Disaster, Crisis and Emergency Management in the Middle East and North Africa

Edris Alam
Editor

Disaster, Crisis and Emergency Management in the Middle East and North Africa

Springer

Editor
Edris Alam
Faculty of Resilience
Rabdan Academy
Abu Dhabi, United Arab Emirates

ISBN 978-3-032-24435-2 ISBN 978-3-032-24436-9 (eBook)
https://doi.org/10.1007/978-3-032-24436-9

This Springer imprint is published by the registered company Springer Nature Switzerland AG
The registered company address is: Gewerbestrasse 11, 6330 Cham, Switzerland

Preface

A substantial body of academic articles and books have been published globally over the past two decades on hazards, disasters, and risk. However, despite the Middle East and North Africa (MENA) region being confronted with a complex constellation of hazards, disasters, crises, and risks arising from both natural and anthropogenic drivers, there remains a notable paucity of academic books focusing specifically on this region. Given its central geopolitical position, rich and ancient cultural heritage, and its significance as the birthplace of major world faiths and belief systems, hazards, disasters, and conflicts in the MENA region warrant analysis grounded in their specific historical, cultural, and socio-political contexts. The region is highly exposed to earthquakes, droughts, sand and dust storms, and extreme heat. These hazards are further intensified by climate change, which accelerates desertification, exacerbates water scarcity, and increases the frequency and severity of flash and coastal flooding. Rapid urbanization, population growth, and inadequate or ageing infrastructure further amplify disaster risk, rendering communities increasingly vulnerable to environmental and socio-political shocks. Beyond natural hazards, the MENA region continues to experience recurrent crises associated with armed conflict, political instability, terrorism, and large-scale population displacement. Such crises place considerable strain on national and regional emergency response systems and generate cascading impacts, including economic disruption, food insecurity, and public health emergencies.

This book addresses a wide range of interrelated themes, including climate change-induced extreme events and food security; water scarcity and compounding risks; sand and dust storms; climate adaptation and disaster risk reduction within development planning; the integration of climate change adaptation into urban planning and the strengthening of early warning systems; community engagement in emergency response in conflict-affected contexts; seismogeodynamics and seismic risk; desert earthquakes and biblical narratives; disaster management approaches across nine case-study countries; and detailed analyses of recent major events, including the Dubai floods (2024), the Beirut explosion (2020), and the Syria–Türkiye earthquakes (2023). The governance of disaster risk reduction (DRR) in the MENA region remains constrained by fragmented policy frameworks, limited early warning capacities, insufficient cross-border cooperation, and weak institutional arrangements. Although technological advances, including AI-enabled early warning systems and remote sensing, offer considerable potential, effective risk reduction in the region ultimately depends on coherent policies, meaningful community engagement, and sustainable development strategies. This volume underscores the urgent need for integrated, multi-sectoral approaches to disaster and crisis management in the MENA region, highlighting the central role of policy reform, regional cooperation, and sustained investment in innovation to strengthen resilience in one of the world's most hazard-prone regions.

The book offers valuable insights for students, researchers, educators, consultants working in climate change and disaster risk management, geopolitical analysts, conflict resolution specialists, and general readers with an interest in the region. We are grateful to Caitlin Kerr, Naanmiap Naanshep Madaki, Lethukubon Gamatsaba, Retshepile P. Motlou, Kimani Mayers, Mantsoi Alice, and Genevieve Berg for their careful review of language and grammar across different chapters of the book. Finally, we extend our sincere appreciation to Arumugam Deivasigamani and Robert Doe for their outstanding support throughout the publication process.

Abu Dhabi, United Arab Emirates Edris Alam

About This Book

The Middle East and North Africa (MENA) region is confronted with a complex constellation of hazards, disasters, crises, and risks arising from both natural and anthropogenic drivers. The region is highly exposed to earthquakes, droughts, sand and dust storms, and extreme heat. These hazards are further intensified by climate change, which accelerates desertification, exacerbates water scarcity, and increases the frequency and severity of flash and coastal flooding. Rapid urbanization, population growth, and inadequate or ageing infrastructure further amplify disaster risk, rendering communities increasingly vulnerable to environmental and socio-political shocks. Beyond natural hazards, the MENA region continues to experience recurrent crises associated with armed conflict, political instability, terrorism, and large-scale population displacement. Such crises place significant strain on national and regional emergency response systems and generate cascading impacts, including economic disruption, food insecurity, and public health emergencies.

This book comprises 23 chapters addressing a wide range of interrelated themes, including climate change-induced extreme events and food security; water scarcity and compounding risks; sand and dust storms; climate adaptation and disaster risk reduction within development planning; the integration of climate change adaptation into urban planning and the enhancement of early warning systems; community engagement in emergency response in conflict-affected contexts; seismogeodynamics and seismic risk; desert earthquakes and biblical narratives; disaster management approaches across nine case-study countries; and in-depth analyses of recent major events, including the Dubai floods (2024), the Beirut explosion (2020), and the Syria–Türkiye earthquakes (2023). The governance of disaster risk reduction (DRR) in the MENA region remains constrained by fragmented policy frameworks, limited early warning capacities, insufficient cross-border cooperation, and weak institutional arrangements. Although technological advances, including AI-enabled early warning systems and remote sensing, offer significant potential, effective risk reduction in the region ultimately depends on coherent policies, meaningful community engagement, and sustainable development strategies. This volume underscores the urgent need for integrated, multi-sectoral approaches to disaster and crisis management in the MENA region, highlighting the central role of policy reform, regional cooperation, and sustained investment in innovation to strengthen resilience in one of the world's most hazard-prone regions.

Contents

Editor and Contributors

About the Editor

Edris Alam is Professor of Integrated Emergency Management at Rabdan Academy, Abu Dhabi, United Arab Emirates. He is an internationally recognized expert in disaster and emergency management and climate change adaptation, with extensive experience working with universities, national governments, non-governmental organizations, and United Nations agencies. He has published over 120 Scopus-indexed articles in leading international journals on disaster risk management and sustainability science. As a lead international consultant for several UN agencies, he has identified strategic gaps in climate change adaptation and disaster risk reduction programming. His work promotes cross-sectoral integration of environmental, climate, and disaster risk considerations, alongside active service as a journal editor, reviewer, and grant assessor.

Contributors

Bouchra Belhaj Abdallah General Department of Environmental Quality, Environmental Agency, Ministry of Environment, Water and Agriculture of Saudi Arabia, Riyadh, Saudi Arabia

Hamad Abushehab Faculty of Resilience, Rabdan Academy, Abu Dhabi, United Arab Emirates

Mohamed Mustaf Ahmed School of Medicine and Health Sciences, SIMAD University, Mogadishu, Somalia

Edris Alam Faculty of Resilience, Rabdan Academy, Abu Dhabi, United Arab Emirates

MHD Bahaa Aldin Alhaffar Disaster Medicine Research Group, Department of Global Public Health, Karolinska Institute, Stockholm, Sweden;
Diseases Program, Directly Transmitted Diseases, European Center for Diseases Prevention and Control, Stockholm, Sweden

Raed Al-Husain Information Systems and Operations Management Department, College of Business Administration, Kuwait University, Kuwait City, Kuwait

Ghalya Almansoori Faculty of Resilience, Rabdan Academy, Abu Dhabi, United Arab Emirates

Mariam Almansoori Faculty of Resilience, Rabdan Academy, Abu Dhabi, United Arab Emirates

Shamma Almheiri Faculty of Resilience, Rabdan Academy, Abu Dhabi, United Arab Emirates

Raghad Alshamsi Faculty of Resilience, Rabdan Academy, Abu Dhabi, United Arab Emirates

S. Al Zaabi Muscat, Oman

Tom Ansell The Hague Humanitarian Studies Centre, International Institute of Social Studies, The Hague, Netherlands

Mahra Binhammad Faculty of Resilience, Rabdan Academy, Abu Dhabi, United Arab Emirates

Zohra Boutaraa Hassiba Benbouali University of Chlef, Chlef, Algeria

Shubham Chaudhary Academy of Scientific and Innovative Research (AcSIR), Ghaziabad, India

Mohamad Diab Syria Civil Defense—The White Helmets, Sarmada, Syria

Ahmed Ekzayez Syria Civil Defense—The White Helmets, Sarmada, Syria

Lamia Elbezzari Department of Geography, Ibn Tofail University, Kenitra, Morocco

Fatna Ikrame El Fanne Department of Process and Environmental Engineering, Faculty of Sciences and Technologies-Hassan II University, Mohammedia, Morocco;
Youth for Climate Morocco, Casablanca, Morocco

Lobna Ahmed Elgamal MOHP, Cairo, Egypt

Maroun El Rayess Rabdan Academy, Abu Dhabi, United Arab Emirates

Bijan Fallah German Climate Computing Center (DKRZ), Hamburg, Germany

Farahnaz Fazel-Rastgar School of Science, University of KwaZulu-Natal, Scottsville, South Africa;
S.V. Raman Researchers Roadmap, Westville, Durban, South Africa

Glenn Fernandez Faculty of Resilience, Rabdan Academy, Abu Dhabi, United Arab Emirates

Ricardo Fonseca ENGEOS Lab, Earth Science Department, Khalifa University, Abu Dhabi, UAE

Diana Francis ENGEOS Lab, Earth Science Department, Khalifa University, Abu Dhabi, UAE

Amos Frumkin Institute of Earth Sciences, The Hebrew University of Jerusalem, Jerusalem, Israel

Bassim M. Hashim Scientific Research Commission, Baghdad, Iraq

Sakineh Khansalari Atmospheric Science and Meteorological Research Center, Tehran, Iran

David MacLeod School of Earth and Environmental Sciences, Cardiff University, Cardiff, United Kingdom

Rinku Mahanta Indian Institute of Technology (IIT) Roorkee, Roorkee, Uttarakhand, India

Rodrigo Mena The Hague Humanitarian Studies Centre, International Institute of Social Studies, The Hague, Netherlands

Amany Mounir Helal Egyptian Ambulance Organization (EAO), Giza, Egypt

Stern Mwakalimi Kita Disaster Management Program, Faculty of Resilience, Rabdan Academy, Abu Dhabi, UAE

Som Nath Indian Institute of Technology (IIT) Roorkee, Roorkee, Uttarakhand, India

Olalekan John Okesanya Department of Public Health and Maritime Transport, University of Thessaly, Volos, Greece

Zhinya Kawa Othman Department of Pharmacy, Kurdistan Technical Institute, Sulaymaniyah, Kurdistan Region, Iraq

Abdelghani Qadem Departement of Geography, Faculty of Arts and Humanities—Sais, Sidi Mohamed Ben Abdellah University, Fez, Morocco

Zohair Qadem Department of Geography, Ibn Tofail University, Kenitra, Morocco

Elnazir Ramadan Department of Geography and Urban Sustainability, United Arab Emirates University, Al Ain, United Arab Emirates

Masoud Rostami Potsdam Institute for Climate Impact Research (PIK), Potsdam, Germany;
Laboratoire de Météorologie Dynamique (LMD), IPSL, ENS-PSL Université, Ecole Polytechnique-Institut Polytechnique de Paris, Sorbonne Université, CNRS, Paris, France;
State Key Laboratory of Deep Oiland Gas, China University of Petroleum (East China), Qingdao, China

Venkataraman Sivakumar S.V. Raman Researchers Roadmap, Westville, Durban, South Africa

V. Svalova Sergeev Institute of Environmental Geoscience, Russian Academy of Sciences, Moscow, Russian Federation;
Geophysical Institute, Vladikavkaz Scientific Center, Russian Academy of Sciences, Vladikavkaz, Russian Federation

Major Hamad Swar Disaster Management, Manama, Kingdom of Bahrain

Chapter 1
Introduction: Complexity of Disaster, Crisis, and Risk in the MENA

Edris Alam

Abstract This introductory chapter conceptualizes the intricate landscape of hazards, crises, and disasters within the Middle East and North Africa (MENA), a region uniquely vulnerable to compounding natural and anthropogenic threats. Synthesizing contemporary literature and international frameworks, including the United Nations Office for Disaster Risk Reduction (UNDRR), the chapter establishes analytical distinctions between distinct risk concepts while rejecting the term natural disaster to highlight how risk is socially constructed through structural vulnerabilities.

Empirically, the chapter charts the region's elevated risk profile, underscored by severe geopolitical instability, high fragile-state indices, and devastating casualties from intersecting crises ranging from the 2023 Turkey–Syria earthquakes and extreme climate-induced temperature events to protracted armed conflicts and industrial failures like the 2020 Beirut explosion. To address these complex, cascading scenarios, the chapter introduces a comprehensive five-part conceptual framework designed to evaluate regional disaster history, analyze institutional and legal governance structures, diagnose operational and coordination challenges, identify research gaps, and propose forward-looking, data-driven mitigation and preparedness strategies. Ultimately, this chapter outlines the structural layout of the volume's 23 chapters, establishing a foundational baseline for shifting the MENA region from reactive emergency response toward proactive, resilient, and collaborative disaster risk management.

E. Alam (✉)
Disaster Management Program, Faculty of Resilience, Rabdan Academy, Abu Dhabi, United Arab Emirates
e-mail: ealam@ra.ac.ae

E. Alam (ed.), *Disaster, Crisis and Emergency Management in the Middle East & North Africa*, https://doi.org/10.1007/978-3-032-24436-9_1

1.1 MENA: A Region of Conflict, Hazard, Emergency, Crisis, and Disaster

The Middle East and North Africa (MENA) region, characterised by its diverse geography, rich cultural heritage, and complex socio-political landscape, is one of the most vulnerable areas in the world to a wide array of natural and human-induced hazards. From devastating earthquakes and prolonged droughts to political instability and economic crises, the region has faced numerous challenges that have shaped its disaster risk profile and management strategies. Its geographical location, climatic conditions, and socio-economic dynamics make it susceptible to a variety of hazards, including seismic activity, floods, droughts, sandstorms, and epidemics, as well as human-induced crises such as armed conflicts, urbanisation pressures, and environmental degradation. In 2025, several notable intra-state conflicts, Syria, Sudan, Yemen and inter-state war, Hamas-Israel exists. In 2024, from an economically fragile state perspective, the first three top fragile states stood from MENA, including Somalia, Sudan, and Syria, superseding all other countries (https://www.theglobaleconomy.com/rankings/fragile_state_index/). From the point of security threats, the region is placed among the elevated countries in the list. A number of MENA countries among lowest global peace index, by hierarchy, are Sudan, Yemen, Syria, South Sudan, Israel, Somalia, Iraq, Palestine, and Iran (Institute for Economics & Peace, 2025). Although the Middle East recorded a 7% reduction in terrorist attacks in 2024, both Israel and Syria ranked among the ten most impacted countries globally (https://www.aap.com.au/aapreleases/cision20250305ae29327/). The resurgence of violence between Israel and Palestine has been ongoing. These hazards often intersect, creating complex disaster scenarios that challenge the resilience of communities, infrastructure, and governance systems. Thus, understanding the context of crises and disasters in the region is crucial for identifying patterns, vulnerabilities, and opportunities for improved risk management.

The MENA has experienced a wide array of high-casualty disasters spanning geological, hydrometeorological, technological, and conflict-driven hazards. Table 1.1 consolidates ten of the most significant events, from the 2023 Turkey–Syria Earthquakes that killed around 59,000 people (Turkey ~ 53,537; Syria ~ 5,951–8,476) to the protracted Iran–Iraq War (1980–1988) with an estimated minimum of ~500,000 killed across both sides. While the global community was facing the most lethal biological hazards, the COVID-19 pandemic was bringing the world to collapse claiming over 1 million lives; human negligence induced the Beirut explosion in 2020 taking 218 lives, around 7,000 injuries, and displacing over 300,000 people reminding us that disasters are often man-made. Within a very short span of time, on 26 April 2025, inappropriate handling of flammable goods stored in containers ignited a fire at a container port in Bandar Abbas claiming at least 70 lives and over 1,200 injured. The conflict between Hamas and Israel began on **7** October 2023 taking over 67 938 lives in Gaza and approximately 1,600 in Israel. Since 2020, the region has faced disasters including the Dubai floods of 2024, Syria–Turkey earthquakes in 2023, and the Libyan Dam failure in 2003, all of which attracted global attention (Francis et al., 2025). The region consistently faced extreme temperature events with 54.0 °C in Kuwait and 51.8 °C in the United Arab Emirates (UAE). These cases reflect how the region's hazards are not confined to earthquakes or floods; they include industrial disasters (e.g., the 2020 Beirut explosion), conflict wars, and storms/flooding, illustrating the complex disaster landscape of MENA.

Table 1.1 Top ten major events from selected categories causing deaths in MENA

Year	Location	Hazard type	Deaths (best-supported estimate)	Affected	Sources
2011–2025	Syria (nationwide)	Civil war (conflict)	~618,000		The Syrian Observatory for Human Rights (SOHR)
2025	Iran	Industrial explosion	~65	1,023,609	Zarafshan et al. (2025)
2023–2025	Israel-Gaza	War	~67,938	~533 000	WHO (2026)
2023	Southern Türkiye and Northern Syria	Earthquake sequence (Mw 7.8/ 7.5)	~59,000 (Turkey ~53,537 and Syria ~5,951–8,476)	23 million	CRED, UNOCHA (2023) and Galasso and Opabola (2023)
2023	Derna and Eastern Libya	Storm Daniel/ floods/dam failures	**~5,923**	250,000	UNOCHA (2024)
2020–2021	MENA	Pandemic COVID-19	~351,822	~23,406,897 confirmed cases	WHO (2023)
2020	Beirut, Lebanon	Industrial explosion	~200	300 000	El Sayed (2022)
2015	Saudi Arabia	Crowd crush (pilgrimage event)	~2,431		Ganjeh and Einollahi (2016)
2003	Bam, Iran	Earthquake (Mw 6.6)	~40,000		Sadeghi-Bazargani et al. (2015)
1990	Rudbar–Manjil, Iran	Earthquake (M_W ~ 7.4)	~40,000	500,000	Berberian et al. (1992)
1980–1988	Iran–Iraq	Interstate war	~188,015 to 217,489 in Iran ~200,000 in Iraq		Salamati et al. (2013), Li et al. (2016)
1978	Iran—Tabas earthquake	Earthquake	~25,000+		Berberian (1978)

Source Compiled by the authors

Table 1.2 Hazard cluster by UNDRR and ISC (2025)

Hazard cluster	Examples
Meteorological and hydrological hazards	Tropical cyclones, drought, riverine floods, and heatwaves
Extraterrestrial hazards	Asteroid and meteorite impacts or solar flares
Geohazards	Three-seismogenic, volcanogenic, and erosion or some type of mass movement
Environmental hazards	Biodiversity loss, land salination, loss of permafrost, and the marine equivalent
Biological hazards	Bacteria, viruses, parasites, venomous animals, and mosquitoes carrying disease-causing agents, COVID-19
Chemical hazards	Corrosive, flammable, and toxic chemicals, Beirut explosion 2020
Technological hazards	Cyber hazards
Societal hazards	Violence and conflict

Source UNDRR and ISC (2025)

1.2 Hazard Types

A variety of classifications are used to describe hazard types and their sources of origin. A commonly used typology categorises hazards into three types: natural, technological and complex or conflict-based events. Natural hazards can be divided based on origin/sources: geological (i.e., earthquakes and tsunamis), hydrological (floods) and climatological (tropical cyclones), and biological (COVID-19).

The three major classifications fail to address a clear understanding of disasters. Since hazard identification and understanding is the first step for disaster risk management, the United Nations Disaster Risk Reduction (UNDRR) and International Science Congress (ISC) (United Nations Office for Disaster Risk Reduction and International Science Council, 2025) developed a robust categorisation of hazards providing eight clusters: (1) Meteorological and hydrological hazards, (2) Extraterrestrial hazards, (3) Geohazards, (4) Environmental hazards, (5) Biological hazards, (6) Chemical hazards, (7) Technological hazards, and (8) Societal hazards. Examples for each hazard cluster have been presented in Table 1.2.

1.3 Conceptualising Hazard, Crisis, Emergency, and Disaster

A variety of terminologies are used to describe hazard types which can lead to a disaster. On the surface, the terms hazard, emergency, disaster, and crisis are used interchangeably. However, contemporary scholarship and international practice differentiate these related but analytically distinct concepts. In the following sections, we review these terminologies including hazards, emergency, disaster, and crisis used in the discipline.

The term 'hazard' denotes a potential source of a 'disaster'. The UNDRR has defined a 'hazard' as follows:

> A dangerous phenomenon, substance, human activity or condition that may cause loss of life, injury or other health impacts, property damage, loss of livelihoods and services, social and economic disruption, or environmental damage.

Emergency and Crisis

Although the term 'emergency' appears similar to the term 'disaster', it is conceptually distinct in risk studies. An *emergency* denotes an event requiring immediate operational action (medical, rescue, containment) to prevent or reduce harm and is the trigger for tactical response systems (e.g., public-health emergency declarations). For example, in the context of biological and technological hazards or health emergencies, which can also relate to hazardous events that do not result in the serious disruption of community or societal functioning (United Nations Office for Disaster Risk Reduction, 2017). A *crisis* is defined in crisis-management literature as a situation in which actors perceive an urgent threat to core values or life-sustaining systems; crises are characterised by high uncertainty, political salience, and time pressure and therefore foreground governance and legitimacy questions (Boin et al., 2005). Generally, a crisis is an event of severity and complexity rather than an emergency that disrupts normal societal functions and hinders government authorities from carrying out their essential responsibilities. These categories frequently overlap in practice (for example, an earthquake can be simultaneously an emergency, a disaster, and a crisis), but keeping the analytical distinctions is valuable for theory, policy, and practice.

Disaster

The concept of disaster has evolved internationally over the last seventy years. In the early 1950s, disaster meant external sources from mostly natural forces causing impact on 'social order', generating 'physical destruction and death' that thus mostly focused on negative consequences (Burton & Kates, 1963; Burton et al., 1993; Killian, 1954). A disaster can be understood as a high-impact occurrence that results from the interaction between a physical hazard and the vulnerabilities inherent within a social system. Susman et al. (1983) developed the political economy perspective on disasters, arguing that vulnerability—not the physical hazard alone—explains why disasters occur. Alexander (1993) explains that natural disasters are typically "*sudden onset physical events*" that create substantial disruption to both the natural environment and human socio-economic systems. Finally, each of these definitions retains the hazard origins of disasters, but also moves to examine them in social terms, particularly of vulnerability and resilience.

Quarantelli (2000, p. 682) identifies disasters in terms of a variety of defining features. They (1) are sudden-onset occasions, (2) seriously disrupt the routines of collective units, (3) are the adoption of unplanned courses of action to cope with the disruption, (4) exhibit unexpected developments shaped by social and temporal contexts, and (5) pose danger to valued social objects. Quarantelli (2000) subsequently emphasised that disasters represent vulnerability, reflecting "weakness in social structures or social systems". It was emphasised as neither an event nor a physical place or time as relevant to disasters. Instead, the entire concept is social, vulnerability is socially constructed by relationships in the social system, and disasters are based on the notion of social change. The definition may have roots in classical era research but clearly departs in significant ways. Quarantelli (2005) points out that it is rooted in two fundamental ideas. First, disasters are inherently social. It is not the hurricane wind or the storm surge that causes disasters; these are the source of damage. Second, the disaster is rooted in the social structure and reflects the processes of social change. It is from these features of the social system that we find vulnerability to the source. The purpose of discussion is to be aware of the conceptual change in defining disasters, presenting how the focus of risk management changes over time. Reflecting on these changes, the United Nations International Strategy for Disaster Risk Reduction (UNISDR) states:

> A serious disruption of the functioning of a community or a society involving widespread human, material, economic or environmental losses and impacts, which exceeds the ability of the affected community or society to cope using its own resources.

The EM-DAT international disaster database (www.emdat.be) categorises a hazard event as a disaster when at least one or more of the following are reported:

- Ten or more people reported killed.
- One hundred or more people reported killed.
- Declaration of a state of emergency.
- Call for international assistance.

To agree and accept that disasters are not 'natural' (O'keefe et al., 1976), we do not use the term 'natural disasters' in this book. If humans did not live in hazardous places, disaster would not occur. A simple example, developing settlement in steep slopes by cutting forestation caused a significant increase of landslide-associated deaths in recent decades (Alam, 2020). Even intensity of tropical cyclones was attributed to anthropologic induced climate change (Sobel et al., 2016). The problematic use of the term natural disasters which often neglect human reasons leaving a solution to control or manage nature has long been widely discussed in the discipline. In 1976, O'keefe et al. (1976) argued this in their seminal research article in Nature titled "Taking the Naturalness out of Natural Disasters". Blaikie et al. (2004) argue that disasters are not caused solely by natural hazards. Instead, they result from the interaction between a hazard and the vulnerability of exposed populations. According to their conceptualisation, a disaster occurs when a hazard interacts with vulnerability, rather than because of the hazard alone. Hazards are natural; disasters are socially constructed. In 2006, Hartman and Squires's (2006) book about Hurricane Katrina was titled "*There Is No Such Thing as a Natural Disaster: Race, Class, and Hurricane Katrina*". This is now not confined to academic discipline. On 17 March 2021, the United Nations Office of Disaster Risk Reduction stated: "Time to recognise there is no such thing as a natural disaster, we're doing it to ourselves".

Risk

In the discipline of natural hazards and disasters, the core idea of risk is defined as Risk = Hazard × Vulnerability (Blaikie et al., 1994; Wisner et al., 2004). UNDRR defined disaster risk reduction as "*The combination of the probability of an event and its negative consequences*". In this thesis, the term 'risk' is used to denote the consequences of hazards.

On the surface, disaster risk management and disaster risk reduction may seem similar. Although, within the discipline, these terms are used to denote administrative and scientific perspectives, respectively, a closer look at the definitions developed by the UNDRR can be used to further understand these perspectives. The UNDRR defines disaster risk reduction as follows:

> The concept and practice of reducing disaster risks through systematic efforts to analyse and manage the causal factors of disasters, including through reduced exposure to hazards, lessened vulnerability of people and property, wise management of land and the environment, and improved preparedness for adverse events.

Disaster Risk

> The potential disaster losses, in lives, health status, livelihoods, assets and services, which could occur to a particular community or a society over some specified future time period.

Emergency Management

> The organization and management of resources and responsibilities for addressing all aspects of emergencies, in particular preparedness, response and initial recovery steps.

Disaster Risk Management

> The systematic process of using administrative directives, organizations, and operational skills and capacities to implement strategies, policies and improved coping capacities in order to lessen the adverse impacts of hazards and the possibility of disaster.

Disaster risk management involves several activity areas from the disaster management cycle. The activities include prevention, protection, mitigation, preparedness, response, relief, recovery, rehabilitation, and reconstruction. These activities are again divided based on before, during, and after a disaster. Prevention, protection, mitigation, and preparedness are considered pre-disaster activities. Response is typically during the disaster period and rescue, relief, recovery, rehabilitation, and reconstructions are post-disaster activities. A different type of framework is seen to be used by different governments. The UAE national response framework (NRF) provides four phases including prevent, prepare, response, and recovery for all types of incidents/disasters. National Preparedness goals of the Federal Emergency Management Authority (FEMA) emphasise building and sustaining core capabilities across 'five mission areas': prevention, protection, mitigation, response, and recovery. The FEMA has specific definitions for each mission area. The UNDRR has developed and provided definitions for each part involved in disaster risk management. A brief discussion of each activity is provided in the following.

The complete avoidance of the harmful effects of hazards and disasters reflects the principle of **prevention**, which involves deliberate measures taken beforehand to eliminate potential risks entirely. Examples include constructing dams or levees to remove flood threats, enforcing land-use policies that prohibit development in hazard-prone areas, and applying seismic-resistant engineering techniques to ensure that essential structures remain functional during earthquakes. However, achieving total prevention is often impractical; consequently, the focus frequently shifts towards **mitigation**, which aims to minimise rather than eliminate potential losses. For this reason, the terms *prevention* and *mitigation* are sometimes used interchangeably in general discourse.

Mitigation refers to the process of reducing or constraining the negative consequences of hazards and associated disasters. While it is rarely possible to eliminate the impacts of hazards entirely, their magnitude and severity can be significantly reduced through targeted actions and strategies. Such measures include the application of engineering solutions, hazard-resistant construction practices, effective environmental management, and the promotion of public awareness. It is important to note that within the context of climate change policy, the term *mitigation* carries a distinct meaning—specifically, the reduction of greenhouse gas emissions responsible for driving climate change.

Preparedness refers to the collective knowledge, skills, and capacities developed by governments, professional response and recovery agencies, communities, and individuals to anticipate, respond to, and recover effectively from potential, imminent, or ongoing hazardous events or conditions. Conducted within the broader framework of disaster risk management, preparedness aims to strengthen the capabilities required to manage emergencies efficiently and to ensure a smooth transition from immediate response to long-term recovery. Effective preparedness relies on comprehensive risk analysis and strong integration with early warning systems. It involves activities such as contingency planning, maintaining essential equipment and supply reserves, establishing coordination mechanisms, planning for evacuation and public communication, and conducting regular training and simulation exercises. These actions must be underpinned by robust institutional, legal, and financial arrangements. The closely related concept of *readiness* refers to the capacity to act promptly and appropriately when a hazard occurs.

Response refers to the immediate actions and assistance provided during or directly after a disaster to protect lives, minimise health impacts, maintain public safety, and meet the essential needs of affected populations. This phase, often termed *disaster relief*, primarily addresses urgent and short-term requirements. The transition between the response and recovery phases is not always distinct, as certain response measures—such as providing temporary shelter, clean water, and other essential services—may continue well into the recovery period.

Recovery refers to the process of restoring—and, where possible, enhancing—facilities, livelihoods, and living conditions in communities affected by disasters, while simultaneously addressing underlying risk factors. The activities of rehabilitation and reconstruction typically commence once the immediate emergency phase concludes and should be guided by pre-established strategies and policies that define institutional roles and promote community involvement. Recovery initiatives, supported by increased public awareness and engagement following a disaster, provide a critical opportunity to integrate disaster risk reduction measures and apply the principle of *building back better* to create safer and more resilient systems.

1.4 Conceptual Framework

This book, titled *Disaster, Crisis, and Emergency Management in the Middle East and North Africa*, seeks to provide a comprehensive overview of the historical and contemporary disaster landscape in the region, examining the interplay between natural and anthropogenic factors that contribute to its unique risk profile. The MENA region is no stranger to disasters. This book is structured around five key questions that aim to unravel the disaster history and risk profile of the MENA region: (1) disaster history and hazard types, crises, and disasters that are most prevalent in the region; (2) institutional framework for disaster and crisis management; (3) the major challenges faced by disaster management authorities in dealing with previous disasters; (4) the key areas lacking sufficient research in disaster management; (5) measures to be taken to reduce disaster-related deaths and the improvements needed in disaster risk management policies and practices.

First, it explores the major disasters that have affected the jurisdiction, providing a historical account of significant events and their impacts. This includes an analysis of both natural hazards, such as floods and droughts, and human-induced crises, such as conflicts and economic instability. By examining past disasters, we can identify recurring themes and lessons learned that inform current and future risk management strategies. Understanding the prevalence and distribution of these hazards is essential for developing targeted mitigation and preparedness measures.

The authors investigate the impacts of these disasters on communities, infrastructure, and the economy. Disasters in the MENA region have far-reaching consequences, often exacerbating existing vulnerabilities and inequalities. For instance, prolonged droughts can lead to food insecurity and displacement, while earthquakes can destroy critical infrastructure and disrupt economic activities. By examining these impacts, we can better understand the cascading effects of disasters and the need for integrated approaches to disaster risk reduction and emergency management.

The MENA region's disaster risk profile is further complicated by its socio-political context. Many countries in the region face challenges related to governance, resource management, and political instability, which can hinder effective disaster response and recovery efforts. Additionally, rapid urbanisation and population growth have increased exposure to hazards, particularly in densely populated cities with inadequate infrastructure. Climate change adds another layer of complexity, as rising temperatures, changing precipitation patterns, and sea-level rise are expected to exacerbate existing hazards and create new ones.

The authors attempted to explore the evolution and institutional structure of crisis and disaster management across the Middle East and North Africa (MENA). It examines how national authorities responsible for disaster and crisis management have developed over time and identifies the current institutional arrangements that guide their operations. The discussion also highlights the key agencies and stakeholders involved in disaster response and mitigation efforts, outlining their respective roles and coordination mechanisms. Furthermore, the legal and policy frameworks that shape disaster governance in the region, providing an understanding of how laws, regulations, and policies influence preparedness, response, and recovery activities within MENA countries are discussed.

The authors examine the key strengths, challenges and weakness that have shaped disaster management across the Middle East and North Africa. It explores the major difficulties authorities have encountered in managing past disasters and considers how political or institutional shortcomings may have affected their capacity to respond effectively. The discussion evaluates the performance of response, recovery, and mitigation efforts, assessing whether existing systems operate more reactively than proactively. Attention is also given to identifying structural weaknesses within current disaster management frameworks and uncovering gaps in coordination among governmental bodies, humanitarian organisations, and other non-governmental actors that limit the overall effectiveness of disaster risk governance in the region.

The authors identify the major research gaps in disaster management across the Middle East and North Africa. It highlights areas where existing studies remain limited or fragmented and where further investigation is needed to strengthen understanding and practice. Particular attention is given to emerging threats and region-specific challenges that have received insufficient scholarly focus. The discussion also considers how targeted research can enhance disaster resilience by informing policy development, guiding institutional reforms, and improving preparedness, response, and recovery capacities within the region.

Despite these challenges, the MENA region has made significant strides in disaster risk management. Many countries have established national disaster management frameworks, early warning systems, and community-based initiatives to enhance resilience. International cooperation and partnerships have also played a crucial role in supporting disaster preparedness and response efforts. However, there is still much work to be done to address the underlying drivers of risk and build long-term resilience.

Finally, the authors outline recommendations and forward-looking strategies to strengthen disaster management in the Middle East and North Africa. It focuses on measures that can help reduce disaster-related fatalities and improve the overall effectiveness of disaster risk management policies and practices. The discussion emphasises the importance of community participation and public awareness in fostering a culture of preparedness. It also explores how technological innovation, data-driven decision-making, and modern communication tools can be leveraged to enhance early warning systems, improve coordination, and build long-term resilience across the region.

In the following sections, we will discuss deeper into disaster history of the MENA region, examining significant events and their impacts. We will then explore the types of hazards and crises that are most prevalent, highlighting the unique characteristics of the region's risk landscape. Finally, we will analyse the impacts of disasters on communities, infrastructure, and the economy, drawing on case studies and empirical evidence to illustrate the multifaceted nature of disaster risk in the MENA region. Through this comprehensive analysis, we hope to provide valuable insights that can inform future efforts to enhance disaster resilience and emergency management in the region.

1.5 Organisation of the Book

The book is organised into 23 chapters comprising specific case chapter countries and chapters focusing on cross-country and regional context. Following the first and introductory chapter, the second chapter presents that MENA faces compounding climate, environmental, and hazard risks that intersect with water scarcity, urbanisation, and institutional fragility, threatening development and human security. This chapter examines how Egypt, Jordan, Morocco, Lebanon, and the UAE integrate disaster risk reduction and climate adaptation into development planning. Policy frameworks are advancing, yet implementation remains fragmented due to weak mandates, limited finance, and uneven capacities, underscoring the need for stronger coordination, legislation, and risk-informed investment.

Chapter 3 presents that rising temperatures, droughts, and extreme rainfall in the MENA region are accelerating soil degradation, erosion, and sedimentation, undermining food systems. Increased evaporation reduces soil moisture and vegetation, while sea-level rise and salinity damage coastal soils, contributing to a 34–37% decline in agricultural productivity since 1961. Evidence from North Africa highlights the need for integrated adaptation combining soil conservation, agroecology, water management, governance, and climate finance.

Chapter 4 analyses urban planning as a vehicle for integrating climate adaptation into city development across the GCC, drawing on cases from the UAE, Oman, and Qatar. While policy innovation is emerging, adaptive planning remains constrained by weak coordination, limited climate data use, and capacity gaps. The chapter highlights priorities including risk-informed spatial planning, nature-based solutions, geospatial tools, participatory processes, and alignment of national and municipal adaptation strategies.

Chapter 5 positions Early Warning Systems as central to preparedness for floods, heatwaves, wildfires, and droughts in MENA. While EWS capacity is expanding, fragmented and non-inclusive data, technology gaps, and limited technical capacity constrain effectiveness and Sendai Framework integration. The analysis reviews regional hazard exposure and EWS performance, identifying governance, cooperation, research, and technology deficits as priorities for strengthening warning-to-action systems.

Chapter 6 shows that classical, state-centred disaster response models often fail in conflict-affected settings, where governance and access are fragmented. Drawing on cases from Syria, Yemen, Gaza, Lebanon, Sudan, Morocco, and Iraq, it demonstrates how community-led responses save lives and sustain resilience despite constraints. The chapter argues for institutionalising and financing community-based mechanisms as a core pillar of disaster preparedness and response in MENA and beyond.

Chapter 7 examines north-western Syria's complex disaster risk landscape shaped by climatic, environmental, conflict-related, and socio-economic pressures. Drawing on climate evidence, field observations, and humanitarian reporting, it shows how floods, heat extremes, and water stress intersect with displacement and fragmented governance. Persistent data gaps and weak coordination constrain planning, underscoring the need for context-sensitive, community-based risk reduction, stronger institutions, and rebuilt monitoring and early warning systems.

Chapter 8 analyses the evolution of disaster management in the UAE, highlighting the institutional and organisational capacities underpinning national resilience. It examines NCEMA's centralised yet adaptive governance, supported by emirate-level institutions, academic bodies, and volunteer organisations. Drawing on official sources, the chapter shows how integrated policy, legal frameworks, education, and community engagement have strengthened preparedness and response, offering transferable lessons for professionalising disaster management in diverse risk contexts.

Chapter 9 examines the evolution of Oman's Emergency Management System amid global shifts towards more collaborative and inclusive governance. Using a qualitative single-case study, it analyses legal frameworks and interview evidence to trace institutional change. Findings show a largely hierarchical system, with growing cross-sectoral coordination and non-state engagement. The chapter highlights localisation and adaptive governance as strengths, while noting institutional constraints that limit deeper reform in centralised states.

Chapter 10 examines Bahrain's evolution in disaster management, highlighting robust governance, inter-institutional coordination, and rapid adoption of advanced technologies across preparedness, response, and recovery. The chapter shows alignment with Vision 2030, the Sendai Framework, and SDGs, and underscores engagement with international partners and the GCC to enhance regional resilience. Case studies illustrate effectiveness, lessons learned, and strategic pathways for sustaining disaster resilience amid evolving global and regional risks.

Chapter 11 assesses Kuwait's disaster and crisis management, examining historical events, institutional structures, and operational challenges. Findings highlight fragmented governance, weak inter-agency coordination, limited public engagement, and inadequate data infrastructure. The chapter identifies gaps in legislative planning, preparedness, and risk communication, and offers recommendations aligned with the Sendai Framework and Kuwait DRR Declaration (2025). It concludes that sustained institutional reform and strategic investment are critical for an integrated, risk-informed disaster governance model.

Chapter 12 examines Iraq's acute vulnerability to climate change and conflict, highlighting water scarcity, drought, floods, desertification, dust storms, and COVID-19 impacts. These crises drive population displacement, land degradation, and weakened infrastructure, exacerbated by corruption and conflict. The chapter emphasises the urgent need for proactive disaster risk management, strengthened national capacities, coordinated governance, enhanced research, and increased funding to shift from reactive response towards resilience and stability.

Chapter 13 examines Algeria's exposure to earthquakes, floods, heatwaves, and forest fires, highlighting their social, economic, and developmental impacts. Since 1980, the government has pursued prevention-focused disaster risk reduction through technical and legislative measures to reduce vulnerabilities. The chapter analyses the evolution and effectiveness of Algeria's disaster management system, assessing preparedness, response, and mitigation, and concludes with recommendations to strengthen national resilience and guide future disaster management strategies.

Chapter 14 assesses Egypt's disaster management amid rising risks from natural hazards, climate change, and urbanisation. Since the 1992 Cairo earthquake, institutional capacities have grown, including the National Centre for Crisis and Disaster Risk Reduction and expanded early warning systems. Yet, the system remains largely reactive, with limited community engagement and vulnerable infrastructure. Economic losses reached $23.1 billion (2000–2024). The chapter calls for proactive strategies, institutional modernisation, community participation, infrastructure resilience, and greater investment in prevention.

Chapter 15 examines Somalia's disaster management evolution from post-1991 collapse to the establishment of SoDMA, the 2020 DRM Policy, and multi-hazard early warning systems. Despite progress, gaps remain in financing, last-mile warning, data, coordination, and local ownership. The chapter recommends institutionalising whole-of-government DRM, linking early warning to anticipatory action, enhancing risk finance, strengthening federal-state coordination, investing in resilient infrastructure, and embedding inclusive, community-led preparedness to transition from reactive relief to lasting resilience.

Chapter 16 examines the 2023 Türkiye–Syria earthquake sequence (M_W 7.8, M_W 7.5), one of the most severe seismic disasters of the century. Using seismological, engineering, and humanitarian evidence, it analyses rupture dynamics, damage patterns, response performance, and conflict-related vulnerabilities. Findings show widespread building code non-compliance, weak risk-sensitive planning, and unequal response capacities, underscoring the need for stronger regulation, local response investment, and regional seismic cooperation.

Chapter 17 examines the extreme heatwave June 2024 that coincided with the Hajj pilgrimage, offering a detailed meteorological and societal analysis of the event. Drawing on atmospheric and synoptic data, the study identifies a combination of anomalies—including a deepened and drier Red Sea Trough, Anticyclonic Rossby Wave Breaking, and weakening of the subtropical jet stream—that amplified surface heating across Makkah and the surrounding region. It further reveals how these shifts, compounded by declining wind speeds and moisture influx, produced hazardous thermal stress. The case underscores mounting climate-induced risks to crowd safety and public health management during mass religious gatherings in the region.

Chapter 18 presents a compelling examination of the 16 April 2024 flood event in Dubai, United Arab Emirates, a phenomenon of first-order significance given the arid, desert-based context. They detail how a record-breaking precipitation episode—driven by tropical-extratropical interactions, elevated atmospheric water vapour, and potential dust loading—catalysed widespread infrastructure disruption, flight cancellations, school and business closures, a dengue outbreak, and four fatalities. Satellite data reveal links among urbanisation, population density, and flood impacts, while desert greening peaked two to three months post-event. Crucially, attribution analysis suggests anthropogenic climate change is very likely increasing the frequency of such extreme rainfall episodes. The chapter offers practical recommendations: impact-based forecasting, public awareness campaigns, early-warning systems, infrastructural adaptation and enhanced drainage and flood-defence investments.

Chapter 19 examines Lebanon's vulnerability to natural and human-induced hazards, emphasising how political instability, weak institutions, mismanagement, and unregulated urban growth amplify risk. The 2020 Beirut port explosion exemplifies institutional failure, regulatory gaps, and poor risk management. While progress includes the Disaster Risk Management Unit and alignment with international standards, disaster response remains largely reactive. The chapter underscores the need for stronger, coordinated, and accountable systems to enhance prevention, preparedness, and resilience in Lebanon and the MENA region.

Chapter 20 investigates possible geological bases for environmental hazards in the biblical Exodus narrative, using textual analysis, paleoseismology, hydrogeology, and historical records. Findings indicate parallels between the narratives and seismic phenomena, including surface ruptures and groundwater changes in the southern Araba valley during the second millennium BCE. Events such as Korah's punishment, water from Kadesh, the Jordan's flow, and Jericho's walls may reflect earthquake-related hazards, suggesting the narratives preserve ancient memories of real environmental events.

Chapter 21 examines drought variability in Morocco's Sebou basin using SPI indices and Mann–Kendall trends, revealing alternating wet and dry periods, significant drying in the Atlas and Middle Atlas, and spatial contrasts in plains. Exceptional wet years occurred in 1996–1997 and 2009–2012. Morocco's response includes inter-basin transfers, modern legal frameworks, and participatory water governance, yet challenges in sustainability, equity, and coordination highlight the need for integrated, multi-scalar approaches under climate change.

Chapter 22 examines seismic hazards in the MENA region, situated at the convergence of the Arabian, Eurasian, and African plates. The 2025 Sea of Marmara earthquake (M 6.2) highlights ongoing risk. Using systematic seismic analyses, geodynamic and geothermal models, and historical earthquake records, the study links earthquake generation to fault activation and mantle diapirism. Despite westward propagation reducing immediate risk near Istanbul, the recurrence interval (~250 years) indicates a major earthquake remains plausible in the near future.

Chapter 23, titled "*Reflecting from Crisis to Resilience Pathways*", synthesises the key findings of the book by examining the challenges and weaknesses of disaster risk management in the region. It highlights significant and concerning research gaps and distils the central takeaways of the volume, emphasising the need for proactive, community-based, data-driven approaches to disaster risk management to strengthen resilience pathways in the region.

1.6 Conclusion

The MENA region faces significant challenges in disaster, crisis, and emergency management, but these challenges also present opportunities for innovation and improvement. By implementing the recommendations outlined in this chapter, the region can reduce disaster-related deaths, enhance risk management policies, strengthen community engagement, and leverage technology to build resilience. A proactive, inclusive, and collaborative approach is essential for addressing the complex and evolving risks faced by the MENA region. Ultimately, these efforts will contribute to safer, more resilient communities and a sustainable future for the region.

References

Alam, E. (2020). Landslide hazard knowledge, risk perception and preparedness in Southeast Bangladesh. *Sustainability*.

Alexander, D. (1993). *Natural disasters* (1st ed.). Routledge. https://doi.org/10.4324/9781315859149

Berberian, M. (1978). Tabas-e-Golshan (Iran) catastrophic earthquake of 16 September 1978. A preliminary field report. *Disasters, 2*(4), 207–219.

Berberian, M., Qorashi, M., Jackson, J. A., Priestley, K., & Wallace, T. (1992). The Rudbar-Tarom earthquake of 20 June 1990 in NW Persia: Preliminary field and seismological observations, and its tectonic significance. *Bulletin of the Seismological Society of America, 82*(4), 1726–1755.

Blaikie, P., Cannon, T., Davis, I., & Wisner, B. (1994). *At risk: Natural hazards, people's vulnerability and disasters*. Routledge.

Blaikie, P., Cannon, T., Davis, I., & Wisner, B. (2004). *At risk: Natural hazards, people's vulnerability and disasters*. Routledge.

Boin, A., Hart, P. T., Stern, E., & Sundelius, B. (2005). *The politics of crisis management: Public leadership under pressure*. Cambridge University Press.

Burton, I., & Kates, R. W. (1963). The perception of natural hazards in resource management. *National Research Journal, 3*, 412.

Burton, I., Kates, R. W., & White, G. F. (1993). *The environment as hazard*. Guilford Press.

El Sayed, M. J. (2022). Beirut ammonium nitrate explosion: A man-made disaster in times of the COVID-19 pandemic. *Disaster Medicine and Public Health Preparedness, 16*(3), 1203–1207.

Francis, D., Fonseca, R., Nelli, N., Cherif, C., Yarragunta, Y., Zittis, G., & Jan de Vries, A. (2025). From cause to consequence: examining the historic April 2024 rainstorm in the United Arab Emirates through the lens of climate change. *npj Climate and Atmospheric Science, 8*(1), 183.

Galasso, C., & Opabola, E. A. (2023, February 6). The 2023 Kahramanmaraş earthquake sequence: Finding a path to a more resilient, sustainable, and equitable society. *Communications Engineering, 3, 24*. https://doi.org/10.1038/s44172-024-00170-y

Ganjeh, M., & Einollahi, B. (2016). Mass fatalities in Hajj in 2015. *Trauma Monthly, 21*(5), e43253.

Hartman, C. W., & Squires, G. D. (2006). *There is no such thing as a natural disaster: Race, class, and Hurricane Katrina*. Taylor & Francis.
Institute for Economics & Peace. (2025, June). *Global peace index 2025: Identifying and measuring the factors that drive peace, Sydney*. Retrieved November 03, 2025, from http://visionofhumanity.org/resources
Killian, L. M. (1954). Some accomplishments and some needs in disaster study. *Journal of Social Issues, 10*(3), 66–72.
Li, S. J., Flaxman, A., Lafta, R., Galway, L., Takaro, T. K., Burnham, G., & Hagopian, A. (2016). A novel method for verifying war mortality while estimating Iraqi deaths for the Iran-Iraq war through operation desert storm (1980–1993). *PLoS ONE, 11*(10), e0164709.
O'keefe, P., Westgate, K., & Wisner, B. (1976). Taking the naturalness out of natural disasters. *Nature, 260*(5552), 566–567.
Quarantelli, E. L. (2000). Disaster. In E. F. Borgatta & R. J. V. Montgomery (Eds.), *Encyclopedia of sociology* (pp. 682–688). Macmillan Reference USA.
Quarantelli, E. L. (2005). What is a disaster? In R. W. Perry, & Q. E. L. (Eds.), *What is a disaster? New answers to old questions* (pp. 39–55). Xlibris.
Sadeghi-Bazargani, H., Azami-Aghdash, S., Kazemi, A., & Ziapour, B. (2015). Crisis management aspects of bam catastrophic earthquake: Review article. *Health Promotion Perspective, 5*(1), 3–13.
Salamati, P., Razavi, S. M., Shokraneh, F., Mohazzab Torabi, S., Laal, M., Hadjati, G., Khaji, A., & Rahimi Movaghar, V. (2013). Mortality and injuries among Iranians in Iraq-Iran war: A systematic review. *Archives of Iranian Medicine, 16*(9), 542–550.
Sobel, A. H., Camargo, S. J., Hall, T. M., Lee, C.-Y., Tippett, M. K., & Wing, A. A. (2016). Human influence on tropical cyclone intensity. *Science, 353*(6296), 242–246.
Susman, P., O'Keefe, P., & Wisner, B. (1983). Global disasters: A radical interpretation. In K. Hewitt (Ed.), *Interpretations of calamity from the viewpoint of human ecology* (pp. 263–283). Allen & Unwin.
United Nations Office for Disaster Risk Reduction. (2017). The Sendai framework terminology on disaster risk reduction. *Disaster*. Retrieved October 30, 2025, from https://www.undrr.org/terminology/disaster
United Nations Office for Disaster Risk Reduction and International Science Council. (2025). UNDRR–ISC hazard definition and classification review: 2025 update of the technical report. United Nations Office for Disaster Risk Reduction, International Science Council. https://doi.org/10.24948/2025.04
UNOCHA. (2023, May 17). Türkiye earthquake 2023 humanitarian response overview. UN Office for the Coordination of Humanitarian Affairs (UNOCHA), 7. https://reliefweb.int/report/turkiye/turkiye-earthquake-2023-humanitarian-response-overview-17-may-2023
UNOCHA. (2024). Libya flood response flash appeal final report Sept 2023–June 2024 (Issued November 2024) [EN/AR]. UN Office for the Coordination of Humanitarian Affairs (UNOCHA). Accessed 13 November 2025.
WHO. (2023). COVID-19 situation updates for epi week 48 (26 November–02 December 2023). Retrieved November 13, 2025, from https://www.emro.who.int/fr/pandemic-epidemic-diseases/covid-19/
WHO. (2026). *Hostilities in the occupied Palestinian territory (oPt): Public health situation analysis (PHSA)*. Retrieved February 23, 2026, from https://cdn.who.int/media/docs/default-source/emergencies-trauma-care/who-phsa-opt-111125-final.pdf?download=true&sfvrsn=efa97590_1
Wisner, B., Blaikie, P., Cannon, T., & Davis, I. (2004). *At risk: Natural hazards, people's vulnerability and disasters*. Routledge.
Zarafshan, H., Fathi, H. R., Seddigh, B., Raeisi, A., Bahramnejad, A., Mirzadeh, Y., Asadi, A., Ghaffarzadeh Razzaghi, M. R., & Shalbafan, M. (2025). Managing the psychosocial aftermath of the Rajaee Port Blast in Southern Iran: Quick emergency response in the first month after the blast. *Medical Journal of the Islamic Republic of Iran, 39*, 99.

Chapter 2
Integrating Resilience: Climate Adaptation and Disaster Risk Reduction in MENA Development Planning

Stern Mwakalimi Kita

Abstract The Middle East and North Africa (MENA) region faces converging risks from climate change, environmental degradation, and natural hazards that increasingly interact with water scarcity, rapid urbanisation, and institutional fragility. Intensifying droughts, floods, heatwaves, and ecosystem decline are creating cascading impacts that threaten development gains and human security. Global frameworks call for integrated approaches that embed disaster risk reduction (DRR) and climate change adaptation (CCA) into development planning. This chapter analyses how five MENA countries—Egypt, Jordan, Morocco, Lebanon, and the United Arab Emirates—are institutionalising disaster risk reduction (DRR) and climate change adaptation (CCA) through risk-informed development. It draws on a structured review of national development plans, adaptation strategies, DRR policies, voluntary national reviews, and peer-reviewed literature to assess the coherence of policy frameworks, the clarity of institutional mandates, and the extent of their operational implementation. Findings show strong progress in policy articulation but persistent fragmentation in implementation, driven by weak legal mandates, parallel institutional arrangements, limited finance, and uneven sub-national capacities. Cross-country comparison highlights emerging innovations including climate-responsive budgeting, nature-based solutions, and expanding digital and geospatial platforms. The chapter concludes with recommendations to strengthen legislative frameworks, risk-sensitive budgeting, institutional coordination, and regional cooperation to advance climate- and disaster-resilient development pathways in MENA.

Keywords Disaster risk reduction · Climate change adaptation · Risk-informed development · Resilience governance · Institutional capacity · MENA region · Climate finance · Mainstreaming integration

---

S. Mwakalimi Kita (✉)
Disaster Management Program, Faculty of Resilience, Rabdan Academy, Abu Dhabi, UAE
e-mail: ealam@ra.ac.ae

E. Alam (ed.), *Disaster, Crisis and Emergency Management in the Middle East & North Africa*, https://doi.org/10.1007/978-3-032-24436-9_2

2.1 Introduction: Framing the Challenge

The Middle East and North Africa (MENA) region stands at a critical juncture in its pursuit of sustainable development. Intensifying climate risks, rapid urbanisation, water scarcity, and social fragility are converging to create complex, cascading threats. Extreme heat events, prolonged droughts, flash floods, sea-level rise, and ecosystem degradation increasingly interact with poverty, displacement, and weak institutions to generate systemic risk (Chehbouni, 2024; Qasem & Scholz, 2025).

Recent, country-specific events illustrate the scale of these risks: The 2023 Derna floods in Libya caused more than 5,000 deaths and widespread infrastructure collapse (Armon et al., 2025); the UAE's April 2024 flash floods disrupted transport systems and exposed drainage limitations (Hussein et al., 2025); Morocco's 2023 Al Haouz earthquake revealed weakness in local preparedness (Achbani et al., 2024; Malhouni & Mabrouki, 2025); and recurrent droughts in Egypt and Jordan continue to strain agriculture and water resources (Abdelaty et al., 2023; Abu-Hamatteh et al., 2025). These shocks increasingly interact with poverty, displacement, and weak institutions to generate systemic risk across the region.

Recent assessments show that the Middle East records some of the world's highest multi-hazard risk levels, driven by rapid urbanisation, fragile health systems, and pollution burdens, while surface-temperature increases across the region are rising at more than twice the global average (Danish & Isaifan, 2024; IPCC, 2022; UNDRR, 2025). North Africa is projected to face longer and more frequent droughts, escalating water stress, and heat extremes, compounding vulnerabilities in countries such as Egypt and Morocco (IPCC, 2022; Shrimpton et al., 2025).

Despite growing recognition of these challenges, development planning across MENA often remains fragmented and reactive. Climate change adaptation (CCA) and disaster risk reduction (DRR) are still approached through parallel mandates or short-term projects, rather than through the governance systems needed to manage compounding risks effectively (UNDRR, 2021). Institutional capacities have not kept pace with escalating hazards, and governments continue to struggle with translating high-level climate commitments into operational planning, budgeting, and monitoring systems (Al-Hajj et al., 2020; Jum'a & Hamdan, 2024).

A deliberate shift towards risk-informed development is therefore essential. Decisions made today on land use, water, infrastructure, and urban growth will determine whether societies move towards resilience or maladaptation. This chapter illustrates how five MENA states—Egypt, Jordan, Morocco, Lebanon, and the United Arab Emirates (UAE)—are addressing this integration challenge through differing institutional pathways and levels of capacity.

2.2 Global Commitments and Regional Relevance

The integration of DRR and CCA into development planning is grounded in the global policy architecture established by the Sendai Framework for Disaster Risk Reduction (2015–2030), the Paris Agreement, and the 2030 Agenda for Sustainable Development. These frameworks call for risk-informed, climate-resilient development and have been widely endorsed by MENA governments through Nationally Determined Contributions (NDCs), National Adaptation Plans (NAPs), and national DRR strategies.

However, translating these global ambitions into coherent national systems remains a major challenge. The region's unique combination of hydro-climatic threats, rapid urbanisation, and fragile governance creates both constraints and opportunities for integration. Donor-supported climate and DRR initiatives have expanded, yet they often operate outside national planning and budgeting structures, limiting institutional ownership.

This chapter situates the MENA experience within this global context and examines how far the selected five countries have moved from commitment to implementation. In doing so it assess how far these global commitments have been reflected in national laws, institutions, and financing frameworks.

2.3 Purpose and Scope of the Chapter

This chapter provides a comparative, evidence-based assessment of how DRR and CCA are being integrated into national development planning across five MENA countries. Its purpose is to identify both promising practices and systemic barriers that shape the uptake of risk-informed development in the region. Specifically, the chapter (i) assesses the extent and quality of DRR/CCA integration within national and sectoral planning frameworks, and (ii) generates actionable recommendations for governments, regional bodies, and partners to advance coherent, inclusive, and sustainable resilience strategies.

The analysis focuses on Egypt, Jordan, Lebanon, Morocco, and the United Arab Emirates, which together represent a spectrum of institutional models and stages of mainstreaming. These range from long-established climate strategies to more recent donor-supported initiatives. The five countries were selected because they represent a spectrum of governance capacities, institutional models, and stages of mainstreaming DRR and CCA. Egypt and Morocco exemplify comprehensive, policy-led frameworks (Abou-Ali et al., 2023; Chehbouni, 2024); Jordan illustrates a governance-reform model shaped by demographic and water pressures (Jum'a & Hamdan, 2024); Lebanon presents a highly fragile, donor-dependent context (Al-Hajj et al., 2020); and the UAE demonstrates a high-capacity, innovation-driven system (Al Marzooqi, 2017; Danish & Isaifan, 2024). This diversity enables comparative insights across different integration pathways within the MENA region.

The comparative assessment is guided by three overarching questions: (1) How are DRR and CCA embedded in national development strategies, sectoral plans, and investment frameworks? (2) What institutional mechanisms enable or constrain coordination across climate, disaster, and development domains? and (3) To what extent are high-level commitments translated into budget systems, monitoring frameworks and sub-national implementation? These dimensions, policy integration, institutional alignment, and operational coherence, anchor the chapter's structure and guide the cross-country comparisons that follow.

This chapter is organised into eight sections. Following the introductory Sect. 2.1 and 2.2 outlines the conceptual framework, and Sect. 2.3 describes the methods. Section 2.4 provides the regional context, while Sect. 2.5 presents five country case studies. Section 2.6 synthesises comparative trends, and Sect. 2.7 offers recommendations for strengthening risk-informed development. Section 2.8 concludes by identifying future pathways for coherent, inclusive, and sustainable resilience governance in the MENA region.

2.4 Conceptual Framework—From Risk Reduction to Integrated Resilience

Integration of DRR and CCA into development planning is best understood as a process of governance transformation that embeds risk awareness in laws, budgets, and investment decisions. This chapter applies the lens of risk-informed development (RID); a concept that positions resilience as the organising principle of sustainable development. RID refers to the systematic incorporation of disaster and climate risk information into laws, policies, planning, budgeting, and investment decisions, with the aim of preventing the creation of new risks while reducing existing ones (UNDP, 2022; UNDRR, 2021).

2.4.1 The Rationale for Integration

Globally, frameworks such as the Sendai Framework for DRR, the Paris Agreement, and the 2030 Agenda all converge on the need to embed risk reduction and adaptation within the core machinery of development. Yet, as the five MENA country cases reveal, achieving such integration depends less on adopting these frameworks and more on how domestic institutions internalise them, through planning systems, budget processes, and inter-ministerial coordination.

The conceptual challenge is therefore one of institutional coherence: how climate and disaster governance are linked to national development planning, fiscal management, and local implementation. Integration succeeds when these systems operate in synergy rather than in isolation. These institutional and governance challenges create a practical need for an analytical structure that can assess not only the presence of policy commitments but also the mechanisms through which they are coordinated and executed. The following section outlines the analytical dimensions used to assess these dynamics.

2.5 Methodology

This chapter applies a structured comparative desk review to assess how DRR and CCA are being integrated into national development planning across five MENA countries: Egypt, Jordan, Morocco, Lebanon, and the UAE.

2.5.1 Approach and Sources

Given the cross-country scope and the absence of primary fieldwork, the analysis draws exclusively on secondary data, including national policy and planning documents, official reports to global frameworks, and peer-reviewed or institutional analyses. This includes NAPs, climate strategies, DRR frameworks, national development plans, and Voluntary National Reviews (VNRs) prepared under the 2030 Agenda. Complementary regional and global assessments from United Nations Economic and Social Commission for Western Asia (ESCWA), United Nations Development Program (UNDP), United Nations Office for Disaster Risk Reduction (UNDRR), and the World Bank were used to contextualise national findings within broader resilience trends.

Academic and grey literature provided additional insight into institutional reform, budget integration, and ecosystem-based approaches. These sources allowed for triangulation of evidence, ensuring analytical consistency despite differences in country data quality and disclosure.

Considerable variation exists in the detail and frequency of national reporting across the five countries. To address these differences, all documents were assessed for publication date, institutional origin, and consistency with other national and regional sources. Priority was given to the most recent and official documents to ensure data reliability.

2.5.2 Country Selection

Countries were selected based on two methodological criteria:

1. Availability of recent, accessible policy and institutional documents on DRR, CCA, and development planning; and

2. Representation of diverse governance and integration models, enabling comparison across different administrative and fiscal capacities.

Contextual justification for the relevance of each country is provided in Sect. 1.3. The assessment applies the three analytical lenses outlined in Sect. 2.2 to ensure a consistent basis for comparison.

2.5.3 Applying Analytical Lens

The analysis throughout this chapter is organised around three interrelated dimensions of integration that together define risk-informed development:

1. Policy Integration—the extent to which DRR and CCA are embedded in national and sectoral strategies and linked to sustainable-development priorities.
2. Institutional Alignment—the coherence of mandates, coordination structures, and inter-ministerial linkages that support or hinder mainstreaming.
3. Operational Coherence—how policy commitments are translated into practice through budget tagging, monitoring systems, and sub-national implementation.

The three analytical dimensions were selected because they are widely recognised in resilience-governance literature as core determinants of effective DRR–CCA mainstreaming and are consistent with global frameworks such as the Sendai Framework and UNDRR guidance. Together, they capture the full cycle of integration from policy intent to institutional coordination and practical implementation. These dimensions allow comparison across countries with different governance capacities and risk profiles, highlighting where integration is rhetorical, partial, or fully institutionalised. By applying this framework, the chapter examines how Egypt, Jordan, Morocco, Lebanon, and the United Arab Emirates have sought to operationalise risk-informed development. The focus is not on definitional debates but on how integration manifests in real policy and institutional practice, whether through legal mandates, budget tagging, local planning, or ecosystem-based interventions. A structured coding matrix was used to organise evidence across the three analytical dimensions. The matrix included specific indicators for each dimension: for example, the presence of legal mandates, coordination mechanisms, budget instruments, monitoring systems, and sub-national delivery arrangements. This structure enabled systematic comparison across countries and ensured consistent interpretation of heterogeneous sources.

2.5.4 Limitations

Reliance on secondary data introduces several constraints. Formal documents may not capture informal coordination dynamics or the political economy of implementation. Variations in terminology, reporting frequency, and disclosure levels also limit comparability. To mitigate these issues, information from national policy documents, regional assessments, and academic literature was triangulated, with key findings cross-checked across at least two independent sources wherever possible. This approach helped strengthen analytical validity and reduce interpretation bias arising from uneven documentation across countries. Despite these caveats, the structured synthesis offers a credible evidence base for cross-national learning and a baseline understanding of how resilience is being institutionalised in the region's development frameworks.

Before turning to the country cases, the next section situates the analysis within MENA's shared risk drivers and regional coordination landscape.

2.6 Regional Context: Risk and Resilience in MENA

2.6.1 A Region Facing Intersecting Risks

The MENA region faces a convergence of natural and climate-related hazards that compound deep-rooted socio-economic fragilities. Droughts, floods, sand and dust storms, heatwaves, and occasional seismic events increasingly threaten lives, infrastructure, and livelihoods (Bousquet et al., 2013; Eltinay & Harvey, 2019; Osman, 2024). Recent country-specific events illustrate this trend, including the 2023 Derna floods in Libya, severe flash floods in Amman and Beirut, and prolonged multi-season droughts affecting Egypt and Jordan (Al-Qinna et al., 2022; Ramat et al., 2023). Population growth, rapid urbanisation, and chronic water scarcity heighten exposure and strain already fragile ecosystems (ESCWA, 2017; Verner, 2012).

Arid and semi-arid zones dominate much of the region, rendering it acutely sensitive to temperature rise and rainfall variability. Worsening drought and desertification continue to undermine agricultural productivity and water security, particularly in Yemen, Sudan, and Syria (Bousquet et al., 2013; Qasem & Scholz, 2025). Recent analyses show temperature increases of 1.2 °C since 1970, projected to rise by 1.5–3 °C by 2050 under high-emission scenarios. These shifts drive frequent droughts, and reduce groundwater recharge and rising heat-related health risks, with implications for food security and rural livelihoods (Gebrael et al., 2024; Qasem & Scholz, 2025; Verner, 2012; Wodon et al., 2014).

Coastal zones such as Egypt's Nile Delta and Morocco's Atlantic corridor face increasing sea-level rise, saline intrusion, and erosion (ESCWA, 2017; Maboudi & D'Amico, 2025). The 2023 Derna floods in Libya and prolonged droughts in Iraq and Tunisia exemplify the increasing frequency and complexity of climate-linked disasters (Cooper, 2020; Eltinay & Harvey, 2019; Ramat et al., 2023).

Cascading and transboundary risks now define much of MENA's risk landscape. Environmental stress interacts with conflict, economic fragility, and weak governance, producing cross-border crises in food security, energy systems, and migration

(Bousquet et al., 2013; Eltinay & Harvey, 2019; Mena & Hilhorst, 2021). Climate-induced displacement has affected more than 1.5 million people in the past decade, and up to 19 million additional people could be displaced by 2050 without effective adaptation (Gebrael et al., 2024; Verner, 2012; Wodon et al., 2014).

The region's population is projected to approach approximately 700 million by 2050, magnifying exposure. Water scarcity already affects ~60% of people and food insecurity affected ≥55 million in 2019. Agricultural output is particularly sensitive, with ≈30% yield losses at 1.5 °C and ≈60% at 3 °C of global warming (Gebrael et al., 2024). These pressures are intensified by weak legal and institutional frameworks for adaptation, as most national climate laws prioritise mitigation over resilience (Maboudi & D'Amico, 2025). Adaptive capacity remains highly uneven: Gulf states such as the UAE and Qatar have invested in digital infrastructure and green cities, while conflict-affected or low-income countries depend on donor-supported short-term projects (Al-Sarihi & Luomi, 2019; Cooper, 2020; Luomi, 2021).

2.6.2 Regional Institutions and Coordination Mechanisms

The integration of climate adaptation and DRR in MENA has been advanced through several intergovernmental and technical initiatives. The League of Arab States (LAS) anchors political coordination through the Arab Permanent Committee on Meteorology and the Arab Committee for Disaster Risk Reduction. The Arab Strategy for Disaster Risk Reduction (2010–2020), later aligned with the Sendai Framework, represented a pioneering attempt to standardise DRR policy across Arab states, although implementation has remained uneven (Bousquet et al., 2013; Eltinay & Harvey, 2019; GFDRR, 2021; Verner, 2012).

The Regional Initiative for the Assessment of Climate Change Impacts on Water Resources and Socio-Economic Vulnerability (RICCAR), led by ESCWA with LAS, UNEP, and ACSAD, is the region's principal technical platform for climate-risk assessment. Its hydro-climatic modelling and vulnerability maps have informed national adaptation and water plans in Jordan, Tunisia, Iraq, and Egypt (ESCWA, 2017). Complementary networks, including the Arab Water Council, Arab Federation for Disaster Risk Management (AFADs), and Arab Climate Outlook Forums, promote sectoral coordination and technical exchange (Cooper, 2020; Wodon et al., 2014).

Despite these advances, binding regional monitoring, peer-review, or enforcement mechanisms remain limited, and governance asymmetries among Gulf, Maghreb, and Mashreq countries continue to affect coherence (Bousquet et al., 2013; ESCWA, 2017). Nonetheless, emerging cooperation on climate diplomacy, transboundary water governance, and nature-based solutions (NbS) provides a foundation for stronger regional resilience (Eltinay & Harvey, 2019; Gebrael et al., 2024).

2.6.3 *Challenges to Integration*

Institutional fragmentation remains the principal barrier to resilience mainstreaming. DRR functions are typically anchored in civil-protection agencies while adaptation sits within environment ministries, producing parallel policy streams and weak coordination (Bousquet et al., 2013; Eltinay & Harvey, 2019; Osman, 2024; Verner, 2012). Coordination challenges are particularly visible during large-scale events, such as flood response in Jordan and infrastructure failures revealed during the 2024 UAE floods (Hussein et al., 2025). The absence of shared data systems and integrated planning frameworks further constrain cross-sectoral action (ESCWA, 2017).

Regulatory and legislative gaps persist. Many countries possess climate or environmental laws but lack enforceable provisions linking adaptation and DRR to public investment and budgeting (Maboudi & D'Amico, 2025; Verner, 2012). At the local level, limited technical expertise and outdated hazard data continue to impede effective risk-informed planning (Cooper, 2020; ESCWA, 2017; Wodon et al., 2014).

Financial constraints further hamper implementation. Most countries rely on external finance such as the Green Climate Fund, AfDB, and the World Bank, with Egypt, Morocco, and Jordan receiving the largest shares, while fragile contexts remain under-supported (Cooper, 2020; GFDRR, 2021; Verner, 2012). Climate-budget tagging and risk-screening tools remain nascent (Eltinay & Harvey, 2019).

Persistent conflict and governance instability continue to undermine national resilience. Yemen, Syria, and Libya illustrate how institutional fragility limits national DRR–CCA integration and shifts priorities towards humanitarian response (Mena & Hilhorst, 2021). These differences in governance capacity directly shape the feasibility and consistency of risk-informed development across the region.

Against this backdrop, the five cases illustrate varied pathways from commitment to implementation.

2.7 Country Case Studies: Integrating Resilience in National Planning

2.7.1 *Egypt*

2.7.1.1 Strategic Framework and Policy Architecture

Egypt provides a strong example of how national climate strategies serve as vehicles for mainstreaming DRR and CCA into development planning. Earlier Organisation for Economic Co-operation and Development (OECD) analysis warned that integrating climate-risk considerations into planning was vital to address vulnerabilities in the Nile Delta's coastal and agricultural systems (Agrawala et al., 2004). The 2022 National Climate Change Strategy (NCCS) 2050 remains the principal policy instrument for climate resilience (Ministry of Environment, 2022). It aligns low-emission development with adaptive capacity across five strategic goals: green economic growth, strengthened climate governance, public awareness, scientific research, and climate finance (Abou-Ali et al., 2023). The NCCS is positioned as a comprehensive resilience-development blueprint linking mitigation, adaptation, and finance through coordinated governance reforms. Recent climate impacts reinforce the strategic emphasis on integration. Increasing saline intrusion in the Nile Delta, periodic flooding in Alexandria, and recurrent heatwaves affecting rural water demand highlight the urgency of linking climate policy with development planning.

2.7.1.2 Institutional Arrangements and Governance

General Directions 2 and 3 of the NCCS (2022) emphasise mainstreaming climate action into national planning and integrating sustainability criteria into budgeting processes. These measures underpin Egypt's broader goal of enhancing adaptive capacity and resilience within national development systems (Ministry of Environment, 2022). Implementation is led by the Ministry of Environment and the Egyptian Environmental Affairs Agency (EEAA), coordinating with sectoral ministries and governorates.

Despite this coordination, recent research highlights persistent gaps between national strategies and local implementation, particularly in agriculture and water. Field studies in the Nile Delta reveal weak horizontal coordination between frameworks such as the NCCS, the National Water Resources Plan (NWRP), and the Nexus for Water, Food, and Energy (NWFE) initiative (Raafat, 2025). Urban planning remains similarly fragmented, with coastal-city case studies indicating limited integration of DRR and flood risk management in local plans (Abdelaty et al., 2023; Dabaieh et al., 2022).

2.7.1.3 Operational Integration and Implementation

The NCCS adopts several cross-sectoral approaches, including climate-resilient infrastructure (Direction 4) and innovative finance instruments such as green bonds and results-based financing (Goal 4). Agricultural adaptation is supported through measures that enhance water-use efficiency and reduce rural vulnerability (Ministry of Environment, 2022). Vulnerability assessments consistently identify agriculture, coastal zones, and water resources as priority sectors requiring improved data systems (Elsehamy, 2015).

Aboelnaga (2021) developed an Egyptian Climate Change Adaptation Index to guide governorate-level planning, demonstrating how localised economic policies can operationalise national resilience goals. The Strategy for Adaptation to Climate Change and Disaster Risk Reduction (2011) and the National Strategy for Disaster Risk Reduction (2011, updated 2017) reinforce this integration by calling for multisectoral coordination, improved risk information, and community-based approaches (Dabaieh et al., 2024). Examples of operational uptake include early warning improvements in coastal areas, expanded use of remote-sensing tools for agricultural monitoring, and the gradual introduction of climate-sensitive investment screening in selected infrastructure projects.

2.7.1.4 Progress, Challenges, and Emerging Lessons

Historical analyses of disaster governance underline that crises such as the 2006 Avian Influenza outbreak exposed the costs of reactive planning and the need for proactive DRR integration in local frameworks (Arram, 2009). The NWFE initiative demonstrates Egypt's movement towards nexus-based financing, aligning adaptation, DRR, and development priorities (Al-Mailam et al., 2023). Egypt's policy architecture now shows increasing convergence between DRR and CCA under a unified resilience framework.

Implementation, however, remains uneven across sectors. Recent assessments note that green-infrastructure and finance initiatives under Vision 2030 are strengthening the DRR–CCA nexus (Abdelmelek et al., 2022), but coordination gaps among climate and disaster-management institutions, limited sub-national capacity, and incomplete data systems continue to hinder coherent implementation (Abdelaty et al., 2023).

Overall, Egypt demonstrates clear strengths in national strategy development, climate-finance innovation, and cross-sectoral policy alignment. Remaining gaps relate primarily to inconsistent local delivery, fragmented coordination between national and governorate institutions, and the need for more integrated monitoring systems to translate policy intent into measurable outcomes.

2.7.2 Jordan

2.7.2.1 Strategic Framework and Policy Architecture

Jordan has progressively integrated DRR and CCA within development planning, driven by chronic water scarcity, rapid population growth, and recurrent climate shocks. The country remains among the most water-stressed globally, and climate projections indicate intensified drought, rising temperatures, and greater hydrological variability. These pressures have reinforced alignment between development planning and resilience objectives.

Jordan's long-term climate governance framework is anchored in its NDC and the Climate Change Policy (CCP) 2022–2050 (MoE & UNDP, 2022), which updated the 2013–2020 policy and provided a comprehensive foundation for mainstreaming climate priorities across all sectors.

The CCP is aligned with the Economic Modernisation Vision (2022–2033) and establishes a results-based framework with monitoring indicators to guide implementation.

Complementing the CCP, the Climate Change Bylaw No. 79 of 2019 formalises national governance structures, including the National Climate Change Committee (NCCC), and mandates sectoral mainstreaming of climate objectives. These instruments collectively guide the policy coherence, NDCs updates, and formulation of the National Adaptation Plan (MoE, 2025). The NAP translates climate objectives into sectoral programmes, with dedicated coverage of water, agriculture, urban systems, biodiversity, health, and socio-economic resilience. The inclusion of "Urban Resilience and Disaster Risk Reduction" as a dedicated programme area highlights Jordan's intent to link adaptation and DRR within a risk-informed development pathway.

The CCP 2022–2050 is notable for being the first in the Arab region to mainstream gender considerations and emphasise the water–energy–food nexus. It builds on earlier frameworks such as the Green Growth Plan (2017). Financing needs identified in the *Country Climate and Development Report,* estimated at US$ 9.5 billion through 2030, reinforce the linkage between climate policy and Jordan's economic transformation agenda.

2.7.2.2 Institutional Arrangements and Governance

Jordan's governance framework seeks to strengthen both horizontal and vertical coordination. The Ministry of Environment leads national efforts, while line ministries are responsible for integrating adaptation within sectoral strategies. Local governance mechanisms exist but remain unevenly implemented. Research shows persistent discrepancies between national adaptation frameworks and municipal practice, largely due to limited fiscal decentralisation, unclear mandates and insufficient risk data systems (Jum'a & Hamdan, 2024).

Municipal case studies, such as Jerash, show that centralised decision-making and a reactive emergency-response culture continue to constrain proactive planning (Al-Nammari & Al-Zaghal, 2015). Jordan is, however, among the few states in the region that have embedded climate change considerations within poverty-reduction and sustainable-development-planning frameworks through successive iterations of the CCP and NAP (MoE, 2025; MoE & UNDP, 2022).

Other national strategies, including the National Water Strategy 2023–2040, NAP, and Groundwater Sustainability Policy (2016), recognise demographic pressures, particularly refugee population growth, as key drivers of water scarcity and infrastructure strain. These frameworks adopt a water–energy–food nexus approach, although refugee considerations are not yet consistently reflected in sub-national adaptation plans (Ministry of Water & Irrigation, 2016, 2023; World Bank, 2022a).

2.7.2.3 Operational Integration and Implementation

Operational integration remains limited. While the CCP outlines broad priorities, such as climate-resilient infrastructure and sustainable land use, few binding mechanisms currently enforce risk-informed investments or budget allocations. Existing synergies between DRR and adaptation appear most clearly in climate-resilient agriculture and technology-transfer initiatives that support livelihoods in arid and semi-arid zones (Al-Rusheidat et al., 2016).

Climate financing remains heavily donor-dependent, and implementation tends to follow a project-based logic, leading to fragmented and limited continuity between planning and delivery. Macro-economic modelling estimates that water scarcity could reduce GDP by 6.6% by 2050 without effective intervention, whereas a US$ 9.5 billion resilience investment pathway could enhance long-term growth if diversified financing is secured (World Bank, 2022a).

Urban evidence from Amman indicates that preparedness for flash floods has improved, yet adaptive and recovery measures remain weakly embedded in urban planning (Dalgamoni & Khwaileh, 2025). Spatial analyses along the Amman–Jarash corridor link increased flood and landslide risks to unregulated expansion and vegetative loss, emphasising the need for risk-sensitive land-use planning (Qutieshat & Al-Assaf, 2022). Energy-sector modelling under high-emission scenarios indicates a 23 per cent efficiency decline, reinforcing the case for embedding risk-assessment tools within infrastructure planning processes (Abu-Hamatteh et al., 2025).

2.7.2.4 Progress, Challenges, and Emerging Lessons

Evidence from 2025 highlights several priorities for strengthening implementation: activating the CCA Technical Advisory Body under Bylaw 79 (2019), integrating CCA and DRR representatives within the NCCC and NCSCM boards, establishing a national flood-management committee, and introducing budget tracking for CCA/DRR investments (Mercy Corps & LSE, 2025). Nationwide surveys indicate growing public concern over floods and heatwaves, signalling the need for locally led adaptation and financial accountability (Mercy Corps & LSE, 2025).

Overall, Jordan demonstrates strong policy coherence and administrative commitment to resilience. However, operational uptake continues to be constrained by centralised governance, donor dependence, limited municipal capacity, and the absence of binding mechanisms for risk-informed investment. Strengthening institutional coordination, expanding domestic financing, and clarifying local mandates remain essential for translating national commitments into effective local action.

2.7.3 *Morocco*

2.7.3.1 Strategic Framework and Policy Architecture

Morocco has emerged as a regional leader in integrating climate adaptation and DRR within national development frameworks. This progress is supported by institutional reforms and a growing portfolio of climate-resilient investments. The governance framework is anchored in a set of strategic documents designed to mainstream resilience across planning, budgeting, and implementation. The updated Nationally Determined Contributions (NDCs) and the Plan National Stratégique d'Adaptation (PNSA 2020–2030) form the core of this architecture, aligning adaptation and mitigation objectives with long-term sustainable-development goals (Ministère de la Transition Énergétique et du Développement Durable [MTEDD], 2022).

These strategies are complemented by findings from the Country Climate and Development Report (CCDR), which identifies critical resilience priorities in water and energy (World Bank, 2022a). The CCDR calls for stronger policy reforms, improved planning instruments, and expanded public investment to address high exposure to climate and disaster risks. Hydrological modelling confirms increasing aridity and rising sensitivity of the energy sector to drought, underscoring the need for integrated planning of water resources and renewable energy (Chehbouni, 2024). The CCDR also reports that rainfall shocks contribute to 37 per cent of national output volatility, while a 25 per cent reduction in water supply could lower GDP by 6.5 per cent by mid-century (World Bank, 2022a). The National Water Plan (2020–2050) allocates around US$41 billion for infrastructure such as dams, desalination, and irrigation modernisation, with an emphasis on demand management to complement large-scale investments (World Bank & IMF, 2022).

2.7.3.2 Institutional Arrangements and Governance

Morocco has strengthened cross-sectoral governance by enhancing coordination between ministries, parliament, and local authorities. The High Commission for Water, Forests, and the Fight against Desertification, alongside the Ministry of Energy Transition and Sustainable Development, plays a key role in guiding resilience planning. Climate and risk considerations have increasingly been embedded within the national Finance Law process, reflecting progress towards climate-responsive budgeting.

Institutional collaboration has improved both vertical and horizontal coherence, though local-level participation and technical capacity remain limited (Bachegour et al., 2023). Ongoing decentralisation reforms aim to address these gaps, but challenges persist in aligning ministerial strategies with territorial plans. Citizen engagement in adaptation planning also remains inconsistent, particularly in rural and mountainous areas where vulnerabilities are most acute (Skrimizea et al., 2022).

2.7.3.3 Operational Integration and Implementation

Operational integration has expanded through initiatives that localise risk-informed development. A notable example is the Stratégie de Résilience de Mohammedia et Ain Harrouda (2022–2027), which incorporates spatial risk assessments, infrastructure upgrades, and early warning mechanisms into municipal planning (World Bank, 2022b). This model demonstrates how urban resilience strategies can translate national priorities into actionable local interventions.

Sectoral progress is also evident. Agricultural water management, watershed restoration, and early recovery from droughts and floods have been implemented in regions such as Settat and El Jadida, where recurrent drought is a primary driver of vulnerability. Fuzzy Cognitive Mapping workshops highlight how unsustainable irrigation practices, soil degradation, and limited participation of women farmers continue to constrain adaptive capacity and reinforce systemic risk (Shrimpton et al., 2025).

The private sector is increasingly recognised as a key actor. Moroccan SMEs face significant exposure to climate hazards yet lack standardised risk management tools. The flowchart approach developed by Jaouhari et al. (2024) offers a practical model for integrating climate-risk assessment into company operations. Despite such innovations, implementation remains fragmented and often donor-driven, with limited cross-ministerial coordination and weak vertical linkages between national and sub-national actors (Bachegour et al., 2023).

2.7.3.4 Progress, Challenges, and Emerging Lessons

Despite strong regulatory frameworks, the translation of national adaptation strategies into territorial plans remains uneven, particularly in rural and mountain regions where data gaps and limited institutional capacity hinder implementation. Findings from the ADAPTtoDEVELOP initiative indicate high exposure to drought, land degradation, and livelihood risks among women and youth, exacerbated by migration and limited access to finance and education (Skrimizea et al., 2022).

The 2023 Al Haouz earthquake demonstrate both institutional strengths and persistent challenges. The Whole-of-Government response enabled coordinated military–civilian action, yet gaps in local preparedness, civil-society participation, and information management were evident. Scholars have recommended institutionalising a national emergency-management agency and adopting a Whole-of-Society approach to enhance inclusiveness and adaptability (Malhouni & Mabrouki, 2025).

Overall, Morocco reflects a transition from policy formulation to delivery. Empirical evidence consistently highlights the economic costs of drought and water scarcity: Rainfall variability contributes significantly to macroeconomic instability, and water rationing could reduce GDP by up to 6.5% (World Bank/IMF, 2022). Innovations in farmer-led adaptation (Shrimpton et al., 2025), private-sector resilience (Jaouhari et al., 2024), and disaster-governance reform (Malhouni & Mabrouki, 2025) demonstrate emerging strengths. However, systemic gaps remain in decentralised capacity, sustained financing, and community-level engagement.

A more consolidated approach, that links local governance, private-sector adaptation, and institutionalised Whole-of-Society coordination, will be essential for maintaining Morocco's leadership in risk-informed development.

2.7.4 Lebanon

2.7.4.1 Strategic Framework and Policy Architecture

Lebanon's efforts to integrate DRR and CCA into national development planning have gained new momentum with the adoption of the National Adaptation Plan (2025–2035), launched in 2025 alongside the updated NDC 3.0 and the Long-Term Low Emission Development Strategy (LT-LEDS) (Ministry of Environment & UNDP, 2025). The NAP represents the country's first comprehensive framework for climate-resilient development, building on earlier GCF- and UNDP-supported readiness work initiated in 2017. It identifies cross-sector priorities spanning water, agriculture, health, forestry, biodiversity, urban systems, and disaster risk reduction, with a strong emphasis on institutional coordination, risk-data management, and municipal capacity development (MoE, UNDP & GCF, 2025).

These priorities align with Lebanon's well-documented vulnerabilities to water scarcity, heat stress, and flood hazards. Long-term hydrological analysis indicates declining precipitation, reduced soil moisture, and groundwater depletion, supporting the case for no-regret measures such as watershed protection and more efficient irrigation systems (Bou-Zeid & El-Fadel, 2002). The World Bank's climate-recovery assessment similarly highlights climate-proofing of infrastructure and agriculture as potential drivers of economic recovery and job creation (Ahmad & Ranade, 2021). Other analyses underscore the role of resilience planning in strengthening fiscal stability by embedding adaptation within macroeconomic and investment frameworks (Zea-Reyes et al., 2021).

Lebanon's Fourth National Communication (FNC) to the UNFCCC provides updated emissions profile, climate projections, and sectoral adaptation needs (Ministry of Environment, 2022). However, these technical contributions have yet to catalyse broader planning reforms or sustained mainstreaming, largely due to institutional fragmentation, political instability, and economic crisis.

2.7.4.2 Institutional Arrangements and Governance

Institutional arrangements for adaptation and DRR remain fragmented. The Prime Minister's Disaster Risk Management Unit serves as the national lead for DRR, yet coordination with sectoral ministries, including environment, infrastructure, and finance, remains inconsistent. The absence of a binding climate law and the lack of climate-responsive budgeting mechanisms further hinder integration (Al-Hajj et al., 2020). Earlier institutional reviews stressed the need for a unified climate-governance platform and the establishment of climate-budget tagging to address fragmentation between the DRM Unit and sectoral ministries (Reyes, 2017).

Some progress has been achieved at the sectoral level. The Ministry of Environment, with support from IUCN and UNEP, developed biodiversity-inclusive Environmental Impact Assessment (EIA) guidelines for marine and coastal projects, providing one of the few formal mechanisms that integrate ecosystem considerations into development approvals (Khoury et al., 2021). The MoE also leads the Climate Change Resilience and Ecosystem Connectivity (CC REC) project, implemented by IFAD, which promotes nature-based solutions and sustainable livelihoods across the Shouf–West Beqaa–Mount Hermon corridor. This initiative illustrates how ecological restoration and community resilience can be advanced through landscape-level planning.

2.7.4.3 Operational Integration and Implementation

Operational integration is advancing slowly and unevenly. Several municipalities in Mount Lebanon and the Bekaa region have piloted local resilience strategies focused on flood management, solid waste systems, and early warning mechanisms. These initiatives demonstrate growing local engagement, yet they often remain disconnected from national frameworks, and risk data continues to be dispersed across institutions. Earlier DRR initiatives in Beirut, Saida, and Tripoli showed strong potential for municipal leadership but also exposed persistent shortcomings in coordination, data accessibility, and technical capacity (Haddad et al., 2014).

Agricultural remains a sector of concern. While farmers increasingly recognise climate-related threats, weak inter-ministerial coordination and insufficient knowledge-transfer systems restrict uptake of climate-smart practices (Zeid et al., 2023). The health sector is similarly underprepared. Rising temperatures, water scarcity, and pollution are heightening risks related to food security and vector-borne diseases, yet institutional capacity to manage these intersecting pressures is limited (Al Khatib et al., 2025).

These gaps highlight the need for integrated adaptation approaches that connect environmental, health, and social systems and that strengthen local delivery models.

2.7.4.4 Progress, Challenges, and Emerging Lessons

Lebanon's resilience landscape reflects a combination of donor-supported innovation and systemic governance challenges. International projects have helped lay the foundation for adaptation governance, notably in the water and ecosystem sectors, but weak institutional mandates and the absence of climate-budgeting mechanisms continue to limit national ownership. Fragmentation between the DRM Unit and line ministries persists, and the lack of a centralised risk-information system undermines policy coherence.

Recent studies highlight the urgency of strengthening the DRM Unit's mandate, institutionalising climate-budget tagging, and embedding adaptation into public investment planning (Al Khatib et al., 2025; Al-Hajj et al., 2020). Integrating resilience into the green-recovery agenda could also help align environmental priorities with economic stabilisation efforts, supporting more sustainable investment pathways (Ahmad & Ranade, 2021).

Moving forward, Lebanon's transition from project-based initiatives to systemic, nationally driven resilience planning will require a legally anchored climate framework, consolidated risk-information systems, and stronger institutional coordination across sectors and governance levels.

2.7.5 United Arab Emirates

2.7.5.1 Strategic Framework and Policy Architecture

The UAE is widely regarded as a regional leader in integrating climate resilience and DRR into national development frameworks. Its approach combines high-level political commitment, significant investment, and alignment with global frameworks such as the Sendai Framework, the Paris Agreement, and the SDGs. The UAE's leadership in resilience-building also reflects its broader sustainable-development transition. National strategies such as Centennial 2071 place climate action at the centre of national transformation, linking economic diversification, innovation, and resilience-building (Al Rustamani, 2014). Earlier regional assessments highlight the Gulf region's exposure to extreme aridity, water scarcity, and limited access to international climate finance, conditions that have encouraged self-financed and innovation-driven climate action (Shahid & Behnassi, 2014).

At COP28, hosted in Dubai in 2023, the UAE launched the UAE Framework for Global Climate Resilience, a landmark international declaration that positions adaptation and resilience as central development priorities. The framework focuses on four pillars: (i) national risk assessments, (ii) integrated resilience planning, (iii) transparent implementation and monitoring, and (iv) accountability through globally agreed indicators. The UAE–Belém Work Program, launched concurrently with Brazil, aims to develop resilience indicators that bridge local implementation and global reporting under Article 7 of the Paris Agreement (UNFCCC, 2025). Comparative analyses indicate that only a few Arab states, notably the UAE, Saudi Arabia, and Qatar, have introduced binding measures to reduce energy consumption, expand renewable-energy deployment, and promote sustainable urban design as part of resilience and low-carbon development strategies (Danish & Isaifan, 2024).

The UAE's climate diplomacy further reflects the strategic use of clean-energy partnerships to advance both domestic diversification and international cooperation. Through Masdar, IRENA, and its COP28 presidency, the UAE has positioned itself as a regional policy entrepreneur that leverages energy transition as an instrument of soft power (Al-Sarihi & Mason, 2023).

2.7.5.2 Institutional Arrangements and Governance

Domestically, the UAE's resilience governance is guided by the National Climate Change Plan 2017–2050, which integrates mitigation, adaptation, and private-sector engagement under one policy framework. Emirate-level plans serve as key operational units within this structure. The country's cross-sectoral coherence stands out within the Gulf, where institutional fragmentation often separates adaptation from mitigation (Danish & Isaifan, 2024).

The Abu Dhabi Climate Change Adaptation Plan for the Environment Sector 2025–2050 outlines multiple adaptation actions across sectors such as groundwater management, soil conservation, and biodiversity protection. These actions follow ecosystem-based adaptation (EbA) principles and align with national development goals (Environmental Agency—Abu Dhabi, 2025). Complementary initiatives, such as Masdar City, the Mohammed bin Rashid Al Maktoum Solar Park, and emerging green hydrogen projects, illustrate the integration of adaptation and mitigation within a unified development pathway (Al Rustamani, 2014).

The Ministry of Climate Change and Environment (MOCCAE) leads national coordination, supported by the Environment Agency—Abu Dhabi and the Dubai Supreme Council of Energy. Increasingly, these bodies embed risk considerations into permitting, infrastructure planning, and preparedness systems. Earlier reviews identified challenges including weak federal–Emirate coordination, unclear Civil Defence mandates, and a reactive disaster-management culture. Evidence from public-awareness surveys also highlights the influence of cultural interpretations of disasters, underscoring the need for ongoing community engagement and simulation-based preparedness (Dhanhani, 2010). Urban-planning frameworks such as *Estidama* show progress in integrating sustainability into building codes, linking urban resilience with mitigation and adaptation (Eltinay & Egbu, 2017).

2.7.5.3 Operational Integration and Implementation

Operational integration has accelerated in response to recent extreme-weather events. The April 2024 flash floods in Dubai and Sharjah exposed serious drainage deficiencies and catalysed large-scale infrastructure investment in stormwater management. The Tasreef Mega Project, valued at US$ 8.2 billion, aims to expand the national rainwater drainage network by 700 per cent, reflecting a risk-informed infrastructure upgrade consistent with Sendai Priority 4 (Build Back Better) and the UAE's circular-economy goals (Varma, 2024).

In 2024, the UAE approved a nationwide package of flood-management projects, including new dams, water canals, and embankments in Sharjah, Ajman, Ras Al Khaimah, Abu Dhabi, and Fujairah, designed to strengthen climate resilience and expand rain- and flood-water storage capacity in line with the UAE Water Security Strategy 2036 (Mair, 2024). Meanwhile, data-driven decision-making has become a cornerstone of the UAE's resilience model: the Ministry of Climate Change and Environment now publishes open geospatial and climate-data platforms, and the UAE Space Data Center's Geo-Spatial Analytics Platform (GIQ) offers satellite- and AI-driven tools for real-time monitoring and early warning (Ministry of Climate Change & Environment, 2017; Mohamed Bin Zayed University of Artificial Intelligence, 2025; UAE Space Agency, 2023).

Assessments of Abu Dhabi's disaster-management system highlight the need for integrated risk databases, strengthened inter-agency data sharing, and greater inclusion of business-continuity management in DRR frameworks (Al Marzooqi, 2017). The adoption of the *UAE Framework for Global Climate Resilience (FGCR)* at COP28 has also influenced domestic thinking on monitoring, evaluation, and learning systems (Beauchamp, 2024).

2.7.5.4 Progress, Challenges, and Emerging Lessons

The UAE's resilience model is characterised by strong political leadership, strategic investment, and technology-enabled implementation. These features position the country as a regional benchmark for integrating DRR and climate adaptation within a unified development framework. Ongoing national initiatives aim to ensure that resilience is embedded consistently across all Emirates and that local-level capacities continue to expand.

Remaining challenges include the need to diversify economic drivers, expand Emirate-level adaptation data, and strengthen linkages between disaster-risk management and broader development programmes. These priorities are reflected in ongoing initiatives under the National Climate Change Plan 2017–2050 and the UAE Framework for Global Climate Resilience (Al Marzooqi, 2017; Danish & Isaifan, 2024).

Overall, the UAE demonstrates how resilience and low-carbon development can be integrated within high-capacity governance. Anchored in strong leadership, robust data systems, and continuous innovation, its approach showcases how climate and disaster-risk integration can advance sustainable development in arid environments (Danish & Isaifan, 2024).

2.8 Comparative Synthesis: Trends, Gaps, and Patterns

Table 2.1 provides a consolidated summary of DRR–CCA integration across the five country cases. The subsequent subsections elaborate these patterns in more detail.

Table 2.1 Summary of DRR–CCA integration across five MENA countries

Country	Policy integration	Institutional alignment and governance	Operational coherence/ Implementation
Egypt	• NCCS 2050 aligned with Vision 2030 • DRR–CCA framed within unified resilience strategy • Linked to NWRP and NWFE initiatives	• MoE & EEAA coordinate with sectoral ministries • DRR strategies in place but coordination remains partial • Limited vertical integration with governorates	• Cross-sectoral measures (resilient infrastructure, green bonds) • Governorate-level adaptation index • Early-warning improvements • Uneven implementation due to capacity and data gaps
Jordan	• CCP 2022–2050 aligned with Economic Modernisation Vision • NAP includes Urban Resilience & DRR programme • Strong WEF nexus and gender integration	• Climate Change Bylaw 79/2019 formalises governance • Clear ministerial mandates but centralised decision-making • Weak municipal capacity and limited fiscal decentralisation	• Synergies in climate-resilient agriculture and urban flood management • Heavy donor dependence • Absence of binding risk-screening tools • Fragmented, project-based delivery
Morocco	• Updated NDCs, PNSA 2020–2030, and Water Plan 2050 integrate DRR–CCA • CCDR designates adaptation as national priority	• Improved cross-sectoral coordination • Climate priorities embedded in Finance Law • Decentralisation reforms ongoing but uneven local capacity	• Urban resilience strategies (e.g. Mohammedia–Ain Harrouda) • Agricultural water management and watershed restoration • SME climate-risk tools • Fragmented delivery and data constraints

(continued)

Table 2.1 (continued)

Country	Policy integration	Institutional alignment and governance	Operational coherence/ Implementation
Lebanon	• New NAP 2025–2035 aligned with NDC 3.0 and LT-LEDS • FNC provides risk data but limited influence on planning	• DRM Unit leads DRR but coordination inconsistent • No climate law or climate-budget tagging • Fragmentation worsened by political and economic crisis	• Local pilots in flood management, solid waste, early warning and NbS • Initiatives disconnected from national systems • Agriculture and health adaptation under-resourced • Risk data dispersed across institutions
UAE	• National Climate Change Plan 2017–2050, Centennial 2071 and COP28 outcomes embed DRR–CCA • Strong integration with climate diplomacy	• MOCCAE, EAD, and Dubai Supreme Council of Energy coordinate federal–Emirate action • Earlier Civil Defence and awareness gaps addressed through reforms	• Major stormwater and flood-management investments (Tasreef, dams, canals) • Open geospatial and climate-data systems • EbA integrated into Emirate-level plans

2.9 Country Typologies

Egypt demonstrates a policy-driven approach that integrates CCA/DRR within overarching development planning. The National Climate Change Strategy 2050, supported by the National Water Resources Plan and the NWFE program, aligns climate action with Egypt Vision 2030 and public investment systems, though local delivery remains limited by data coordination and data gaps (Abou-Ali et al., 2023; Dabaieh et al., 2024; Raafat, 2025).

Jordan offers a governance-reform model in which Climate Change Bylaw No. 79 (2019) formalises adaptation mandates and cross-sector coordination, yet operations remain donor-dependent and centralised, with persistent municipal capacity constraints and weak integration of refugee/gender dimensions (El-Anis & Poberezhskaya, 2023; Jum'a & Hamdan, 2024; Mercy Corps & LSE, 2025).

Morocco exemplifies an investment-driven model anchored in reforms and climate-responsive budgeting; the PNRC and the CCDR link adaptation to the Finance Law and the National Water Plan 2020–2050, while decentralisation and local participation lag and many projects remain donor-financed (Bachegour et al., 2023; Chehbouni, 2024; Skrimizea et al., 2022; World Bank, & International Monetary Fund, 2022). Lebanon reflects a nascent, donor-driven model where initiatives are fragmented and institutionalised mainly through pilots; NAP readiness and biodiversity-focused programmes (e.g., CC REC) build foundations, but coordination between the Prime Minister's Disaster Risk Management Unit and line ministries remains limited amid economic and governance constraints (Al Khatib et al., 2025; Al-Hajj et al., 2020; Khoury et al., 2021; Zea-Reyes et al., 2021).

At the other end of the spectrum, the UAE represents a high-capacity, innovation-driven model, where political leadership, integrated planning, and strong financing mechanisms enable rapid translation into practice, via the National Climate Change Plan 2017–2050, Emirate-level adaptation strategies, and flagship projects such as Tasreef (Danish & Isaifan, 2024; Al Marzooqi, 2017; Rustmanai et al., 2023).

2.10 Dimensions of Mainstreaming Across the Region

2.10.1 Policy Integration

Across the MENA region, disaster risk reduction (DRR) and climate change adaptation (CCA) are increasingly being embedded within national development frameworks, albeit with varying degrees of institutionalisation and enforcement. Egypt aligns its National Climate Change Strategy 2050 with Vision 2030, signalling a whole-of-government approach to climate-resilient planning (Abou-Ali et al., 2023). Jordan's Climate Change Bylaw No. 79 (2019) legally anchors adaptation mandates and links them to the Economic Modernisation Vision 2022–2033 (El-Anis & Poberezhskaya, 2023). Morocco's updated NDCs, Plan National Climate (PNRC), and National Water Plan 2020–2050 integrate adaptation priorities into fiscal and public-investment policies (Chehbouni, 2024; World Bank, & International Monetary Fund, 2022). Lebanon's NAP remains largely pilot-driven with limited institutionalisation (Al-Hajj et al., 2020). The UAE, by contrast, embeds adaptation within the National Climate Change Plan 2017–2050 and positions resilience as a core diplomatic priority (Danish & Isaifan, 2024).

2.10.2 Institutional Alignment

Countries have formalised coordination mechanisms with mixed effectiveness. Egypt's environment institutions lead integration but face persistent local-level challenges (Raafat, 2025). Jordan's structures are codified but remain centralised (Jum'a & Hamdan, 2024). Morocco's vertical coordination is improving, though sub-national participation is limited (Bachegour et al., 2023). Lebanon continues to experience fragmented mandates (Reyes, 2017), and the UAE shows strong federal–Emirate cooperation, yet it still requires stronger community inclusion (Al Marzooqi, 2017; Dhanhani, 2010).

2.10.3 Operational Coherence

Implementation capacity varies widely. Egypt and Morocco have begun integrating climate considerations into public investment and budgeting systems (Aboelnaga, 2021; World Bank, & International Monetary Fund, 2022). Jordan and Lebanon remain dependent on project-based and donor-funded models (Jum'a & Hamdan, 2024; Khoury et al., 2021). The UAE illustrates advanced operationalisation through infrastructure and data-driven systems such as the Tasreef national drainage initiative and the Geo-Spatial Analytics Platform (GIQ) (Al Marzooqi, 2017; UAE Space Agency, 2023).

2.10.4 Cross-Cutting Trends and Innovations

Across the five MENA countries, institutional and technical innovations indicate a gradual transition from policy intent to operational practice, reflecting a more mature regional approach to risk-informed development.

First, policy convergence is strengthening. Egypt, Jordan, and the UAE have established coherent policy architectures that link climate and development strategies, supported by legislative or executive mandates for implementation (Abou-Ali et al., 2023; El-Anis & Poberezhskaya, 2023; Danish & Isaifan, 2024).

Second, resilience localisation is advancing through sub-national planning processes. Morocco's Stratégie de Résilience de Mohammedia et Ain Harrouda (2022–2027) and Egypt's governorate-level climate-risk indexing system illustrate how national visions are being translated into place-specific interventions (Aboelnaga, 2021; Shrimpton et al., 2025; World Bank, 2022b).

Third, infrastructure-led adaptation remains a hallmark of higher capacity systems. The UAE's Tasreef programme exemplifies large-scale, technology-driven flood-management planning that combines engineering innovation with sustainability standards (Danish & Isaifan, 2024).

Fourth, data-driven and digital tools are reshaping governance processes. The UAE's open geospatial and climate-information platforms, together with Jordan's expanding monitoring and evaluation systems, mark a shift towards evidence-based and anticipatory planning (Al Marzooqi, 2017; Jum'a & Hamdan, 2024; UAE Space Agency, 2023).

Finally, fiscal innovation is emerging as a key enabler of climate-resilient development. Morocco's climate-sensitive budgeting framework, Egypt's pilot initiatives on budget tagging, and Jordan's medium-term investment pathway collectively signal movement towards more risk-responsive public finance (Abou-Ali et al., 2023; World Bank, & International Monetary Fund, 2022; World Bank, 2022a).

Together, these trends underscore a growing institutional coherence in the region: Policy integration, decentralised implementation, infrastructure resilience, data innovation, and fiscal reform are increasingly reinforcing one another.

2.10.5 Common Gaps and Constraints

Across the five cases, broader governance and political-economy factors shape the depth of DRR–CCA integration. Countries facing higher institutional fragmentation or conflict-related pressures, most notably Lebanon, and in governance terms Jordan, tend to rely on donor-driven, project-based approaches with limited continuity. By contrast, higher capacity and more stable systems such as the UAE and, increasingly, Morocco demonstrate how political cohesion and institutional clarity can accelerate risk-informed development.

Persistent constraints include fragmented mandates, weak vertical coordination, and limited decentralisation (Al-Hajj et al., 2020; Jum'a & Hamdan, 2024; Raafat, 2025). Financing dependence and fragile budget systems limit sustainability in Jordan and Lebanon, while Egypt and Morocco continue to pilot budget-tagging mechanisms (Abou-Ali et al., 2023; World Bank, & International Monetary Fund, 2022; World Bank, 2022a). Across all cases, monitoring gaps, data inconsistencies, and limited community participation persist (Al-Hajj et al., 2020; Al Marzooqi, 2017; Jum'a & Hamdan, 2024).

2.10.6 Conditions for Effective Integration

The comparative analysis of the five countries confirms that successful integration of climate adaptation and disaster risk reduction depends on more than the existence of strong policy frameworks. It requires a constellation of institutional, financial, informational, and social conditions that reinforce one another in practice. Four mutually supporting conditions stand out across the cases:

1. Institutional coherence between national and sub-national levels to bridge DRR and CCA mandates, clarify responsibilities, and ensure vertical coordination in planning and implementation.
2. Data and evidence systems that provide timely, disaggregated, and actionable information for risk-informed decision-making, monitoring, and learning.
3. Climate-responsive finance that aligns adaptation priorities with public-investment frameworks and mobilises private and concessional resources for resilience.
4. Localised adaptive capacity that embeds inclusion, equity, and sustainability principles into local governance and community action.

Comparative evidence shows differentiated progress. The UAE and Morocco demonstrate how institutionalised policy-finance linkages and strong technology systems enable more consistent operational uptake. In contrast, Lebanon and Jordan illustrate the difficulty of sustaining integration under fiscal stress and donor dependence. Egypt sits between these cases, with strong national frameworks but uneven local delivery (Abou-Ali et al., 2023; Al-Hajj et al., 2020; Al Marzooqi, 2017;

Chehbouni, 2024; Danish & Isaifan, 2024; Jum'a & Hamdan, 2024; Khoury et al., 2021; Mercy Corps & LSE, 2025; Raafat, 2025; World Bank, & International Monetary Fund, 2022). Collectively, these experiences suggest that progress depends on strengthening multi-level governance, scaling innovative financing instruments, and embedding risk considerations into routine planning and budgeting systems.

2.10.7 Operational Levers: Ecosystem-Based Approaches, Tools, and Capacity

Beyond institutional and policy reform, several operational levers are shaping how MENA countries translate resilience commitments into practice.

2.10.7.1 EbA and NbS

Morocco's farmer-led drought responses highlight how irrigation choices, soil health management, and gender inclusion serve as key levers for adaptation in drought-prone zones (Shrimpton et al., 2025; Skrimizea et al., 2022). In Lebanon, biodiversity-inclusive Environmental Impact Assessment (EIA) guidelines and the CC REC programme embed ecosystem considerations into development approvals and landscape management (Khoury et al., 2021). The UAE's Estidama framework links green-building standards to urban resilience and climate-adaptation objectives (Eltinay & Egbu, 2017). Together, these examples demonstrate that EbA and NbS are being mainstreamed through various pathways, though they remain subject to challenges that mirror those noted in governance and finance.

2.10.7.2 Governance Tools and Capacity-Building Mechanisms

Countries are increasingly using risk-screening tools, integrated EIA, vulnerability assessments, and climate-sensitive public investment pilots (Elsehamy, 2015; Jum'a & Hamdan, 2024; Khoury et al., 2021). Budgeting reforms range from Morocco's climate-sensitive Finance Law to Egypt's early budget-tagging efforts and Jordan's quantified investment pathway (Abou-Ali et al., 2023; World Bank, & International Monetary Fund, 2022; World Bank, 2022a), though municipal capacity remains limited in Lebanon and Jordan. The UAE's focus on integrated data governance and business-continuity highlights the role of institutionalised DRR systems (Al-Hajj et al., 2020; Al Marzooqi, 2017; Jum'a & Hamdan, 2024). Coordination mechanisms differ in legal force and vertical coherence across Egypt, Jordan, and Morocco, while public engagement remains a priority, particularly in the UAE (Bachegour et al., 2023; Dhanhani, 2010El-Anis & Poberezhskaya, 2023; Poberezhskaya & El-Anis, 2024).

Collectively, these tools show how policy intent is being operationalised through planning, financing, and capacity-building practices, creating practical pathways towards risk-informed, climate-resilient development.

2.11 Recommendations for Advancing Risk-Informed Development in MENA

Building on the cross-country findings, five priority actions can accelerate the integration of disaster risk reduction and climate adaptation into development planning across the region:

1. **Institutionalise risk screening and data-driven decision-making**: Make risk assessment and the use of climate and disaster data mandatory in national planning and public investment processes. This would close persistent monitoring and evidence gaps and ensure that infrastructure, land-use, and budget decisions are informed by real-time risk information.
2. **Strengthen legal mandates and vertical coordination**: Provide statutory authority for inter-ministerial and multi-level coordination bodies to address fragmentation and clarify institutional responsibilities. Codifying coordination roles can help translate high-level strategies into coherent implementation frameworks across national and local levels.
3. **Advance climate-responsive public finance**: Embed climate risk into budget classification systems, appraisal procedures, and investment tracking. Scaling climate-budget tagging and blended finance mechanisms will help align fiscal policy with adaptation and resilience priorities while maintaining macroeconomic stability.
4. **Resource and empower local implementation**: Expand fiscal and technical support to municipalities and governorates so that they can translate national resilience strategies into local land-use plans, infrastructure projects, and service delivery. Decentralised implementation is key to linking national policy intent with community-level impact.
5. **Scale up ecosystem-based and nature-based approaches**: Move beyond pilot projects by embedding EbA and NbS within national planning and financing instruments. This would enhance cost-effectiveness, generate co-benefits for biodiversity and livelihoods, and strengthen long-term resilience.

Collectively, these recommendations highlight that advancing risk-informed development in MENA is not a question of policy design alone but of institutional practice. Sustained progress will require political commitment, inter-sectoral coordination, and investment in the human and data systems that turn plans into measurable resilience outcomes.

2.12 Conclusion

The five country experiences illustrate a clear regional transition from policy formulation to implementation, albeit through different institutional and capacity pathways. Egypt demonstrates robust national frameworks but uneven localisation; Jordan has formalised governance structures yet remains constrained by donor dependence and limited municipal capacity; Morocco is advancing fiscal and territorial integration but requires deeper decentralisation; Lebanon continues to rely on donor-driven pilots amid fragmented mandates and weak climate budgeting; and the UAE offers a high-capacity, innovation-led model that still needs stronger community inclusion and equitable delivery.

Across these varied contexts, risk-informed development is emerging as both a policy vision and a practical framework for aligning disaster risk reduction, climate adaptation, and sustainable development. Sustained progress will depend on four mutually reinforcing conditions: coherent institutions that link national and subnational action, data and evidence systems that inform decision-making, climate-responsive finance that bridges public and private investment, and localised capacity that anchors resilience in everyday planning and service delivery.

The MENA region now faces a decisive opportunity: to consolidate these gains by institutionalising integrated resilience governance, scaling financing instruments, and embedding risk awareness in all stages of development planning. Doing so will determine whether the region's shift towards risk-informed, climate-resilient development becomes systemic and enduring, rather than episodic and project-based.

References

Abdelaty, H., Weiss, D., & Mangelkramer, D. (2023). Climate policy in developing countries: Analysis of climate mitigation and adaptation measures in Egypt. *Sustainability, 15*, 9121. https://doi.org/10.3390/su15119121

Abdelmelek, S., El Ghandour, S., Ibrahiem S, Madi, T. H., & Nadeem, Z. (2022). Mitigating the impact of climate change on Egyptian cities: Sustainable building and construction as a strategy. *Policy Brief 38*. The Policy Hub–GAPP School. https://fount.aucegypt.edu/studenttxt/106/

Aboelnaga, S. (2021). *Local economic development policies to adapt climate change hazards in Egypt* (Doctoral dissertation, Hungarian University of Agriculture and Life Sciences). https://doi.org/10.54598/001230

Abou-Ali, H., Elayouty, A., & Mohieldin, M. (2023). Climate action in Egypt: Challenges and opportunities. In A. Bhattacharya, H. Kharas, & J. W. McArthur (Eds.), *Keys to climate action* (pp. 57–83). Brookings Institution Press.

Achbani, A., Bouchriti, Y., Sine, H., et al. (2024). Key takeaways from the Al Haouz earthquake, Morocco, 2023. *Disaster Medicine and Public Health Preparedness, 18*, Article e88. https://doi.org/10.1017/dmp.2024.80

Agrawala, S., Moehner, A., El Raey, M., Conway, D., van Aalst, M., Hagenstad, M., & Smith, J. (2004). Development and climate change in Egypt: Focus on coastal resources and the Nile Delta (OECD Working Paper No. 3). Organisation for Economic Co-operation and Development. https://research.fit.edu/media/site-specific/researchfitedu/coast-climate-adaptation-library/africa/egypt-amp-libya/Agrawala-et-al.-2004.-Egypt-Coastal-Development--CC.pdf

Ahmad, E., & Ranade, S. (2021). *Climate-responsive economic recovery: Post-pandemic opportunities in the Mashreq (Lebanon, Iraq, Jordan).* World Bank. https://documents1.worldbank.org/curated/en/869611630485230362/pdf/Climate-Responsive-Economic-Recovery-Post-pandemic-Opportunities-in-Mashreq.pdf

Al Khatib, A., Alsaleh, B., Almari, M., Hassanein, S., & Al Hakawati, N. (2025). Mitigating climate change impacts on health: A comparative analysis of strategies in Saudi Arabia and Lebanon. *Frontiers in Public Health, 13,* 1551559. https://doi.org/10.3389/fpubh.2025.1551559

Al Marzooqi, A. (2017). *Disaster risk reduction in the emirate of Abu Dhabi, United Arab Emirates: Effects of natural and non-natural disasters over business continuity and sustainability* [Doctoral dissertation, Université Sorbonne Paris Cité]. HAL Archives. https://shs.hal.science/tel-01928654/

Al Rustamani, Z. A. (2014). *Impacts of climate change on urban development in the UAE: The case of Dubai* [Master's thesis, UAE University]. https://scholarworks.uaeu.ac.ae/all_theses/29

Al-Hajj, S., Abou-El-Hassan, H., Khalill, L., Kaafarani, H. M. A., & Sayed, M. (2020). Hospital disaster and emergency preparedness (HDEP) in Lebanon: A national comprehensive assessment. *International Journal of Disaster Risk Reduction, 51,* Article 101889. https://doi.org/10.1016/j.ijdrr.2020.101889

Al-Mailam, M., Arkeh, J., & Hamzawy, A. (2023). *Climate change in Egypt: Opportunities and obstacles.* Carnegie Endowment for International Peace. https://carnegieendowment.org/2023/10/climate-change-in-egypt

Al-Nammari, F., & Al-Zaghal, M. (2015). Towards local disaster risk reduction in developing countries: Challenges from Jordan. *International Journal of Disaster Risk Reduction, 12,* 34–41. https://doi.org/10.1016/j.ijdrr.2014.11.005

Al-Qinna, M. I., Hammouri, N. A., Obeidat, M. M., et al. (2011). Drought analysis in Jordan under current and future climates. *Climatic Change, 106,* 421–440. https://doi.org/10.1007/s10584-010-9954-y

Al-Rusheidat, J., Hammour, W. A., & Al-Jaafreh, S. (2016). Climate change adaptation and technology transfer: the path to disaster risk reduction in the arid and semi-arid zones. The case of Jordan. *New Medit 15*(1), 2–6. https://newmedit.ciheam.org/edizioni_new_medit,229,229,2016,162,1048,climate-change-adaptation-and-technology-transfer:-the-path-to-disaster-risk-reduction-in-the-arid-and-semi-arid-zones-the-case-of-jordan.htm

Al-Sarihi, A. & Luomi, M. (2019). Climate change governance and cooperation in the Arab Region. EDA Insights; Emirates Diplomatic Academy.

Al-Sarihi, N., & Mason, M. (2023). Challenges and opportunities for climate policy integration in oil-producing countries: The case of the UAE and Oman. *Climate Policy, 20*(10), 1226–1241. https://doi.org/10.1080/14693062.2020.1781036

Armon, M., et al. (2025). Anatomy of a foreseeable disaster: Lessons from the 2023 dam-breaching flood in Derna, Libya. *Science Advances, 11,* eadu2865. https://doi.org/10.1126/sciadv.adu2865

Arram, S. (2009). *Integrating disaster risk reduction into development planning in Egypt: The impact of avian influenza on livelihoods in Fayoum* (Master's thesis). American University in Cairo.

Bachegour, H., Moussaid, F. Z., Qafas, A., & Jerry, M. (2023). Integrating climate policy, social inclusion, and indigenous knowledge: Toward a holistic approach for Morocco's sustinable future. *International Journal of Accounting, Finance, Auditing, Management and Economics, 4*(5–1), 45–59.

Beauchamp, E. (2024, February 15). *What is the UAE Framework for Global Climate Resilience, and how can countries move it forward?* International Institute for Sustainable Development (IISD). https://www.iisd.org/articles/deep-dive/uae-framework-global-climate-resilience

Bousquet, F., et al. (2013). Natural disasters in MENA: A regional overview. *MENA Knowledge and Learning, No. 15.*

Bou-Zeid, E., & El-Fadel, M. (2002). Climate change and water resources in Lebanon and the Middle East. *Journal of Water Resources Planning and Management, 128*(5), 343–355. https://doi.org/10.1061/(ASCE)0733-9496(2002)128:5(343)

Chehbouni, G. (2024). Assessing the alignment and implementation challenges of the Sustainable Development Goals (SDGS) 2030 in Morocco: A Relevant Inquiry. *The Journal of Sustainable Development, Law and Policy, 15*(2), 186–223. https://doi.org/10.4314/jsdlp.v15i2.8

Combaz, E. (2019). Jordan's environmental policies and engagement on climate change. K4D Helpdesk Report. UK Department for International Development. https://opendocs.ids.ac.uk/opendocs/handle/20.500.12413/14551

Cooper, R. (2020). Donor support for climate change initiatives in the Middle East and North Africa. K4D Helpdesk Report.

Mercy Corps, & London School of Economics. (2025). *Improving climate policy in Jordan: Evidence paper.* Zurich Climate Resilience Alliance. https://zcralliance.org/resources/item/improving-climate-policy-in-jordan

Dabaieh, M., Maguid, D., Abodeeb, R., & El Mahdy, D. (2022). The practice and politics of urban climate change mitigation and adaptation efforts: The case of Cairo. *Urban Forum, 33*(1), 83–106. https://doi.org/10.1007/s12132-021-09444-6

Dabaieh, M., Maguid, D., El Mahdy, D., & Al Hegazi, S. (2024). Climate change adaptation at the margins: The case of Cairo Egypt. *Cities, 154*, Article 105390. https://doi.org/10.1016/j.cities.2024.105390

Dalgamoni, A., & Khwaileh, M. (2025). Policy gaps and urban flooding: A critical evaluation of flash flood risk management and resilience in Amman, Jordan. *SSRN*. https://ssrn.com/abstract=5392440

Danish, M., & Isaifan, R. J. (2024). Strategic plans to support action against climate change in the Arab countries in the Middle East region. *Journal of Sustainable Development Law and Policy, 15*(1), 304–335. https://doi.org/10.4314/jsdlp.v15i1.11

Dhanhani, H. A. G. (2010). *Evaluation of the response capability of the United Arab Emirates (UAE) to the impact of natural hazards* [Doctoral dissertation, University of Bedfordshire]. https://uobrep.openrepository.com/handle/10547/241787

El-Anis, I., & Poberezhskaya, M. (2025). Responding to climate change in Jordan: Understanding institutional developments, political restrictions and economic opportunities. *British Journal of Middle Eastern Studies, 52*(2), 335–353. https://doi.org/10.1080/13530194.2023.2279332

Elsehamy, M. I. (2015, July 27–31). *Climate change vulnerability and adaptation in Egypt and the role of research and science in informing adaptation to climate change in food production and water resources.* Ministry of Environment (Egypt).

Eltinay, N., & Egbu, C. (2017). Sustainable futures: Making cities resilient in the MENA region. In G. Aouad, A. Al-Hajj, & C. Egbu (Eds.), *Proceedings of the International Conference on Sustainable Futures 2017* (pp. 23–32). http://hdl.handle.net/2436/621230

Eltinay, S., & Harvey, M. (2019). Building urban resilience in the Arab Region: Implementing the Sendai Framework for Disaster Risk Reduction 2015–2030 at the local level. Contributing Paper to GAR 2019. UNDRR. https://www.undrr.org/publication/building-urban-resilience-arab-region-implementing-sendai-framework-disaster-risk

Environmental Agency–Abu Dhabi. (2025). *Abu Dhabi climate change adaptation plan for the environment sector 2025–2050.* https://www.ead.gov.ae/-/media/Project/EAD/EAD/Documents/EN_Abu-Dhabi-Climate-Change-Adaptation-Plan-for-the-Environment-v4.pdf

ESCWA. (2017). Climate Change and Disaster Risk Reduction in the Arab Region. *ESCWA Water Development Report 7.* https://www.unescwa.org/publications/escwa-water-development-report-7-climate-change-and-disaster-risk-reduction-arab

Gebrael, R., Mitri, G., & Kalantzi, O. (2024). Overview of nature-based solutions for climate resilience in the MENA region. *Nature-Based Solutions, 6*, 100159. https://doi.org/10.1016/j.nbsj.2024.100159

GFDRR. (2021). *Natural disasters in the Middle East and North Africa: A regional overview.* Global Facility for Disaster Reduction and Recovery.

Haddad, E. A., Farajalla, N., Camargo, M., Lopes, R. L., & Vieira, F. V. (2014). Climate change in Lebanon: Higher-order regional impacts from agriculture. *The Journal of ERSA, 1*(1), 9–24.

Hussein, K., Alhosani, N., Al-Areeq, A. M., et al. (2025). Unprecedented rainfall in the United Arab Emirates: Hydrologic and flood impact analysis of the April 2024 event. *Natural Hazards, 121*, 9363–9385. https://doi.org/10.1007/s11069-025-07156-9

IPCC. (2022). Climate change 2022: Impacts, adaptation and vulnerability. In *Contribution of Working Group II to the Sixth Assessment Report of the Intergovernmental Panel on Climate Change*. Cambridge University Press. https://www.ipcc.ch/report/ar6/wg2/

Jaouhari, Y., Stour, L., & Agoumi, A. (2024). Towards a standardized tool for improving the mainstreaming of climate-change adaptation in Moroccan private companies. *Multidisciplinary Science Journal, 6,* e2024062. https://doi.org/10.31893/multiscience.2024062

Jum'a, A., & Hamdan, S. (2024). Disaster risk reduction and resilience building in vulnerable communities: The role of community participation in Jordan. SSRN Preprint. https://papers.ssrn.com/sol3/papers.cfm?abstract_id=4926910

Khoury, R., Atiyah, P., Nader, M., Alhaj, D., & Ismail, D. (2021). *Mainstreaming biodiversity conservation in Lebanon: Guidelines for environmental impact assessment for projects affecting marine and coastal areas.* IUCN Regional Office for West Asia & Ministry of Environment.

Luomi, M. (2021). *Mainstreaming climate policy in the gulf cooperation council states.* OIES Paper: MEP7. https://doi.org/10.26889/9781907555916

Maboudi, T., & D'Amico, N. (2025). Vulnerability, climate laws, and adaptation in the Middle East and North Africa. *Environmental Policy and Governance, 35*(1), 145–158. https://doi.org/10.1002/eet.2134

Mair, F. (2024, October 18). *UAE approves major projects to battle flooding: It will benefit five emirates.* Timeout Abu Dhabi. https://www.timeoutabudhabi.com/news/uae-rain-canals-dams

Malhouni, Y., & Mabrouki, C. (2025). Whole-of-government approach in disaster management: Collaborative case study on the 2023 Morocco earthquake response. *Journal of Humanitarian Logistics and Supply Chain Management, 15*(1), 1–31. https://doi.org/10.1108/JHLSCM-10-2024-0152

Mena, R., & Hilhorst, D. (2021). The transition from development and disaster risk reduction to humanitarian relief: The case of Yemen during highintensity conflict. *Disasters, 46*(4), 1049–1074. https://doi.org/10.1111/disa.12521

Ministry of Climate Change and Environment. (2017). *National Climate Change Plan of the United Arab Emirates 2017–2050.* https://u.ae/en/about-the-uae/strategies-initiatives-and-awards/strategies-plans-and-visions/environment-and-energy/national-climate-change-plan-of-the-uae

Ministry of Environment. (2022). Egypt National Climate Change strategy 2050. Arab Republic of Egypt. https://www.fao.org/faolex/results/details/en/c/LEX-FAOC213202/

Ministry of Water and Irrigation. (2016). Jordan Groundwater Sustainability Policy. https://www.pseau.org/outils/ouvrages/mwi_groundwater_sustainability_policy_2016.pdf

Ministry of Water and Irrigation. (2023). *National Water Strategy 2023–2040.* Government of Jordan. https://www.mwi.gov.jo

MoE & UNDP. (2022). *National Climate Change Policy of the Hashemite Kingdom of Jordan 2022–2050.* https://www.undp.org/jordan/publications/jordans-2022-2050-national-climate-change-policy

MoE, UNDP & GEF. (2022b). *Lebanon's 4th National Communication on Climate Change.* https://unfccc.int/sites/default/files/resource/Lebanon-Fourth%20National%20Communication.pdf

MoE, UNDP, & GCF. (2025). *Lebanon's National Adaptation Plan (NAP) on climate change 2025–2035.*

MoE. (2025). *The National Climate Change Adaptation Plan of Jordan—2022.* https://unfccc.int/documents/645950

Mohamed Bin Zayed University of Artificial Intelligence. (2025, June 4). *MBZUAI to launch lab at UAE's Ministry of Climate Change and Environment.* https://mbzuai.ac.ae/news/mbzuai-to-launch-lab-at-uaes-ministry-of-climate-change-and-environment/

MTEDD. (2022). Plan National Stratégique d'Adaptation (PNSA-2030). Rabat, Morocco: Royaume du Maroc. https://unfccc.int/documents/636730

Osman, M. (2024). Assessing climate adaptation plans in the Middle East and North Africa. Carnegie Endowment for International Peace. https://carnegieendowment.org/research/2024/04/assessing-climate-adaptation-plans-in-the-middle-east-and-north-africa?lang=en

Qasem, R. H. M., & Scholz, M. (2025). Climate change impact on resources in the MENA region: A systematic and critical review. *Physics and Chemistry of the Earth, 139*, Article 103936. https://doi.org/10.1016/j.pce.2025.103936

Raafat, Y. (2025). National efforts in climate change adaptation in agriculture: Assessing policies and practices in Egypt's Nile Delta (Master's thesis). American University in Cairo. https://fount.aucegypt.edu/etds/2403

Ramat, G., Santi, E., Paloscia, S., Fontanelli, G., Pettinato, S., Santurri, L., et al. (2023). Remote sensing techniques for water management and climate change monitoring in drought areas: case studies in Egypt and Tunisia. *European Journal of Remote Sensing, 56*(1). https://doi.org/10.1080/22797254.2022.2157335

Reyes, L. Z. (2017). Which institutional barriers (and opportunities) exist in the climate adaptation planning process and how they interact with each other in the Municipality of Beirut (Lebanon)? MSc thesis. Erasmus University Rotterdam. https://thesis.eur.nl/pub/42843

Shrimpton, E. A., Sarmah, T., Balta-Ozkan, N., Malki, S., Jelloul, A., Ouikhalfan, M., Nilahyane, A., Kouisni, L., & Huo, D. (2025). Unpacking drought impacts and adaptive strategies in Morocco–Perspectives from small-scale farmers. *International Journal of Disaster Risk Reduction, 128*, Article 105732. https://doi.org/10.1016/j.ijdrr.2025.105732

Skrimizea, E., Larrue, C., El Gueddari, M., & Chourabi, H. (2022). *The development–climate change nexus in Morocco: Priorities for climate-resilient development in a changing environment.* ADAPTtoDEVELOP Policy Brief No. 1. Agence Française de Développement (AFD). https://ees.kuleuven.be/project/adapt2develop/results/socio-environmental-roadmap-for-morocco/deliverable-4-policy-brief-ADAPTtoDEVELOP.pdf

UAE Space Agency. (2023, December 3). *UAE Space Agency officially launches the operational phase of Geo-Spatial Analytics Platform (GIQ).* https://space.gov.ae/en/media-center/news/3/12/2023/uae-space-agency-officially-launches-the-operational-phase-of-geo-spatial-analytics-platform-giq

UNDP. (2022). *The UNDP approach to risk-informed development*. https://www.undp.org/publications/undp-approach-risk-informed-development

Ministry of Environment, & UNDP. (2025, September 22). *Lebanon launches climate policy package to advance Paris Agreement goals and accelerate low-carbon transition.* https://www.undp.org/lebanon/press-releases/lebanon-launches-climate-policy-package-advance-paris-agreement-goals-and-accelerate-low-carbon-transition

UNDRR. (2021). *Promoting synergy and alignment between climate change adaptation and disaster risk reduction in the context of national planning.* United Nations Office for Disaster Risk Reduction. https://www.undrr.org/publication/promoting-synergy-and-alignment-between-climate-change-adaptation-and-disaster-risk

UNDRR. (2025). *Global assessment report on disaster risk reduction 2025 (GAR 2025).* United Nations Office for Disaster Risk Reduction. https://www.undrr.org/gar/gar2025

UNFCCC. (2025). *Final list of potential indicators: UAE–Belém work programme on indicators.* https://unfccc.int/documents/649629

Varma, A. (2024, June 24). Dubai's new $8.2 billion "Tasreef" project to enhance rainwater drainage capacity. *Middle East Economy.* https://economymiddleeast.com/news/dubais-new-8-2-billion-tasreef-project-to-enhance-rainwater-drainage-capacity/

Verner, D. (Ed.). (2012). *Adaptation to a changing climate in the Arab countries: A case for adaptation governance.* World Bank. https://documents.worldbank.org/en/publication/documents-reports/documentdetail/740351468299700935/adaptation-to-a-changing-climate-in-the-arab-countries-a-case-for-adaptation-governance-and-leadership-in-building-climate-resilience

Wodon, Q., Burger, N., Grant, A., & Liverani, A. (2014). *Climate change, migration, and adaptation in the MENA region*. MPRA Paper No. 56927. World Bank. https://mpra.ub.uni-muenchen.de/56927/

World Bank, & International Monetary Fund. (2022). Climate change and development in Morocco. In *Morocco's quest for stronger and inclusive growth* (pp. 121–126). World Bank Group.

World Bank. (2022a). *Jordan country climate and development report*. World Bank Group. https://documents.worldbank.org/en/publication/documents-reports/documentdetail/368701682402395190/jordan-country-climate-and-development-report

World Bank. (2022b). *Stratégie de Résilience de Mohammedia et Ain Harrouda 2022–2027 (French)*. World Bank Group. https://documents.worldbank.org/en/publication/documents-reports/documentdetail/099810009272212622

Zea-Reyes, L., Olivotto, V., & Bergh, S. I. (2021). Understanding institutional barriers in the climate change adaptation planning process of the city of Beirut: vicious cycles and opportunities. *Mitigation and Adaptation Strategies for Global Change, 26*(26). https://doi.org/10.1007/s11027-021-09961-6

Zeid, S., Aoun, R., Hassan, L., & Mouawad, J. (2023). *Climate-smart agriculture readiness in Lebanon: Country profile and policy pathways*. Food and Agriculture Organization (FAO). https://www.fao.org/3/cc4480en/cc4480en.pdf

Chapter 3
Climate Change in MENA Region: Soil Degradation, Desertification Dynamics, and Food Systems Risk

Fatna Ikrame El Fanne

Abstract The increasing temperatures, occurrence of droughts, and extreme rainfall events in the MENA region have created challenges to the degradation of soils, further threatening food systems. As evaporation increases due to rising temperatures, the availability of soil moisture and vegetation also decreases, which increases the likelihood of soil erosion. In addition, erratically fluctuating weather patterns are resulting in greater amounts of sediment transport and consequently siltation in water bodies. Rising sea levels and saline groundwater have created additional threats to coastal areas and have compromised the integrity of the soil, which in turn has decreased agricultural production (by 37% since 1961). It has been noted, based on errors identified by several sources, that the regions of Morocco, Algeria, Tunisia, and Egypt present substantial evidence of degradation (e.g., aridity). This loss of agricultural productivity (approximately 34% reduction from 1961) due to climate change indicates there is a need for integrated adaptation strategies, such as soil conservation and agroecological practices, combined with improved water management and enhanced governance, and climate finance to best support the development of tailored resilience strategies for the MENA region.

Keywords Climate change · Soil degradation · Erosion · Desertification · Salinisation · Extreme events · MENA region · Food security · Adaptation · Resilience

F. I. El Fanne (✉)
Department of Process and Environmental Engineering, Faculty of Sciences and Technologies-Hassan II University, Mohammedia, Morocco
e-mail: ikrameelfanne@gmail.com

Youth for Climate Morocco, Casablanca, Morocco

E. Alam (ed.), *Disaster, Crisis and Emergency Management in the Middle East & North Africa*, https://doi.org/10.1007/978-3-032-24436-9_3

3.1 Introduction

Climate change is rapidly affecting the soil systems in the Middle East and North Africa (MENA), which depend on these systems for food production. As a result of rapid warming, increased drought, and rainfall variability, recent evaluations have found the wider Mediterranean region to be a "climate change hotspot" due to severe warming trends and reduced water availability. In North Africa, the increase in the frequency of dry–hot extremes, hydrological droughts and short-duration, high-intensity rainfall events, has compounded the degradation of soils across a region already burdened by significant land-use pressures. Soils, located in semi-arid and Mediterranean climates are particularly sensitive to rapid fluctuations in moisture, temperature, and rainfall intensities. Longer and hotter dry seasons usurp vegetation cover and soil organic matter, while concentrated rainfall on dry or crusty surfaces significantly increases both water erosion and sediment transport. Coastal and irrigated soils are also impacted by salty water tables due to the effects of sea-level rise and the use of marginal quality irrigated water. The combined effects of these soil-system stresses will ultimately impact food production systems across MENA. According to the IPCC AR6, there has been a decline of approximately 34% in African agricultural productivity growth since the early 1960s, which is the highest decline caused by climate change globally. This chapter reviews the climate change processes impacting soil degradation in MENA by collecting quantitative data from North African case studies and identifying resilience and adaptation pathways to improve both soil and food production systems in MENA.

3.2 Research Gaps

As the regional body of climate risk and land degradation continues to increase, researchers still see many gaps in knowledge. The first gap is that researchers often analyse climate change and soil on separate levels and therefore do not fully understand how climate change affects soil degradation processes (i.e., erosion, salinity, organic matter loss). The second gap is that few studies have been conducted that connect links system degradation with food-system vulnerability at the regional level; especially in North Africa, the variability in available data is enormous. The third gap is that there are few comparative analyses among Morocco, Algeria, Tunisia, and Egypt, thus creating inconsistency in the methods by which soil threats and climate signals are analysed. Finally, the majority of the studies that exist do not examine the effect that governance, institutional barriers, and adaptation finance impose on the ability to scale-up technical solutions. This chapter will address these gaps through the development of a comprehensive and integrated cross-country analysis.

3.3 Objectives

This chapter will examine the following objectives:

1. Analysing the most recent climate signals observed in the MENA region that are currently having a direct impact on soil degradation.
2. Identifying the key physical and chemical pathways of soil degradation associated with climate extremes.
3. Comparing the country-level research findings for Morocco, Algeria, Tunisia, and Egypt to identify similarities and differences in patterns of degradation.
4. Examining how the degradation of soils will impact regional food systems.
5. Recommending adaptation and resilience pathways to enhance the stability of soils and food systems through improved governance and policy-making in view of deteriorating climate conditions.

3.4 Methodology

This chapter presents a systematic overview of the published literature, synthesising reviews from peer-reviewed Journal Articles, national-level hydrological assessments, FAO and WMO datasets, and the regional/sub-regional assessments available from IPCC AR6 (2021). Sources for this systematic review will be limited to 2015–2024 literature, thus allowing for analysis of the most recent climate change trends and the most current hydrological and soil data available. As outlined in the chapter, this synthesis identifies four countries (Morocco, Algeria, Tunisia, Egypt) based on their rich data availability and their representation as examples of four very different climate sub-regions (Mediterranean, semi-arid steppes, heavily irrigated plains, and deltaic systems). Quantitative indicator values were extracted for erosion rates (RUSLE estimates), sediment yield, salinity/sodicity (EC/SAR), and gen-groundwater quality as well as for climate data (temperature trends, drought indices, and rainfall intensity). These results demonstrated several key limitations resulting from unbalanced data availability across countries as well as variability in methodology across source studies; yet based on the information received to date, the synthesis provides a clear and uniform regional assessment related to climate change risks imposed on soil/resource systems and food-preparation systems.

3.5 Climate Signals Relevant to Soils in North Africa

North Africa is one of the most sensitive regions of the Mediterranean basin to the impacts of climate change. Due to anthropogenic warming and changes in atmospheric circulation patterns, North Africa has seen some of the most dramatic regional climate signals. North Africa has been impacted by rapidly increasing temperatures, increased frequency and intensity of heat waves, decreased and erratically patterned precipitation, increased incidence of multi-year drought conditions, increased saltwater intrusion into coastal areas, and increased wind and dust eroding the land. Each of these changes has resulted in the degradation of the land, mainly because of the change in the erosion of the land and the loss of soil organic carbon, water erosion from the degradation of the soil structure, and increased soil salinisation. Furthermore, North Africa has been warming faster than the average rate for the rest of the Earth. North Africa is the only region in the world that recorded the largest increase in temperature for the period of January to August 2023. Higher temperatures create higher rates of evapotranspiration in the atmosphere, and, therefore, are making the soils increasingly weak and vulnerable to crusting formation and disintegration. The combinations of increased extreme heat events and widespread drought conditions in the past 50 years are a result of increasing evaporative demand due to the rising temperature, and not solely due to decreasing precipitation amounts. Rainfall trends from 1960 to the present day reveal decreasing total rainfall amounts during the annual rainfall seasons, as well as increasing variability of intra-annual rainfall events. Future predictions based on the CMIP6-AR6 models indicate that the continuation of these trends will result in more intense and longer duration future meteorological droughts, especially under the scenario of high emissions producing a decrease in the number of growing seasons, reduction in vegetative cover, and overall reduction in soil moisture content (Figs. 3.1 and 3.2).

The Atlantic Multidecadal Variability (AMV) and Mediterranean Oscillation are important to the study of rainfall patterns over North Africa (JA, MDOP, AMV). Phases of the positive North Atlantic Oscillation (NAO) have tended to push low-pressure systems to the poles. These shifts have resulted in reduced winter precipitation rates across the northern African countries of Morocco, Algeria, and Tunisia.

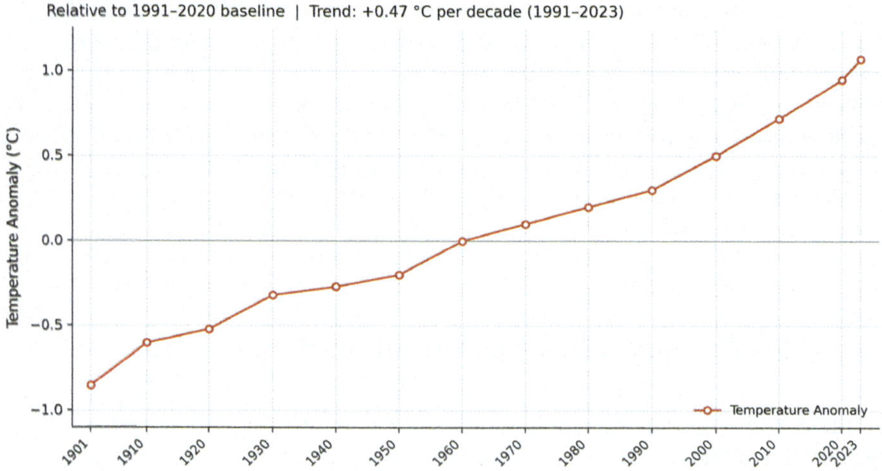

Fig. 3.1 North Africa Annual Temperature Anomalies (1901–2023). *Data source* WMO (2024), State of the Climate in Africa 2023

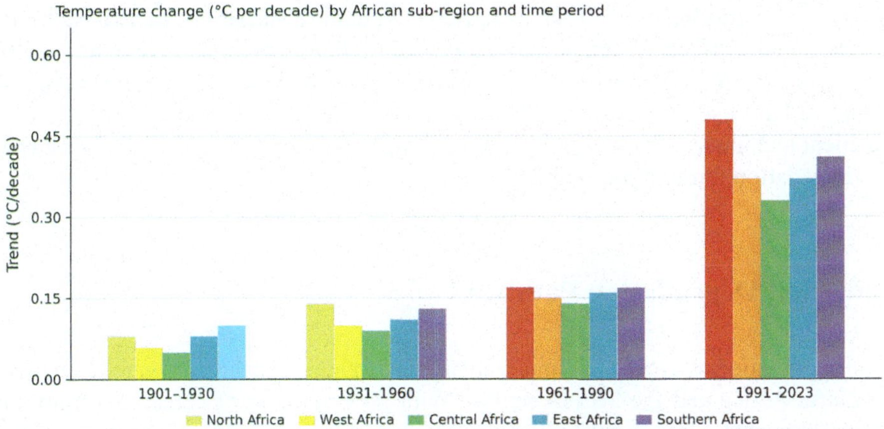

Fig. 3.2 Regional temperature trends across Africa. *Data source* WMO (2024), State of the Climate in Africa 2023. *Key Finding* North Africa shows the highest warming rate in the most recent period (1991–2023: + 0.45 °C/decade), exceeding all other African sub-regions

The AMV has altered wet–dry cycles occurring on a decadal basis. Both movements create soil moisture anomalies, which remain visible for several years, thereby inhibiting plant growth and adding to the generation of dust emissions and increased potential for erosion (Hakam et al., 2025). While some long-term assessments estimate that the total volume of annual precipitation will be reduced, the Mediterranean region is experiencing an increase in the frequency of high-magnitude events (i.e., "flash floods") that occur in areas with mountainous terrain (Adelo et al., 2026; Hakam et al., 2025). Flash floods can promote a greater risk for soil erosion due to increased rainfall and runoff events. The trend of increased flash flood hazards over the northern Mediterranean region has been reported in the literature since at least 2006 and continues to be reported through 2020 (Tramblay et al., 2013, 2024). Current Soil Erosion Data and Future Landscape Erosion Risk using RUSLE and SWAT reveal that, as a whole, the Mediterranean region will continue to be a source of high-risk areas for soil erosion, which will continue to increase due to compound dry–hot extremes (CDHEs) (Ed-Dakiri et al., 2024; Fredj et al., 2024; Serbaji et al., 2023). CDHEs represent a critical component of the CDHE phenomenon because they are the driving force behind soil degradation and are rapidly accelerating soil desiccation. The consequences of CDHEs, which are becoming increasingly common in the Mediterranean region, include rising air temperatures, decreased soil moisture preservation, and a higher chance of soil disintegration. In Southern Morocco, Western Algeria, and the peripheral areas of steppes, wind erosion is the primary land degradation concern. In North Africa, dust emissions are a significant contributor to dust and soil organic carbon exports to the Earth (Darvishi Boloorani et al., 2023). The salinisation of soil is currently impacting large-scale irrigated agriculture and is projected to worsen. Land subsidence, reduced sedimentation rates, and saline intrusion are contributing to increased soil salinity levels in the Nile Delta region

(FAO, 2021; Tarolli et al., 2024). Increased temperature variances, fluctuating water levels, and the effects of human activities (land use/land cover) will continue to alter soil–climate interactions within the Maghreb region. Without the implementation of adaptive land-use management practices, soil conservation measures, and climate-resilient techniques, the North African soils will continue to face an increasing risk of degradation through the year 2100.

3.6 Soil Degradation Pathways

Soil degradation as a result of climate change has a significant impact on food security in North Africa and the MENA region. With 50% or more of cereal consumption being imported, climate-related deterioration will have negative consequences for future food availability. Erosion caused by water runoff during storms has resulted in high loss of soil in Morocco, Algeria, Tunisia, and other Mediterranean regions. GIS mapping has identified areas with the greatest risk of damage due to high slopes, heavy storm frequency, and agri-business expansion.

For each centimetre of topsoil that is lost, wheat production is estimated to decrease by approximately 4–8%. Climate models suggest that future climate patterns will result in increased rainfall and soil erosion during periods of intense storm activity following extended dry periods. Wind erosion is also a concern, especially in the Saharan margins, where crop and forage production take place. Dust produced from this region also causes extensive loss of rangeland due to desertification.

In addition, soil organic carbon (SOC) levels are extremely low in Mediterranean dryland environments and have been reported to be below the critical threshold (2% SOC) for sustainable agricultural production. The loss of SOC caused by climate change makes SOC more vulnerable to extended droughts or further temperature increases. Conservation farming techniques, including crop residue retention and reduced tillage practices, can help prevent SOC losses from occurring. The salt-affected soils that exist in agricultural regions of Egypt's Nile Delta contribute to the ongoing reduction of agricultural output. Heavy rainfall also leaves behind a layer of soil; therefore, water infiltration will not occur and will increase the stress for agricultural productivity (Table 3.1).

These ongoing economic, climatic, and agricultural factors are creating a situation for agricultural producers where multiple vulnerabilities are compounding to increase concerns over food security. These impacts, when coupled, will reduce cereal production in North Africa between 20 and 40% by 2050, making it more difficult for smallholder farmers to meet their increasing consumer demand for food, leading to increased costs, and creating a greater link between the price volatility of food and the potential for civil unrest in North Africa. Consequently, soil degradation will establish significant challenges for maintaining food security in North Africa during a time of climate change (Fig. 3.3).

Table 3.1 Soil degradation pathways and food security impacts in MENA

Degradation pathway	Extent in MENA	Yield/Productivity impact	Food security consequence	Key sources
Water erosion	High-to-very-high in 15–25% of Mediterranean basins	4–8% yield loss per cm of topsoil lost	Reduced cereal production; reservoir siltation (0.5–1%/yr)	Bakker et al. (2004), Ed-Dakiri et al. (2024)
Wind erosion	~ 20 million ha (Algerian steppe alone)	Selective removal of fertile fine fractions	Rangeland degradation; pastoral livelihood loss	Houyou et al. (2016), UNCCD (2022)
SOC depletion	< 1% SOC in most MENA agricultural soils	30–50% reduction in N mineralisation	Increased fertiliser dependency; smallholder vulnerability	Zdruli et al. (2004), Mrabet (2002)
Salinisation	46% of Nile Delta; expanding in irrigated plains	25–50% yield decline in severe areas	$12B global annual losses; increased import dependency	Kotb et al. (2000), Qadir et al. (2014)
Sealing/ Compaction	Widespread in mechanised drylands	50–90% infiltration reduction	Reduced water availability; increased drought vulnerability	

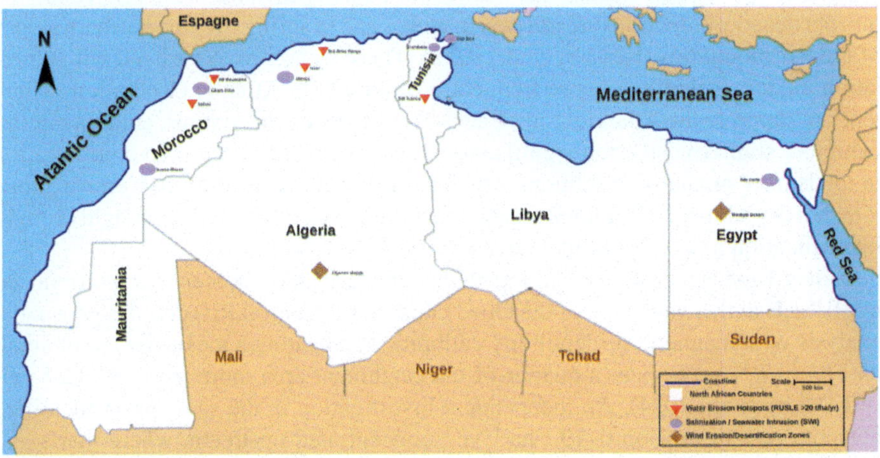

Fig. 3.3 Map showing the four study countries with major soil degradation hotspots. Red triangle: water erosion zones with RUSLE-based estimates (Rif Mountains, Morocco; Tell Atlas, Algeria; SW Tunisia); Brown diamond: wind erosion and desertification fronts (Algerian steppe, 20M ha affected); Purple ellipses: salinisation and seawater intrusion zones (Nile Delta, Egypt 46% affected, Cap Bon and Grombalia, Tunisia; Souss-Massa, Morocco 37 g/L salinity, Gharb Plain, Mitidja)

3.7 Country-Level Perspectives (Morocco, Algeria, Tunisia, and Egypt)

Climate change is modifying hydroclimate regimes across North Africa—shorter and more erratic wet seasons, more frequent dry–hot extremes, and increased coastal hazards. These changes then intensify the pathways of soil degradation (water/wind erosion, salinisation/sodification, organic matter losses) and propagate through food systems. The yield growth of agriculture in Africa at a continental scale has already been reduced by ~34% since 1961 due to climate change, more than in any other continental region, highlighting the magnitude of risk (high confidence).

3.7.1 Morocco

The soils of Morocco lie at the intersection of multiple climate stresses, which include warming, a tendency to have drier winters and springs, and more concentrated autumn rainfall patterns, all mediated by severe land pressures in the Rif, Middle Atlas, and Atlantic plains. Long quality-controlled rain-gauge records indicate a statistically significant decline in annual precipitation patterns over much of the agriculturally important Sebou basin since 1961, with Theil–Sen slopes in the order of −10 to −58 mm per decade; reductions are greatest in winter and in spring over the northern slopes of the Rif mountains (Kessabi et al., 2022). These trends are correlated with circulation modes, including a more frequent positive North Atlantic Oscillation, which tends to suppress the Atlantic storm tracks over Morocco. In contrast, autumn rainfall exhibits a slight (mostly non-significant) upturn, suggesting increased concentration in totals inter-annually, conditions which tend to increase erosion risk on bare soils after hot dry summers. Regional assessments for the Mediterranean highlight drought intensification as a major climate hazard (IPCC, 2023) (Fig. 3.4).

Soil erosion is therefore a major pathway of degradation. Recent work introduced by RUSLE in the watershed of the Oued Ghiss dam (Central Rif) couples geospatial analysis with magnetic susceptibility calibration indicating a mean erosion of about 24 t ha^{-1} yr^{-1}, with about a quarter of the catchment area more than 10 t ha^{-1} yr^{-1} (Ed-Dakiri et al., 2024). An independent investigation in the same basin identified hotspot areas of erosion 0–19 t ha^{-1} yr^{-1} concentrated upstream, where marls and steep slopes coincide (Taher et al., 2022). Such losses translate directly into siltation of reservoirs and reduced operational ages of infrastructures, a chronic problem in Morocco as noted throughout the reservoir systems, including considerable areas such as Al Wahda, areas of the highest recorded annual silt volume in the country (Loudyi et al., 2018; Mosaid et al., 2024) (Figs. 3.5 and 3.6).

Further strain on the region due to flash floods is associated with short-duration, high-intensity storms. The findings from case studies conducted in Morocco suggest that urban areas associated with built environments are at risk of flash flooding due to

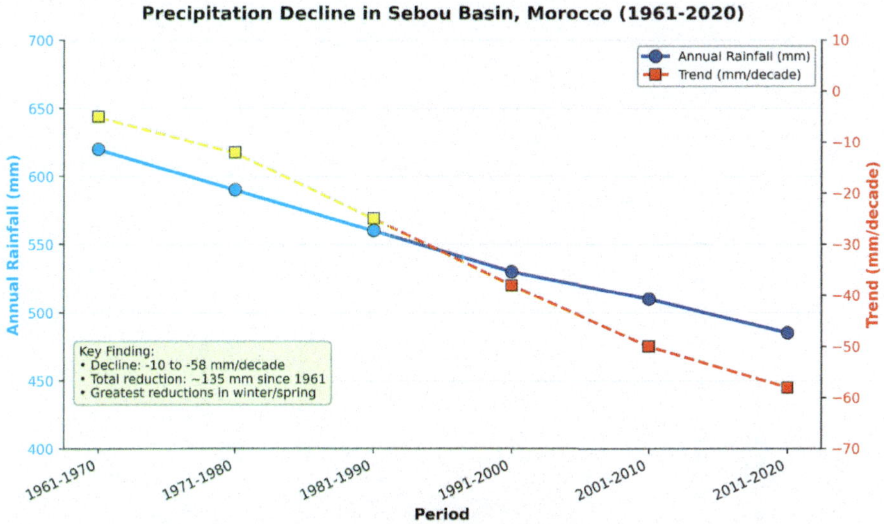

Fig. 3.4 Precipitation decline in Sebou basin (1961–2020). *Key Finding* Statistically significant decline of -10 to -58 mm/decade across the Sebou basin, with greatest reductions in winter and spring (Kessabi et al., 2022). Total reduction ~135 mm since 1961

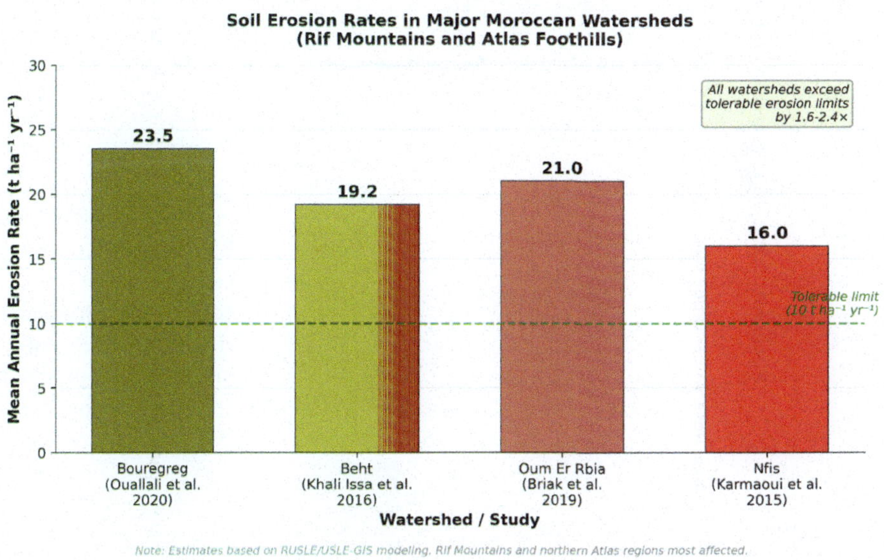

Fig. 3.5 Soil erosion rates in major Moroccan watersheds. *Sources* Ed-Dakiri et al. (2024), Taher et al. (2022), Kessabi et al. (2022)

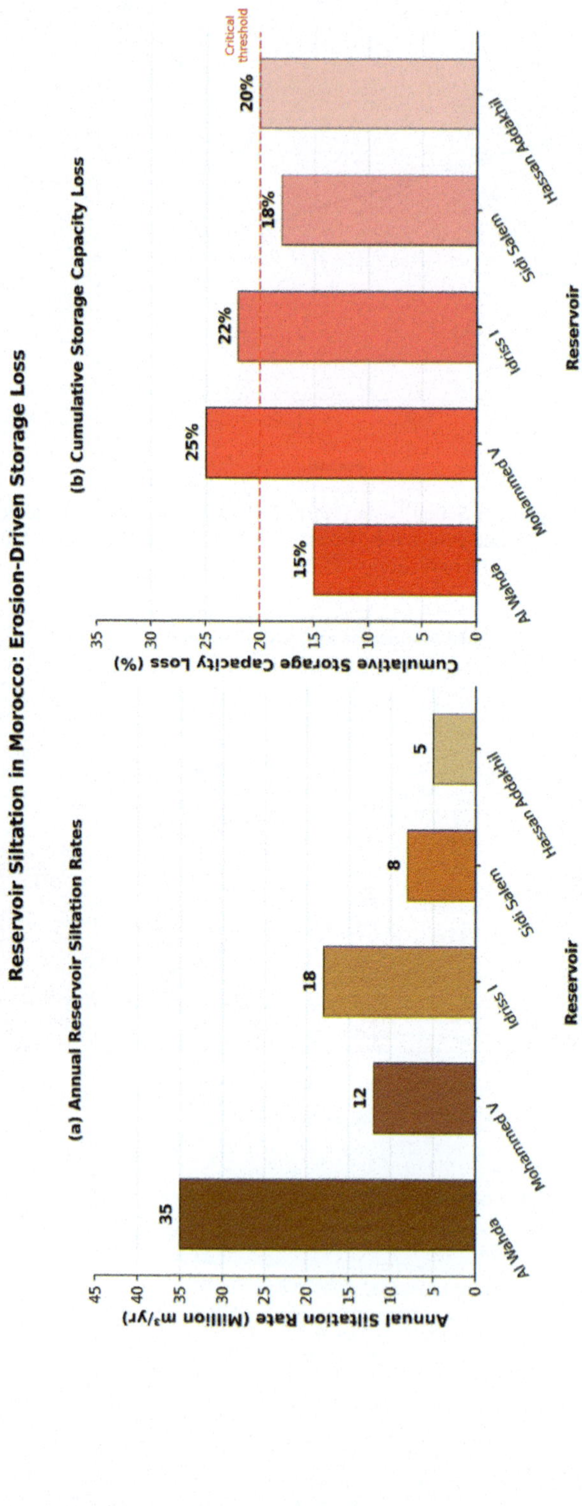

Fig. 3.6 Major reservoir siltation and storage loss. *Sources* Loudyi et al. (2018), Mosaid et al. (2024)

excessive rainfall on both impervious and erodible surfaces. Recent studies demonstrate that a diminishing number of high-intensity storms have occurred over recent years, while the overall trend in North Africa has been towards drying. Consequently, both soil erosion and sediment movement into reservoirs will result from increased storm intensity. Another avenue of degradation on the Atlantic coast of North Africa is through salinity caused by the overabundance of irrigation. Recent reports have documented the salinity of aquifers to have reached 37 g l^{-1}, and, therefore, it has been demonstrated that contemporary seawater influxes are contributingto this, along with the influence of ancient marine deposits. Detecting and demarcating areas at risk of increased salinity is possible through current advances in geophysics, while improvements in management options will require mapping high-risk areas using this technology (Table 3.3).

Agricultural pollution in the Atlantic plains of Morocco is adding to the saline stress caused by saltwater intrusion. Within the Gharb region specifically, the extensive survey undertaken during the period of hydro-geochemistry (2024) provided evidence of elevated nitrate contamination levels in groundwater sources used for drinking water purposes. Unsustainable applications of nutrients have degraded sandy soil types within Gharb through the contamination of groundwater with high levels of nitrates. The review across the wider area corroborated these findings and also documented a widespread vulnerability to nitrate contamination and salinisation, with the Gharb identified as a "hotspot".

A recent sampling within the Doukkala plain confirmed that 23% of groundwater samples collected contained sufficient salinity and sodicity levels to cause severe restrictions on irrigation use and associated impacts on soil function. The impact of climate change has resulted in a 34% decline in overall agricultural productivity between 1961 and 2025 in the nation of Morocco, particularly in terms of rain-fed cereal systems and irrigated fruit and vegetable systems. Management strategies identified to reduce salinisation include the reduction of soil erosion through re-vegetating and conservation practices, sediment retention, controlled groundwater abstraction, and the establishment of improved drainage systems. As noted in specific studies of individual basins, these strategies will also include stabilising slopes and concentrating groundwater abstractions on recharged areas (Fig. 3.7).

Table 3.3 Groundwater salinity in irrigated plains

Location	Salinity (g/L)	Depth (m)	Affected Area (ha)	Status
Souss-Massa Coastal	37	15	12,000	Critical
Souss-Massa Inland	8.5	45	8,000	High
Gharb-Mnasra	4.2	8	15,000	Moderate
Doukkala Plain	6.8	12	9,500	High

Source Hssaisoune et al. (2022), Sanad et al. (2024a, 2024b), Al-Aizari et al. (2024)

Fig. 3.7 Morocco: regional distribution of soil degradation hotspots. *Data from* Ed-Dakiri et al. (2024), Kessabi et al. (2022), Al-Aizari et al. (2024), Hssaisoune et al. (2022)

3.7.2 Algeria

The varied topography of Algeria, with its coastal highlands and semi-arid steppe environments, is facing a land degradation risk as a result of climate change, due to rising temperatures (+0.4°C per decade since 1991) and increased rainfall variability. Regions of Northern Algeria have had longer periods of decades without rain since the late 1900s, and, in the same time frame, the increasing intensity of rainfall events contributes to the runoff of soil and sediment erosion in particular areas such as the Tell Atlas, or "Tell", which define the basin of the Wadi Isser. The amount of soil lost from this area has been estimated at approximately 33.5 tons/ha/year, and this is a major threat to soil fertility and long-term viability of the reservoir (Fig. 3.8 and Tables 3.4).

An additional attempt was made to integrate RUSLE with the long-standing sediment rating curves that had been produced with the station lower down at Lakhdaria showed for the period 1970–2017 an average basin erosion of 25.6 t ha^{-1} yr^{-1} but a sediment delivery ratio of the order of 0.09 which answer the question of how much of the debris detached as solution finally get to the outlets of the catchment area (Bensekhria & Bouhata, 2022; Fredj et al., 2024). There are similar sectors that describe this picture in addition: damming with respect to sedimentation and large suspended loads have been noted in the Wadi Mina system, while in Oued el-Hai (Aurès), the USLE–GIS technique has brought to attention the sort of areas it has either been reported have or shown up the increased as yet undisguised risk

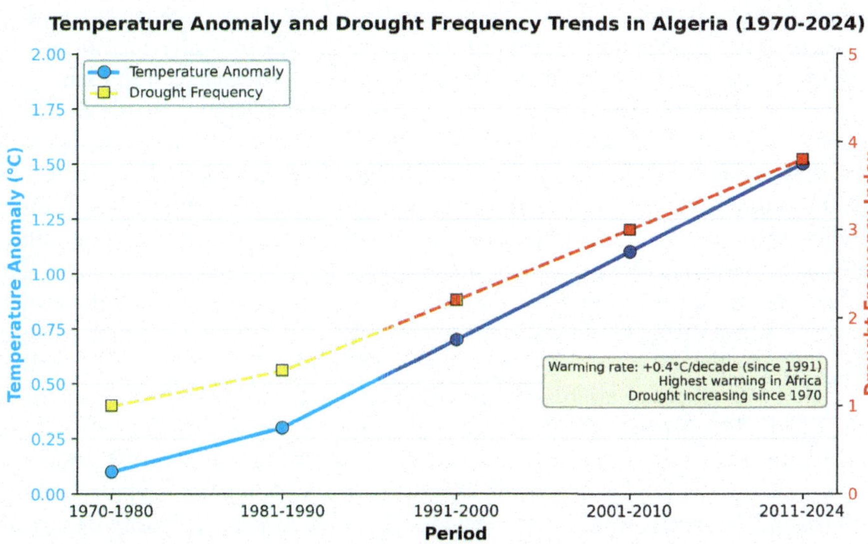

Fig. 3.8 Temperature anomaly and drought frequency trends in Algeria (1970–2024). Blue line (left axis) shows cumulative temperature anomaly based on warming rate of + 0.4°C/decade since 1991. Red dashed line (right axis) shows increasing drought frequency index. *Sources* WMO (2024); Habibi et al. (2018)

Table 3.4 Climate change signals in Algeria (1970–2024)

Climate parameter	Trend	Period	Region most affected	Impact on soil
Temperature	+ 0.40°C/ decade	1991–2024	North Africa (the highest in the continent)	Increased evaporative demand, soil moisture deficit
Precipitation	Declining since the 1970s	1970–2024	Northern basins (Tell Atlas)	Reduced vegetation cover, increased erosion risk
Drought frequency	More frequent	1970–present	Semi-arid interior, steppe zones	Vegetation thinning, wind erosion amplification
Intense rainfall events	Increasing	Recent decades	Tell Atlas, the coastal mountains	Flash floods, high runoff on bare soils

catastrophe owing to steeply inclined slopes, erodible substrata, and poor coverage (Hallouz et al., 2018) (Fig. 3.9 and Table 3.5).

The body of these studies presents a quantitative assessment of the degree of hydric edaphic loss, which takes place under the prevailing climatic and land-use conditions in Algeria. Furthermore, the long series of observations has also explained the sedimentary dynamics in such circumstances, where the surface water runoff is semi-arid. Resulting from 22 years of records (1973–1995), in Wadi Abd (northwest; 2480 km^2), the specific suspended-sediment yield is approximately 136 t km^{-2} yr^{-1} (median), with great inter-annual variation and a preponderance of the fluxes in autumnal-springtime, which accounted for factors indicative of Mediterranean flash-flood seasonalities (Achite & Ouillon, 2007). The record also showed that there was a shift to more irregular fluxes in the late 1980s there was a change to one of more irregular fluxes, as the drought vigour increased, the importance of which, certainly, was both to show, that changes in hydrological climates were directly manifestly, if in a secondary way, propagated in sedimentary balances, which were of significance in respect of soil-protection and water reservoir hydration. Apart from water erosion, desertification processes in the steppic belt (≈20 million ha.) are very intimately associated with climatic variabilities and land use (overgrazing, rain-fed crops) (Mihi et al., 2024) (Fig. 3.10).

A wide range of literature include (in passing and more focused investigations of land degradation in Laghouat), which shows that the transition of steppic to cereal-crop pasture, in sandy, weakly aggregated soils, is of an extraordinarily damaging character, vis-a-vis greatly enhanced rotation of wind-erosion, surface-sealing, and nutrient-losses (Houyou et al., 2016). Remote-sensing studies in NE steppic areas (Tébessa) and the Algerian Green Barrier have also shown great tracts of land prone

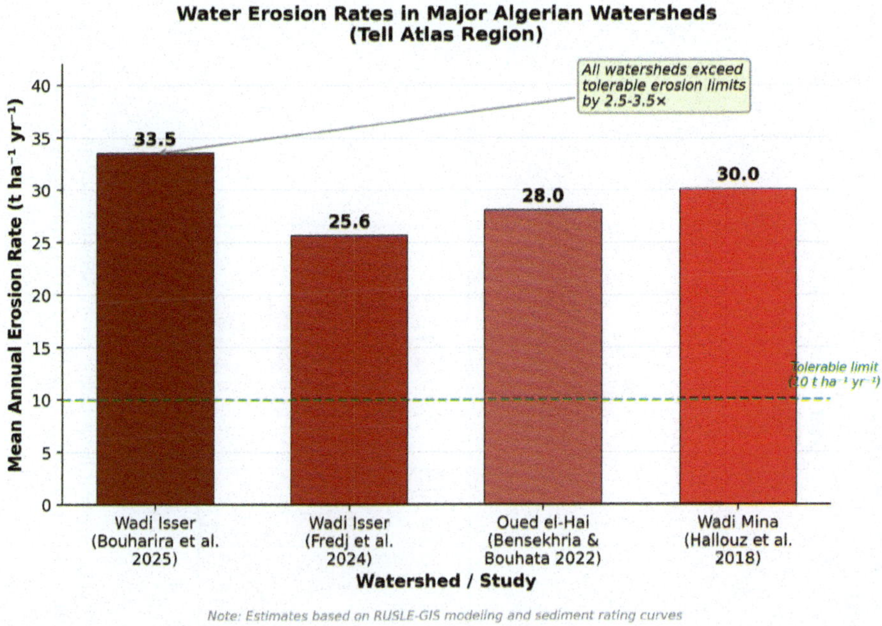

Fig. 3.9 Water erosion rates in major Algerian watersheds in the Tell Atlas region. All watersheds exceed tolerable erosion limits (10 t ha^{-1} yr^{-1}) by 2.5–3.5 × . Estimates based on RUSLE-GIS modelling and sediment rating curves. *Sources* Bouharira et al. (2025), Fredj et al. (2024), Bensekhria and Bouhata (2022), Hallouz et al. (2018)

Table 3.5 Water erosion rates in major Algerian watersheds

Watershed/Basin	Area (km^2)	Mean erosion rate (t/ha/yr)	Sediment delivery ratio	Specific sediment yield (t/km^2/yr)	Source
Wadi Isser	2,480	33.5	0.09	256	Bouharira et al. (2025)
Wadi Abd	2,480	18.2	0.12	136 (median)	Achite and Ouillon (2007)
Oued el-Hai (Aurès)	875	25.6	0.15	385	USLE-GIS studies
Wadi Mina	4,900	22.8	0.11	251	Hallouz et al. (2018)
Mafragh Basin	7,800	28.5	0.10	285	Medjani et al. (2023)

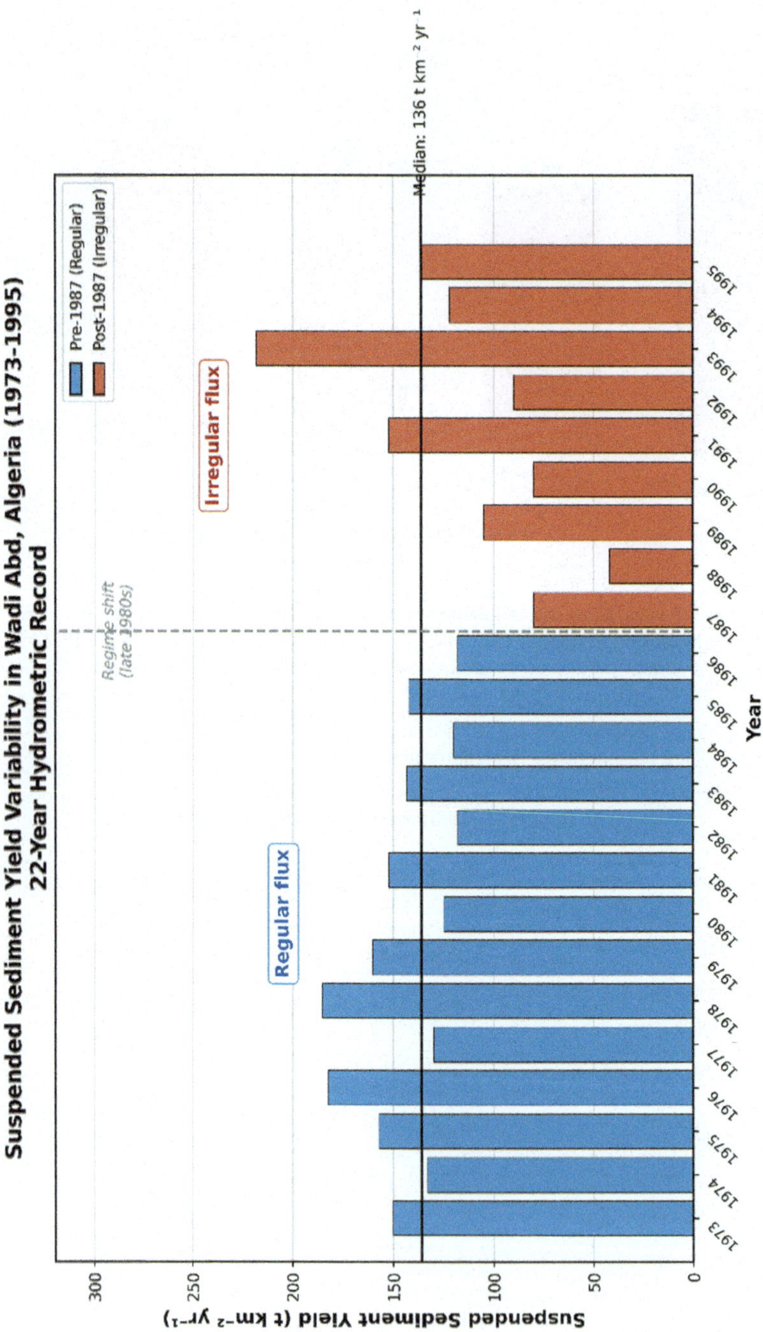

Fig. 3.10 **Suspended sediment yield variability in Wadi Abd, Algeria (1973–1995)**. Twenty-two-year hydrometric record showing median yield of 136 t km^{-2} yr^{-1} with shift to irregular fluxes in the late 1980s associated with drought intensification. Flux dominated by autumn–spring seasonality. *Source* Achite and Ouillon (2007)

to desertification, which were confirmed by field observations of depletion of vegetational cover and sand-encroachment phenomena (Mihi et al., 2022). These results cadenced with the global models of IPCC that desertification, particularly in arid lands, was accelerated by the increase in temperature, and that again human disturbance was a major factor in laying low the protective mantle. Wind-erosion and aeolian sand-sourcing processes were therefore spatially and regionally importantly related to internal Algeria. Research conducted in the steppic region, particularly demonstrating how environmental conditions (such as aridity and extreme translational seasonal winds) were used in conjunction with surface-detection phenomena, so as to give mobility to the fine particulates, whereas geo-spatial modelling of aeolian phenomena, shows the corridors of transport, south of the Tell, were sufficient to imperil infrastructures, cropland, and pasture (Rerboudj et al., 2024) (Fig. 3.11).

In the Laghouat area, hazard-mapping and process-studies were nearly in agreement with the cultivated sandy soils, degenerated rangelands, the key-elements, which, if graphically put forward, would enormously strengthen the case for soil cover, wind-breaks, and in land-use planning (Saadoud et al., 2018). The third route was that of salinisation phenomena, in coastal + irrigated plains, where there was great evaporative, etc., and shallow saline groundwater and over-extraction. In the Annaba coast hydro-chemical survey, records indicate the electric conductivity/influx and Cl^{-}–Na^{+} hallmark connectivity with seawater intrusion accelerated by pumping, which meant obvious results in respect of irrigation value and soil structure (Alfarrah & Walraevens, 2018; Benchaib et al., 2025). In the Mitidja alluvial plain (Great Algiers), hydro-chemical and isotopic record-study philosophies tackled the quality-debilitating phenomena, due to the urban, agricultural dompand on the quality-debilitating phenomena, as a result of urban and agricultural pollutants; nonetheless, overstress of aquifer-groundwater resources is confirmed by restatement-modelling metrophilosophy, which led eventually to a serious undersupply, ultimately calling for integrated surface–groundwater movement (Kastali et al., 2024; Yahiaoui et al., 2023). Vulnerability map, in the GALDIT type in Algerian coastland uriaquifers (in accordance), shows a large extent of high potential reserves to marine intrusion, which again confirms the potential sodicity and crusting form-reserves in the matter of the situation of irrigated soils (Boufekane et al., 2022) (Fig. 3.12 and Table 3.6).

The combination of accelerated warming and rain variability with erodible substrates, together with land-use pressures, leads to a composite risk profile across Algeria of (i) an increased on-site soil loss in mountain and piedmont basins during short, but intense storms in the wet season, (ii) wind erosion in steppes/desertification in areas where vegetation is thinned out, and (iii) salinisation in coastal and irrigated soils due to increased water stress and over-extraction of water resources. The basin-scale quantifications of erosion rates and now-recorded trends in the salinity of renewable groundwater resources provide valuable base line data for the targeting of conservation agriculture, rangeland recovery, "sediment smart" measures in watersheds, and coastal aquifer conservation (Alfarrah & Walraevens, 2018; Bouharira et al., 2025; Fredj et al., 2024).

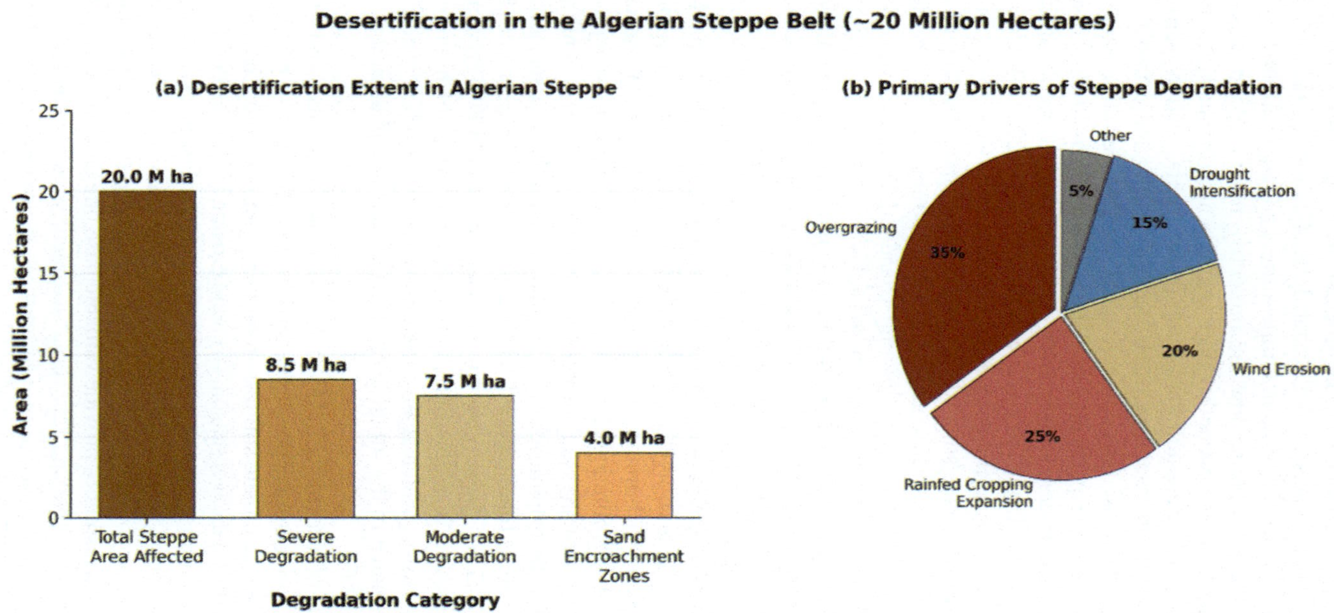

Fig. 3.11 Desertification extent and drivers in the Algerian steppe belt. **a** Area affected by degradation category; **b** Primary drivers of steppe degradation. Total affected area approximately 20 million hectares, with overgrazing and rainfed cropping expansion as dominant drivers. *Sources* Mihi et al. (2024), Houyou et al. (2016), Saadoud et al. (2018)

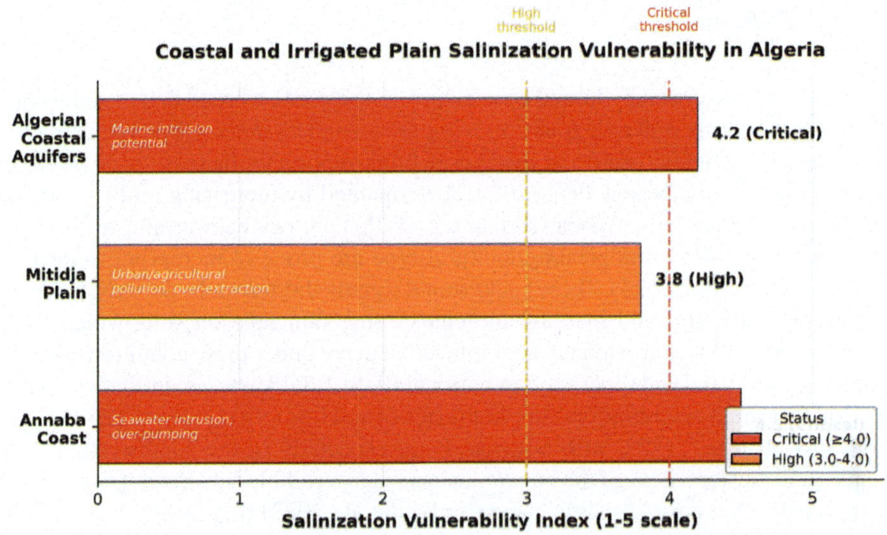

Fig. 3.12 Coastal and irrigated plain salinisation vulnerability in Algeria. Horizontal bars show vulnerability index (1–5 scale) with primary drivers annotated. GALDIT-type assessment indicates high-to-critical risk of marine intrusion in coastal aquifers. *Sources* Benchaib et al. (2025), Kastali et al. (2024), Boufekane et al. (2022)

Table 3.6 Coastal and irrigated plain salinisation in Algeria[a]

Location	Aquifer type	Primary salinity source	EC/Chloride signature	Impact	References
Annaba Coast	Coastal shallow	Seawater intrusion (SWI)	High EC, Cl–Na^+ dominance	Irrigation quality decline, soil structure damage	Benchaib et al., (2025)
Mitidja Plain (Algiers)	Alluvial	Urban + agricultural demand, overstress	Quality deterioration	Groundwater undersupply, need surface–GW integration	Kastali et al., (2024), Yahiaoui et al., (2023)
Multiple Coastal Aquifers	Various coastal	Marine intrusion + over-extraction	GALDIT vulnerability HIGH	Widespread sodicity, crusting risk in irrigated soils	Boufekane et al., (2022)

Annaba Coast: Hydro-chemical surveys reveal elevated electrical conductivity with Cl^-–Na^+ signature indicating seawater intrusion accelerated by pumping. Direct consequences for irrigation water quality and soil aggregate stability. **Mitidja Plain:** Major agricultural zone near Algiers facing groundwater quality degradation from combined urban and agricultural demand. Hydrochemical and isotopic studies confirm aquifer overstress requiring integrated surface-groundwater management. **Regional Vulnerability:** GALDIT-based vulnerability mapping shows extensive Algerian coastal aquifers at high risk for marine intrusion, threatening soil structure through sodicity development and crusting in irrigation-dependent agriculture

3.7.3 Tunisia

Tunisia is characterised by humid coastal hills to the north, a broad belt of semi-arid/steppe in the centre and arid zones, moving towards the Sahara. Observed warming is strong: The climate profile of the World Bank reports an increase of + 0.46 °C per decade over the period 1971–2020, accompanied by increasing aridity, and the IPCC AR6 for the Mediterranean further reports that with global warming above 1.5–2 °C, droughts of hydrological/agricultural type are prone to increasing frequency (IPCC, 2022; WBG, 2025). These background signals shorten moist periods, increase evaporative demand, and increase the chance that rain falls on soils which have been withered. Water erosion is geographically severe under these conditions, which intensify runoff and soil separation. A national RUSLE/GIS assessment shows 6.43% of Tunisia's areal extent suffers very high rates of soil loss (>30 t ha^{-1} yr^{-1}) and 4.20% high (20–30 t ha^{-1} yr^{-1}), with hotspots lying especially in the southwestern, and central and western areas of the country, areas where erodible substrates, pronounced slopes, and scant cover are associated (Serbaji et al., 2023) (Fig. 3.13).

At the scale of basins, therefore, updated applications verify considerable transport of soil. The corresponding sediment yield and fallout radionuclide (^{137}Cs) are 2.6 t ha^{-1} yr^{-1} (SDR calibrated to the relationship of Maner). The event-driven nature of losses under Mediterranean whiplash rains is highlighted by RUSLE evaluations for other basins in the centre and north, which place similarly sizable tracts in the slight–very severe classes (Cheikha et al., 2023; Marghmi et al., 2024).

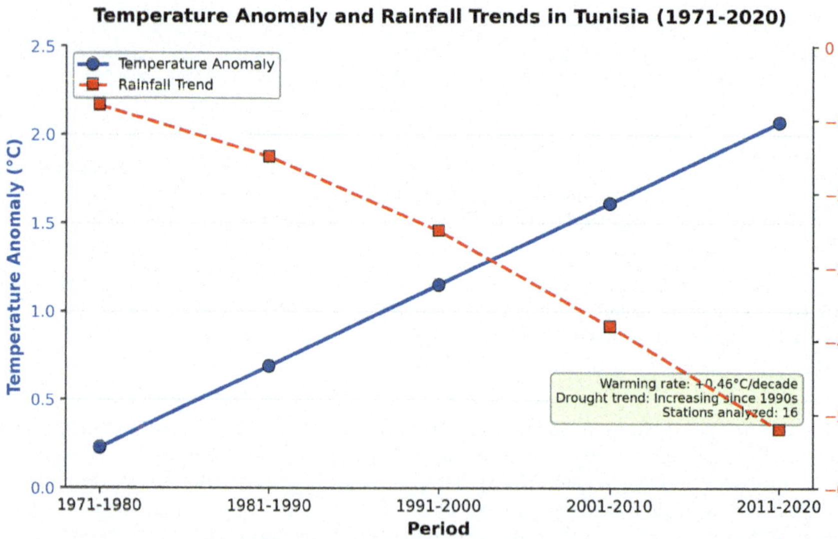

Fig. 3.13 Temperature anomaly and rainfall trends in Tunisia (1971–2020). Blue line (left axis) shows cumulative temperature anomaly based on warming rate of + 0.46 °C/decade. Red dashed line (right axis) shows declining rainfall trend (mm/decade). *Sources* WBG (2025), Mohamed (2023)

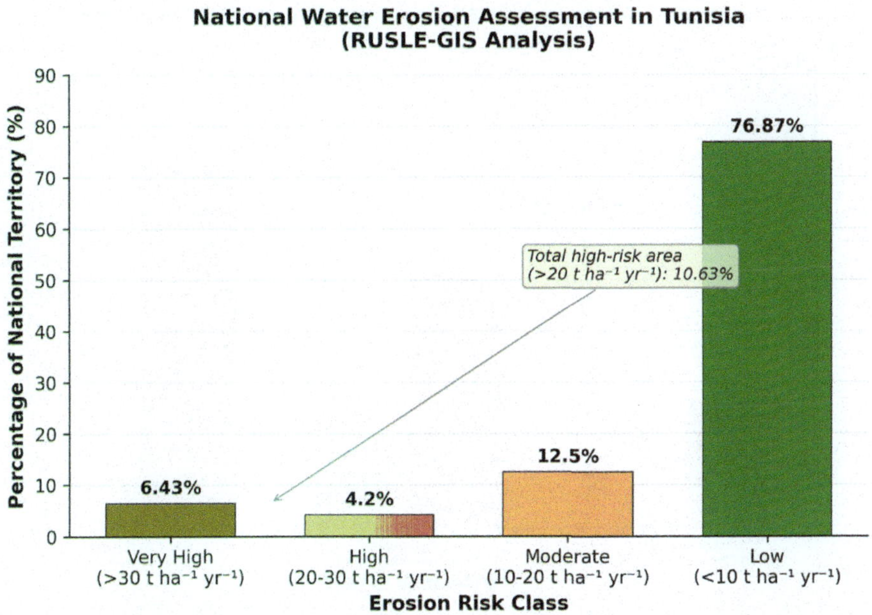

Fig. 3.14 National water erosion assessment in Tunisia based on RUSLE-GIS analysis. Bars represent percentage of national territory in each erosion risk class. Total high-risk area (>20 t ha^{-1} yr^{-1}) comprises 10.63% of Tunisia. *Source* Serbaji et al. (2023)

Long hydrometric series from central Tunisia also show significant inter-annual variation in sediment yield related to fluctuating stores of sediment, and periodic drought sequences. The information presented here together abstracts only a part of Tunisia's erosion menace, and its sensitivity to climatic fluctuations (Ayadi et al., 2010) (Figs. 3.14 and 3.15).

Drought trends refine the erosion signal. An analysis of rainfall for 1958–2020 in a monthly sequence in 16 stations provides evidence of a great deal of drying in many parts, and development of the severe-drought classes of droughts since the late 1990s, agreeing with AR6 reports of frequent droughts of long duration, and increasing intensity over the southern Mediterranean. These recessionary and drought shifts in climate lessen the protective cover of vegetation, promote surface sealing/crusting, and secure the hillslopes for high runoff coefficients, when the live thunderstorms return (IPCC, in press. Mostafa Abd El-Hameed Mohamed, 2023).

The salinisation may bc regarded as an approximately co-equal émeute, particularly the saline effects of salt water intrusion on irrigated and coastal plains. In Cap Bon (Korba System) and Grombalia Plain, the intensive abstraction of Karst waters has, due to years of continued drought, distance, and closeness to salt sea, of varying salinity, the groundwaters have been salinised through salt water intrusion (SWI) and evaporation-concentration (Ben Saad et al., 2023). A more up-to-date hydro-geological synthesis of Grombalia shows that the salinity of the Karst waters

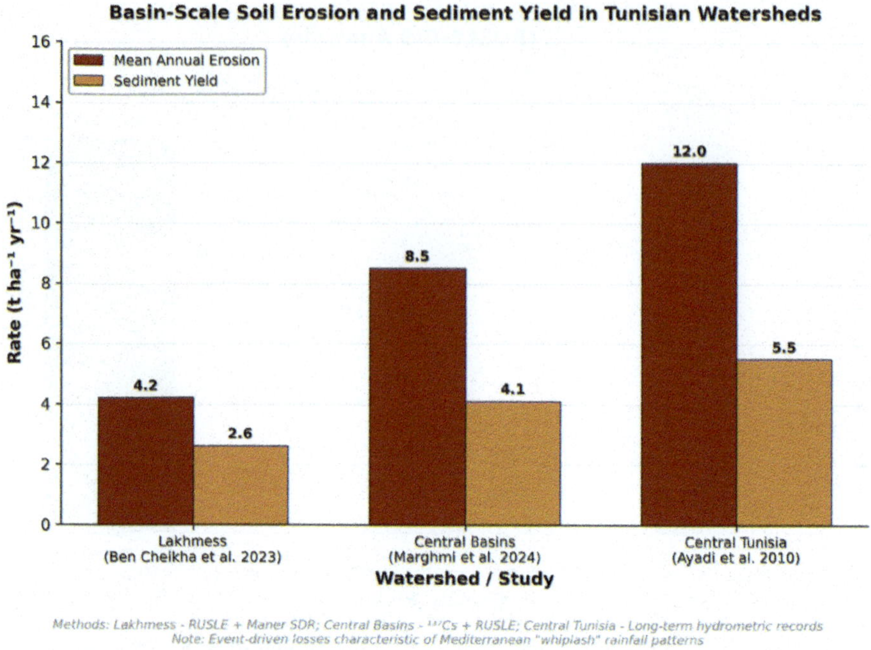

Fig. 3.15 Basin-scale soil erosion rates and sediment yields in Tunisian watersheds. Mean annual erosion (dark bars) and sediment yield (light bars) determined using RUSLE with Maner SDR calibration (Lakhmess), ^{137}Cs fallout with RUSLE validation (Central Basins), and long-term hydrometric records (Central Tunisia). *Sources* Cheikha et al. (2023), Marghmi et al. (2024), Ayadi et al. (2010)

may be traced back to irrigation returns and to the legacy of industrial waters fed into it, and gives an insight into the aquifer's true behaviour in aquifer formation from a new conceptual model of aquifer synthesis. A paired analysis of the farming system's practical and marketing value is conducted. Irrigation uses the zonation of stated efficiency to disrupt the cropping area's width, demonstrating significant and common physical limitations to which refer, directing EC from applied water to the cultivated areas so that irrigation may be effective (Panagiotou et al. 2025).

The results of an investigation of the drought during a three-year period in Korba, specifically the Gantera system in use, demonstrate the severe sensitivity of saltwater intrusion (SWI) impacting, under a prolonged drought and pumped up, with isotopic relationships and chemical wave-tracing, linearly showing up areas of intrusion of these into the SWI (Slama, 2022). Likewise, regional investigations attribute the salinisation of Tunisian coastal aquifers to a combination of seawater intrusion and agricultural land use, reportedly suggesting the need to jointly manage pumping, artificial recharge, and drainage to remedy soil salinity and sodicity (Mohamed Neji 2025) (Fig. 3.16).

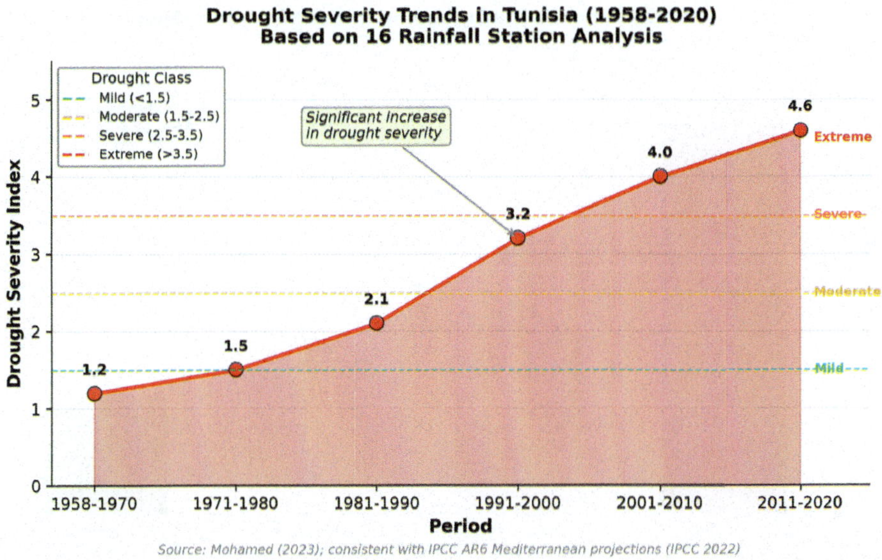

Fig. 3.16 Drought severity trends in Tunisia (1958–2020) based on analysis of 16 rainfall stations. Horizontal dashed lines indicate drought classification thresholds. A significant transition to severe drought conditions occurred in the 1990s, consistent with IPCC AR6 Mediterranean projections. *Source* Mohamed (2023)

3.7.4 Egypt

The high vulnerability of soils in Egypt to climate change is largely influenced by their exposure to multiple stressors in the Nile Delta and the reclaimed irrigation lands, where low relief, intensive irrigation, and closeness to the Mediterranean Sea provide for multiple impacts in terms of land degradation and soil degradation. The Nile Delta is a densely populated agricultural region dependent on irrigation from the Nile River and underground water resources such as aquifers, with rainfall negligible over the delta, such that soil and agriculture are subject to changes in quantities and salinity of water as well as increased groundwater levels (Elshinnawy & Almaliki, 2021, Hagage et al., 2024b).

Research findings have described the problems of the Nile Delta, clearly indicating the presence of widespread salinity and sodicity. It is reported that about 46% of the area of the Nile Delta is characterised as having saline and sodic soils, which have detrimental effects on crop production (Kotb et al., 2000). Soil monitoring in the northeastern portions of the Nile Delta has shown that salinisation and waterlogging are serious environmental problems, with the use of remote sensing, in situ sampling, and hydrological evaluation indicating strong evidence of increased returns of ground water and capillary rise in the delta soils leading to shallow water tables and high surface accumulation of salts which decrease the productive quality of the soils (Hagage et al., 2024c). The undersupply of irrigation water because of

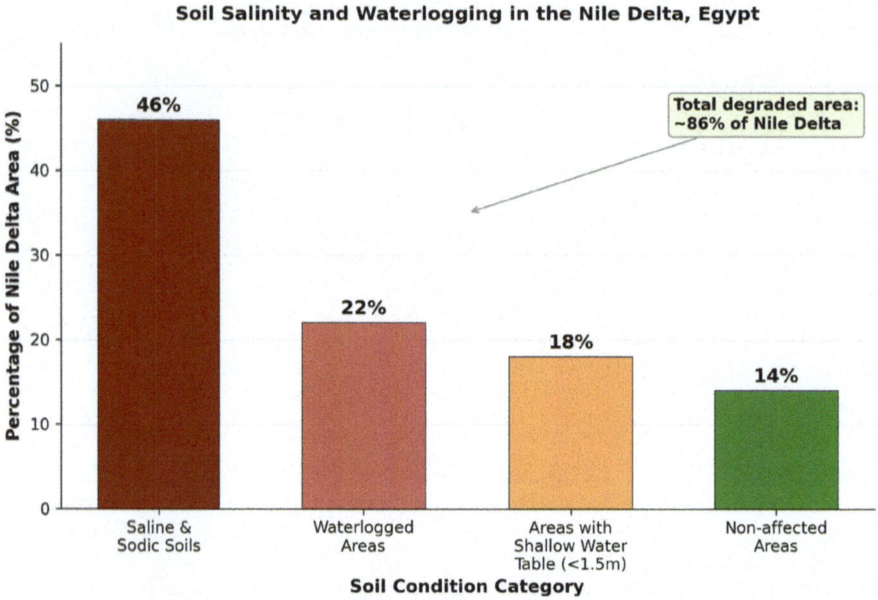

Fig. 3.17 Soil salinity and waterlogging in the Nile Delta, Egypt. Approximately 46% of the delta area is characterised by saline and sodic soils, with an additional 22% affected by waterlogging. *Sources* Kotb et al. (2000), Hagage et al. (2024)

increased irrigation needs and demands from population increases results in excessive utilisation of underground water resources from the aquifers of the delta and the reclaimed desert lands (Fig. 3.17).

In some cases, the aquifers in the delta also have their source waters in direct connection with the Mediterranean Sea, making such waters subject to infiltration of seawater. The development of numerical groundwater models (such as SEAWAT[1]) indicates that the process whereby irrigation is effected through pumping of ground water and the rise of sea level causes a progressive inlet of seawater into the deeper and more inland portions of the Nile Delta aquifers, leading to an increase in the salinity of the aquifers of the Nile Delta and a decreased availability of fresh water for irrigation (Abd-Elaty et al., 2023, Abd-Elhamid et al., 2016). The effect of the presumptive increased sea levels (1.25 m or more) indicates increased inland extent of saline water intrusion, and increased groundwater levels, resulting in increased salinity and waterlogging (Abdrabo et al., 2015, Armanuos et al., 2022, Sefelnasr & Sherif, 2014).

In the delta, irrigation water is sourced from sources such as drainage or marginal-quality water, particularly in lands that are reclaimed or intensively cultivated. One experiment showed that irrigation of a limited area leads to salt accumulation through

[1] SEAWAT is a generic MODFLOW/MT3DMS-based computer program designed to simulate three-dimensional variable-density groundwater flow coupled with multi-species solute and heat transport.

the reuse of drainage water: Nearly 62 tons were added to a 2-ha area in one summer by irrigation and reuse of drainage water, and only about 93% of the salt which was added was leached out during the winter months. A great amount of salt was determined to be still in the soil profile at the time of the hot dry months. This indicates that repeated irrigation, without sufficient leaching or drainage, results in salt accumulation in the root zone (Abbott & El-Quosy, 1996). Other experiments, involving salt mapping by means of groundwater salinity determination in wells, are classified into five classes (normal, moderate, high, very high, extreme). In the western part of the delta area (west of the Cairo–Alexandria highway), many wells draw from deeper confined aquifers (Moghra, Wadi El Natrun aquifer, Oligocene aquifers) and show very high to extreme salinity as well as high sodium adsorption ratios (SARs) (in excess of 10), and consequently the possibility of sodicity and damages to soil structure by the use of this water for irrigation is greater (Figs. 3.18 and 3.19).

The wells in the eastern part of the delta area, on the other hand, are often better because the wells draw from unconfined aquifers, which receive their water from recharge through the canals lying near them (Amer, 2021). The combination of rising water tables (waterlogging) and the use of saline waters for irrigation will

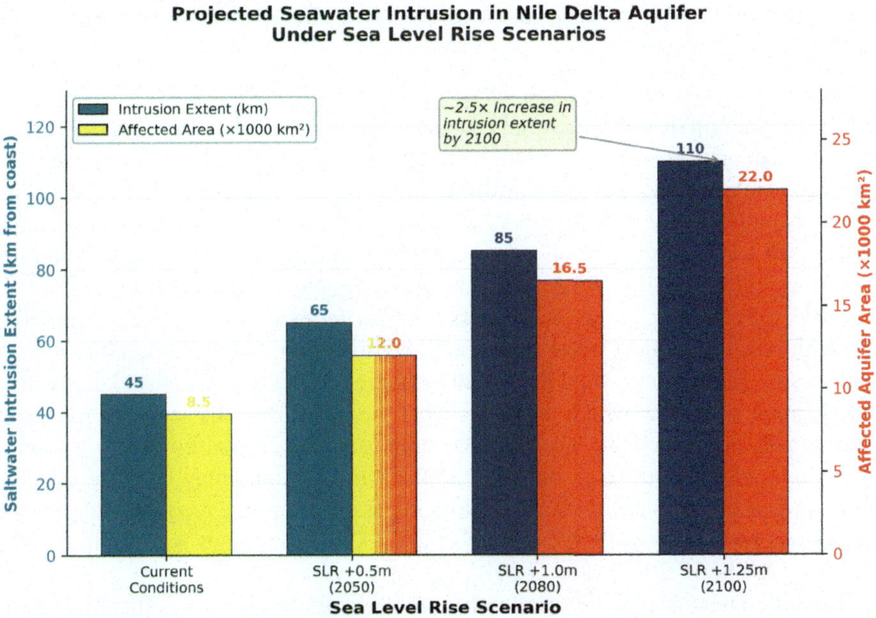

Fig. 3.18 Projected seawater intrusion in the Nile Delta aquifer under sea-level rise scenarios. Dark bars show inland intrusion extent (km from coast); red bars show affected aquifer area. Based on SEAWAT numerical groundwater modelling. *Sources* Abd-Elaty et al. (2023), Sefelnasr and Sherif (2014), Armanuos et al. (2022). *Note* Based on SEAWAT numerical groundwater modelling

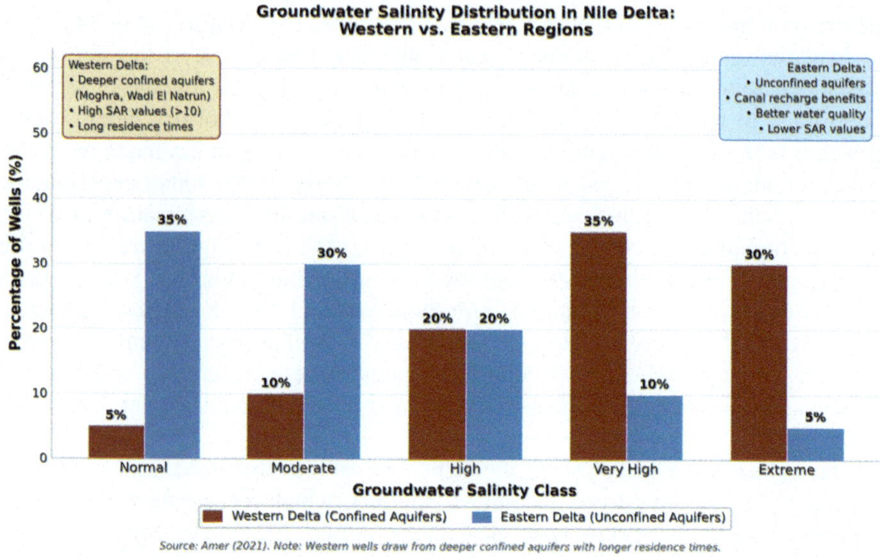

Fig. 3.19 Groundwater salinity distribution in Nile Delta comparing western and eastern regions. Western delta wells draw from deeper confined aquifers (Moghra, Wadi El Natrun) with higher salinity and SAR (>10), while eastern delta wells benefit from canal recharge in unconfined aquifers. *Source* Amer (2021)

cause degradation of soil structure. The salts, especially sodium, will render dispersion of clay particles in soils, and this will produce decreased soil structure stability, increased bulk density, decreased infiltration, and diminished effective rootable zone. The soils may become more impervious or may have surface capillary salt accumulation with waterlogging to the crops, especially during the dry periods (Abbott & El-Quosy, 1996; Elshinnawy & Almaliki, 2021; Hagage et al., 2024c).

Projected climate change (for example, higher temperatures, greater evaporative demand, reduced surface water flows) is likely to compound these pressures. More arid seasons mean lower canal flows and recharge, resulting in increased reliance on groundwater pumping, which promotes what the literature calls seawater intrusion. Additionally, sea-level rise increases the salinity of the coastal boundary, meaning even slight additions of (e.g., decimetres) the salinity inland may increase substantially. Increased soil salinity and waterlogging result from this feedback loop which causes less productive soils (Abd-Elaty et al., 2023; Elshinnawy & Almaliki, 2021; Sefelnasr & Sherif, 2014) (Fig. 3.20).

The Nile Delta is a highly productive area, accounting for a significant share of Egypt's food production. Food security will be seriously threatened by decreased soil productivity brought on by salt, waterlogging, and declining irrigation water quality, the livelihoods of the farmer, and the community, economic analyses put future losses at a high level, such as in a study of the Damietta governorate (which is part of the coastal delta) where it was calculated that by the year 2100 the losses in

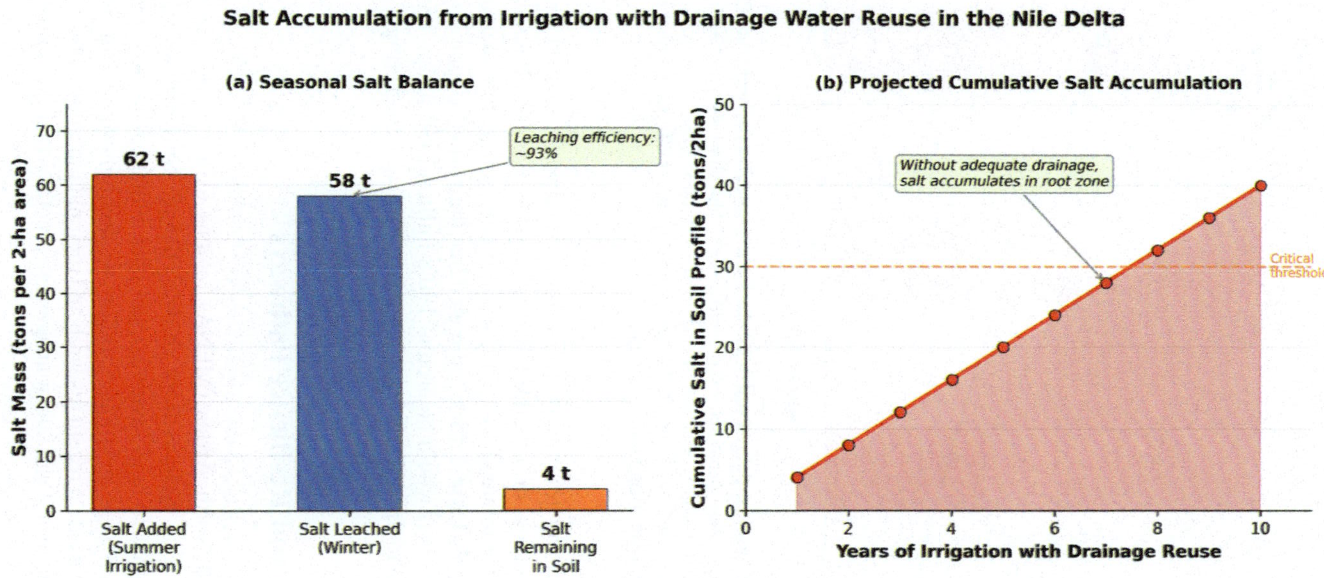

Fig. 3.20 Salt accumulation from irrigation with drainage water reuse in the Nile Delta. a Seasonal salt balance showing 62 tons added during summer irrigation with ~93% leached during winter; **b** Projected cumulative salt accumulation over 10 years without adequate drainage. *Source* Abbott and El-Quosy (1996)

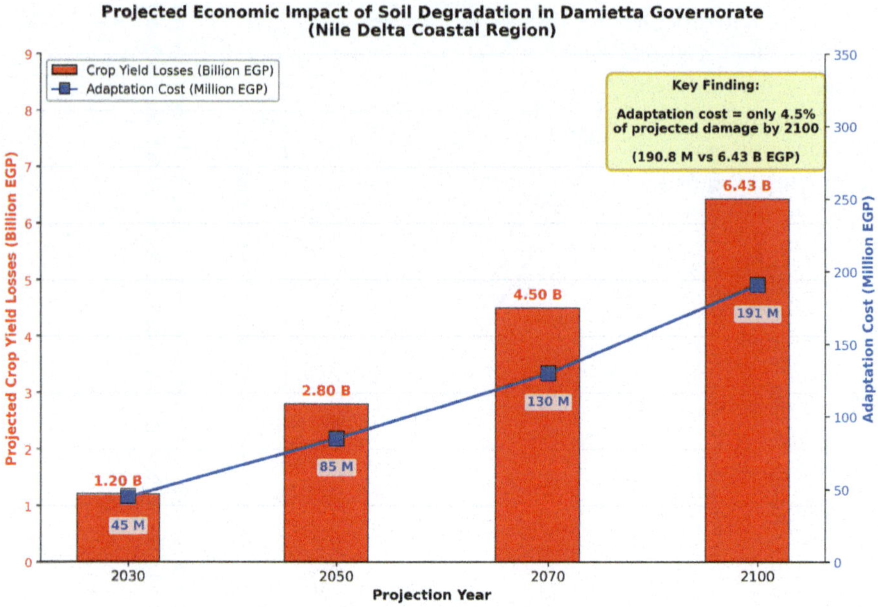

Fig. 3.21 Projected economic impact of soil degradation in Damietta Governorate (Nile Delta coastal region). Red bars show projected crop yield losses; blue line shows adaptation costs (primarily drainage improvements). Adaptation cost represents only 4.5% of projected damage by 2100. *Source* Abdrabo et al. (2015). *Note* Losses exclude indirect effects (employment, food supply); adaptation primally drainage improvement

crop yield due to the rising groundwater (again connected with sea-level rise) would amount to an appraised loss of something like 6.43 billion Egyptian Pounds on crop production of that governorate alone (not giving but indirect effects, such as loss of jobs, general food supply losses) it was estimated that the cost of adaptation (such as in drainage improvements) would be around 190.8 million EGP, about 4.5% of the projected damage due to the effect of rainfall (Abdrabo et al., 2015) (Fig. 3.21).

3.8 Adaptation and Resilience Pathways

The Middle East and North Africa (MENA) regions exhibit severe vulnerabilities in their soil and food systems from climate change. To reduce risk and build resiliency, adaptation has to be assessed around land management, water use, agricultural systems, governance, and finance. Below, we highlight some of the important adaptation pathways based on recent research and regional initiatives. One important strategy is the adoption of soil and land management practices that generate resilience in these areas to drought, erosion, and salinisation. For instance, a recent study highlights that "building soils", meaning adding organic matter, surface cover, and structure, is crucial in the dryland MENA for adaptation to soil-degrading influences of climate change (Deeb et al., 2024). Agroecological management practices such as reduced tillage, retention of residues, agroforestry, contour bunding, and micro catchments have been shown to increase soil moisture, reduce runoff and erosion, and maintain fertility in North Africa. In a detailed survey, Boutagayout et al. (2025) reveal that agroecological management practices in the semi-arid regions of North Africa have improved infiltration, organic carbon, and crop resilience (Boutagayout et al., 2025). Given the prevalence of historic soil degradation and the projected increased incidences of droughts and extreme rainfall "whiplash" events throughout MENA, these soil-centred practices actually represent a first line of defence against rapid degradation. Also, since soils are intrinsically slow to change, early adoption of resilience-promoting practices helps "lock in" improvements before degradation manifests. Water scarcity and salinisation are the twin threats in MENA. Adaptation must therefore include irrigation efficiency, drainage, water quality management, and salt-tolerant crops. For instance, Aryal et al. (2025) mention that salinity in soil and irrigation water is increasing in North Africa due to hotter, drier conditions and poor drainage and that adaptation measures are needed to reduce the amount of salt accumulating and to protect the soils (Aryal et al., 2025). At a systems level, the International Centre for Biosaline Agriculture (ICBA) has developed climate-smart, salt-tolerant crops appropriate to marginal, saline, and water-deficient environments typical of MENA.

In many MENA countries, adaptation involves transitioning to drip or sprinkler irrigation, precision scheduling, use of treated water or recycling water, salt-leaching protocols, and improvements in drainage and aquifer-recharge systems to reduce the salt build-up in soils and maintain agricultural productivity under stressed conditions. Beyond soil and water management, cropping systems must themselves be adapted. In the MENA region, research into farmer adaptation shows wheat growers in certain countries are altering sowing dates, selecting drought-resilient varieties, and modifying fertiliser and irrigation regimes due to climate stress. For example, Mirgol et al. (2025) found that wheat farmers in MENA region adapted many on-farm strategies to cope with rising heat and water stress (Mirgol et al., 2025). Additionally, Boutagayout et al. (2025) underline that agroecological diversification (inter-cropping, integration of livestock, agroforestry) in North Africa improves resilience to both soil degradation and the impact of climatic extremes. Similar to these system-level adaptations, climate-smart agriculture (CSA) systems are becoming more popular in MENA. For instance, a 2025 study on MENA agriculture recommends investment in CSA systems for improving crop and soil resilience (Gallas et al., 2025). Restoration of degraded landscapes and use of nature-based solutions (NbS) are fundamentally important for large-scale resilience. The World Bank's 2019 review of sustainable land management in MENA region points out that restoration of degraded land can yield very high economic benefits as well as prevent desertification and soil loss (Ruiz & Sanz, 2022; Ziadat et al., 2012). NbS examples are afforestation/re-vegetation of eroded hillslopes, restoration of riparian strips to reduce sediment

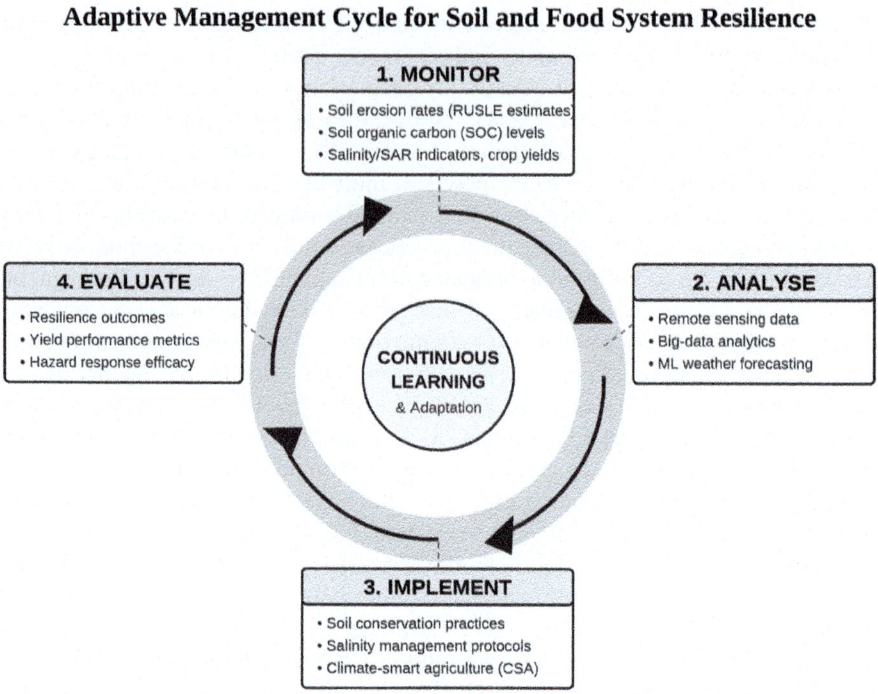

Fig. 3.22 Adaptive management framework for climate-responsive soil and food system resilience in the MENA region

flow, wind-breaks against aeolian erosion, and wetland restoration for salt-buffering in coastal plains (Jaber 2024). These interventions result in improved soil health and more sustainable ecosystems, which in turn protect food systems (Fig. 3.22).

Adaptation in soils and food systems is not only technical, but it is also institutional and financial. To effectively scale up adaptation measures, governance, policy, and investment are needed. The Carnegie Endowment's recent work on climate governance in MENA shows that limited research funding, incoherent institutional structures, and weak local capacities hinder adaptation strategies (Ahmed Eladawy 2025). Similarly, the UNFCCC adaptation knowledge-platform shows that it is important to link knowledge gaps, build capacity, and provide access to climate finance in MENA. In terms of tackling investment, the World Bank's "Climate and Development in MENA" brief (2023) shows that it is scaling up climate-finance operations in the region, including in food systems and natural capital, but that much more is needed (World Bank, 2023). Governance measures include mainstreaming the resilience of soils and food systems into national climate planning instruments (NDCs, adaptation strategies), strengthening of land-use policies designed to incentivise sustainable land practices (payments for ecosystem services, carbon-soil credits), and facilitating public–private partnerships in restoration and climate-smart agriculture. For maximum impact, adaptation measures must move beyond isolated interventions and integrate across scales: field, farm, watersheds, food systems. For example, linkages between soil-health interventions, farmer support, value-chain adaptation, market access for climate-smart crops, and early-warning systems create synergies.

Recent literature highlights that the importance of scaling, e.g., a 2024 multi-country review of adaptation plans in MENA shows that many remain pilot-scale plans with limited institutionalisation and scaling out of soil/land-based adaptation measures (Saber Osman 2023). In the realm of soils and agriculture, this means linking conservation agriculture and salinity-management practices with national programs, incentivising their uptake with subsidy reform, alignment with climate-finance instruments, and monitoring of derived outcomes (C/SOC as well as erosion rates, yields) over time. Adaptation is a dynamic process. In the MENA region, it is necessary to monitor soil health (erosion rates, SOC, salinity/SAR), land cover change, crop yields, and system resilience. Emerging technologies such as remote sensing soil monitoring, big-data analytics, and localised weather forecasting (including machine learning models) are starting to help support adaptation decisions at both the farm and the regional scale. For example, recent investigations into neural weather forecasting tailored to MENA have shown the possibility for localised early warning systems adapted to support agriculture-led soil management (Munir et al., 2024). Coupling such innovations with farmer-led participatory monitoring and adaptive management loops ensures that soil-and-food-system adaptation remains responsive to accelerated climate change and an evolving hazard environment.

3.9 Conclusions

Evidence presented in this chapter indicates that the MENA region is experiencing a confluence of climate indicators—faster warming, increasingly intense drought, combined dry–hot extremes, and episodic heavy rainfall, which foster soil degradation pathways across a wide range of geographic areas. Short wet seasons and high evaporative demand dry terrestrial soils and vegetation in Mediterranean and semi-arid regions, priming slopes for high flow coefficients and event-scale erosivity during rainy seasons. Wind erosion is caused by surface problems and vegetation stress in steppes and desert border regions. In irrigated interiors and low-lying coastal areas, the diminishing high-quality water, shallow saline water tables, and sea-level rise aggravate salinity and sodicity. These phenomena collectively deprive soils of organic matter, structure, and infiltration, which undermine agricultural productivity, ecosystem stability, and rural livelihoods. Country-specific assessments (Morocco, Algeria, Tunisia, Egypt) illustrate how common drivers achieve differentiated, place-based hazards: hotspots of erosion on steep marly flysch terrains and piedmonts; fronts of desertification in grazed steppes; and salinity-waterlogging syndrome in coastal and deltaic lowlands. These patterns correlate with the IPCC AR6 assessments for Africa and the Mediterranean hotspot, as well as those of basin studies mapping erosion and groundwater degradation. This combined signal is clear: Climate change without targeted adaptation will lock prosperity into levels of soil loss and chemical degradation, narrowing agronomic choices and intensifying food-system fragility. Simultaneously, this chapter identifies ways to slow or reverse degradation which work. Conservation and agroecological practices that retain residues, enhance ground cover, and create soil organic carbon are able to diminish sealing/crusting, enhance infiltration, and buffer yields in phases of dry–hot weather. Erosion-control structures and sediment-smart watershed management reduce both on- and off-site harm. In systems endangered with salinity, integrated water-salt management, efficient irrigation, seasonal leaching with adequate drainage, blending/reuse goals in source water-driven aquifer recharge, and use of salt-tolerant varieties of crops stabilise soils and productivity. Solutions provided by nature and restoration of landscapes (windbreaks, riparian buffers, rangeland rejuvenation) complement farm-scale methods and derive co-benefits for biodiversity and regulation of microclimate.

Scaling such solutions will involve enabling institutions and finance means mainstreaming soil health in national climate policies (NDCs, and NAPS), aligning incentives to encourage sustainable land management, and expanding climate-finance pipelines for restoration and climate-smart agriculture. Monitoring systems (pairing remote sensing with ground observations of erosion, SOC, salinity/SAR, and groundwater) are necessary to underpin adaptive management and evaluation of impact. Closing knowledge and capacity gaps (e.g., through extension, participatory co-design with farmers and pastoralists, and applied research on climate–soil–water interactions) will be crucial to whether adaptation remains at the pilot level or is achieved at the landscape and food-system level that the region now requires. In summary, soils in MENA are on the frontline of climate risk—and also at the centre of climate resilience. Policies and investments that protect and rebuild soil functions will deliver benefits in terms of water security, food production, rural incomes, and ecosystem services. There is a narrow window of opportunities to implement at scale, and the technologies and practices are known; what is currently required is coordinated deployment, sustained financing, and monitoring of outcomes.

References

Abbott, C. L., & El-Quosy, D. E. D. (1996). *Soil salinity processes under drainwater reuse in the Nile Delta, Egypt*. HR Wallingford in collaboration with the Water Management Research Institute, Report No. OD/133. Cairo, Egypt.

Abd-Elaty, I., Abdoulhalik, A., & Ahmed, A. (2023). The impact of future hydrology stresses and climate change on submarine groundwater discharge in arid regions: A case study of the Nile Delta aquifer, Egypt. *Journal of Hydrology: Regional Studies*, *47*, 101395. https://doi.org/10.1016/j.ejrh.2023.101395

Abd-Elhamid, H. F., Javadi, A., Abdelaty, I., & Sherif, M. (2016). Simulation of seawater intrusion in the Nile Delta aquifer under the conditions of climate change. *Hydrology Research*, *47*(6), 1198–1210. https://doi.org/10.2166/nh.2016.157

Abdrabo, M. A., Hassaan, M. A., & Selmy, A. R. N. (2015). Economic valuation of sea level rise impacts on agricultural sector in northern governorates of the Nile Delta. *Low Carbon Economy*, *6*, 51–63.

Achite, M., & Ouillon, S. (2007). Suspended sediment transport in a Semiarid Watershed, Wadi Abd, Algeria (1973–1995). *Journal of Hydrology, 343*(3–4), 187–202. https://doi.org/10.1016/j.jhydrol.2007.06.026

Al-Aizari, H. S., Aslaou, F., Mohsen, O., Al-Aizari, A. R., Al-Odayni, A.-B., Abduh, N. A. Y., Al-Aizari, A.-J., & Taleb, E. A. (2024). Assessment of groundwater quality for irrigation purpose using Irrigation Water Quality Index (IWQI). *Journal of Environmental Engineering and Landscape Management, 32*(1), 1–11. https://doi.org/10.3846/jeelm.2024.20598

Alfarrah, N., & Walraevens, K. (2018). Groundwater overexploitation and seawater intrusion in coastal areas of arid and semi-arid regions. *Water, 10*(2), 143. https://doi.org/10.3390/w10020143

Amer, R. (2021). Spatial relationship between irrigation water salinity, waterlogging, and cropland degradation in the arid and semi-arid environments. *Remote Sensing, 13*(6), 1047. https://doi.org/10.3390/rs13061047

Armanuos, A. M., Eldin Ibrahim, M. G., Negm, A., Takemura, J., Yoshimura, C., & Mahmod, W. E. (2022). Investigation of seawater intrusion in the Nile Delta aquifer, Egypt. *Journal of Engineering Research, 6*(1), 10–28. https://doi.org/10.21608/ERJENG.2022.113766.1046

Aryal, J. P., Becerra Lopez-Lavalle, L. A., & El-Naggar, A. H. (2025). Crop loss due to soil salinity and agricultural adaptations to It in the Middle East and North Africa region. *Resources, 14*(9), 139. https://doi.org/10.3390/resources14090139

Ayadi, I., Abida, H., Djebbar, Y., & Mahjoub, M. R. (2010). Sediment yield variability in central Tunisia: A quantitative analysis of its controlling factors. *Hydrological Sciences Journal, 55*(3), 446–558. https://doi.org/10.1080/02626661003741526

Benchaib, A., Djabri, L., & Hani, A. (2025). (PDF) Hydrochemistry of Groundwater Salinity Sources in the Shallow Aquifer: Case of Annaba Plain (Ne Algeria). *ResearchGate*. https://www.researchgate.net/publication/330565775_Hydrochemistry_of_groundwater_salinity_sources_in_the_shallow_aquifer_Case_of_Annaba_plain_Ne_Algeria. Retrieved September 1, 2025.

Ben Saad, E., Ben Alaya, M., Taupin, J. D., Patris, N., Chaabane, N., & Souissi, R. (2023). A hydrogeological conceptual model refines the behavior of a Mediterranean coastal aquifer system: A key to sustainable groundwater management (Grombalia, NE Tunisia). *Hydrology, 10*(9), 180. https://doi.org/10.3390/hydrology10090180

Bensekhria, A., & Bouhata, R. (2022). Assessment and mapping soil water erosion using RUSLE approach and GIS tools: Case of Oued El-Hai Watershed, Aurès West, Northeastern of Algeria. *ISPRS International Journal of Geo-Information, 11*(2), 84. https://doi.org/10.3390/ijgi11020084

Boufekane, A., Maizi, D., Madene, E., Busico, G., & Zghibi, A. (2022). Hybridization of GALDIT method to assess actual and future coastal vulnerability to seawater intrusion. *Journal of Environmental Management, 318*, Article 115580. https://doi.org/10.1016/j.jenvman.2022.115580

Bouharira, A., Achite, M., Hermassi, T., Toubal, A. K., & Ouillon, S. (2025). Water erosion, sediment transport, and sediment delivery ratio in the Wadi Isser Basin (North-Central Algeria): insights from the RUSLE model and sediment rating curves. *Mediterranean Geoscience Reviews*. https://doi.org/10.1007/s42990-025-00186-2

Boutagayout, A., Hamdani, A., & Adiba, A. (2025). Advancing agroecology for sustainable water management: A comprehensive review and future directions in North African countries. *Water Conservation Science and Engineering, 10*(1), 22. https://doi.org/10.1007/s41101-025-00350-7

Chapter 9. 2021. Chapter 9: Africa. https://www.ipcc.ch/report/ar6/wg2/chapter/chapter-9/?utm_source=chatgpt.com. Retrieved August 20, 2025.

Cheikha, L. B., El Asmi, A. M., Gdiri, A., Oueslati, M., Attia, R., Rhouma, A. B., Aouadi, T., Jaoued, M. and Gueddari, M. (2023). Soil erosion rates evaluation via ^{137}Cs and RUSLE modeling in the Sigilil watershed (northeastern Tunisia). *Geoderma Regional, 35*, e00714. https://doi.org/10.1016/j.geodrs.2023.e00714

Darvishi Boloorani, A., Soleimani, M., Papi, R., Neysani Samany, N., Teymouri, P., & Soleimani, Z. (2023). Sources, drivers, and impacts of sand and dust storms: A global view. In A. Al-Dousari & M. Z. Hashmi (Eds.), *Dust and health*, emerging contaminants and associated treatment technologies (pp. 31–49). Springer International Publishing. https://doi.org/10.1007/978-3-031-21209-3_3

Deeb, M., et al. (2024). The urgency of building soils for Middle Eastern and North African countries: Economic, environmental, and health solutions (p. 917).

Ed-Dakiri, S., Etebaai, I., El Moussaoui, S., Tawfik, A., Lamgharbaj, M., El Talibi, H., Dekkaki, H. C., & Taher, M. (2024). Assessing soil erosion risk through geospatial analysis and magnetic susceptibility: A study in the Oued Ghiss Dam Watershed, Central Rif, Morocco. *Scientific African, 26*, e02401. https://doi.org/10.1016/j.sciaf.2024.e02401

FAO. (2021). Global Map of Salt-Affected Soils. https://openknowledge.fao.org/handle/20.500.14283/cb7247en. Retrieved August 20, 2025.

Fredj, A., Ghernaout, R., Dahmani, S., & Remini, B. (2024). Assessing soil erosion through the implementation of the RUSLE model and geospatial technology in the Isser Watershed, Northern Algeria. *Water Supply, 24*(7), 2487–2505. https://doi.org/10.2166/ws.2024.154

Habibi, B., Meddi, M., Torfs, P. J. J. F., Remaoun, M., & Van Lanen, H. A. J. (2018). Characterisation and prediction of meteorological drought using stochastic models in the Semi-Arid Chéliff-Zahrez Basin (Algeria). *Journal of Hydrology: Regional Studies, 16*, 15–31. https://doi.org/10.1016/j.ejrh.2018.02.005

Hagage, M., Abdulaziz, A. M., Elbeih, S. F., & Hewaidy, A. G. A. (2024). Monitoring soil salinization and waterlogging in the Northeastern Nile Delta linked to shallow saline groundwater and irrigation water quality. *Scientific Reports, 14*(1), 27838. https://doi.org/10.1038/s41598-024-77954-x

Hakam, O., Bouras, E. H., Amazirh, A., Ongoma, V., Eddamiri, S., Saidi, L., Zerouali, A., & Chehbouni, A. (2025). Influence of atmospheric and oceanic circulation patterns on precipitation variability in North Africa with a focus on Morocco. *Scientific Reports, 15*(1), 17977. https://doi.org/10.1038/s41598-025-02718-0

Hallouz, F., Meddi, M., Mahé, G., Toumi, S., & Rahmani, S. E. A. (2018). Erosion, suspended sediment transport and sedimentation on the Wadi Mina at the Sidi M'Hamed Ben Aouda Dam, Algeria. *Water, 10*(7), 895. https://doi.org/10.3390/w10070895

Houyou, Z., Bielders, C. L., Benhorma, H. A., Dellal, A., & Boutemdjet, A. (2016). Evidence of strong land degradation by wind erosion as a result of rainfed cropping in the Algerian Steppe: A case study at Laghouat. *Land Degradation and Development, 27*(8), 1788–1796. https://doi.org/10.1002/ldr.2295

Hssaisoune, M., Bouchaou, L., Qurtobi, M., Marah, H., Beraaouz, M., & Messari, J. S. E. (2022). Isotopic and chemical tracing for residence time and recharge mechanisms of groundwater under semi-arid climate: Case from Rif Mountains (Northern Morocco). *Geosciences, 12*(2), 74. https://doi.org/10.3390/geosciences12020074

IPCC. (2022). Cross-Chapter Paper 4: Mediterranean Region. https://www.ipcc.ch/report/ar6/wg2/chapter/ccp4/. Retrieved September 17, 2025.

IPCC. (2023). *Climate Change 2022–Impacts, adaptation and vulnerability: Working Group II contribution to the sixth assessment report of the intergovernmental panel on climate change* (1st ed.). Cambridge University Press. https://doi.org/10.1017/9781009325844.

Kastali, F., Meddi, M., Boufekane, A., & Longobardi, A. (2024). Integrated modeling of surface water and groundwater interactions in the Eastern Mitidja Plain, North Algeria: Investigating the impact of human activities and climate change. *Journal of Water and Climate Change, 15*(11), 5570–5586. https://doi.org/10.2166/wcc.2024.413

Kessabi, R., Hanchane, M., Krakauer, N. Y., Aboubi, I., El Kassioui, J., & El Khazzan, B. (2022). Annual, seasonal, and monthly rainfall trend analysis through non-parametric tests in the Sebou River Basin (SRB), Northern Morocco. *Climate, 10*(11), 170. https://doi.org/10.3390/cli10110170

Kotb, T. H. S., Watanabe, T., Ogino, Y., & Tanji, K. K. (2000). Soil salinization in the Nile Delta and related policy issues in Egypt. *Agricultural Water Management, 43*(2), 239–261. https://doi.org/10.1016/S0378-3774(99)00052-9

Loudyi, D., Chagdali M., & Belmatrik, S. (2018). Reservoirs Silting in Morocco. Reservoir Sedimentation: Challenges and Management Strategies.

Marghmi, A., Ben Cheikha, L., El Asmi, A. M., & Gueddari, M. (2024). Soil erosion risk assessment of the Lakhmess watershed (northwestern Tunisia) via the SEAGIS model: Inferred prioritization of risky sub-watersheds. *International Journal of Sediment Research, 39*(6), 1015–1026. https://doi.org/10.1016/j.ijsrc.2024.08.001

Medjani, F., Derradji, T., Zahi, F., Djidel, M., Labar, S., & Bouchagoura, L. (2023). Assessment of soil erosion by universal soil loss equation model based on geographic information system data: A case study of the mafragh watershed, North-Eastern Algeria. *Scientific African, 21*, Article e01782. https://doi.org/10.1016/j.sciaf.2023.e01782

Mihi, A., Ghazela, R., & Wissal, D. (2022). Mapping potential desertification-prone areas in North-Eastern Algeria using logistic regression model, GIS, and remote sensing techniques. *Environmental Earth Sciences, 81*(15), 385. https://doi.org/10.1007/s12665-022-10513-7

Mihi, A., Zerroug, K., Kouachi, M. E., & Benarfa, N. (2024). Spatiotemporal changes of desertification degree in the Algerian green barrier over the last four decades (1984–2020). *Arid Land Research and Management, 38*(3), 362–381. https://doi.org/10.1080/15324982.2024.2316657

Mirgol, B., Dieppois, B., Northey, J., Eden, J., Jarlan, L., Khabba, S., Le Page, M., & Mahé, G. (2025). Understanding wheat farmers' climate change adaptation in the middle East and North africa: The role of socioeconomic factors and perceptions in decision-making. *Journal of Agriculture and Food Research, 23*, 102290. https://doi.org/10.1016/j.jafr.2025.102290

Mosaid, H., Barakat, A., Bouras, E. H., Ismaili, M., El Garnaoui, M., Abdelrahman, K., & Kahal, A. Y. (2024). Dam siltation in the mediterranean region under climate change: A case study of Ahmed El Hansali Dam, Morocco. *Water, 16*(21), 3108. https://doi.org/10.3390/w16213108

Munir, M. A., Khan, F., & Khan, S. (2024). Efficient localized adaptation of neural weather forecasting: A case study in the mena region. arXiv preprint arXiv:2409.07585. Accepted at NeurIPS 2024 Workshop on Tackling Climate Change with Machine Learning.

Qadir, M., Quillérou, E., Nangia, V., Murtaza, G., Singh, M., Thomas, R. J., Drechsel, P., & Noble, A. D. (2014). Economics of salt-induced land degradation and restoration. *Natural Resources Forum, 38*(4), 282–295. https://doi.org/10.1111/1477-8947.12054

Rerboudj, A., Guettouche, M.-S., & Callot, Y. (2024). Geospatial modeling of aeolian dynamics in the algerian steppe from Zahrez Chergui to Hodna. *Engineering, Technology and Applied Science Research, 14*(6), 18695–18701. https://doi.org/10.48084/etasr.9095

Ruiz, I., & Sanz, M. J. (2022). Sustainable land management for rural adaptation in the Mediterranean and middle eastern watersheds. In: W. Leal Filho, & E. Manolas (Eds.), *Climate change in the Mediterranean and Middle Eastern Region. Climate change management*. Springer. https://doi.org/10.1007/978-3-030-78566-6_21

Saadoud, D., Hassani, M., Peinado, F. J. M., & Guettouche, M. S. (2018). Application of fuzzy logic approach for wind erosion hazard mapping in laghouat region (Algeria) using remote sensing and GIS. *Aeolian Research, 32*, 24–34. https://doi.org/10.1016/j.aeolia.2018.01.002

Sanad, H., Mouhir, L., Zouahri, A., Moussadek, R., El Azhari, H., Yachou, H., Ghanimi, A., Lhaj, M. O., & Dakak, H. (2024a). Assessment of groundwater quality using the Pollution Index of Groundwater (PIG), Nitrate Pollution Index (NPI), Water Quality Index (WQI), Multivariate Statistical Analysis (MSA), and GIS approaches: A case study of the mnasra region, Gharb Plain, Morocco. *Water, 16*(9), 1263. https://doi.org/10.3390/w16091263

Sanad, H., Lhaj, M. O., Zouahri, A., Saafadi, L., Dakak, H., & Mouhir, L. (2024b). Groundwater pollution by nitrate and salinization in morocco: a comprehensive review. *Journal of Water and Health, 22*(10), 1756–1773. https://doi.org/10.2166/wh.2024.024

Sefelnasr, A., & Sherif, M. (2014). Impacts of seawater rise on seawater intrusion in the Nile Delta aquifer, Egypt. *Groundwater, 52*(2), 264–276. https://doi.org/10.1111/gwat.12058

Serbaji, M. M., Bouaziz, M., & Weslati, O. (2023). Soil water erosion modeling in tunisia using RUSLE and GIS integrated approaches and geospatial data. *Land, 12*(3), 548. https://doi.org/10.3390/land12030548

Slama, F., Nasri, N., & Bouhlila, R. (2022). Delineating the origins and processes of groundwater salinization and quality degradation in a coastal irrigated plain, Korba (Northeastern Tunisia). *Marine Pollution Bulletin, 181*, 113914. https://doi.org/10.1016/j.marpolbul.2022.113914

Taher, M., Mourabit, T., Bourjila, A., Saadi, O., Errahmouni, A., El Marzkioui, F., & El Mousaoui, M. (2022). An estimation of soil erosion rate hot spots by integrated USLE and GIS methods: A case study of the Ghiss Dam and Basin in Northeastern Morocco. *Geomatics and Environmental Engineering, 16*(2), 95–110. https://doi.org/10.7494/geom.2022.16.2.95

Tarolli, P., Luo, J., Park, E., Barcaccia, G., & Masin, R. (2024). Soil salinization in agriculture: Mitigation and adaptation strategies combining nature-based solutions and bioengineering. *IScience, 27*(2), 108830. https://doi.org/10.1016/j.isci.2024.108830

Tramblay, Y., El Adlouni, S., & Servat, E. (2013). Trends and variability in extreme precipitation indices over Maghreb Countries. *Natural Hazards and Earth System Sciences, 13*(12), 3235–3248. https://doi.org/10.5194/nhess-13-3235-2013

Tramblay, Y., Khalki, E. M. E., Khedimallah, A., Sadaoui, M., Benaabidate, L., Boulmaiz, T., Boutaghane, H., et al. (2024). Regional flood frequency analysis in North Africa. *Journal of Hydrology, 630*, Article 130678. https://doi.org/10.1016/j.jhydrol.2024.130678

WBG (World Bank Group). (2025). *Climate risk country profile: Tunisia*. World Bank Group/ Climate Change Knowledge Portal. https://climateknowledgeportal.worldbank.org

WMO. (2024). *State of the Climate in Africa 2023*. Erscheinungsort nicht ermittelbar.

World Bank. (2023). *Tunisia country climate and development report*. World Bank Group. https://documents.worldbank.org

Yahiaoui, S., Meddi, M., Razack, M., Boufekane, A., & Bekkoussa, B. S. (2023). Hydrogeochemical and isotopic assessment for characterizing groundwater quality in the Mitidja Plain (Northern Algeria). *Environmental Science and Pollution Research, 30*(33), 80029–80054. https://doi.org/10.1007/s11356-023-27952-9

Zhou, W., Wang, H., & Ge, Q. (2024). Contributions of Climatic Factors and Vegetation Cover to the Temporal Shift in Asian Dust Events. *NPJ Climate and Atmospheric Science, 7*(1), 328. https://doi.org/10.1038/s41612-024-00887-9

Ziadat, F. M., Zdruli, P., Christiansen, S., Caon, L., Abdel Monem, M., & Fetsi, T. (2022). An overview of land degradation and sustainable land management in the Near East and North Africa. *Sustainable Agriculture Research, 11*(1). https://doi.org/10.5539/sar.v11n1p11

Chapter 4
Integrating Climate Change Adaptation into Urban Planning: Pathways for Resilient Cities in the MENA Region

Elnazir Ramadan

Abstract The Middle East and North Africa (MENA) regions face escalating climate hazards including heatwaves, sea-level rise, water scarcity, and flash floods which are exacerbated by rapid urban growth, demographic pressures, and fragmented governance. This chapter examines how urban planning can function as a strategic mechanism for embedding climate adaptation into city development across the region. Drawing on case studies from GCC countries, including the United Arab Emirates, Oman, and Qatar, it assesses the extent to which adaptation is integrated into land use planning, zoning, infrastructure, and urban design. The analysis identifies persistent gaps in adaptive planning, driven by limited inter-agency coordination, weak climate data integration, and insufficient local technical capacity. The chapter underscores the need to mainstream climate risks into spatial plans, expand nature-based solutions, employ geospatial and decision-support technologies, and strengthen inclusive and participatory planning processes. It also reviews emerging policy innovations—such as climate-responsive building codes and advanced water management systems—and considers their potential for replication across MENA. The chapter concludes with practical recommendations to enhance institutional coherence, build planning capacities, and align national and municipal adaptation strategies with global frameworks.

Keywords Urban planning · Climate adaptation · Resilience · MENA region · Nature-based solutions

E. Ramadan (✉)
Department of Geography and Urban Sustainability, United Arab Emirates University, Al Ain, United Arab Emirates
e-mail: elnazirr@uaeu.ac.ae

E. Alam (ed.), *Disaster, Crisis and Emergency Management in the Middle East & North Africa*, https://doi.org/10.1007/978-3-032-24436-9_4

4.1 Introduction: Framing the Challenge

The Middle East and North Africa (MENA) region finds itself at a critical juncture, where the intertwined forces of climate change and rapid urban expansion are converging to create unprecedented climate hazards such as floods, heatwaves, and extreme storms which are classified within a globally standardized hazard system developed through the Hazard Information Profiles (HIPs) initiative (UNDRR & ISC, 2025). Integrating hazard information into development plans is essential for reducing risk and building resilience (World Bank, 2022). This chapter posits that urban planning is not merely a technical or administrative function but a fundamental strategic tool for navigating this complex landscape. It moves beyond a simplistic identification of climate hazards to critically examine the institutional frameworks, policy mechanisms, and practical tools that can systematically embed climate adaptation into the very fabric of urban development. By drawing on a comparative analysis of case studies from both the oil-rich Gulf states and the North African urban centers, the chapter seeks to provide a nuanced understanding of the region's diverse challenges and opportunities. Ultimately, it aims to offer an actionable roadmap for policymakers, planners, and researchers, advocating for a paradigm shift from reactive disaster management to proactive, long-term resilience building. This transition is framed as essential not only for safeguarding economic assets and infrastructure but also for ensuring social equity and sustainable urban futures across the MENA region, in alignment with global agendas like the Paris Agreement and the Sendai Framework.

4.2 Urbanization Trends and Climate Risks in MENA

The MENA region is experiencing one of the most rapid and transformative urbanization processes in the world, a trend that dramatically increases its exposure to climate-related hazards. Cities across the region are expanding both in population and physical footprint, with a significant proportion of the population residing in cities that are often ill-prepared for such stresses (Sowers et al., 2011), through informal settlements and unplanned development that encroach upon ecologically sensitive areas such as floodplains, coastal zones, and arid peripheries. This rapid, and often sprawling, urban growth places immense strain on already limited resources, particularly water, and creates vast expanses of heat-absorbing concrete that exacerbate the urban heat island effect (Waha et al., 2017). Consequently, the inherent climate vulnerabilities of the region (IPCC, 2022), which include extreme heat, water scarcity, and flash floods, are not just environmental issues but are intensely mediated by this urban context. The concentration of people, economic assets, and critical infrastructure in high-risk zones means that a single climate event can trigger cascading failures, disrupting essential services, economies, and public health. Coastal cities, such as Alexandria, Casablanca, and Doha, are threatened by sea-level rise and saltwater intrusion, jeopardizing infrastructure, housing, and freshwater resources (Dasgupta et al., 2009; Reimann et al., 2018). Urbanization patterns compound these biophysical risks. Many cities in the region are characterized by informal settlements, which are often located in high-risk zones like floodplains or steep slopes and lack basic services, making their inhabitants disproportionately vulnerable (Al-Naber & Lwasa, 2017). Furthermore, the urban heat island (UHI) effect, where built-up areas are significantly warmer than surrounding rural areas, intensifies heat-related risks in densely populated cities (Estoque et al., 2017; Gago et al., 2013). Water governance challenges, including over-extraction of groundwater and inefficient irrigation, further strain the capacity to respond to climate-induced droughts (Molle & Berkoff, 2006). This dangerous synergy between demographic pressure, spatial expansion, and intensifying climatic extremes defines the unique and urgent challenge facing MENA metropolises.

4.3 Why Urban Planning Is Central to Climate Change Adaptation

Historically, urban planning in the MENA region has largely followed a master-planning approach focused on economic development and physical expansion, often with limited consideration for environmental sustainability and climate risks (Alawadi & Benkraouda, 2019). This reactive, rather than proactive planning model, has created a significant adaptation deficit. Urban planning is central to climate change adaptation because it provides the primary set of instruments for proactively shaping the physical, social, and functional structure of a city to withstand and respond to climatic shocks. Unlike reactive disaster response, which addresses impacts after they occur, planning allows for the systematic integration of resilience into long-term decision-making about where and how to build. Research highlights a disconnect between high-level national climate policies, such as Nationally Determined Contributions (NDCs), and local urban planning implementation (Chu et al., 2016). While many MENA countries have mentioned urban adaptation in their NDCs, the translation into enforceable local regulations, building codes, and land-use plans remains weak (UN Habitat, 2025). The dominant focus has often been on mitigation (e.g., through renewable energy projects) rather than on adapting urban form and infrastructure to inevitable climate impacts (Atallah, 2023). This is partly due to the perception of adaptation as a cost rather than an investment, and a lack of locally specific climate data to inform planning decisions (Grafakos et al., 2019). Through strategic spatial planning and land-use regulation, cities can steer development away from high-risk zones like coastal erosion corridors or wadi beds prone to flash flooding. Zoning ordinances and building codes can mandate climate-sensitive design, requiring features like heat-reflective materials, passive cooling, and water-efficient fixtures. Furthermore, planning facilitates the deployment of green infrastructure such as urban parks, green corridors, and restored wetlands, which simultaneously mitigates heat, manages stormwater, enhances biodiversity, and improves citizen well-being. By governing the location, density, and quality of infrastructure and housing, urban planning determines a city's long-term vulnerability or resilience. It is the key mechanism for translating broad national climate policies into concrete, on-the-ground actions that directly protect communities, economies, and ecosystems from the escalating impacts of climate change.

4.4 Climate Change Hazards and Urban Vulnerabilities

The urban centers of the MENA region are confronted by a multifaceted suite of climate hazards that interact with built environments to produce acute and chronic vulnerabilities. These are not isolated threats but a compound crisis where one hazard can exacerbate another, creating complex chains of risk. The region is a global hotspot for extreme heat and heatwaves, which elevate mortality and morbidity rates, strain energy grids through soaring cooling demand, and reduce outdoor labor productivity. Concurrently, acute water scarcity, driven by low precipitation, high evaporation rates, and growing demand, threatens the very viability of cities, forcing a reliance on energy-intensive desalination and often leading to social conflict over allocation. This scarcity interacts with desertification, degrading arable land on urban fringes and increasing dust storms that impact air quality and health. Meanwhile, sea-level rise poses an existential threat to coastal cities which house a significant portion of the region's population and economic activity through permanent inundation, saltwater intrusion into freshwater aquifers, and worsening coastal erosion. Paradoxically, despite overall aridity, the intensity of flash floods is increasing; when rains do occur, they are often torrential, and the impermeable surfaces of urban areas prevent infiltration, leading to devastating runoff that overwhelms drainage systems and damages infrastructure built in dry riverbeds (wadis).

4.5 Overview of Region-Specific Hazards

The MENA region's climate hazards are distinctly severe and deeply interconnected. Water scarcity is arguably the most defining challenge, with most countries in the region experiencing absolute water stress. This is driven by the world's lowest levels of renewable freshwater resources per capita, which is now being compounded by longer drought periods and rising temperatures that increase evaporation from reservoirs and soils. This scarcity directly fuels desertification, the process by which fertile land typically loses its productivity because of drought, deforestation, or inappropriate agriculture. As vegetation recedes on the outskirts of cities, soil erosion increases, and dust and sandstorms become more frequent and intense, impacting air quality, respiratory health, and even disrupting air travel. Another critical, region-specific hazard is the rapid rise in sea level, which particularly threatens the numerous low-lying coastal cities and the vast investments in infrastructure along the coastlines of the Gulf, Egypt, and North Africa. Saltwater intrusion into deltaic agricultural lands, like the Nile Delta, simultaneously threatens food security and urban water supplies. These slow-onset changes create a baseline of chronic vulnerability upon which acute shocks, such as a record-breaking heatwave or a sudden catastrophic flood, can have a devastating impact.

4.6 Spatial Distribution of Risk in Major MENA Cities

Within major MENA cities, climate risks are not distributed evenly but are heavily influenced by socio-economic factors and the quality of urban planning, creating stark spatial inequalities. The most vulnerable populations are often concentrated in informal settlements and low-income neighborhoods, which are frequently located on the most hazardous steep slopes prone to landslides, low-lying areas susceptible to flooding, or on the urban periphery where the urban heat island effect intersects with limited access to green space and water. These areas typically lack robust infrastructure, adequate drainage, and climate-resilient housing, making residents disproportionately exposed. In coastal cities like Alexandria, Casablanca, or Doha, risk is spatially concentrated along the shoreline and low-lying reclaimed areas, where valuable assets are exposed to sea-level rise, storm surges, and erosion. Furthermore, the urban core often experiences the most intense heat island effect, with temperatures significantly higher than the surrounding countryside due to the density of heat-absorbing materials and lack of vegetation. Conversely, affluent districts often have greater access to green spaces, reliable water supplies, and air-conditioned buildings, illustrating how urban climate vulnerability is a direct function of spatial inequality and planning priorities.

4.7 The Escalating Climate Vulnerability of MENA Cities

The Middle East and North Africa (MENA) region stands on the front lines of the global climate crisis, facing a confluence of intensifying and interlinked hazards that pose an existential threat to its urban centers (World Bank, 2022). The region's inherent aridity is being drastically amplified by more frequent and severe heatwaves, prolonged droughts, and accelerating desertification, which collectively strain already scarce water resources and agricultural systems (IPCC, 2022; Waha et al., 2017). Simultaneously, coastal megacities and burgeoning Gulf metropolises are confronting the insidious threat of sea-level rise, which threatens critical infrastructure, salinizes freshwater aquifers, and erodes shorelines (Dasgupta et al., 2009;

Reimann et al., 2018). These slow-onset crises are punctuated by acute, high-impact events like flash floods, which are exacerbated by erratic rainfall patterns and the region's often dry and impermeable terrain (Sowers et al., 2011). This complex risk profile is not unfolding in a vacuum; it intersects dangerously with powerful socio-economic trends, primarily rapid and often unplanned urbanization. The concentration of people and assets in high-risk areas, such as floodplains and coastal zones, coupled with governance challenges, outdated infrastructure, and existing social inequities, creates a potent recipe for disaster (Al-Naber & Lwasa, 2017; Chu et al., 2016). Consequently, MENA cities are becoming increasingly vulnerable hotspots where climate change acts as a threat multiplier, deepening environmental stresses and potentially undermining economic stability and social cohesion (Sowers, 2013).

4.8 Institutional and Technical Barriers to Climate-Resilient Planning

Despite the clear and present danger, the integration of climate adaptation strategies into urban planning across the MENA region is hampered by a suite of persistent institutional and technical barriers (Aylett, 2015). A primary institutional challenge is the pervasive fragmentation of governance, where responsibilities for land use, water management, infrastructure, and disaster response are often siloed across different ministries and agencies with limited coordination (Betsill & Bulkeley, 2006; Sowers, 2013). This lack of policy coherence leads to contradictory agendas and prevents a holistic, city-wide approach to resilience building (Dorkenoo et al., 2022). Furthermore, there is a frequent significant disconnect between national climate policies, such as Nationally Determined Contributions (NDCs), and their implementation at the municipal level, where local governments often lack the mandate, financial resources, and technical capacity to enact meaningful change (Chu et al., 2016; Atallah et al., 2023). On the technical front, a critical barrier is the inadequate integration of reliable, high-resolution climate data and risk assessments into the spatial planning process (Grafakos et al., 2019). Planners often operate with outdated information or lack access to localized climate projections needed to make informed decisions about zoning, infrastructure design, and land-use allocation (Johnson et al., 2012). This is compounded by a traditional planning model that often prioritizes short-term economic development and reactive disaster response over long-term, proactive risk reduction and resilience investment (Alawadi & Benkraouda, 2019), leaving cities perpetually playing catch-up with the escalating climate threat.

4.9 A Multi-Scalar Framework for Integrated Adaptation

To overcome these systemic challenges, a transformative and multi-scalar framework is essential for embedding climate adaptation into the very fabric of urban development in the MENA region. This framework operates simultaneously at the regional, national, city, and community levels to ensure coherence and effectiveness. At its core, it advocates for the mainstreaming of climate risk assessments into all land-use planning, zoning regulations, and infrastructure development codes, ensuring that every new planning decision is evaluated through climate resilience lens. This integrated approach actively promotes Nature-based Solutions (NbS) as a central pillar of urban strategy, advocating for the restoration of urban aquifers, the creation of green corridors to mitigate heat island effects, and the preservation of natural floodplains instead of concrete channelization. To support evidence-based decision-making, the framework emphasizes the utilization of advanced geospatial technologies and multi-criteria decision-making tools, which can model climate scenarios, map vulnerabilities, and prioritize interventions with the highest cost-benefit returns. Finally, and crucially, this model insists on enhancing participatory governance, moving beyond technocratic approaches to actively engage communities, civil society, and the private sector in the planning process. This ensures that adaptation strategies are not only technically sound but also socially inclusive, equitable, and locally owned, thereby enhancing their legitimacy and long-term sustainability.

4.10 Emerging Practices and Policy Innovations for Urban Resilience

Amid the significant challenges, promising emerging practices and policy innovations across the MENA region offer valuable, replicable models for building urban resilience. In the Gulf Cooperation Council (GCC) states, cities like Abu Dhabi and Dubai are pioneering the implementation of mandatory climate-sensitive building codes and rating systems such as Estidama and the Dubai Green Building Regulations, which mandate energy and water efficiency, specify reflective building materials to reduce heat absorption, and encourage the integration of green roofs and walls (Birge & Berger, 2019). In North Africa, countries like Morocco and Tunisia are innovating in water management through decentralized and sustainable infrastructure, including large-scale wastewater recycling for irrigation and urban greening, and the revival of traditional water conservation techniques like *khettaras* (qanats) to combat scarcity (Boutagayout et al., 2025). Furthermore, several cities are beginning to explore the strategic use of geospatial technology to create detailed urban heat island maps or to identify neighborhoods most vulnerable to flooding, allowing for targeted interventions. There is also a growing, though nascent, movement toward participatory planning projects that co-design green spaces with local communities, ensuring these spaces meet social needs while providing critical ecosystem services like shade, stormwater absorption, and biodiversity habitat. These innovations, though often isolated, demonstrate a shifting mindset and provide a critical foundation upon which to build a comprehensive regional resilience strategy.

4.11 Policy and Planning Reform

Effective integration of CCA into urban planning is fundamentally a governance challenge. It requires multi-level coordination, stakeholder participation, and institutional capacity-building (Betsill & Bulkeley, 2006). Studies on environmental governance in the MENA region often point to centralized decision-making structures that can hinder local innovation and context-specific responses (Sowers, 2013). However, there is a growing recognition of the need for polycentric governance models that involve various actors (Djalante et al., 2011).

The role of local governments is critical. Research on city networks like C40 and ICLEI demonstrates how transnational municipal networks can facilitate knowledge transfer, capacity-building, and the mainstreaming of climate action (Bulkelye & Betsill, 2005; Heikkinen et al., 2020). Financing remains a major barrier. Accessing international climate finance (e.g., the Green Climate Fund) is often challenging for MENA municipalities due to complex procedures and a lack of bankable project proposals (Atela et al., 2018; Michaelowa & Michaelowa, 2011). Innovative financial mechanisms, such as green bonds and resilience bonds, are still nascent in the region (Flammer, 2021).

To support a widespread transition toward climate-resilient urban futures, a concerted effort is required by planners, policymakers, and international partners to implement a set of targeted recommendations. First, there is an urgent need to strengthen policy coherence by establishing inter-ministerial climate task forces and mandating the integration of climate adaptation objectives into all national urban policies and local master plans, ensuring alignment with global commitments like the Paris Agreement and the Sendai Framework. Second, significant investment must be made in local capacity-building, providing municipal authorities with training, tools, and dedicated funding streams to develop, implement, and monitor local adaptation action plans. Third, governments should incentivize nature-based solutions and climate-sensitive design through fiscal mechanisms, revised procurement rules, and updated planning regulations that make sustainable practices the mainstream action not the exception. Fourth, it is critical to foster transboundary collaboration for knowledge exchange and to manage shared resources, particularly water basins and coastal zones, through regional platforms and partnerships. Finally, advancing applied research to generate localized climate data and develop context-specific solutions is paramount. By adopting these recommendations, MENA nations can empower their urban planning systems to become proactive agents of change, shifting from a cycle of disaster recovery to a pathway of long-term, sustainable, and inclusive resilience.

4.11.1 Assessment of Current Urban Planning Practices

Urban planning practices across the MENA region are characterized by a significant disconnect between the escalating climate risks and the institutional frameworks designed to manage urban growth (Moukarzel & Azoury, 2025). Historically, planning systems in many MENA countries have been heavily influenced by master planning approaches, which are often rigid, centralized, and focused on physical expansion and infrastructural development to accommodate rapid population growth. While effective for directing large-scale urbanization, these traditional models are not designed to handle the dynamic, uncertain, and cross-cutting nature of climate change. The institutional landscape is typically fragmented, with responsibilities for water, environment, housing, transportation, and civil defense siloed across different ministries and levels of government. This lack of coordination often results in contradictory policies; for instance, a housing ministry may promote development in a coastal zone that an environment ministry is trying to protect from sea-level rise (Chabou et al., 2025). Furthermore, planning mandates are frequently weak at the municipal level, where the implementation ultimately occurs, leading to a gap between national climate strategies and local land-use decisions. The regulatory tools, such as zoning ordinances and building codes, have largely been designed for conventional purposes of controlling density and use, not for managing complex climate risks like urban heat or pluvial flooding.

4.11.2 Planning Tools and Institutional Frameworks in MENA

The predominant planning tools in the MENA region remain conventional land-use plans and building regulations, which are often not leveraged for climate adaptation (Babapoorkamani & Ricci, 2025). Zoning codes, for example, rarely include overlays that prohibit or restrict development in high-risk areas such as floodplains, coastal erosion zones, and extreme heat hotspots. Where more advanced tools exist, such as Environmental Impact Assessments (EIAs), they are often applied project-by-project rather than strategically at the plan or policy level, limiting their ability to influence broader urban forms. A World Bank (2022) report, Climate and Development in the Middle East and North Africa, explicitly notes that "institutional fragmentation and centralized decision-making" are key barriers to climate action. Institutionally, governance is highly centralized in most MENA nations. National ministries hold the primary authority for setting planning standards and approving major urban projects, while local municipalities often lack the technical capacity, financial resources, and political autonomy to innovate and implement context-specific climate resilience measures (Dabaieh et al., 2024).

This top-down structure stifles local innovation and makes it difficult to respond nimbly to localized climate impacts. The legal frameworks supporting planning are also often outdated, not explicitly mandating the integration of climate risk assessments or the consideration of future climate scenarios into the urban planning process, leaving adaptation as an optional add-on rather than a core requirement.

4.11.3 Limitations in Integrating Risk-Informed Decision-Making

The integration of risk-informed decision-making into urban planning is severely limited by several critical barriers. Firstly, there is a profound data gap. Planners often lack access to reliable, high-resolution, and localized climate data (e.g., downscaled precipitation models, detailed flood maps, future heat projections) that is essential for understanding and quantifying risks. The data that does exist is often held by different agencies (meteorological, hydrological, geological) and is not synthesized in a format usable for spatial planning. Secondly, there is a technical and capacity gap. Many planning departments lack staff trained in climate risk assessment, geospatial analysis, or the design of nature-based solutions (Jack et al., 2020). This leads to a reliance on conventional engineering approaches (e.g., concrete drainage channels) over more adaptive, integrated strategies. Thirdly, a financial and political gap persists. Gray infrastructure projects are often more politically visible and easier to finance than long-term resilience planning or green infrastructure, which may have benefits that are distributed over time and across sectors. The short-term political cycle prioritizes immediate, tangible results over long-term risk reduction, making it difficult to justify investments in proactive adaptation measures whose primary success is a disaster that does *not* happen (Menoni, 2025).

4.12 Best Practices and Case Studies

Despite the overarching challenges, pioneering cities and nations in the MENA region are demonstrating leadership through innovative projects and policies that offer valuable replicable models for climate-resilient urban planning.

4.12.1 Case Examples from the UAE, Morocco, Tunisia, and Egypt

United Arab Emirates: Abu Dhabi's Estidama Pearl Rating System is a mandatory sustainability framework that has successfully embedded climate adaptation into the building approval process (Alawadi & Benkraouda, 2019). It requires developers to integrate features such as shaded walkways, high-performance building envelopes to reduce cooling demand, water-efficient irrigation, and the use of recycled materials. The city of Al Ain has been praised for its traditional *aflaj* irrigation system and its modern efforts to integrate green spaces, which help mitigate the extreme desert heat through evapotranspiration.

Morocco: In response to critical water scarcity, Morocco has invested heavily in large-scale wastewater treatment and reuse for irrigating urban green spaces and agricultural peripheries. This decouples urban greening from potable water supplies, a critical adaptation strategy. Cities are also exploring the rehabilitation of traditional water management systems (Ortega-Pozo et al., 2022).

Tunisia: The Agricultural Development Groups model in rural and peri-urban Tunisia is a successful example of community-based natural resource management. While focused on agriculture, it provides a template for participatory governance that could be applied to managing urban green corridors, restoring degraded lands, and implementing local-scale NbS to reduce erosion and flood risk.

Egypt: Facing severe threats from sea-level rise in the Nile Delta, Egypt has initiated large-scale coastal protection works combining hard infrastructure (seawalls) with softer measures like the restoration of sand dunes and mangroves. There are also ongoing projects to improve drainage infrastructure in informal settlements in Cairo to reduce flood vulnerability (Sharaan et al., 2022).

4.13 Successful Models of NbS, Green Infrastructure, and Community-Based Adaptation

Successful models are emerging that move beyond gray infrastructure. Integrating green spaces, urban forests, and permeable surfaces can mitigate the UHI effect, manage stormwater runoff, and enhance biodiversity (Kabisch et al., 2017). Research on cities like Cairo and Dubai shows the potential of green infrastructure, though implementation is often fragmented (Elshater et al., 2022). In the UAE, the creation of urban parks and green corridors (e.g., Dubai's Safa Park), though often reliant on desalinated water, provides critical relief from the urban heat island effect and recreational space that improves public health during heatwaves. Green building codes, as seen in Abu Dhabi and Dubai, indirectly promote NbS by mandating landscaping with native, drought-tolerant species and the inclusion of green roofs and walls (Najini et al., 2020; Jung et al., 2023; Alhamlawi et al., 2021). In North Africa, there are growing efforts in rainwater harvesting in public buildings and parks, reducing pressure on municipal water systems and mitigating flood runoff. A key lesson from successful cases is the importance of blending modern technology with traditional knowledge, such as reviving ancient water channel systems for modern urban greening, demonstrating that effective adaptation is often context-specific and culturally informed.

4.14 Proposed Framework for Climate-Resilient Urban Planning

A transformative, multi-scalar framework is required to systematically embed climate resilience into the urban planning in MENA cities. This framework must be iterative, flexible, and participatory.

4.14.1 *Integrating Climate Data into Planning*

The foundation of resilient planning is actionable information. This requires a dedicated effort to downscale global climate models to produce high-resolution projections for temperature, precipitation, sea-level rise, and storm intensity specific to each urban area. The use of Geographic Information Systems (GIS), remote sensing, and climate modeling can help identify risk hotspots and simulate the impacts of different planning scenarios, supporting evidence-based decision-making (Johnson et al., 2012). This data must then be translated into hazard maps (flood, heat, storm surge) and vulnerability assessments that identify exposed populations, critical infrastructure, and economic assets. These maps and assessments cannot be static documents; they must be integrated directly into planning support systems and legally mandated as part of the evidence base for all new comprehensive plans, zoning updates, and major infrastructure investments. Planners need to shift from planning based on historical data to planning for a range of future climate scenarios.

4.14.2 *Multi-Criteria Decision Analysis (MCDA) and GIS Tools*

To navigate the complex trade-offs inherent in urban planning, planners need robust decision-support tools. Geographic Information Systems (GIS) provides the platform for layering climate hazard maps with socio-economic data, infrastructure networks, and land-use plans to visually identify areas of highest risk and priority for intervention. Multi-criteria decision analysis (MCDA) is a method that works within GIS to evaluate different adaptation options (e.g., building a seawall vs. restoring a mangrove wetland) against a set of weighted criteria, such as cost-effectiveness, social equity, environmental benefit, and technical feasibility. This structured approach moves decision-making away from political whims or conventional practice and toward evidence-based, transparent, and justifiable choices for resilience investments (Malczewski, 2006).

4.14.3 Adaptive and Flexible Urban Design Approaches

Given the deep uncertainty of climate future, planning must embrace adaptive and flexible design. This includes

- **Adaptive Zoning**: Creating planning regulations that can be triggered by changing conditions, such as prohibiting certain types of development after a sea-level rise threshold is reached.
- **Flexible Public Spaces**: Designing parks, plazas, and urban rights-of-way to serve dual purposes as recreational spaces most of the time and as water retention areas during extreme rainfall events (e.g., "sponge city" concepts).
- **Modular and Green Infrastructure**: Prioritizing infrastructure that can be easily expanded or adapted, and favoring distributed green infrastructure (permeable pavements, bioswales, urban forests) over single, massive gray infrastructure projects.

4.15 Policy Recommendations

To operationalize the proposed framework, a concerted effort is needed at all levels of governance. Key policy recommendations include the following.

Planning Reform for Climate Resilience

National governments must undertake fundamental planning laws and policy reforms. This includes mandating that all urban plans be subjected to a Climate Resilience Impact Assessment and legally embedding the use of future climate scenarios and risk maps into the plan-making process. Zoning codes must be updated to include climate adaptation overlays that restrict development in high-risk areas and incentivize climate-sensitive design through density bonuses or fast-track permits.

Cross-sectoral Integration and Regional Cooperation

Silos must be broken down. Establishing inter-ministerial climate resilience task forces can ensure policy coherence among water, energy, environment, and urban development sectors. At the regional level, MENA countries must strengthen trans-boundary collaboration on shared challenges, particularly water basin management, dust storm mitigation, and knowledge sharing on adaptation technologies and policies, potentially through a dedicated regional knowledge hub.

Capacity Development for Local Governments

Empowering local authorities is critical. National governments and international partners must invest in long-term capacity development programs for municipal planners and engineers. This includes training on GIS and MCDA tools, climate risk assessment, participatory planning methods, and the design of NbS. Establishing dedicated funding streams (e.g., climate adaptation grants) for cities to pilot innovative resilience projects is essential to translating knowledge into action.

4.16 Conclusion

The climate crisis presents a profound threat to the sustainable future of cities in the MENA region, but it also presents an unprecedented opportunity to rethink and reform urban planning practices. The chapter establishes the urgent need to integrate climate change adaptation into urban planning in the MENA region. While the vulnerabilities are well-documented, and a suite of potential strategies exists, the primary challenges lie in governance, financing, and the mainstreaming of adaptation into core urban planning processes. Key research gaps persist. There is a need for more empirical studies evaluating the effectiveness and cost-benefit of specific adaptation interventions in MENA cities (Gouldson et al., 2015). Further research is required on context-specific governance models that can bridge the national-local divide and foster multi-stakeholder collaboration (Dorkenoo et al., 2022). As this chapter has argued, the path to resilience requires a fundamental paradigm shift from a reactive, siloed, and static approach to urban development to a proactive, integrated, and adaptive one. This transition is not merely technical; it is institutional, legal, and political. It demands that governments move beyond symbolic commitments and undertake the hard work of reforming planning systems, investing in local capacity, and fostering cross-sectoral collaboration. The emerging best practices from across the region prove that innovation is possible, even in the most challenging environments. By leveraging spatial planning as a strategic tool, mainstreaming nature-based solutions, harnessing technology for decision-making, and placing equity at the center of adaptation efforts, MENA cities can navigate the uncertainties of the twenty-first century. They can evolve into more livable, sustainable, and resilient urban centers that are not only capable of withstanding climatic shocks but also of providing a higher quality of life for all their inhabitants. The roadmap is clear; what is needed now is political will and collective action to follow it.

References

Ahern, J. (2011). From fail-safe to safe-to-fail: Sustainability and resilience in the new urban world. *Landscape and Urban Planning, 100*(4), 341–343. https://doi.org/10.1016/j.landurbplan.2011.02.021

Alawadi, K., & Benkraouda, O. (2019). The debate over neighborhood density in Dubai: Between theory and practicality. *Journal of Planning Education and Research, 39*(1), 72–88. https://doi.org/10.1177/0739456X17720459

Alhamlawi, F., Alaifan, B., & Azar, E. (2021). A comprehensive assessment of Dubai's green building rating system: Al Sa'fat. *Energy Policy, 15*, 7. https://doi.org/10.1016/j.enpol.2021.112503

Al-Masri, N., & Abu-Hijleh, B. (2022). The effectiveness of passive cooling design strategies in reducing cooling energy demand in a residential building in Dubai. *Energy and Buildings, 255*, 111667.

Al-Naber, P., & Lwasa, S. (2017). The role of informal settlements in building urban resilience: The case of Amman. In *Urban resilience in a global context* (pp. 145–162). *Springer*.

Atallah, J. (2023). The best path forward for low and middle-income countries in the MENA region. MEP_231124_Atallah Brief_FINAL.pdf

Atela, J. O., Quinn, C. H., & Minang, P. A. (2018). Implementing REDD+ at the local level: Assessing the key enablers for credible mitigation and sustainable livelihood outcomes. *Journal of Environmental Management, 220*, 1–13.

Aylett, A. (2015). Institutionalizing the urban governance of climate change adaptation: Results of an international survey. *Urban Climate, 14*, 4–16.

Babapoorkamani, A., & Ricci, L. (2025). Decision-making strategies for climate change adaptation in coastal regions of Africa. *Environmental Development, 55.*

Betsill, M. M., & Bulkeley, H. (2006). Cities and the multilevel governance of global climate change. *Global Governance, 12*(2), 141–160.

Birge, D., & Berger, A. (2019). Transitioning to low-carbon suburbs in hot-arid regions: A case-study of Emirati villas in Abu Dhabi. *Building and Environment, 147.*

Boutagayout, A., Hamdani, A., Boutgayout, I., & Adiba, A. (2025). Towards sustainable water management in Morocco: Assessing resources, challenges, and adaptation strategies. *Frontiers in Sustainable Food Systems, 9*, 1666555. https://doi.org/10.3389/fsufs.2025.1666555

Bulkeley, H., & Betsill, M. (2005). Rethinking sustainable cities: Multilevel governance and the 'urban' politics of climate change. *Environmental Politics, 14*(1), 42–63. https://doi.org/10.1080/0964401042000310178

Chabou, O., Lazri, Y., Mannucci, S., Ciardiello, A., Rosso, F., & Ferrero, M. (2025). Advancing flood resilience in north African coastal cities: A contextual analysis of Jijel, Algeria. *City and Environment Interactions, 28.*

Chu, E., Anguelovski, I., & Carmin, J. (2016). Inclusive approaches to urban climate adaptation planning and implementation in the Global South. *Climate Policy, 16*(3), 372–392. https://doi.org/10.1080/14693062.2015.1019822

Dabaieh, M., Maguid, D., El Mahdy, D., & Al Hegazi, S. (2024). Climate change adaptation at the margins: The case of Cairo, Egypt. *Cities, 154.*

Dasgupta, S., Laplante, B., Murray, S., & Wheeler, D. (2009). Climate change and the future impacts of storm-surge disasters in developing countries. *World Bank Policy Research Working Paper, 5001*. https://openknowledge.worldbank.org/handle/10986/4024

Djalante, R., Holley, C., & Thomalla, F. (2011). Adaptive governance and managing resilience to natural hazards. *International Journal of Disaster Risk Science, 2*(4), 1–14.

Dorkenoo, K., Scown, M., & Boyd, E. (2022). A critical review of disproportionality in loss and damage from climate change. *Wiley Interdisciplinary Reviews: Climate Change, 13*(4), e770. https://doi.org/10.1002/wcc.770

Elshater, A., Aboelata, A., & Afifi, S. (2022). The impact of green infrastructure on thermal comfort in a hot-arid climate: A case study of new Cairo, Egypt. *Urban Forestry & Urban Greening, 67*, 127441.

Estoque, R. C., Murayama, Y., & Myint, S. W. (2017). Effects of landscape composition and pattern on land surface temperature: An urban heat Island study in the megacities of Southeast Asia. *Science of the Total Environment, 577*, 349–359.

Faruqui, N. I., Biswas, A. K., & Bino, M. J. (Eds.). (2001). *Water management in Islam*. United Nations University Press.

Few, R., Brown, K., & Tompkins, E. L. (2007). Public participation and climate change adaptation: Avoiding the illusion of inclusion. *Climate Policy, 7*(1), 46–59.

Flammer, C. (2021). Corporate green bonds. *Journal of Financial Economics, 142*(2), 499–516.
Fletcher, T. D., Shuster, W., Hunt, W. F., Ashley, R., Butler, D., Arthur, S., & Viklander, M. (2015). SUDS, LID, BMPs, WSUD and more–the evolution and application of terminology surrounding urban drainage. *Urban Water Journal, 12*(7), 525–542.
Gago, E. J., Roldan, J., Pacheco-Torres, R., & Ordóñez, J. (2013). The city and urban heat islands: A review of strategies to mitigate adverse effects. *Renewable and Sustainable Energy Reviews, 25*, 749–758.
Gouldson, A., Colenbrander, S., Sudmant, A., McAnulla, F., Kerr, N., Sakai, P., & Hall, S. (2015). Exploring the economic case for climate action in cities. *Global Environmental Change, 35*, 93–105.
Grafakos, S., Trigg, K., Landauer, M., Chelleri, L., & Dhakal, S. (2019). Analytical framework to evaluate the level of integration of climate adaptation and mitigation in cities. *Climatic Change, 154*(1–2), 87–106.
Heikkinen, M., Ylä-Anttila, T., & Juhola, S. (2020). Incremental, reformistic or transformational: What kind of change do C40 cities advocate to deal with climate change? *Journal of Environmental Policy & Planning, 22*(1), 90–102.
IPCC. (2022). Climate change 2022: Impacts, adaptation, and vulnerability. In *Contribution of working group II to the sixth assessment report of the intergovernmental panel on climate change*. Cambridge University Press.
Jack, C. D., Jones, R., Burgin, L., & Daron, J. (2020). Climate risk narratives: An iterative reflective process for co-producing and integrating climate knowledge. *Climate Risk Management, 29.*
Johnson, K. A., Dana, G., Jordan, N. R., Draeger, K. J., Kapuscinski, A., Olabisi, L. K. S., & Reich, P. B. (2012). Using participatory scenarios to stimulate social learning for collaborative sustainable development. *Ecology and Society, 17*(2).
Jung, C., Awad, J., Ismail, M. A., & Chohan, A. H. (2023). Enhancing urban sustainability through green roofs: A thermal performance evaluation in Dubai. *Future Cities and Environment, 9*(1), 1–12. https://doi.org/10.5334/fce.206
Kabisch, N., Korn, H., Stadler, J., & Bonn, A. (Eds.). (2017). *Nature-based solutions to climate change adaptation in urban areas: Linkages between science, policy and practice,.*
Larice, M., & Macdonald, E. (Eds.). (2013). *The urban design reader* (Vol. 384). London: Routledge.
Leeshawn, Y. (2012). *The evolving Arab City: Tradition, modernity and urban development*. Routledge.
Malczewski, J. (2006). GIS-based multicriteria decision analysis: A survey of the literature. *International Journal of Geographical Information Science, 20*(7), 703–726. https://doi.org/10.1080/13658810600661508
Meerow, S., Newell, J. P., & Stults, M. (2016). Defining urban resilience: A review. *Landscape and Urban Planning, 147*, 38–49. https://doi.org/10.1016/j.landurbplan.2015.11.011
Menoni, S. (2025). Urban planning for disaster risk reduction and climate change adaptation: A review at the crossroads of research and practice. *Sustainability, 17*(20), 9092. https://doi.org/10.3390/su17209092
Michaelowa, A., & Michaelowa, K. (2011). Coding error or statistical embellishment? The political economy of reporting climate aid. *World Development, 39*(11), 2010–2020.
Molle, F., & Berkoff, J. (2006). Cities versus agriculture: A review of intersectoral water re-allocation. *Natural Resources Forum, 30*(1), 1–12.
Moukarzel, A., & Azoury, N. (2025). AI and smart cities: Transforming urban living in the MENA. In N. Azoury & G. Yahchouchi (Eds.), *AI in the Middle East for growth and business*. Palgrave Macmillan. https://doi.org/10.1007/978-3-031-75589-7_11
Najini, H., Nour, M., Al-Zuhair, S., & Ghaith, F. (2020). Techno-economic analysis of green building codes in United Arab Emirates based on a case study office building. *Sustainability, 12*(21), 8773. https://doi.org/10.3390/su12218773
Ortega-Pozo, J. L., Alcalá, F. J., Poyatos, J. M., & Martín-Pascual, J. (2022). Wastewater reuse for irrigation agriculture in Morocco: Influence of regulation on feasible implementation. *Land, 11*(12), 2312. https://doi.org/10.3390/land11122312

Reimann, L., Vafeidis, A. T., & Brown, S. (2018). Mediterranean UNESCO world heritage at risk from coastal flooding and erosion due to sea-level rise. *Nature Communications, 9*(1), 4161.
Sharaan, M., Iskander, M., & Udo, K. (2022). Coastal adaptation to sea level rise: An overview of Egypt's efforts. *Ocean & Coastal Management, 218*. https://doi.org/10.1016/j.ocecoaman.2021.106024
Shi, L., Chu, E., & Debats, J. (2016). The global justice implications of urban climate adaptation. *Climatic Change, 133*(3), 327–339.
Sowers, J. (2013). Environmental politics in Egypt: Activists, experts, and the state. *The Journal of Environment & Development, 22*(2), 139–144.
Sowers, J., Vengosh, A., & Weinthal, E. (2011). Climate change, water resources, and the politics of adaptation in the Middle East and North Africa. *Climatic Change, 104*(3–4), 599–627.
Taleb, H. M., & Sharples, S. (2011). Developing sustainable residential buildings in Saudi Arabia: A case study. *Applied Energy, 88*(1), 383–391.
UN Office for Disaster Risk Reduction & International Science Council. (2025). *Hazard definition and classification review: Technical report*. https://www.undrr.org/publication/documents-and-publications/hazard-definition-and-classification-review-technical-report
UN-Habitat. (2018). *The new urban agenda*. https://habitat3.org/the-new-urban-agenda/
UN-Habitat/UNDP/UNESCO. (2022/2024). *The urban content of the NDCs: Global review/local climate action explored through in-depth country analyses*. UN-Habitat/UNDP report.
United Nations. (2015). *Paris Agreement*. https://unfccc.int/process-and-meetings/the-paris-agreement.
United Nations Office for Disaster Risk Reduction (UNDRR). (2015). *Sendai framework for disaster risk reduction 2015–2030*. https://www.undrr.org/publication/sendai-framework-disaster-risk-reduction-2015-2030
Waha, K., Krummenauer, L., Adams, S., Aich, V., Baarsch, F., Coumou, D., & Schleussner, C. F. (2017). Climate change impacts in the Middle East and Northern Africa (MENA) region and their implications for vulnerable population groups. *Regional Environmental Change, 17*(6), 1623–1638.
World Bank. (2018). *Beyond scarcity: Water security in the Middle East and North Africa*. World Bank Group. https://www.worldbank.org/en/topic/water/publication/beyond-scarcity-water-security-in-the-middle-east-and-north-africa
World Bank. (2021). *Integrating green and gray: Creating next generation infrastructure*. World Bank Group. https://www.worldbank.org/en/topic/water/publication/integrating-green-and-gray-creating-next-generation-infrastructure
World Bank. (2022). *Climate and development: An agenda for action—Middle East and North Africa*. World Bank Group.
Ziervogel, G., Pelling, M., Cartwright, A., Chu, E., Deshpande, T., Harris, L., & Zweig, P. (2016). Inserting rights and justice into urban resilience: A focus on everyday risk in cities. *Environment and Urbanization, 29*(1), 123–138. https://doi.org/10.1177/09562478166869

Chapter 5
Enhancing Early Warning Systems for Better Resilience against Extreme Weathering Events in the Middle East and North Africa (MENA) Region

Bouchra Belhaj Abdallah

Abstract Climate change, biodiversity loss, land degradation, waste, and pollution are global threats. The United Nations Framework Convention on Climate Change (UNFCCC) and the Intergovernmental Panel for Climate Change Impact (IPCCC) revealed intriguing science-based evidence in its sixth report that rang the alarm bell about serious climate and environmental crisis impacts on countries and communities, with a strong emphasis on the Middle East and North Africa (MENA) region. At the heart of an unprecedented era of extreme weathering events, data availability is key for correct vulnerability identification and for driving projection translated into convenient mitigation and adaptation actions. Early warning systems (EWS) is a key tool which countries are empowered and equipped with to gather data and increase their preparedness in facing floods, severe heatwave, wildfires, and droughts, among other extreme weathering conditions. However, data which lack coherence, inclusiveness, and integration remain invalid with interpretation errors and significant barriers towards resilience and successful action plans to extreme weathering conditions. The MENA region still suffers from readiness disparity, technology shortage, lack of technical capacity, and knowledge acquisition in facing climate environmental crisis manifestation, which highlight a weakness in integrating the Sendai Framework in government resilience plans. This chapter examines the current extreme weathering events that frequently impact the MENA region, and the level of sophistication and the performance of EWS. Furthermore, challenges and weaknesses in policies, and gaps in governance framework, multilateralism-multicountry-level and cooperation, research, and technology are identified as key areas which need to be addressed further in mitigating the effects of extreme weathering events in the MENA region.

Keywords Extreme weathering events · Climate change · Early warning systems · Sendai framework · Resilience · Adaptation · Preparedness

---

B. B. Abdallah (✉)
General Department of Environmental Quality, Environmental Agency, Ministry of Environment, Water and Agriculture of Saudi Arabia, Riyadh, Saudi Arabia
e-mail: bouchra.belhajabdallah@eui.eu

E. Alam (ed.), *Disaster, Crisis and Emergency Management in the Middle East & North Africa*, https://doi.org/10.1007/978-3-032-24436-9_5

5.1 Introduction

The climate risk 2025 report (Adil et al., 2025) revealed that total inaction and the escalation in human and economic cost for the time frame 1993-by 2022 has reached 765,000 deaths worldwide with a direct associated loss of about 4.2 trillion USD resulting from more than 9400 extreme weather events. There was a clear increase of the climate anomalies and risks contributing to transforming uncommonly extreme climate events into continual threats, with a great number of weathering events particularly noticed in 2024, demonstrating a shift in the frequency and the intensity to a point of creating a new normal (Adil et al., 2025). Torrential rains and flooding in the United Arab Emirates (UAE), Libya, and recently in Brazil, which hosted the United Nations Climate Change Conference (COP30), displaced over 600,000 individuals from their homes in the whole state and claimed many lives (Simoes et al., 2025). Deadly heat waves swept across Asia, the Mediterranean, and Mexico, resulting in numerous fatalities and devastating impacts on the population and ecosystems (Adil et al., 2025). In September 2024, deadly rains hit West and Central Africa, also leading to multiple fatalities and displacing hundreds of thousands. In the United States, hurricanes Helene and Milton, fuelled by record-breaking Gulf Mexico temperatures, killed hundreds and left a trail of destruction. In October 2024, in just minutes heavy downpours caused flash floods in eastern Spain, sweeping away nearly everything in their path. More than 200 people died and thousands of livelihoods were devastated.

The UN Secretary-General launched the *Early Warnings for All* (EW4All) initiative at the 2022 Conference of the Parties (COP27) to the United Nations Framework Convention on Climate Change (UNFCCC), setting a clear and ambitious goal that every person on Earth be protected by life-saving early warning systems. Multi-hazards early warning systems by 2027. As the impact of the climate crisis accelerates, there is a need to support communities left unprotected from its most devastating consequences. Thus, scaling up EWS isn't just a choice, it's a necessity. By making the best use of technology, strategic investments, and working hand in hand with national governments, local stakeholders, the United Nations Development Programme (UNDP), World Meteorological Organization (WMO), United Nations Office for Disaster Risk Reduction (UNDRR), International Telecommunications Union (ITU), International Federation of Red Cross and Red Crescent Societies (IFRC) and other partners, the UN will be able to translate its commitment to action to leave no one behind in facing the impact of extreme weathering cause associate to climate change.

Combined actions and efforts—including upgrading weather observation networks, integrating indigenous knowledge into early warning systems, strengthening community-based preparedness, improving climate risk communication, and engaging the private sector (such as the telecommunications industry)—could deliver a tailored and sustainable early warning model with potential for scalability. Such an approach would enhance disaster preparedness and adaptation strategies (Climate Information and Early Warning Systems, Climate Action, 2025) and transform climate extremes from persistent challenges into opportunities for building resilience and safeguarding the future of populations most affected by the climate crisis, particularly those exposed to extreme weather events. In this chapter, we (i) highlight the most devastating extreme weathering events that are becoming the "new normal" and are threatening communities across the Middle East and North Africa (MENA) region, and (ii) suggest recommendations to improve preparedness via early warning systems and increase the communities' resilience.

5.2 Extreme Weathering Events in MENA region

The MENA region is witnessing a notable rise in the frequency and severity of extreme weather events. This trend underscores the region's growing vulnerability to the adverse effects of climate change, which manifests itself in various forms such as droughts, heatwaves, and floods. As a result, communities will be exposed to inevitable economic and social consequences. The increasing risk of extreme weathering events in the MENA region clearly shows that better preparedness and adaptation instruments to such events must be integrated in national planning (Mezyen, n.d.). The warming trend has triggered changes in both global and regional climate conditions (Arnell et al., 2019; Olonscheck et al., 2021). Projections under high-emission scenarios are reporting alarming predictions, with a potential warming increase by 7.6 °C $\pm$ 1.53 °C in the Arabian Peninsula. More locally, some inland areas could experience a 9 °C warming (Zittis et al., 2022). Therefore, it is expected that a large part of the MENA population will be exposed to this accelerated heating —potentially 600 million people—with recurring "super- and ultra-extreme heatwaves" with temperatures soaring to 56 °C and higher for extended periods (Malik et al., 2024). According to the Intergovernmental Panel for Climate Change (IPCC), the MENA region (Bangar, 2024; Blunden et al., 2021; Caloiero et al., 2018; Ceccherini et al., 2017) is now identified as the most vulnerable regions across in the world; threatened by extreme heat with unpreceded breaking records, water scarcity, extended drought, and other extreme weathering events more frequently and severely occurring. These science-based facts have been echoed first by the IPCC's six reports (IPCC, 2021).

The ratification of the 2015 Paris Agreement is a concrete commitment for countries to limit the rise in global mean surface temperature to 1.5 °C (IPCC, 2018), although projections suggest a potential rise of 2.1–3.5 °C under intermediate warming scenarios (IPCC, 2021). The warming response of the MENA region to GHG forcings is intricately tied to the global response to the Paris Agreement. The strong warming in the region (Almazroui et al., 2022; Zittis et al., 2021) threatens the future human habitability of especially in areas closer to coastal zones (e.g., the Arabian Gulf and the Red Sea (Andrews et al., 2018; Raymond et al., 2020). Under the threats of extreme weathering events which occur in different ways and patterns across the MENA, understanding their dynamic is insufficient on its own, solution on how to strengthen the resilience of communities is capital.

5.2.1 Extreme Heat

In the past decades, the extreme weather events' characteristics, including the frequency, duration, and intensity of heatwaves, have amplified across the Eastern Mediterranean and Middle East (EMME) region (Abbasnia & Toros, 2019; Tolika, 2019; Collins et al., 2013; Cook et al., 2016; Coumou & Rahmstorf, 2012; Donat, 2014; Dust storms sweep across Arab countries, Examining the Historic, 2024; First Global Early Warnings, 2025; Founda, & Giannakopoulos, 2009). Particularly in the Middle East, the observed trend of cumulative heat is + 50% per decade, being among the highest in the world (Perkins-Kirkpatrick & Lewis, 2020). Parts of the region were affected by some of the most severe, record-breaking hot weather events of the past decade (Coumou & Rahmstorf, 2012; Founda & Giannakopoulos, 2009). Temperature records were broken for other locations in the Middle East (e.g., Syria, Jordan, and Lebanon) during this exceptional 2020 autumn heatwave (Blunden & Boyer, 2021). Further west of the Eastern Mediterranean and Middle East (EMME) region, the Sicilian Meteorological Information Service for Agriculture reported a record. In the summer of 2021, the Kuwaiti city of Nawasib recorded 53.2 °C, the highest temperature in the inhabited world. Previously, on 21 July 2016, in the hottest year on record so far, the temperature in Mitribah, Kuwait, reached a record 54 °C (Merlone et al., 2019). In the following summer, on 29 June 2017, a temperature of 53.7 °C was recorded at Ahwaz, southwest Iran. In the arid regions of the Arabian Peninsula and eastern Iran, summer land surface temperature, usually derived by satellite sensors, routinely soared above 60 °C (Mildrexler et al., 2011). In 2018, the world highest land surface temperature of 80.8 °C was observed in the Lut Desert in Iran (Zhao et al., 2021). The Arabian Gulf recently experienced record-breaking sea surface temperatures (SSTs). On 30 July 2020, sea surface temperatures in Kuwait Bay reached 37.6 °C (Alosairi et al., 2020). Cold-weather extremes in the region (e.g., cold spell duration, number of ice, and frost days) were also subject to warming trends; however, these have not been as significant as the trends for hot-weather extremes (Donat Kuglitsch, 2014; Abbasnia & Toros, 2020; Ntoumos et al., 2020; Yosef et al., 2019). There is strong evidence that the frequency and intensity of droughts have increased in many countries of the region (Caloiero et al., 2018; Spinoni et al., 2019; IPCC, 2018; IPCC, 2021; Islam et al., 2025; Kelley et al., 2015; Knaepen, 2024; Kostopoulou, & Jones, 2005; Kuglitsch et al., 2010). Recent mega-droughts in the Eastern Mediterranean and the Levant have received great attention, and their magnitude cannot be explained by natural climate variability alone (Cook et al., 2016; Mathbout et al., 2018; Lelieveld et al., 2016; Majdi et al., 2022; Malik et al., 2024). It is worth noting that the IPCC Fifth Assessment Report (Collins et al., 2013). projected a 5–7 °C rise in Saudi Arabian temperatures by the end of the twenty-first century under the high-emission scenario, with coastal zones near the Arabian Gulf warming more than areas near the Red Sea (Lelieveld et al., 2016). Corroborating these findings, the World Bank Group (World Bank Group, 2014). indicated stronger warming in summer than in winter, especially in Algeria and in the Arabian Peninsula. Additionally, Majdi et al. (Majdi et al., 2022). predicted lower future warming in North Africa (except Algeria) than the Middle East and less warming over the coastal zones than over the inland areas of the Arabian Peninsula. Over a broader perspective, Hamed et al. (2023) (Hamed et al., 2023). demonstrated that the temperature increases in the MENA region respond nonlinearly to global temperature rises.

Around the world, extreme weather events are progressively becoming the frequent manifested anomalies of climate change (Singh & Zommers, 2014). This is especially the case for parts of the EMME region, where the increasing severe weather events include droughts, heatwaves, hydro-climate extremes, storm surges, wind—and dust storms. To better cope with more severe and frequent extreme events, the development of accurate EWS is vital. Early warning and timely response play a major role in reducing vulnerability, limiting the casualties caused by disasters and enhancing the awareness and resilience of communities (Seng, 2012). This will be important for the EMME, where the mitigation of climate extreme weathering event is still limited. These are key elements of climate change adaptation and disaster risk reduction and aim to avoid or reduce the damage caused by hazards, including extreme weather events and geoclimatic hazards.

5.2.2 *Flooding Events*

The classic reputation for hot and dry weather have changed for the MENA region and it's now experiencing an increase in heavy rain events leading to floods in recent years (Oxford Analytica, 2024). It is the case of the UAE where the type of April 2024 storm is 30 times more likely to occur (Francis & Fonsec, 2024; Examining the Historic, 2024). In April 2024, heavy rains swept across extensive areas of the Arabian Peninsula, sparking devastating flash floods throughout the region. Some areas received nearly a year's worth of rain in just one day, resulting in widespread flooding and significant damage in Oman and the UAE (Singh, 2024). During this event, the UAE experienced an unprecedented deluge, marking the most substantial rainfall event ever recorded in this desert nation. Heavy rains caused floods that primarily affected Dubai, Sharjah, Ajman, Fujairah, Ras Al Khaimah, and Umm Al Quwain (Bangar, 2024). The rarity of rainfall in the UAE underscores the region's vulnerability to such extreme weather events. The storm was characterized by a total rainfall of over 250 mm within 24 h, an amount that shattered 75-year records, being above the annual average in some regions and paralyzing cities like Dubai, causing massive infrastructure damage, with an estimated $ 2.9–3.4 billion losses. Flash floods occurred at the Dubai International Airport, causing extensive disruptions to flights at the world's busiest international airport. With standing water impeding taxiways and halting arrivals, the infrastructure struggled to cope, leaving passengers stranded on waterlogged roads (Francis et al., 2025). While this event caused massive disruption, it paled in comparison to the human cost of the Storm Daniel in Libya. Daniel, which hit Libya back in 2023, recorded a total precipitation of 379.08 mm per day, with the city of Derna experiencing the highest rainfall anomaly (Haifa et al., 2023). Furthermore, this extreme rainfall generated a catastrophic loss of over 5000 lives amplified nearly 20-fold by the failure of two upstream dams (Armon et al., 2025). Releasing an estimated 30 million cubic meters of water that swept away entire sections of the city. The floods in Derna left a devastating impact and generated an estimated $19 billion in total damage and losses, equivalent to 3.6% of Libya's 2022 GDP, devastating the city's water and sanitation infrastructure and leading to severe public health crises (Post-disaster reconstruction, 2024). Such loss was avoidable and foreseeable underscoring the failing infrastructure and the absence of an effective EWS, associated with the aging infrastructure amid the era of fatal risks extremely vulnerable to intensified climate shocks. It also highlights the need for improved flood forecasting and urban resilience to cope with recurrent extreme events (Hussein et al., 2025). Communities in the region are increasingly feeling the impact and distress caused by such events. Recent floods in Dubai, UAE, and in Derna, Libya, have unveiled both countries' lack of preparedness and adaptive capacity, respectively. While both countries have starkly different political contexts, facilities and adaptation capacities, more frequent extreme weather events have become an environmental reality in the region.

5.2.3 *Sand-Dust Storm Events*

On the other hand, the IPCC has noted that the frequency and intensity of dust storms have exacerbated in recent decades due to human-related major changes in land use and land cover. The MENA region is heating up twice as fast as the global average, leading to drier conditions that fuel these storms. A 2 °C temperature increase in Iraq over sixty years has contributed to a jump in the number of annual dust storms rising from 75 to 200 (Special Report on Climate Change and Land — IPCC ste, n.d.). By analyzing 40 years of data, recent studies suggest that rising temperatures and decreased precipitation are exacerbating dust events in Saudi Arabia, Iraq, Yemen, parts of Iran, Egypt, and countries around the Arabian Gulf. The area between the Tigris and Euphrates rivers, in particular, has seen a sharp increase in dust sources over the last 20 years (Mousavi et al., 2024).

In May 2022, exceptional dust storms in terms of their intensity and duration, battered the Middle East and resulted in serious effects on human health and disrupted daily activities due to worsening air quality over several days. Over Iraq, the first storm occurred on 16 May and the second on 23 May. The dust fronts were seen to propagate downstream into the southeastern Arabian Peninsula on the days following initial emission (17 and 24 May, respectively). The dust emission was promoted by density currents emanating from deep convection over Turkey. The convective systems were triggered by cut-off lows from mid-latitudes fed by moisture from African atmospheric rivers. The dust clouds were transported southward at 4 km in altitudes but sunk to ground levels when they reached the southern Arabian Peninsula due to strong subsidence. At a station in coastal UAE, the dust caused a 350 W m−2 drop in the surface downward shortwave flux and a 70 W m−2 increase in the longwave one during the dust episodes. This contributed to a 9°C increase in nighttime temperatures which exacerbated the effects of the heat for the population. The newly highlighted mechanism for dust emission in the Middle East, in which a cut-off low interacts with an atmospheric river, as well as direct observations of the dust impact on the radiative budget can contribute to reducing associated uncertainties in climate models (Francis et al., 2023). On May 5, 2025, a severe dust storm swept across Iraq, turning skies orange and sickening numerous people. According to ReliefWeb, nearly 550 people were hospitalized in the country following the storm, despite being urged by authorities to stay indoors and take precautionary measures. Airborne dust is respiratory irritating and can trigger severe breathing issues in people with respiratory illness, the elderly, and very young (MODIS Web, 2025). On May sixth 2025, a large dust storm originating in Iraq swept across several western Iranian provinces, confirmed by the Iran's meteorological authorities, forcing the closure of schools and offices to protect an estimated 13 million people from dangerously poor air quality (Dust storms sweep across Arab countries, n.d.). The storm caused extremely low visibility and high levels of air pollution and the Anadolu Agency reported official warnings about low visibility and urging residents to avoid outdoor and maritime activities in the UAE, Saudi Arabia, Qatar, Kuwait, Syria and Iraq with turbulent conditions, driven by strong winds carrying dust and debris, according to meteorological authorities (Dust storms sweep across Arab countries, n.d.). For example, in the UAE, the NCM reported two dust waves impacting the region—one affecting the eastern and northern Arabian Gulf and another hitting Syria, Iraq, and northern Saudi Arabia—causing significantly reduced visibility. Additionally, the Saudi Arabia's NCM flagged a low-level yellow alert for light rain, active winds and reduced visibility in Al Baha, alongside dust particles in Riyadh and Qassim issuing therefore an alerts and warning of dust-stirring winds continuing in several regions like Mecca and Medina with active winds, while orange-level alerts in Riyadh and the Eastern Province cautioned about dust particles. Furthermore, the World Bank (World Bank, 2019) estimates that dust storms cost the MENA region's economies over $150 billion annually. The storms halt air travel, impact air quality, damage critical infrastructure, reduce the efficiency of solar energy installations, and devastate agriculture. On Thursday, July 10, 2025, a violent sandstorm swept across southern Tunisia, particularly affecting the governorates of Tozeur and Kébili (In video: Spectacular sandstorm hits southern Tunisia - Tunisie; Tunisie Numerique, n.d.). The storm caused a sharp drop in visibility and led to significant material damage in several areas. According to data from the National Institute of Meteorology, exceptionally strong wind gusts were recorded up to 144 km/h in Tozeur, and 107 km/h in Kébili. These intense winds swept through roads, homes, and farmland, forcing many residents to remain indoors for safety (In video: Spectacular sandstorm hits southern Tunisia - Tunisie; Tunisie Numerique, n.d.). Earlier, in mid-May 2025, a powerful storm system carried heavy dust from the Sahara across Tunisia and the central Mediterranean, causing severe weather that led to flooding in coastal cities like Kelibia (MODIS Web, 2025). Extreme weather events, such as sand-dust storms, heatwaves, droughts, and flooding, could manifest (Abadi et al., 2025). differently and in distinct ways and pattern, drawing therefore a critical challenge for the MENA region. The problem lies in the fact that MENA

Table 5.1 Summary of the most occurring climate extreme weathering events in MENA region

Climate extreme event type	Specific locations
Extreme heatwaves	The overall Arabian Peninsula and the coastal zones near the Arabian Gulf including Iraq, Kuwait, Saudi Arabia, Oman, Syria, Jordan, Lebanon, Iran, and Algeria
Multi-Year drought	Morocco, Algeria, Tunisia, Egypt, Libya
Flash floods	The overall Arabian Peninsula including United Arab Emirates, Oman, Yemen, Qatar, Kuwait, Bahrain, Saudi Arabia; Libya, and Tunisia
Sand-Dust-Storms	The overall Arabian Peninsula and the coastal zones near the Arabian Gulf including Iran, United Arab Emirates, Kuwait, Saudi Arabia, Iraq, Yemen, Egypt, Qatar, Syria; and Southern Tunisia.

countries are not ready and resilient enough to face the consequences of such accelerating extreme events, which is aggravated by persistent challenges related to EWS (Table 5.1).

5.3 Early Warning Systems: Benefit and Lead Institutions

The MENA region is in the front lines of a climate crisis characterized by a dangerous combination of extreme heat and unpredictable and intense precipitation. Such an unequivocal affirmation by the scientific community confirms that the region already faces the world's highest levels of water stress to be severally exacerbated by the ongoing climatic shifts (Rezaei et al., 2024). The MENA region faces a future of escalating humanitarian and economic catastrophes. Additionally, recent scientific publications by the World Bank highlighted recent catastrophic weather events that further place the MENA region as a super-hot spot for climate change, facing a dual threat of extreme heat and floodwater that is outpacing global averages and overwhelming unprepared nations (World Bank Group, 2023). In fact, the immense human and economic toll of these events have exposed a critical gap in regional preparedness. While nations like the UAE had EWS that issued alerts ahead of the April 2024 storm, the disaster in Derna underscored the fatal consequences of their absence, a challenge echoed across other vulnerable nations. Despite the fact that the technology for early warnings is improving, its effectiveness is often hindered by a lack of institutional coordination and the capacity to predict and communicate risks to vulnerable communities on time (Islam et al., 2025). This gap's growing recognition is fueling a drive to implement more effective, inclusive, accessible EWS across the region (World Bank Group, 2023). In an era of dramatic cuts to humanitarian aid and overseas development assistance, one truth remains: the cost of inaction is far greater than the cost of preparedness. Few investments in disaster preparedness deliver as much impact for so little cost as EWS. A $1 investment in early warning can save up to $10 in disaster losses. Globally, these systems offer a return of 3 to

16 times the cost, according to the Global Commission on Adaptation and UNDRR. (Why early warning systems deserve more attention in Asia-Pacific climate security, people in need, 2025). Otherwise, the cost of inaction would be catastrophic. Building resilience requires immediate and sustained investment in climate-adapted infrastructure, robust EWS, and transboundary cooperation to manage a transnational crisis that recognizes no borders. In response to this escalating threat, a region-wide push to bolster defenses is underway, centered on the urgent need for integrated EWS which are defined, by the UNDRR, as an integrated system of hazard monitoring, forecasting and prediction, disaster risk assessment, communication and preparedness activities systems and processes that enable individuals, communities, governments, businesses and others to take timely action to reduce disaster risks in advance of hazardous events.

The EWS includes four interrelated key elements: (1) disaster risk knowledge based on the systematic collection of data and disaster risk assessments; (2) detection, monitoring, analysis and forecasting of the hazards and possible consequences; (3) dissemination and communication, by an official source, of official, timely, accurate and actionable warnings and associated information on likelihood and impact; and (4) preparedness at all levels to respond to the warnings received. These four interrelated components need to be coordinated within and across sectors and multiple levels for the system to work effectively and to include a feedback mechanism for continuous improvement. Failure in one component or a lack of coordination across them could lead to the failure of the whole system (United Nations Office for Disaster Risk Reduction (UNDRR), 2017). They are put in place to analyze and identify potential weather- and climate-related risks and hazards enabling early action to save and protect lives, livelihoods and assets of people at risk. In fact, following the 2024 floods, the UAE is accelerating investments in artificial intelligence (AI) and machine learning (ML) to create sophisticated models that can better predict flooding hotspots and analyze near real-time satellite data for disaster management and impact assessment (Sun et al., 2024). However, the persistence and the manifestation of recent extreme weathering events revealed that gaps remain, and technology alone is insufficient to address them. An effective system requires not only accurate forecasting but also robust risk communication and the capacity for communities to respond—elements missing in the Derna floods, with which its effects would have been minimized. Well-tech equipped nations are committing to increasing resilience and reducing vulnerability investing in improved climate predictions and projections, high-quality weather and climate data, flood defences, and urban planning. Yet, significant gaps remain in communicating risks and ensuring community-level preparedness. On the other hand, many MENA countries lack proper tools to manage the consequences of likely destructive extreme weathering events due to ineffective policies, dysfunctional warnings (Mezyen, n.d.) and slow funding mechanism much needed to catalyse and accelerate sustainable financial support. In fragile and conflict-ridden states, challenges such as fragmented governance, technology, and lack of resources severely hinder the implementation of such life-saving systems. Only half of the countries worldwide are equipped with the capability to timely alert their citizens about impending hazardous weather conditions. In fact, less than 50%

of the least developed countries and only one-third among small island developing states are covered. Moreover, the Arab countries have the lowest regional coverage score. By identifying upcoming hazards including as early as possible, communities can take measures to increase its preparedness in time against the impact of hazardous events and try to minimise disruption and damage. Furthermore, through real-time monitoring, more accurate forecasting and prediction, disaster risk assessment, and effective communication tools, equipped stakeholders and decision-makers can carry out early plan preparedness actions to boost the nation and communities' resilience. Such climate information encompasses a wide range of data and knowledge related to past, present, and future climate conditions. This includes historical weather records, climate model forecasts, projections, and information on climate variability and trends. Accessing high-quality weather and climate data is essential for understanding climate risks, informing adaptation planning, and developing effective early warning systems.

According to the UN Environment Programme (UNEP), a 24-hour notice of impending hazardous events could reduce confirmed damage by 30%. Additionally, investing 800$ US million in integrated EWS in developing countries would prevent a $3 to $16 billion loss annually (Climate Information and Early Warning Systems, Climate Action, United Nation Environment Programme, Climate Information and Early Warning Systems | UNEP - UN Environment Programme, n.d.). EWS are put in place to analyze and identify potential weather-and climate-related hazards, enabling early action to save and protect lives, livelihoods and assets of people at risk.

Countries with functional and effective EWS could play a crucial role by significantly lowering disaster-related losses and protect public infrastructure; forecasting extreme weather enables businesses, farmers and governments to mitigate financial risks by protecting assets and infrastructure. By issuing advanced notice of potential dangers, authorities can also take pre-emptive measures to safeguard critical infrastructure such as bridges, roads, energy, and healthcare facilities, ensuring that essential services remain accessible and operational to assist people during and after a hazardous event. Furthermore, by fostering a greater awareness of risks and promoting preparedness, EWS are instrumental in building long-term resilience to the escalating impacts of climate change. Investing in climate information and EWS have a humanitarian and economic return providing up to tenfold return on investment. EWS are an effective tool by which climate change adaptation measures can effectively be implemented. There is a growing consensus, supported by international bodies like the UNDRR and WMO, that every nation must have access to effective early warnings. To strengthen and expand the efficiency and the benefits of EWS and to adhere to the principle of 'leaving no one behind', the United Nations Secretary General has launched in the Cop 27 a global initiative; EW4All, which aims to cover all countries worldwide by an early warning system by the end of 2027 (Why early warning systems deserve more attention in Asia-Pacific climate security, people in need, 2025).

Driving transformational changes for communities through EWS components is only possible via, initially, addressing all elements of EWS in an integrated manner to ensure that accurate, timely, and actionable information reaches everyone that needs it; through its components there is a need to strengthen institutional frameworks for climate information and MHEWS, enhance capacity for monitoring, analysis and forecasting of climate and its impacts through promoting the role of AI/ML models in MENA region, improve dissemination and communication of risk information and early warnings, and enhance climate risk management capacity. Furthermore, via investing in the entire chain of climate services, from national institutional strengthening to building preparedness capabilities at the community level, engaging multiple stakeholders by bringing on board a wide range of agencies and communities to enhance cooperation and synergies between both producers and users, and addressing cross-cutting issues through an emphasis on social inclusion and gender responsiveness (First Global Early Warnings for All Multi-Stakeholder Forum launches with call to accelerate universal protection from disasters, 2025, United Nations Office for Disaster Risk Reduction (UNDRR), Launch of First Global Early Warnings for All Multi-Stakeholder Forum | UNDRR, n.d.), is also a step forward toward effective EWS.

5.4 Challenges and Enhancement Opportunities for EWS

5.4.1 Recent Investments and Pilot EWS Projects

As global warming threatens to exceed the 1.5 °C threshold, a change that would make multiple climate hazards and risks more frequent and intense to humans and ecosystems, the need for proactive adaptation has never been more urgent. Projections of economic losses due to climate change are also highly alarming. By 2050, climate change could cost the global economy up to 38 trillion annually USD. This cost is primarily through losses in agriculture, infrastructure, and public health. In absolute and relative terms, the global south is particularly affected. Increasing climate change impacts primary result in increased costs for adaptation and addressing loss and damage (L and D), therefore the UNEP adaptation Gap report 2024 indicates that 215–387 billion USD are needed annually for effective adaptation including EWS. Moreover, current estimates of needs for residual L and D in developing countries jump from USD 116–435 billion in 2020, to 290–580 billion USD in 2030, to 5551–1016 billion USD in 2040, and to 1132–1741 billion USD in 2050 (Mezyen, n.d.). Early Warning Systems, if functional, integrated, and accessible, can save lives and avoid social and economic losses. However, critical implementation gaps remain unbridged, as these life-saving systems currently only cover half of the countries in the world. Despite significant progress, this shortfall persists as the effectiveness of these systems is hindered by interconnected challenges, particularly within the MENA region and least developed countries where vulnerability is most acute. Primarily, the shortages and the limitation of financial resources remain a foundational barrier towards empowering nations and countries with efficient and inclusive

EWS. This issue goes far beyond instant investment in technology to sustain funds covering updating, upgrading, maintaining, and integrating operational EWS in the long run.

The GCF is the world's largest dedicated climate fund. GCF's mandate is to foster a paradigm shift towards low emission, climate-resilient development pathways in developing countries. The GCF has a portfolio of 286 projects worth USD 15.9 billion (USD 61.3 billion including co-financing) delivering transformative climate action in 133 developing countries. GCF is an operating entity of the financial mechanism of the UNFCCC and serves the 2015 Paris Agreement, supporting the goal of keeping average global temperature rise well below 2 °C.

The GCF has approved an unprecedented US$103.2 million in grant financing to scale up life-saving EWS in aiming to reach the most vulnerable communities and strengthen communities' resilience seven of the world's most climate-vulnerable countries. Led by the UNDP, the five-year project will bolster multi-hazard EWS in Antigua and Barbuda, Cambodia, Chad, Ecuador, Ethiopia, Fiji, and Somalia, and directly benefit more than 26 million people representing the three groups, Least Developed Countries, Small Island Developing States, African States. With a total budget of $114.6 million including $11.3 million in co-financing from the IFRC, ITU, WMO, and national governments—the initiative will improve risk assessment, enhance forecasting capabilities, and ensure that critical early warnings reach the most remote and at-risk communities. By improving production, dissemination and effective use of Multi-Hazard Early Warning Systems (MHEWS). The dedicated investment will accelerate access to global and regional technical and operational support for these countries to develop comprehensive EWS, address policy gaps, inter-agency coordination and strategies, build technical and scientific capacities, and promote effective local action. Early Warning Systems will enable impeding shocks and enhancing climate resilience of governments and communities by providing critical information to guide risk management, including anticipatory action and making informed long-term decisions that help avoid otherwise crippling damage and losses to communities and livelihoods. National projects tailored to each participating country's needs will establish early warning systems infrastructure, set up systems for its long-term operation, support local capacities in the generation and dissemination of early warnings products, and help institute relevant policies and governance mechanisms for end-to-end MHEWS. It will be implemented through a multi-agency approach, with UNDP as the Accredited Entity, working alongside the World Meteorological Organization (WMO), the United Nations Office for Disaster Risk Reduction (UNDRR), the International Telecommunication Union (ITU), the International Federation of Red Cross and Red Crescent Societies (IFRC), national governments, and local stakeholders to develop tailored, sustainable early warning solutions.

Financial support comes at a crucial moment as least developed countries and small island developing states bear the brunt of accelerating climate impacts. Prolonged droughts in Ethiopia and Somalia are driving food and water insecurity, while Fiji and Antigua and Barbuda face intensifying storms and rising sea levels. In Chad, Cambodia, and Ecuador, communities are increasingly exposed to extreme weather and environmental degradation (Green Climate Fund approves $103 million to strengthen early warning systems for vulnerable countries, 2025, United Nation Development programme, Green Climate Fund approves $103 million to strengthen early warning systems for vulnerable countries | United Nations Development Programme, n.d.). Although the UNDP is making progress with the most vulnerable countries being present with strong initiatives in the least developed countries, in small islands developed states in Africa and Latin America, its efforts in the MENA region is non-conclusive and scattered to mainly the Northern Africa. The MENA region is not receiving a similar attention and follow up from the UNDP as other regions (United Nations Office for Disaster Risk Reduction (UNDRR), 2017).

5.4.2 Technical Capacity

The aforementioned financial constraints fuel the second major challenge: insufficient technical capacity shortages. Many nations face unsustainable financial needs to operate EWS and maintain them in working. There is a lack of skilled personnel in areas including meteorology, hydrology, risk communication and climate data analysis, to correctly operate EWS and interpret climate projections and data. Additionally, EWS suffers from inaccurate predictions and forecasting significantly impacted by the lack of quality data leading to inaccurate warnings that erode public trust. Furthermore, ensuring credible, inclusive, equitable, and understandable warnings by particularly marginalized communities and vulnerable groups clustered in remote areas remains a real challenge toward translating forecast into life-saving action. Addressing these challenges requires sustained efforts and collaboration among governments, international organizations, financial institutions, civil society, private sector and indigenous people. These efforts should be aligned with the frequency and the complexity of climate change including extreme weathering events like wildfires flooding, droughts, heatwaves, pollution, and storms. In fact, since the UN Secretary-General International Decade for Natural Disaster Reduction (IDNDR) Programme Forum in Geneva, July 1999, native people started being recognized for their role in protecting the planet and themselves. This will lead to involving citizens in all stages of resilience and mitigation projects life cycle, such as flood data analysis and its interpretation rather than only data collection (Perera et al., 2020). Kofi Annan, the UN Secretary-General from 1997 to 2006, emphasized the importance of the indigenous people's role, stating "Much has been learnt from the creative disaster prevention efforts of poor communities in developing countries. Prevention policy is too important to be left to governments and international agencies alone. To succeed, it must also engage civil society, the private sector, and the media." A prerequisite to creating a dynamic early-warning system is the ability to gather data on a constant or highly frequent basis. By leveraging new technologies, scientific knowledge, tools, methods and principles, communities are influencing policy and decision-making alongside professionals in disaster risk management and climate change communities. Citizen science is facilitated by digital and mobile apps through accessible platforms—for example, citizen mapping of floods in real-time via social media or through NGOs, youth groups and local experts (e.g., community leaders, agricultural extension workers and local or national meteorological and hydrological services officers). It enables field and household data collection via open data digital survey tools and crowdsourcing, and it supports sharing of local geospatial data on hazards, exposure, vulnerabilities, capacities and risks. Citizen science complements data collected by sensors described above, offering various entry points for public engagement to help improve the early warning system value chain. A combination of technology measures was used, namely: sensor data (hydrometeorological equipment, geological monitoring equipment), geographic information system (GIS), models, high-performance computer and citizen science (i.e., social media, mobile app, survey tools). As a result, more than 100 of the most vulnerable communities across the 11 river basins were identified in 45 (in western Georgia) implementing community-based DRR measures (UNFCCC & GEO, 2024). According to the Global risks' report 2025, which is survey-based results, extreme weathering events are classified as the second current global risk that is anticipated to become even more severe in a warming climate, with the risk ranked first over the next decade and second in the upcoming two years. Biodiversity loss and ecosystem collapse jump up to rank #2, and with a significant deterioration compared to its two-year ranking (#21). Critical change to Earth systems is at #3, natural resource shortages at #4, and pollution at #10 complete the very bleak outlook for environmental risks over the long term (10 years) (The Global Risks Report 2025, 20th Edition, insight report, world economic forum, n.d.).

Similarly, the report revealed that extreme weathering events are yet to be a priority and are notably absent in the national and local treaties, while assessing the top global risks addressed by national and local regulations where only pollution (air, soil, water, etc.), and biodiversity loss and ecosystem collapse succeeded to receive considerable attention. Additionally, extreme weathering conditions are not on the table with the risks addressed by multi-stakeholders' engagement. Furthermore, extreme weathering conditions are on the tale of interest while assessing the top risk addressed by financial instruments and not a priority for research and development.

To better manage the generated risks by extreme weather events (floods, heatwaves, etc.) there is a need at first for a financial instrument 12%, strengthened by national and local regulations 13%, multi-lateral treaties and agreements 3%, global treaties and agreements 15%, development assistance 17%, corporate strategies 5%, research and development 18%, public awareness and education 10%, and multi-stakeholder engagement 8% (The Global Risks Report 2025, 20th Edition, insight report, world economic forum, n.d.). Governance and Mitigation Policies.

Effective climate action is enabled by political commitment, well-aligned multi-level governance, institutional frameworks, laws, policies, and strategies and enhanced access to finance and technology. Clear goals, coordination across multiple policy domains, and inclusive governance processes facilitate effective climate action. Regulatory and economic instruments can support deep emissions reductions and climate resilience if scaled up and applied widely. Climate resilient development benefits from drawing on diverse knowledge (Intergovernmental Panel on climate change, Climate change, 2023). The Sendai Framework works hand in hand with the other 2030 Agenda agreements, including The Paris Agreement on Climate Change, The Addis Ababa Action Agenda on Financing for Development, the New Urban Agenda, and ultimately the Sustainable Development Goals. Endorsed by the UN General Assembly following the 2015 Third UN World Conference on Disaster Risk Reduction (WCDRR), the Sendai Framework aims for the substantial reduction of disaster risk and losses in lives, livelihoods, health, and in the economic, physical, social, cultural and environmental assets of people, businesses, communities and countries. This framework recognizes that the State has the primary role to reduce disaster risk, but that responsibility should be shared with other stakeholders including local governments, the private sector, landowner, indigenous people, policymakers, scientists, and other stakeholders. To achieve this ambition, the integration of Sendai Framework into local and national policies for mitigation and adaptation to climate change, including extreme weathering events, is beyond essential. This action would enable the governments to strengthen their evaluation efforts and progress towards resilience after the occurring of extreme weathering and the assessment of readiness roadmaps which should be planned before and after occurring extreme weathering events.

The recent Global reporting status on Sendai Framework (Snapshot of Sendai Framework Monitoring, Global reporting status, 2024) outlined that 163 countries (84% of countries in the world) reported on the global targets through the Sendai Framework Monitor by October 2024, 81 countries have reported on all seven targets, and 107 countries have reported on all Sendai Development Goals (SDG)-related targets (A-E). Specifically, for the target G, 44 out of 113 countries, which reported having early warning, achieved a self-scored "substantial to comprehensive" EWS. Sendai framework is a set of 38 indicators, used to track progress in implementing seven priority targets of the Sendai Framework and its associated dimensions of the Sustainable Development Goals 1, 11 and 13 (Target A: Disaster-related deaths and missing persons (excluding COVID-19), Target B: Disaster-affected population, Target C: Disaster-related economic losses, Target D: Disaster damage to critical infrastructure and disruption of basic services, Target E: National and local disaster risk reduction (DRR) strategies, Target F: International cooperation for disaster risk reduction, and Target G: Multi-hazard early warning systems (MHEWS). The UNDRR is tasked to support the implementation, follow-up, and review of the Sendai Framework. Yet, to realize the best outcomes of the adaptation of Sendai Framework and help the most vulnerable communities globally, the UNDRR is called to follow-up closely on the integration of the Sendai Framework with the local policies and treaties prioritizing extreme weathering events as the new normal phenotype of climate anomalies caused by climate change. The adherence of the Sendai framework and its integration at local policy across sectors would harmonize policies across countries while customizing them according to each country's ecosystems and the lack of high-quality weather and climate related data, including indicators related to economic and social, and updating it with new targets as environmental loss and sustainable finances, allocated national budget dedicated as emergency plan to remediate a given situation before and after the occurring of a given crisis, number of benefiting people and sustainable resilient infrastructure and habitat for example). Governance deficits, such as centralized power structures, exclusion of relevant groups from decision-making, and inefficient resource allocation, weaken resilience-efforts at both national and regional levels. Many MENA countries operate under highly centralized systems. While centralization can streamline responses to climate impacts, it also presents challenges. Governments frequently delay urgent climate action and the inclusion of civil society, women, youth, and other vulnerable groups in the decision-making processes groups that are often at the forefront of identifying adaptive solutions and highlighting maladaptation-would help to minimize the effects of extreme weathering events.

In response to the escalating threat associated with extreme weathering events, a region-wide push to bolster defenses is underway, centered on the urgent need for an EWS that should take into account the environmental, financial, the social and economic dimensions of extreme weathering conditions associated with climate change. Key priorities should include generated social and economic losses upon forecasting risk scenarios, allocated finances dedicated for resilience for a given town, which enables individuals, communities, governments, and businesses to reduce their vulnerabilities and increase their preparedness against the potential impact of the occurring hazards. The enhancement of the current EWS framework with a clear financial component, standing on its own, would help better report Sendai Framework indicators and track the government effort translated into actions in the ground. Such holistic approach would be a better comprehensive framework to see the technical and financial efforts combined in one place. The 'new pillar', the financial flow, should cluster all mobilized funds, initiatives, and outcomes across sectors and come up with innovative instruments from international partners, financial institutions, and domestic resources to support mitigation and adaptation to transboundary extreme weathering risks and strengthen systemic resilience in the MENA region. Continuing the strong emphasis on Target E and G would foster the effectiveness of EWS and mitigate the effects of extreme weathering events. With respect to the Sendai Framework, particularly Target F on *international cooperation for disaster risk reduction*, the financial dimension should be separated from international cooperation and treated as a distinct target in order to allow the development of more precise and measurable indicators. Target F should instead focus explicitly on science, innovation, and diplomacy, strengthening cooperation between developed and developing countries to foster innovation and facilitate transparent knowledge exchange. This process should be guided by regional scientific boards that feed into a transboundary platform addressing cross-border environmental and climate crises. Knowledge, tools, and innovation are required to understand and manage the complexity and the uncertainty of transboundary extreme weathering risks across sectors and scales. This revised target will foster the engagement of countries and catalyse partnerships, exchanges, and collaboration among stakeholders to have a better outcome on nations.

Multi-stakeholder collaboration between governments, the private sector, academia, and civil society is essential for effective EWS, yet the limitations and weaknesses of the current cooperation efforts regarding collaboration and data sharing remain a challenge across the MENA region. Many databases are not updated regularly and the existing observational networks in the MENA region are still sparse to cover the remote areas. For example, in Egypt, national disaster databases and vulnerability and exposure data are considered incomplete and outdated. Additionally, the efforts of some institutions overlap and thus might hamper governments effort in anticipating extreme weather events. In Tunisia, the (INM—National Institute of Meteorology) operates a network of weather data collection stations across the country. At the same time, the Ministry of Agriculture operates agrometeorological stations and the General Directorate for Water Resources (DGRE) has established precipitation and surface water stations, resulting in scattered data and a lack of coordination. This limits collaboration across the region around risk information sharing, forecasting and early warning, despite the transboundary nature of extreme weather events (Mezyen, n.d.). According to the intergovernmental panel for climate change in its 2023 (Perera et al., 2020) report, Society, livelihoods, and economies, policy mixes that include weather and health insurance, social protection and adaptive social safety nets, contingent finance and reserve funds, and universal access to EWS combined with effective contingency plans, can reduce vulnerability and exposure of human systems (Fig. 5.1).

5.4.3 *Quality Data and Coverage*

As the world approaches the +1.5 °C temperature threshold sooner than anticipated by the latest IPCC report, the associated irreversible impacts and risk are increasingly evident and disproportionally affecting the most vulnerable people and countries. These climate impacts both from extreme weathering events and slow-onset processes could fundamentally reshape societal trajectories (Mezyen, n.d.). In fact, less than half of least developed countries have a multi-hazard EWS, about only 40% of small island developed states have adequate systems, many countries especially those in Africa, the Americas, and the Caribbean, have significant gaps in their EWS, critically remaining in surface and upper air meteorological observations, particularly across Africa, parts of the Pacific, and West of Latin America (Intergovernmental Panel on climate change, Climate change, 2023). Although a slight progress has been registered for the Arab countries, the multi-hazard monitoring and forecasting systems coverage reached only 41%. The EWS dashboard reports also that only 41% of these countries acquired a comprehensive coverage of local or national dissemination mechanism, and only 27% of local governments having a plan to act on early warning systems. Some countries are actively working to expand their early warning coverage, but progress is not happening fast enough to protect vulnerable populations, according to the UNODRR (Knaepen, 2024). Additionally, the nations that are not covered by EWS are not able to accurately generate weather and climate warning-related data. This situation may incur geographical bias in the Emergency Events Database (EM-DAT) resulting from unequal reporting quality data and coverage across space. There are data gaps particularly for global south

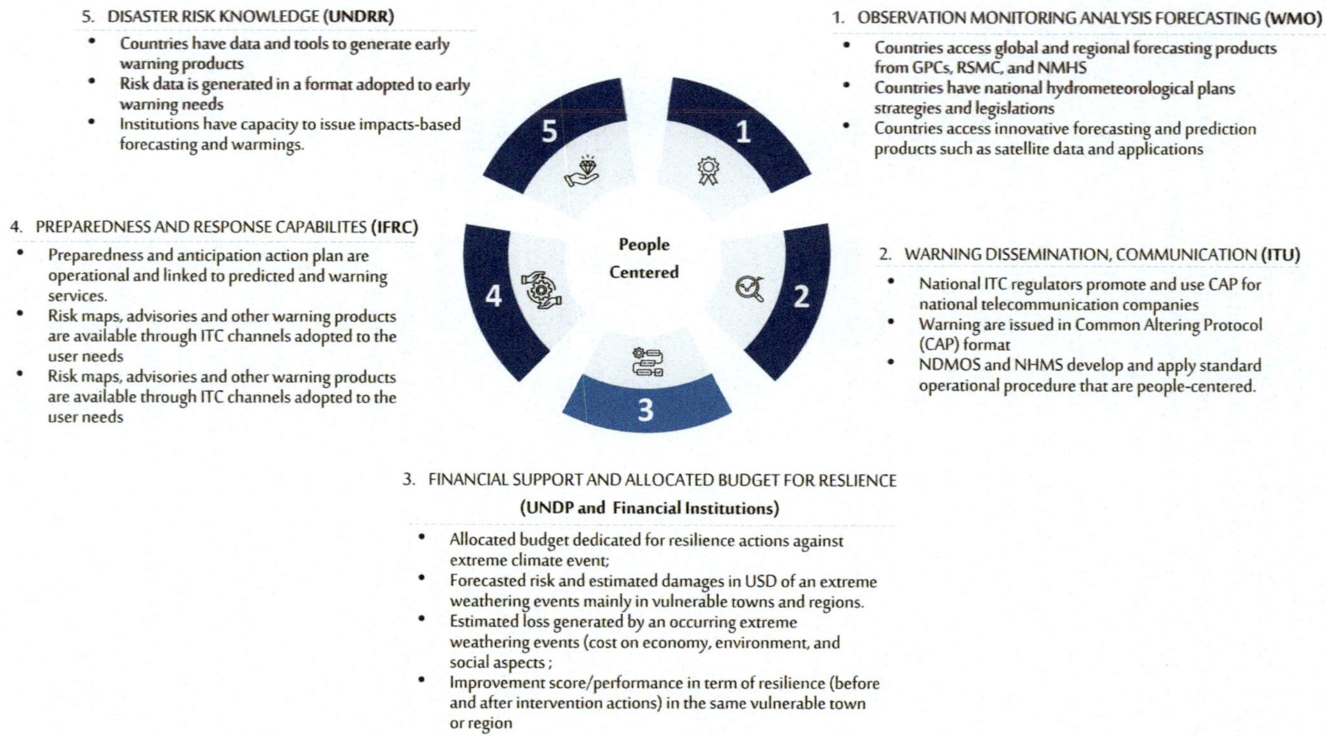

Fig. 5.1 The enhanced early warning system framework

countries, which might lead to these countries' misrepresentation in the Climate Risk index (Adil et al., 2025). This issue is particularly pronounced for heat waves, also with a view to EM-DAT. Heat waves are not well recorded for sub–Saharan Africa. Extreme weather event databases, such as EM-DAT, report no significant heat waves impacts in sub-Saharan Africa since 1990, although the region has experienced several heat waves (Spyrou et al., 2020). About 52% of heat waves events in EM-DAT occurred in nine countries: Japan, India, Pakistan, and the United States, followed by western European countries (France Belgium United Kingdom, Spain, and Germany). Additionally, the MENA region is now experiencing new pattern of extreme heat, floods, and sand and dust storms, which become more unpredictable generating data gap about the dynamics of extreme climates. Additionally, once occurred, the reporting and the interpretation of extreme weathering events are slow and often underrepresented in reports and platforms. The disparity in coverage of EWS created a geographical disparity. The existence of data gaps becomes therefore well-known and clearly revealed and is mainly due to many reasons including the distribution of meteorological stations, the insufficient systematic data collection and operational warning systems, and cataloguing or profiling, the incoherent data collection techniques and the methodology boundaries of data collection. Countries within the MENA region suffer from dispar challenges: financial and technical issues dominate in Northern Africa, while strategic and technical gaps are more manifested in Western Asia. Some countries initiated serious initiatives in mainstreaming adaptation into planning, for example Egypt, Tunisia and Morocco have carried out downscaling and adjustment of climate change scenarios, and of vulnerability assessments on the major sectors, particularly water and agriculture. Algeria is in the process of preparing a vulnerability assessment for coastal zones. Saudi Arabia has also carried out vulnerability assessments and developed climate scenarios focusing on water resources, desertification, agriculture and health. For the first time, Jordan introduced on its Third National Communication Report (TNC) climate projections with dynamic downscaling until the year 2100 with moderate-high levels of confidence for all climate parameters using 43 grid points (Regional briefing on National Adaptation Plans, 2019). In terms of institutional arrangements MENA countries are also strengthening their institutional arrangements to coordinate their adaptation efforts, although there are still areas in which they need support, particularly in strengthening the committees and legal frameworks. Several countries have set up National Climate Change Committees, but for many these have either been inactive or lack a strong legal mandate. Many are in the process of strengthening the institutional coordination arrangements. For example, Egypt was looking to reform its National Climate Change Committee, composed of representatives of 26 ministries and institutions, into a national council for climate change. Tunisia is also reforming the National Committee for Climate Change with the establishment of a more comprehensive coordination and implementation structure to oversee mitigation, adaptation, climate finance capacity building and partnerships. Morocco has a National Committee on Adaptation, as well as a specific Committee for the NAP process, and is working to strengthen their mandates through a legal text. Jordan, Algeria and Bahrain, also have Climate Change Committees. On the other hand, in terms of plans and policies, several countries in the region are integrating climate change and disaster risk reduction considerations into national frameworks. Tunisia, for example, has a National Strategy for Climate Change, a portfolio for adaptation projects, an early warning system feasibility study, and plans for the development of the green economy and sectorial adaptation strategies for agriculture, coastal protection, tourism and health. Morocco has integrated climate change and adaptation as one of the key pillars of its National Sustainable Development Strategy. In Saudi Arabia, the Ministry of Environment is leading the development of a national adaptation strategy, and the Ministry of Planning is steering the integration of adaptation activities into the national development plan (Regional briefing on National Adaptation Plans, 2019). The UNDP could lead with ownership of national authorities and leadership the harmonization of countries efforts toward functional and operational EWS and regional action agenda related in the MENA region to avoid disparity in reporting, accurately assess the countries profiles and leave no one behind supporting the MENA region.

The close follow-up with local governments, and the incorporation of ground observations and citizen science measurements can effectively adjust the investigation of rare weather extremes, their spatial variability, drivers, and their impacts on arid environments that remain a significant challenge in socio-hydrological research. This can be mitigated by Improved public preparedness, better flood mitigation strategies, and interaction with disaster response agencies for more effective emergency interventions. By deepening our understanding of the MENA's vulnerability to extreme weathering events, such as the extreme rainfall occurred in UAE, we can enhance public awareness and support disaster agencies in planning more effective responses (Regional briefing on National Adaptation Plans, 2019).

A lack of good data makes it difficult to provide meaningful and specific assessments other than to acknowledge that, without investment in adaptation and resilience measures, the intensification or the increased frequency of extreme weather events is bound to impact Gross National Product (GDP) growth in the future (Wilbanks, 2007). Additionally, data generated by institutions at local levels and collected by citizen science should be easily understandable and accessible across communities, which facilitate the knowledge exchange and the translation to action to reduce the vulnerability in today's new normal of extreme weathering events.

5.4.4 Technology

Ensuring that EWS reach all people, including vulnerable groups, requires a combination of financial resources, community engagement, local knowledge, and technological innovation. Local leaders and organizations must be actively involved in designing and disseminating warnings to ensure that alerts are culturally appropriate and accessible. Advancements in mobile alerts, social media, and AI-driven forecasting have improved the reach and precision of warnings. AI is a powerful tool that provides better information by exploiting relationships between existing data sets alongside its ability to learn from new knowledge, generating hindcasts and forecasts.

With a set of mathematical and computer science techniques, AI encompasses large language models, advanced big data and predictive analytics, and machine and deep learning, which have emerged as critical tools that play a key role at each stage of the MHEWS value chain. In the context of risk knowledge, AI may be used to process large volumes of raw satellite imagery to create standardized baseline data on exposure (e.g., built-up areas and population distribution), some of which are available as open and free global data products. The accuracy and reliability of statistical data, including high-resolution satellite imagery and field data to provide updated estimates of the status of the country's major coastal and marine ecosystems (coral reefs and mangroves). The results of this exercise have yielded valuable insights for improving risk knowledge, particularly regarding vulnerability and exposure of environmental areas, specifically in coastal and marine ecosystems. Furthermore, under the UNFCCC Technology Mechanism Initiative on AI for Climate Action, a series of activities are underway that explore the role of AI as a tool for advancing and scaling up transformative climate solutions in developing countries, with a focus on LDCs and SIDS. This includes the AI Innovation Grand Challenge which has attracted ideas for AI-powered climate action, around 20 per cent of which are on EWS (UNFCCC & GEO, 2024). Similarly, AI is playing a significant role in the development and implementation of conservation actions that address urgent threats to biodiversity across the globe requiring large amounts of historic and current data. Machine learning approaches offer the tools to process, analyse and interpret large data sets, giving insights into trends and guiding evidence-based allocation of limited resources to maximise positive biodiversity outcomes. Data acquired from remote sensing, citizen science and other monitoring approaches could feed in near-real time into an early warning system for biodiversity that integrates automated red-listing of species with the identification of priority areas for conservation. Recent advances in computing, combined with increased processing power and data availability, have enabled the approach to considerably take off (Antonelli et al., 2023). Although the role of AI in equipping countries with a well-functioning early warning system, challenges remain. Many nations do not have the resources to run traditional weather models for operational purposes, with the computationally cheaper AI/ML models, collecting data quality and effective disseminating warnings to communities in mainly remote areas remains a major issue.

There is a need for more inclusive initiatives towards the equal access to technology and network coverage equipping the most vulnerable and remote nations with homogenous data collection tools. These efforts should be complemented by multi-channel communication strategies, such as radio broadcasts, mobile phone notifications, visual signals, and door-to-door alerts, to ensure broad coverage, particularly in areas with limited digital access notifications, limited climate monitoring stations, lack of electronic data management, monitoring networks unevenly distributed in countries, data continuity and frequency of production, Ground-truthing remote sensing–derived information.. Additionally, the accurate tracking of consecutive extreme weathering events, which become more frequent with the intensification of other climate change impacts, would require more transformative action of EWS. Warnings should be fed with high quality data, mainly from the most impacted areas to ensure high preparedness and effective response at global, regional, or local levels. Moreover, warnings should be translated into real time multi-resilience maps and economic loss avoided at short-, medium-and long-term run. The consideration of warnings at micro-level still needs more attention; ensuring the delivery of high-quality data to the most affected areas is very crucial to ensure effective preparedness against extreme weathering conditions. Utilizing big data, predictive analytics, and geographic information systems can enhance monitoring, analysis, and visualization of risks. In the UAE for example, a recent study highlights the complexity of urban hydrology and the challenges of predicting flood extents in urban areas. The findings underscore the need for improved flood forecasting systems in the UAE, emphasizing enhanced satellite rainfall estimation accuracy and advanced modelling approaches for better urban flood management and mitigation (Hussein et al., 2025). Integrating new technologies and methodologies in the MENA urban flood forecasting is crucial for enhancing resilience against future extreme weather events, ensuring the safety and preparedness of vulnerable regions like the UAE.

The share of extreme weathering impacts in the economic losses should be dashboarded and conducted in real time based on the degree of vulnerability, preparedness, and resilience of the affected countries. This upgraded tool would have a significant impact globally but more noticeable on the least developed countries and small island states, Africa, MENA region, and Latin America that are threatened by the most by extreme weather conditions including floods, rise of sea levels, droughts, heatwaves, and tropical cyclones.

Similarly, the native and indigenous people should be recognized for their significant role in global climate mitigation and adaptation efforts. Special attention should be given to these communities to reduce their vulnerabilities caused by extreme weathering events. The distribution, identification, exchange and follow-up on the data would allow for a better tracking of the impact and threat of the extreme weathering conditions on such communities. Moreover, indigenous knowledge and community-based observations can complement scientific forecasting, enhancing accuracy and relevance for at-risk populations. An inclusive approach ensures that vulnerable populations, including persons with disabilities, women and rural communities, receive timely warnings that they can act upon effectively.

5.4.5 International Cooperation

Alongside international and local partners, the UNDP is a key player in EWS worldwide, supporting governments with funding (including fund management), capacity building, and technology to develop and improve existing EWS. By working together with governments, the private sector, and local communities, the UNDP is helping to ensure that climate information and EWS are accessible, actionable and lifesaving for all.

The UNDP plays a leading role in several global initiatives aimed at accelerating and integrating SDGs achievement, the development, and implementation of EWS.

The efforts include the Systematic Observations Financing Facility (SOFF) which supports least developed countries and small island developing states in generating and using essential climate and weather observations data, and in implementing the UN Secretary-General's EW4All initiative. This initiative is a global effort led by the WMO to ensure that everyone on Earth is protected by EWS by 2027. Under this global action, the UNDP is leading a multi-agency effort to advance multi-hazard EWS benefiting 26 million people in seven of the world's most climate-vulnerable countries: Antigua and Barbuda, Cambodia, Chad, Ecuador, Ethiopia, Fiji and Somalia, with more than $100 million in financing from the GCF. The WMO (Simoes et al., 2025). underscored the need for EWS to cover the global nations to better increase their readiness and resilience by identifying potential weather-and climate-related risks and hazards. The number of disasters has increased five-fold in the past 50-year period (1970–2019), with 50% of all disasters associated with weather, climate, and water hazards, which make EWS increasingly important for mitigating the effects of extreme weathering events and better prepare the countries for adaptation actions (Simoes et al., 2025). In fact, 30% of the damage caused by a disaster can be reduced if an EWS is issued within 24 h, preventing a global annual loss of 3–16 billion USD. Yet, most of nations particularly in small island developing states and least developed countries– those who are most vulnerable to the consequences of extreme weathering events and climate change remain deprived of early action that can save lives, livelihoods, and assets of people at risk, with only 30% of the global population covered by EWS.

Within this framework, the UNEP co-founded the SOFF to address the long-standing problem of missing weather and climate observations from least developed countries and small island developing states. The SOFF supports both by providing grant financing and technical assistance for the sustained collection and international exchange of surface-based weather and climate observations in alignment with the international Global Basic Observing Network (GBON) regulations, i.e. higher resolution, real-time observations that feed into global forecasting and prediction systems for the global public good. This initiative, among others, aims to support the most vulnerable countries to bridge critical weather and climate data gaps, which helps to develop better weather forecasts, EWS, and climate information services. By enhancing weather and climate observation networks to facilitate compliance with the international GBON standard, strengthened human and institutional capacity is necessary to deliver on the GBON requirements and the ICT infrastructure for weather and climate data collection, storage and sharing. Early Warning Systems are urgently needed as climate change is causing more frequent, extreme, and unpredictable climate-related hazards.

Furthermore, participants from governments, civil society, academia, the private sector, and international organizations under the theme, and in collaboration with the United Nations Office for Disaster Risk Reduction (UNDRR) and the League of Arab States, Kuwait held the " The Sixth Arab Regional Platform for Disaster Risk Reduction toward "Building Resilient Arab Communities with a declaration that affirmed the urgent need to strengthen resilience across the Arab region in the face of increasing disaster risks and underscored the importance of accelerating of Sendai Framework, enhancing governance for risk reduction, increasing investments in disaster resilient infrastructure, and leveraging science, technology, artificial intelligence, and early warning systems. It also emphasizes the need for comprehensive disaster loss databases and risk assessments to support evidence-based policymaking, greater regional cooperation, particularly in supporting countries affected by crises, inclusive and sustainable approaches that engage governments, civil society, and the private sector in reducing disaster risks and protecting communities. This forum paved the way for the "First Arab Regional Forum on Early Warning Systems and Disaster Preparedness Together for Enhanced Coordination and Effective Response" which was held in December 2025 in Egypt. This forum is the first of its kind within the Early Warning for all initiative (EW4All) framework to gather MENA region to develop a Joint Arab Roadmap anchored around the following key pillars; (i)Strengthening risk knowledge and emergency telecommunications and EWS capacities, (ii)fostering inter-agency coordination, aligning national and regional priorities under the EW4All initiative and (iii)enhancing the effectiveness and inclusivity of early warning systems in addressing increasingly complex and cascading risks, and (iv)Promoting preparedness and resilience across Arab societies to protect lives, livelihoods, and critical infrastructure. These efforts should be completed by translating these key takeaways into concrete action by setting up at first place a transboundary regional data-sharing or joint forecasting centres for MENA.

On another hand and in terms of international cooperation, Saudi Arabia is leading the regional efforts. The country affirmed its commitment to supporting climate and environmental action and to promoting scientific and technical cooperation to specifically address the impacts of extreme weathering including sand and dust storms at the local, regional, and international levels. These efforts can be made possible, through key global, regional, and local initiatives including the Global Riyadh Drought Resilience Partnership launched under the leadership of Saudi Arabia and the presidency of UN Convention to Combat Desertification (UNCCD) conference (COP16), the Saudi Green Initiative, the Middle East Green Initiative, and the establishment of specialized centres such as the Regional Centre for Sand and Dust Storms, the Climate Change Centre, the Cloud Seeding Program, and the National Centre for Vegetation Cover Development. Similarly, Saudi Arabia aims to enhance early warning systems for dust storms, including a financial contribution of $ten million to develop global models and build capacities, as well as partnerships with the UNCCD and the UNEP to formulate a comprehensive plan to reduce dust sources.

5.5 Conclusion

As extreme weathering events become more frequent and intense, many communities are left unprotected as they are in the frontline of climate change, with substantial impacts on lives and lead to major economy losses. Preparedness and resilience are key towards mitigation and adaptation climate action. From extreme heats, floods, to sand dust storms, EWS are now crucial warning tools toward a timely and accurate protection of lives and infrastructure. Accurate diagnosis of the cause-impact on extreme events and its equal reporting among nations, coherent and homogeneous national policies across countries, good and transparent quality data generation, coverage of early warning system coverage mainly in remote areas, sustainable finance, and consistent international cooperation bridging the most vulnerable nations and the ones which succeed so far to minimize the impact of extreme weathering events on people and the economy remains challenging and yet to be achieved. To this end, we emphasized on the need of an updated EWS framework tailored to community's needs and the climate impacts anomalies associated to climate change. Enriching the actual early warning systems framework with a financial component with key enablers indicators would help, for the first time, integrating the economical dimension of sustainability. With integrated indicators, the enhanced EWS would give a bigger and clearer picture on the level of policy implementation tracking the results of issued warming, and the early detection of weaknesses and gaps to remediate accordingly on time. In parallel, ensuring the integration of Sendai Framework indicators into local policy agenda the same way across countries and with respect to each nations' ecosystem, is a key enabler towards resilience and mitigating the impact of extreme weathering events. Scaling up successful stories in managing risks, promoting, and sharing lessons learned with nations could also enhance communities' resilience through effective active cooperation and leave no one behind.

References

Abadi, A. R. S., Hossein Hamzeh, N., Kaskaoutis, D. G., Xu, B., Li, L., & Ghassabi, Z. 2025. Future projections of dust storm dynamics and sources in the Middle East: Insights from CMIP6 models, Atmospheric Pollution Research. Available online 9 October, 102775. https://www.sciencedirect.com/science/article/abs/pii/S1309104225003770

Abbasnia, M., & Toros, H. (2019). Analysis of long-term changes in extreme climatic indices: A case study of the Mediterranean climate, Marmara Region, Turkey. In I. Vilibić, K. Horvath, & J. L. Palau (Eds.), (pp. 141–153). Springer International Publishing.

Abbasnia, M., & Toros, H. (2020). Trend analysis of weather extremes across the coastal and non-coastal areas (case study: Turkey). *Journal of Earth System Science, 129*(1), 95.

Adil, L., Eckstein, D., Kunzel, V., & Schafer, L. 2025. Climate risk index 2025, who suffers most from extreme weather events? By GERMANWATCH. 1–72. https://www.germanwatch.org/sites/default/files/2025-02/Climate%20Risk%20Index%202025.pdf.

Almazroui, M., Alowaibdi, T., & Hasanean, H. (2022). Dynamical downscaled CMIP5 scenario–based future climate changes over the Arabian Peninsula. *Arabian Journal of Geosciences, 15*(10), 951. https://doi.org/10.1007/s12517-022-10247-7

Alosairi, Y., Alsulaiman, N., Rashed, A., & Al-Houti, D. (2020). World record extreme sea surface temperatures in the northwestern Arabian/Persian Gulf verified by in situ measurements. *Marine Pollution Bulletin, 161*, 111766.

Andrews, O., Le Quéré, C., Kjellstrom, T., Lemke, B., & Haines, A. (2018). Implications for workability and survivability in populations exposed to extreme heat under climate change: A modelling study. *The Lancet Planetary Health, 2*, 1–8.

Antonelli, A., Dhanjal-Adams, K. L., & Silvestro, D. (2023). Integrating machine learning, remote sensing and citizen science to create an early warning system for biodiversity. *Plants, People, Planet, 5*, 1–10.

Armon, M., Shmilovitz, Y., & Dente, E. (2025). Anatomy of a foreseeable disaster: Lessons from the 2023 dam-breaching flood in Derna Libya. *Sci Adv, 11*, 1–9.

Arnell, N. W., Lowe, J. A., Bernie, D., Nicholls, R. J., Brown, S., Challinor, A. J., & Osborn, T. J. (2019). The global and regional impacts of climate change under representative concentration pathway forcings and shared socio-economic pathway socio-economic scenarios. *Environmental research letters, 14*, 1–20.

Bangar, N. (2024). April 17 Look: Top UAE places hit by floods after record rains. *Khaleej Times*.. https://www.khaleejtimes.com/uae/dubai-rains-6-most-flood-hit-areas-after-record-downpour

Blunden, J., & Boyer, T. (2021). State of the climate in 2020. *Bulletin of the American Meteorological Society, 102*, 1–481.

Caloiero, T., Veltri, S., Caloiero, P., & Frustaci, F. (2018). Drought analysis in Europe and in the Mediterranean basin using the standardized precipitation index. *Water, 10*, 1–13.

Ceccherini, G., Russo, S., Ameztoy, I., Francesco Marchese, A., & Carmona-Moreno, C. (2017). Heat waves in Africa 1981–2015, observations and reanalysis. *Natural Hazards and Earth System Sciences, 17*, 115–125.

Climate Information and Early Warning Systems, Climate Action 2025. United Nation Environment Programme., https://www.unep.org/topics/climate-action/climate-transparency/climate-information-and-early-warning-systems

Climate Information and Early Warning Systems, Climate Action, United Nation Environment Programme, Climate Information and Early Warning Systems | UNEP - UN Environment Programme.

Collins, M., Knutti, R., Arblaster, J., Dufresne, J.-L., Fichefet, T., Friedlingstein, P., et al. (2013). Long-term Climate Change: Projections, Commitments and Irreversibility. In T. F. Stocker, D. Qin, G.-K. Plattner, M. Tignor, S. K. Allen, J. Boschung, et al. (Eds.), *Climate Change 2013: The Physical Science Basis. Contribution of Working Group I to the Fifth Assessment Report of the Intergovernmental Panel on Climate Change* (pp. 1029–1136). Cambridge University Press.

Cook, B. I., Anchukaitis, K. J., Touchan, R., Meko, D. M., & Cook, E. R. (2016). Spatiotemporal drought variability in the Mediterranean overthe last 900 years. *Journal of Geophysical Research: Atmospheres, 121*, 2060–2074.

Coumou, D., & Rahmstorf, S. (2012). A decade of weather extremes. *Nature Climate Change, 2*, 491–496.

Donat, M. G., Peterson, T. C., Brunet, M., King, A. D., Almazroui, M., Kolli, R. K., et al. (2014). Changes in extreme temperature and precipi-tation in the Arab region: Long-term trends and variability related to ENSO and NAO. *International Journal of Climatology, 34*, 581–592.

Dust storms sweep across Arab countries. https://www.aa.com.tr/en/middle-east/dust-storms-sweep-across-arab-countries/3558188

Examining the Historic. April 2024. Rainstorm in the UAE Through the Lens of Climate Change-Khalifa University. https://www.ku.ac.ae/examining-the-historic-april-2024-rainstorm-in-the-uae-through-the-lens-of-climate-change

First Global Early Warnings for All Multi-Stakeholder Forum launches with call to accelerate universal protection from disasters, 2025, United Nations Office for Disaster Risk Reduction (UNDRR), Launch of First Global Early Warnings for All Multi-Stakeholder Forum | UNDRR.

Founda, D., & Giannakopoulos, C. (2009). The exceptionally hot summer of 2007 in Athens, Greece—A typical summer in the future climate? *Global and Planetary Change, 67*, 227–236.

Francis, D., & Fonsec, R. (2024). Recent and projected changes in climate patterns in the Middle East and North Africa (MENA) region. *Scientific Reports, 14*, 1–25.

Francis, D., Fonseca, R., Nelli, N., Bozkurt, D., Cuesta, J., & Bosc, E. (2023). On the Middle East's severe dust storms in spring 2022: Triggers and impacts. *Atmospheric Environment, 296*, 1–15.

Francis, D., Fonseca, R., Nelli, N., Cherif, C., Yarragunta, Y., Zittis, G., & Jan de Vries, A. (2025). From cause to consequence: examining the historic April 2024 rainstorm in the United Arab Emirates through the lens of climate change. *npj Climate and Atmospheric Science, 8*, 1–14.

Green Climate Fund approves $103 million to strengthen early warning systems for vulnerable countries, 2025, United Nation Development programme, Green Climate Fund approves $103 million to strengthen early warning systems for vulnerable countries I United Nations Development Programme.

Güner Bacanli, Ü. (2017). Trend analysis of precipitation and drought in the Aegean region Turkey. *Meteorological Applications, 24*, 239–249.

Haifa, M., Miloud, B., & Elmaremi, R. (2023). A diagnostic study of storm daniel's meteorological and environmental characteristics in Derna, Eastern Libya, Sebha University. *Journal of pure and Applied Sciences, 2024*, 1–15.

Hamed, M. M., Nashwan, M. S., & Shahid, S. (2023). Projected changes in thermal bioclimatic indicators over the Middle East and North Africa under Paris climate agreement. *Stochastic Environmental Research and Risk Assessment, 37*, 577–594.

Hochman, A., Marra, F., Messori, G., Raveh-Rubin, S., Yosef, I., Zittis, G., & Pinto, J. G. (2022). ESD reviews: Extreme weather and related societal impacts over the Eastern Mediterranean: A systematic review. *Earth System Dynamics, 19*, 1–29.

Hoerling, M., Eischeid, J., Perlwitz, J., Quan, X., Zhang, T., & Pegion, P. (2012). On the increased frequency of Mediterranean drought. *Journal of Climate, 25*, 2146–2161.

Hussein, K., Alhosani, N., Al-Areeq, A. M., Al Aghbari, A. A., Elkamali, M., Alsumaiti, T., O., S.,. H., Ahmed, M. G., & Abdalati, A. W. (2025). Unprecedented rainfall in the United Arab Emirates: Hydrologic and flood impact analysis of the April 2024 event. *Natural Hazards, 121*, 1–24.

In video: Spectacular sandstorm hits southern Tunisia—Tunisie; Tunisie Numerique. https://news-tunisia.tunisienumerique.com/in-video-spectacular-sandstorm-hits-southern-tunisia/

Intergovernmental Panel on climate change, Climate change. 2023. Synthesis report, summary for policymakers. (pp. 1–42). https://www.ipcc.ch/report/ar6/syr/downloads/report/IPCC_AR6_SYR_SPM.pdf

IPCC. (2018). Global warming of 1.5°C. An IPCC special report on the impacts of global warming of 1.5°C above pre-industrial levels and related global greenhouse gas emission pathways. In V. Masson-Delmotte, P. Zhai, H.-O. Pörtner, D. Roberts, J. Skea, P. R. Shukla, et al. (Eds.), *The context of strengthening the global response to the threat of climate change, sustainable development, and efforts to eradicate poverty* (p. 616). Cambridge University Press. https://doi.org/10.1017/9781009157940

IPCC. (2021). Climate Change 2021: The Physical Science Basis. In V. Masson-Delmotte, P. Zhai, A. Pirani, S. L. Connors, C. Péan, S. Berger, et al. (Eds.), *Contribution of working group I to the Sixth Assessment Report of the Intergovernmental Panel on Climate Change (p. 2391).* Cambridge University Press. https://doi.org/10.1017/9781009157896

Islam, M. M., Mehedi Hasan, M. D., Mia, S., Al Masud, A., Abu Reza, M. D., & Islam, T. (2025). Early warning systems in climate risk management: Roles and implementations in eradicating barriers and overcoming challenges. *Natural Hazards Research, 5*, 1–16.

Kelley, C. P., Mohtadi, S., Cane, M. A., Seager, R., & Kushnir, Y. (2015). Climate change in the Fertile Crescent and implications of therecent Syrian drought. In *Proceedings of the National Academy of Sciences of the United States of America* (Vol. *112*, pp. 3241–3246).

Knaepen, H., 2024. Tackling transboundary climate risks in the Middle East and North Africa. Adaptation without borders policy brief ECDPM, adaptation without borders 1–12. https://adaptationwithoutborders.org/wpcontent/uploads/2024/10/AWBPolicyBrief4_AFD_finalpdf.pdf.

Kostopoulou, E., & Jones, P. D. (2005). Assessment of climate extremes in the Eastern Mediterranean. *Meteorology and Atmospheric Physics, 89*, 69–85.

Kuglitsch, F. G., Toreti, A., Xoplaki, E., Della-Marta, P. M., Zerefos, C. S., Trke, M., & Luterbacher, J. (2010). Heat wave changes in the Eastern Mediterranean since 1960. *Geophysical Research Letters, 37*, 1–5.

Lelieveld, J., Proestos, Y., Hadjinicolaou, P., Tanarhte, M., Tyrlis, E., & Zittis, G. (2016). Strongly increasing heat extremes in the Middle East and North Africa (MENA) in the 21st century. *Climatic Change, 137*, 245–260. https://doi.org/10.1007/s10584-016-1665-6

Majdi, F., Asaad, S., Alireza, H., Maryam, K., & Sara, K. (2022). Future projection of precipitation and temperature changes in the Middle East and North Africa (MENA) region based on CMIP6. *Theoretical and Applied Climatology, 147*, 1249–1262.

Malik, A., Stenchikov, G., Mostamandi, S., Parajuli, S., Lelieveld, J., Zittis, G., Ahsan, M. S., Atique, L., & Usman, M. (2024). Accelerated historical and future warming in the Middle East and North Africa. *Journal of Geophysical Research: Atmospheres, 129*, 1–31.

Mathbout, S., Lopez-Bustins, J. A., Royé, D., Martin-Vide, J., Bech, J., & Rodrigo, F. S. (2018). Observed changes in daily precipitation extremesat annual timescale over the Eastern Mediterranean during 1961–2012. *Pure and Applied Geophysics, 175*, 3875–3890.

Merlone, A., Al-Dashti, H., Faisal, N., Cerveny, R. S., AlSarmi, S., Bessemoulin, P., et al. (2019). Temperature extreme records: World Mete-orological Organization metrological and meteorological evaluation of the 54.0°C observations in Mitribah, Kuwait and Turbat, Pakistan in 2016/2017. *International Journal of Climatology, 39*, 5154–5169.

Mezyen, H. In the eye of the storm: Resilience against extreme weather events in the MENA region, KAS-REMENA Policy Paper Series No. 4/2024, Konrad-Adenauer-Stiftung, Regional programme energy security and climate change Midldle East and North Africa.1–11

Mildrexler, D. J., Zhao, M., & Running, S. W. (2011). Satellite finds highest land skin temperatures on Earth. *Bulletin of the American Meteoro-logical Society, 92*, 855–860.

MODIS Web. 2025. https://modis.gsfc.nasa.gov/gallery/individual.php?db_date=2025-05-09#:~:text=The%20storm%20turned%20skies%20orange%20and%20sickened,a%20true%2Dcolor%20image%20of%20the%20dust%20storm.

Mousavi, H., Panahi, D. M., & Kalantari, Z. (2024). Dust and climate interactions in the Middle East: Spatio-temporal analysis of aerosol optical depth and climatic variables. *Science of the Total Environment, 927*, 1–14.

Nashwan, M. S., Shahid, S., & Abd Rahim, N. (2018). Unidirectional trends in annual and seasonal climate and extremes in Egypt. *Theoretical and Applied Climatology, 136*, 1–17.

Nastos, P. T., Politi, N., & Kapsomenakis, J. (2013). Spatial and temporal variability of the aridity index in Greece. *Atmospheric Research, 119*, 140–152.

Ntoumos, A., Hadjinicolaou, P., Zittis, G., & Lelieveld, J. (2020). Updated assessment of temperature extremes over the Middle East–North Africa (MENA) region from observational and CMIP5 data. *Atmosphere, 11*, 1–27.

Olonscheck, D., Schurer, A. P., Lücke, L., & Hegerl, G. C. (2021). Large-scale emergence of regional changes in year-to-year temperature variability by the end of the 21st century. *Nature Communications, 12*, 1–10.

Oxford Analytica. (2024). Flash floods will increase in frequency in the Gulf. *Emerald Expert Briefings*. https://doi.org/10.1108/oxan-db286682

Pashiardis, S., & Michaelides, S. (2008). Implementation of the standardized precipitation index (SPI) and the reconnaissance drought index (RDI) for regional drought assessment: A case study for Cyprus. *European Water, 23*, 57–65.

Perera, D., Seidou, O., Agnihotri, J., Mehmood, H., & Rasmy, M. (2020). *Untitled Challenges and Technical Advances in Flood Early Warning Systems (FEWSs)* (pp. 1–18). Flood Impact Mitigation and Resilience Enhancement.

Perkins-Kirkpatrick, S. E., & Lewis, S. C. (2020). Increasing trends in regional heatwaves. *Nature Communications, 11*, 1–8.

Post-disaster reconstruction, 2024. Tackling water security in Derna after Storm Daniel Middle East Institute. https://www.mei.edu/publications/post-disaster-reconstruction-tackling-water-security-derna-after-storm-daniel.

Raymond, C., Matthews, T., & Horton, R. M. (2020). The emergence of heat and humidity too severe for human tolerance. *Science Advances, 6*, 1–8.

Regional briefing on National Adaptation Plans. 2019. Focus on middle East and North Africa. (pp. 1–4). https://www.adaptationundp.org/sites/default/files/resources/regional_briefing_on_naps_mena_in_focus.pdf.

Rezaei, A., Karami, K., Tilmes, S., & Moore, J. C. (2024). Future water storage changes over the Mediterranean, Middle East, and North Africa in response to global warming and stratospheric aerosol intervention. *Earth system dynamics, 15*, 1–18.

Seager, R., Osborn, T. J., Kushnir, Y., Simpson, I. R., Nakamura, J., & Liu, H. (2019). Climate variability and change of mediterranean-type climates. *Journal of Climate, 32*, 2887–2915.

Seng, D. C. (2012). Improving the governance context and framework conditions of natural hazard early warning systems. *Journal of Integrated Disaster Risk Management, 2*, 1–25.

Simoes-Sousa, I. T., Camargo, C. M. L., Tavora, J., Piffer-Braga, A., Farrar, J. T., & Pavelsky, T. M. (2025). The May 2024 flood disaster in southern Brazil: Causes, impacts, and SWOT-based volume estimation. *Geophysical Research Letters, 52*, 112442.

Singh, A., & Zommers, Z. 2014. *Reducing disaster: Early warning systems for climate change, Springer, Reducing Disaster: Early warning systems for climate change springer nature Link (formerly Springer Link).*

Singh, M. (2024). *April 18 Unusual weather system brings rain to Saudi Arabia.* Gulf Business. https://gulfbusiness.com/unusual-weather-system-rain-saudi-arabia-bahrain/

Snapshot of Sendai Framework Monitoring, Global reporting status, 2024. United nation. https://www.undrr.org/monitoring-sendai-framework/snapshot

Special Report on Climate Change and Land — IPCC ste. https://www.ipcc.ch/srccl/

Spinoni, J., Barbosa, P., De Jager, A., McCormick, N., Naumann, G., Vogt, J. V., et al. (2019). A new global database of meteorological drought events from 1951 to 2016. *Journal of Hydrology: Regional Studies, 22*, 1–24.

Spyrou, C., Varlas, G., Pappa, A., Mentzafou, A., Katsafados, P., Papadopoulos, A., et al. (2020). Implementation of a nowcasting hydrome-teorological system for studying flash flood events: The Case of Mandra. *Greece, Remote Sensing, 12*, 1–21.

Suarez-Gutierrez, L., Wolfgang, G., Chao, A. M., & Jochem, L. (2020). Hotspots of extreme heat under global warming. *Climate Dynamics, 55*, 429–447.

Sun, Y. Q., Hassanzadeh, P., Shaw, T., & Pahlavan, A. H. 2024. Predicting Beyond Training Data via Extrapolation versus Translocation: AI Weather Models and Dubai's Unprecedented Rainfall. arXiv, 1–36.

The Global Risks Report 2025, 20th Edition, insight report, world economic forum. https://reports.weforum.org/docs/WEF_Global_Risks_Report_2025.pdf

Tolika, K. (2019). Assessing heat waves over Greece using the excess heat factor (EHF). *Climate, 7*, 1–15.

UNFCCC & GEO. (2024). *Realising early warnings for all: Innovation and technology in support of risk-informed climate resilience policy and action* (pp. 1–52). United Nations Climate Change Secretariat.

United Nations Office for Disaster Risk Reduction (UNDRR). 2017. The Sendai Framework Terminology on Disaster Risk Reduction, "Early warning system". https://www.undrr.org/terminology/early-warning-system

Why early warning systems deserve more attention in Asia-Pacific climate security, people in need. 2025. Why Early Warning Systems Deserve More Attention-People in Need. https://www.peopleinneed.net/why-early-warning-systems-deserve-more-attention-in-asia-pacific-climate-security-12727gp

Wilbanks, T. J. (2007). Scale and sustainability. *Climate Policy, 7*, 278–287. https://doi.org/10.1080/14693062.2007.9685656.149

World Bank. (2019). *Sand and dust storms in the Middle East and North Africa region—sources, costs, and solutions* (pp. 1–50) https://documents1.worldbank.org/curated/en/483941576489819272/pdf/sand-and-dust-storms-in-the-middle-east-and-north-africa-mena-region-sources-costs-and-solutions.pdf

World Bank Group. 2014. *Turn down the heat: Confronting the new climate normal* vol. (275). World Bank. https://hdl.handle.net/10986/20595.

World Bank Group. (2023). *Climate change action in the Middle east and North Africa-key insights from country climate and development reports* (pp. 1–5). World Bank Group.

Yosef, Y., Aguilar, E., & Alpert, P. (2019). Changes in extreme temperature and precipitation indices: Using an innovative daily homogenized database in Israel. *International Journal of Climatology, 39*, 5022–5045.

Zhao, L., Oleson, K., Bou-zeid, E., Krayenhoff, E. S., Bray, A., Zhu, Q., Zheng, Z., Chen, C., & Oppenheimer, M. (2021). Global multi-model projections of local urban climates. *Nature Climate Change, 11*, 152–157.

Zhao, Y., Norouzi, H., Azarderakhsh, M., & Agha Kouchak, A. (2021). Global patterns of hottest, coldest and extreme diurnal variability on Earth. *Bulletin of the American Meteorological Society, 102*, 1–23.

Zittis, G., Almazroui, M., Alpert, P., Ciais, P., Cramer, W., Dahdal, Y., Fnais, M., Francis, D., Hadjinicolaou, P., Howari, F., Jrrar, A., Kaskaoutis, D. G., Kulmala, M., Lazoglou, G., Mihalopoulos, N., Lin, X., Rudich, Y., Sciare, J., Stenchikov, G., Xoplaki, E., & Lelieveld, J. (2022). Climate change and weather extremes in the Eastern Mediterraneanand Middle East. *Reviews of Geophysics, 40*, 1–48.

Zittis, G., Hadjinicolaou, P., Almazroui, M., Bucchignani, E., Driouech, F., El Rhaz, K., Kurnaz, L., Nikulin, G., Ntoumos, A., Ozturk, T., Yiannis, P., Stenchikov, G., Zaaboul, R., & Jos Lelieveld, J. (2021). Business-as-usual will lead to super and ultra-extreme heatwaves in the Middle East and North Africa, npj Climate and Atmospheric. *Science, 4*, 1–9.

Chapter 6
Community Engagement as a Tool for Emergency Response in Conflict-Affected Countries

MHD Bahaa Aldin Alhaffar

Abstract Classical disaster response models—reliant on stable governance, centralized coordination, and secure operating environments—often fail in conflict-affected settings, where governance is fragmented, humanitarian access is restricted, and public trust in authorities is minimal. In such contexts, communities themselves emerge as the true first responders, leveraging local knowledge, networks, and resources to meet urgent needs. This chapter argues that community engagement is not only a practical necessity in fragile and conflict-affected countries but also a proven way to enhancing the effectiveness, relevance, and sustainability of emergency response. Through case studies from Syria, Yemen, Palestine (Gaza), Lebanon, Sudan, Morocco, and Iraq, the chapter demonstrates how community-led initiatives—ranging from crowdsourced disaster mapping to neighborhood relief committees—have saved lives, filled operational gaps, and built local resilience amid insecurity. While challenges like safety risks for volunteers, resource constraints, and integration with formal systems remain, evidence shows that community engagement delivers tangible benefits. The conclusion calls for institutionalizing and investing in community-based mechanisms as a core component of disaster preparedness and response in the MENA region and beyond.

Keywords Disaster · Emergency · Conflict · Safety · Community engagement

MHD B. A. Alhaffar (✉)
Disaster Medicine Research Group, Department of Global Public Health, Karolinska Institute, Stockholm, Sweden
e-mail: bhaa.alhafar@gmail.com

Diseases Program, Directly Transmitted Diseases, European Center for Diseases Prevention and Control, Stockholm, Sweden

E. Alam (ed.), *Disaster, Crisis and Emergency Management in the Middle East & North Africa*, https://doi.org/10.1007/978-3-032-24436-9_6

6.1 Introduction: Conflict and Disaster Response in the MENA Region

The Middle East and North Africa (MENA) region faces escalating conflict and disaster risks. With at least 45 active armed conflicts, it is among the most conflict-prone regions globally (Geneva Academy of International Humanitarian L, Human R, 2025). In 2023, conflict-related deaths more than doubled from the previous year reaching over 35.000 conflict-related deaths, driven by protracted wars in Syria and Yemen and new crises such as Sudan and the Gaza war (Stockholm International Peace Research I, 2024). The region now accounts for a disproportionate share of high-intensity conflicts, a trend that has steadily worsened over the past decade (Stockholm International Peace Research I, 2024).

Conflict has deepened communities' vulnerability to disasters. Years of warfare have destroyed infrastructure, displaced millions, and dismantled governance, leaving populations exposed to natural hazards and public health emergencies (Alhaffar & Janos, 2021). In Syria and Yemen, for example, basic services are so degraded that disasters or disease outbreaks meet almost no local response capacity. Yet disaster management frameworks have long assumed stability, overlooking the conflict–disaster nexus (Mena, 2020). The omission of "conflict" from the Sendai Framework illustrates this gap and has hindered adaptation of strategies to fragile settings (Overseas Development I, 2025).

Consequently, classical disaster response—relying on stable governance, centralized coordination, and secure humanitarian access—often fails in conflict zones (Mena, 2020). Governments may be absent, fractured, or actively obstructive; large areas may be outside state control; and insecurity can block international responders (Börzel & Risse, 2010). Aid can be politicized, while the absence of unified systems forces local communities to become the primary responders (Rosvold, 2023).

This chapter examines community engagement as a vital tool for emergency response in conflict-affected countries. It first outlines the limits of traditional approaches, then defines community engagement and its theoretical foundations. It presents case studies from Syria, Yemen, Palestine (Gaza), Lebanon, Sudan, Iraq, and Morocco, showing how community-led responses deliver more effective, context-sensitive outcomes. It concludes with an analysis of challenges and practical recommendations for integrating and strengthening community engagement in disaster preparedness and response across MENA.

6.2 Limitations of Classical Disaster Response in Fragile and Conflict Settings

Traditional disaster response models assume functioning governments, stable institutions, and secure access—conditions that rarely exist in conflict zones (Yang et al., 2023). In fragile and conflict-affected states (FCAS), governance is fragmented, often split between governments and armed groups. In places like Yemen and Syria,[1] no single authority can enforce laws or coordinate aid at scale (Abbara et al., 2021; Dureab et al., 2021). Ongoing violence makes humanitarian access dangerous or impossible. Relief teams face shelling, blocked roads, and attacks, forcing agencies to suspend operations, as seen in Sudan and Gaza in 2023 and onwards. Standard approaches—airlifting aid or deploying search-and-rescue teams—often fail under such conditions (Haar et al., 2021). At the same time, needs are enormous. Conflict zones host displaced populations and overlapping crises, but insecurity and politics block effective aid delivery. Centralized relief efforts struggle to operate across frontlines, leaving many communities without support (George & Adelaja, 2022).

Global disaster frameworks rarely account for these realities. The Sendai Framework, for example, omits any mention of conflict (United Nations Office for Disaster Risk R, 2015). Humanitarian actors often avoid long-term preparedness work in war zones, focusing instead on short-term relief, with little investment in resilience or early warning systems (Spiegel, 2017). Distrust also limits classical responses. In places like Lebanon, years of state failure have led to widespread skepticism toward government-led aid. Communities often prefer to rely on their own networks, ignoring official instructions. This weakens the effectiveness of top-down interventions (Shirleyana et al., 2023). Finally, external actors often miss crucial local knowledge. Unfamiliar with the terrain or power dynamics, they may deliver inappropriate aid or set up operations in unsafe areas. Without flexibility, traditional models fail to adapt to the complexities of conflict.

Fragile and conflict-affected contexts expose the limitations of conventional disaster response along multiple dimensions: governance (no central authority to lead), security (unsafe for responders), logistics (access routes cut and infrastructure destroyed), scale (needs overwhelm resources), and trust/ethics (local suspicion of outside actors). These limitations compel a shift in approach. As several analysts have argued, it is neither acceptable nor necessary to treat conflict settings as off-limits for disaster risk reduction and emergency preparedness (Overseas Development I, 2025). Instead, practitioners must adapt and find ways to reduce risk and respond to crises amid conflict, working through and with the communities who are living the reality of those conflicts. One promising approach is community engagement and participation, which the next sections will explore in depth.

---

[1] This remains true in Syria despite the political changes represented by the fall of Assad regime on the eighth of December 2024 which unified large parts of the countries under the control of what is previously known as the Syrian armed opposition; however, certain parts of Syria in northeast and in the south remain outside the new government control.

6.3 Defining Community Engagement and Theoretical Frameworks

Community engagement in disaster and emergency management refers to the process of actively involving local communities in the planning, decision-making, implementation, and evaluation of preparedness and response activities (Sufri et al., 2020). Rather than being passive recipients of aid, community members become equal partners and leaders in creating solutions for the emergencies that affect them (Ryan et al., 2020). According to the World Health Organization, community engagement is about developing relationships and structures that empower communities to share leadership and jointly design responses that are *acceptable and workable* for those impacted (WHO, 2025). In practical terms, this means recognizing community members' local knowledge, capabilities, and needs, and ensuring they have a voice in how aid is delivered and how risks are managed.

Key theoretical frameworks have evolved to guide community-based approaches. One foundational concept is Community-Based Disaster Risk Management (CBDRM) (Van Niekerk et al., 2018). CBDRM is an approach in which at-risk communities themselves take the lead in identifying, analyzing, and reducing disaster risks with support from external facilitators. It "originates from, and is organized by, local communities" rather than outside experts. In CBDRM, community members map their own hazards, vulnerabilities, and resources and develop locally appropriate measures for prevention, preparedness, and response. The process is highly participatory, aiming to build on indigenous knowledge and foster local ownership of risk reduction. Successful CBDRM depends on maintaining community motivation and integrating these local plans into broader policy, so that national disaster management systems support and sustain grassroots initiatives (Yore et al., 2018).

Participatory risk assessment is central to Community-Based Disaster Risk Management (CBDRM) and related frameworks. Tools like Hazard, Vulnerability, and Capacity Assessments (HVCAs) actively involve community members in identifying local hazards, vulnerable groups, and existing capacities such as skills, networks, and resources (Yore et al., 2018). This process ensures that disaster risk reduction (DRR) measures reflect the community's realities rather than external assumptions. Complementing CBDRM, the International Federation of Red Cross and Red Crescent Societies (IFRC) promotes Community Engagement and Accountability (CEA), which positions communities as decision-makers and holds agencies accountable through two-way communication and feedback systems (International Federation of Red C, Red Crescent S, Iraqi Red Crescent S, 2025a).

A range of participatory approaches further strengthen community engagement. These include Participatory Rural Appraisal for mapping resources and hazards, Community-Based Disaster Preparedness training in first aid and rescue skills, Participatory Action Planning for locally led contingency plans, and social capital frameworks that leverage trust and networks during crises (Van Niekerk et al., 2018). The guiding principle across these models is that those most affected are not passive victims but active agents of their own survival—particularly critical in conflict-affected areas where external assistance is limited.

The next sections illustrate these concepts through concrete case studies from MENA. Each case shows a different facet of community engagement in emergency response. These examples demonstrate not only the possibility of community engagement in FCAS, but the benefits in terms of lives saved, aid delivered, and trust built in environments where conventional approaches struggle.

6.4 Case Studies of Community Engagement in Emergency Response

6.4.1 Syria (Northwest): Community-Led Response to the 2023 Earthquake

On February 6, 2023, a 7.8 magnitude earthquake struck southern Türkiye and northwest Syria, causing massive destruction on both sides of the border (Dal Zilio & Ampuero, 2023). In Syria, the hardest-hit areas—Idlib and parts of Aleppo—were under opposition control, inaccessible to the Syrian government, and already devastated by 12 years of conflict (Jabbour et al., 2023). Türkiye, facing its own extensive disaster, was unable to provide immediate large-scale assistance, and strict border controls delayed the entry of international rescue teams. With no functioning state presence, and formal disaster response capacity extremely limited, local communities were left to organize and lead their own emergency response (Jabbour et al., 2023). Figure 6.1 represents the areas mostly affected by the earthquake.

In this context, a remarkable and now documented scientific case emerged: the rapid use of community engagement and crowdsourcing as a tool for disaster risk assessment (Alhaffar et al., 2023). Within hours of the quake, Syrian volunteers and local responders launched a Google Form, shared via social media, to collect real-time reports of collapsed buildings and trapped individuals. Over 1200 reports poured in within the first 24 hours. A volunteer tech team verified and filtered these submissions—cross-checking with local contacts and available communication channels—producing a verified map of 129 locations with more than 1000 collapsed buildings. This map was immediately shared with local authorities and the Syria Civil Defence (White Helmets), allowing scarce rescue teams and heavy equipment to be deployed where they were needed most (Alhaffar et al., 2023).

At the same time, thousands of residents became first responders. The White Helmets coordinated an estimated 3000 spontaneous local volunteers, many without formal training, who dug through rubble, rescued survivors, and provided basic aid. Privately owned excavators and heavy machinery were mobilized by citizens and even local armed groups, reflecting an adaptive civic ecosystem built over years of war. Trust, informal coordination networks, and institutional memory from previous

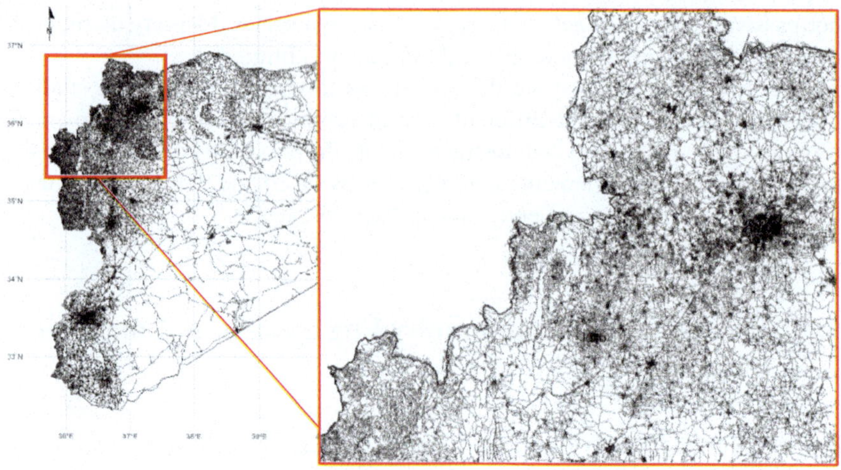

Fig. 6.1 Maps of Syria and earthquake-affected areas. Left: National map of Syria generated in RStudio, displaying administrative boundaries and main road networks. Right: Zoomed-in view of the north-western region most affected by the February 2023 earthquake, encompassing the governorates and major cities of Aleppo, Idlib, and Hama, where structural damage and humanitarian impact were most severe

crises enabled this rapid mobilization. The result was a highly targeted, timely, and effective response (Alhaffar et al., 2023).

The Syria case illustrates that in conflict settings with fragmented governance; community engagement can be an important tool for both immediate lifesaving action and rapid risk assessment. By combining local knowledge, social trust, and accessible technology, communities were able to identify and prioritize needs in real time, bridging the gap until outside help became possible. It also demonstrates the value of institutionalizing such approaches—through training, preparedness planning, and integrating crowdsourcing tools—so that in future disasters, affected populations are not only the first responders but also the first and most reliable source of operational intelligence.

6.4.2 *Yemen: Local Flood Mitigation and Preparedness in a Fragile Conflict Environment*

Since 2015, Yemen's conflict has created one of the world's worst humanitarian crises, while seasonal floods and storms repeatedly devastate displacement camps (United Nations Office for the Coordination of Humanitarian A, 2018). With formal disaster risk reduction (DRR) efforts, which were minimal, due to state fragility, local actors have taken the lead. In early 2024, Yemen Al-Khair for Relief and Development (YARD), with UN support, conducted a community-based flood risk assessment across hundreds of IDP sites in high-risk zones (United Nations Office for the Coordination of Humanitarian A, 2018). Displaced residents identified hazards such as flood-prone locations, lack of drainage, and absence of early warning systems—exposing a gap in humanitarian response, which had focused on post-flood relief rather than prevention.

Working with camp residents, YARD moved quickly from assessment to action. In Al-Jawf's Al-Mahzam Al-Sharqi site, the community co-designed and helped excavate a 2.5 km flood diversion channel, reinforced with local stone, along with earthen berms to redirect water. Volunteer risk reduction committees, composed of men and women from the camps, were trained in early warning, hazard mapping, and maintenance. The impact was immediate: When heavy rains returned, the new infrastructure spared the camp from flooding. One resident, Samaira, who once feared for her children's safety during floods, described sleeping "with peace of mind" after the project. By mid-2024, similar measures had improved safety in multiple sites, combining modest technical interventions with strong local ownership. This case shows that even in active conflict, community-driven DRR can protect lives, reduce displacement, and build resilience—supporting the principle that "preparedness cannot wait for peace." It is now cited internationally as evidence that locally led risk reduction should be mainstreamed into humanitarian operations.

6.4.3 *Palestine (Gaza Strip): Volunteer Networks and First Responders in a Blockaded Conflict*

The Gaza Strip, home to over two million people, operates in a state of chronic crisis marked by repeated wars (2008, 2012, 2014, 2021, and 2023), economic collapse, and fragile infrastructure. The Israeli–Egyptian blockade severely limits the entry of aid workers and supplies, and during conflicts, borders often close entirely. Governance is split between Hamas in Gaza and the Palestinian Authority in the West Bank, leaving disaster management fragmented. In practice, emergencies—whether conflict, disease outbreaks, or floods—are met first and foremost by community-based resources (Alokaily, 2024).

Despite these conditions, Gaza has a strong tradition of volunteerism. The Palestine Red Crescent Society (PRCS) provides the backbone of organized emergency medical services, staffed entirely by local professionals and volunteers. During the 2023 war, PRCS teams delivered trauma care to around 15,000 injured people and recovered over 5000 bodies despite losing several volunteers in the line of duty. Many responders continued working after losing family or homes themselves, a testament to their deep integration in the communities they serve. The IFRC publicly commended their courage and commitment (International Federation of Red C, Red Crescent S, 2023).

Alongside PRCS, informal volunteer networks are critical. Neighborhood teams self-organize after airstrikes to search rubble, distribute supplies, and help the elderly evacuate. Local NGOs, mosque committees, and charities coordinate food and water deliveries to shelters. Community Emergency Teams, activated during past escalations and again in 2023, combine local first aid and firefighting skills to support overstretched civil defense forces. In one instance, volunteer paramedics coordinated via WhatsApp to reach bomb sites, effectively creating a community-based ambulance system when formal services were overwhelmed (United Nations Office for the Coordination of Humanitarian Affairs oPt, 2017).

These efforts have saved countless lives and sustained communities under siege. Immediate local rescues prevent higher casualty rates, and solidarity networks help families share scarce resources like food, water, and fuel. However, the strain is immense. Volunteers work with limited protection and resources, face constant danger, and must make life-or-death triage decisions. Fuel shortages in late 2023 forced PRCS to halt some ambulance operations, and local aid groups struggled as supplies dwindled. Such constraints show that community action, while essential, cannot fully replace secure humanitarian access and external support.

Gaza's experience demonstrates both the necessity and limits of community engagement in protracted conflict. Palestinian communities have repeatedly been the first and often only responders, and experts agree this role will remain crucial. Strengthening these systems—through better equipment, protection for responders, and formalized local emergency plans linked to national and international frameworks—would transform ad-hoc heroism into sustainable capacity. Supporting these "armies of compassion" is not just humanitarian prudence but a prerequisite for an effective, equitable response in Gaza's foreseeable future.

6.4.4 Lebanon: Civil Society-Led Emergency Response to the 2020 Beirut Port Explosion

Lebanon's fragile state, already weakened by economic collapse, political paralysis, and a legacy of conflict, was devastated on August 4, 2020, when improperly stored ammonium nitrate exploded at the Port of Beirut. The blast killed over 220 people, injured 6000, and left 300,000 temporarily homeless (Al-Hajj et al., 2023). With governance fragmented and the official response slow and disorganized, local communities and civil society became the primary responders from day one.

Within hours, volunteers were rescuing the injured, pulling survivors from collapsed buildings, and transporting them to hospitals in private cars. Off-duty medical professionals set up street-side triage clinics, relieving pressure on overwhelmed hospitals. The next morning, thousands of citizens descended on the worst-hit neighborhoods, clearing rubble, sweeping glass, and reopening streets. Social media became a coordination hub, matching displaced families with hosts and organizing food and supply distribution (Chehayeb, 2020).

Civil society organizations quickly structured this volunteer surge into organized aid. Tadamon al-Nass, a collective with 4000 volunteers, deployed teams for door-to-door needs assessments, distributing food, medicine, shelter, and repair materials, often in partnership with larger charities. Other NGOs, youth groups, and informal initiatives repurposed their operations to provide emergency relief. The Dafa Campaign, for instance, reactivated to deliver food boxes and essential supplies, with volunteers openly stating they were doing what the state should have done (Chehayeb, 2020).

The outcomes were significant. Within two weeks, nearly all displaced families had secured temporary shelter, mostly through community connections. Volunteers cleared vast amounts of debris, enabling quicker reopening of roads and reducing secondary hazards. Damage and needs data gathered by NGOs informed later UN and international recovery efforts. The visible absence of government relief contrasted sharply with the speed and scale of community action, prompting calls for greater localization of aid (Chehayeb, 2020).

The Beirut blast response demonstrated the effectiveness of a vibrant civil society and strong volunteer culture in an urban disaster. Years of experience with economic crisis and refugee assistance had built networks and trust that could be mobilized instantly. The case highlights that in fragile states, investing in local networks and empowering community actors is critical—they can act faster, adapt better, and maintain public trust when centralized systems fail. As one volunteer remarked during cleanup: "We saved ourselves."

6.4.5 Sudan: Neighborhood "Resistance Committees" as First Responders in the 2023 Conflict

When heavy fighting erupted in April 2023 between Sudan's national army and the Rapid Support Forces, cities like Khartoum became war zones overnight (United Nations Office for the Coordination of Humanitarian A, 2025). Hospitals shut down, utilities collapsed, and international agencies evacuated. Sudan entered the crisis with a unique advantage: a nationwide network of grassroots "Resistance Committees" originally formed during pro-democracy protests. With governance and services absent, these committees pivoted instantly to humanitarian response, becoming the backbone of relief efforts (Mat, 2023; Abbashar, 2023).

Within days, hundreds of neighborhood committees mobilized medics, engineers, and volunteers to meet urgent needs. They reopened defunct clinics by gathering volunteer doctors and nurses, securing fuel for generators, and sourcing medical supplies through donations. Makeshift first aid stations were set up in homes and schools, and several hospital wings were brought back into service solely through community action. Committees also coordinated evacuations from besieged districts, using scarce fuel to run convoys and real-time WhatsApp updates to guide civilians along safer routes (Mat, 2023; Abbashar, 2023).

Relief distribution was hyper-local. Communities pooled water from private wells, cooked meals for neighbors, and purchased essentials—flour, canned goods, fuel—using funds raised from Sudanese diaspora networks. Aid stations in each neighborhood ensured that basic supplies reached those in shelters. In addition to relief, the committees acted as peacekeepers, running awareness campaigns to prevent ethnic or political divisions from fracturing communities under stress (Bashri, 2025).

The impact was substantial. Community-run health facilities provided care to the wounded and chronically ill who otherwise had no access. Evacuation convoys saved families trapped in combat zones. Real-time information sharing improved civilian safety in a fast-changing battlefield. These committees filled the vacuum left by international agencies, proving that decentralized, trusted local networks can mount an effective response even in extreme insecurity (Mat, 2023; Bashri, 2025).

Constraints remained—supplies were scarce, and international aid could not easily reach them—but their ability to improvise, share resources across neighborhoods, and maintain a functioning "social protection network" kept thousands alive. Sudan's experience shows that existing grassroots movements can rapidly transform into lifesaving emergency responders, suggesting a clear opportunity for humanitarian agencies to partner with and strengthen such networks before crises erupt.

6.4.6 Iraq: Community Engagement and Health Response—The "Building Trust" Project for Disease Outbreak

Although large-scale conflict in Iraq has subsided since ISIS's defeat in 2017, the country remains fragile, with weak public services. In 2023–2024, an outbreak of Crimean-Congo Hemorrhagic Fever (CCHF) threatened rural communities, especially butchers, farmers, and their families. The Iraqi Red Crescent Society (IRCS), supported by the IFRC, launched the *Building Trust Project* to curb transmission by changing high-risk practices and ensuring timely treatment. Recognizing that health orders alone would fail without community buy-in, IRCS began with a participatory assessment, gathering feedback on local perceptions, barriers to safer practices, and needs. This revealed that access to proper protective equipment (PPE) was a major gap (International Federation of Red C, Red Crescent S, Iraqi Red Crescent S, 2025b).

Guided by Community Engagement and Accountability (CEA) principles, the project worked directly with at-risk groups. Actions included distributing PPE kits (gloves, masks, aprons) to high-risk workers in outbreak hotspots, with live demonstrations on correct use; holding interactive awareness sessions in Arabic dialects and Kurdish; and involving religious leaders, tribal elders, and local authorities to endorse prevention measures. Continuous feedback loops—via suggestion boxes, a hotline, and face-to-face volunteer visits—allowed rapid adaptation, such as replacing unsuitable gloves with better ones. Communities also proposed their own solutions, like weekly market cleaning days to reduce tick populations (United Nations Development P, Public Aid O, Alliance of Iraqi M, Fight for H, 2022).

The project reached over 15,000 high-risk individuals across affected regions. Post-session surveys showed 97% reported greater understanding of CCHF prevention, and 91% had adopted safer hygiene and animal handling practices. The intervention turned key community members into advocates—one butcher pledged to "spread awareness through my profession" and enforced stricter hygiene among his workers. Trust between communities and the Red Crescent improved markedly, reducing resistance to health measures and fostering local ownership. Protective behaviors persisted beyond the project's timeline, demonstrating sustained impact. The IFRC evaluation concluded that embedding CEA "enhanced effectiveness and sustainability," making this a model for outbreak control in fragile settings (International Federation of Red C, Red Crescent S, Iraqi Red Crescent S, 2025b; United Nations Development P, Public Aid O, Alliance of Iraqi M, Fight for H, 2022).

In broader perspective, this case study demonstrates how even in a fragile state like Iraq, where trust in government might be low, leveraging the reach and rapport of organizations like the Red Crescent and involving communities directly can overcome suspicion and change behaviors during a public health emergency. It highlights the importance of *building trust* (aptly named) before and during crises. A key takeaway is that technical solutions (PPE, disinfectants, medical treatment) must be paired with social solutions (community engagement, risk communication) to succeed. By getting "buy-in" from the community, the project avoided pitfalls like people rejecting health advice or rumors undermining the response.

6.4.7 Morocco: Community Solidarity and Local Volunteerism in the 2023 Earthquake Response

On September 8, 2023, a 6.8 magnitude earthquake struck Morocco's High Atlas Mountains, with the epicenter in Al-Haouz province south of Marrakech. It was Morocco's deadliest earthquake in over 60 years, killing nearly 3000 people and injuring thousands more. Many affected areas were remote mountain villages, already difficult to reach and further cut off by landslides. While the Moroccan government and military mobilized a national response, limited international assistance was accepted in the early days, leaving many communities to depend on themselves and local resources (Al, 2023).

In the first 24–48 hours—when survival odds were highest—neighbors rescued neighbors from collapsed adobe houses, often using only shovels, pickaxes, or bare hands. In villages like Imi N'Tala, residents worked through aftershocks to pull people from rubble before professional teams arrived. The Moroccan Red Crescent Society (MRCS) rapidly activated local branches, deploying hundreds of volunteers trained in first aid and psychosocial support. They reached remote hamlets on foot or by mule when roads were blocked, evacuated people from unstable buildings, provided first aid, and managed the deceased with dignity (Newsfeed, 2023).

Spontaneous aid convoys from nearby cities brought food, water, blankets, and tents directly to affected villages. Small groups of youth volunteers, businessmen, and diaspora members delivered supplies—sometimes by donkey—to bypass blocked roads. In Moulay Brahim, survivors organized communal kitchens, shared limited water, and arranged collective shelter for about 600 displaced people, restoring the town's water supply and cleaning public buildings for use as shelters within 48 hours (Riley, 2024).

Local action saved lives by enabling immediate rescues and quicker treatment of the injured. MRCS volunteers, often the first organized responders, provided urgent care and psychological support, reducing shock and panic. Domestic volunteer convoys filled the supply gap before larger official and international aid could arrive. Within six months, MRCS—working with community volunteers—had delivered shelter, food, and medical assistance to around 55,000 people. Local knowledge made aid delivery more efficient and culturally appropriate. Community bonds also deepened; survivors cited unprecedented solidarity, and in some villages, reconstruction committees formed to plan safer rebuilding. Women played active roles, including leadership positions—such as Naima from Chichaoua, MRCS's first female shelter officer; her work helped ensure relief met women's needs and encouraging broader female participation (Riley, 2024).

The Morocco case highlights that community engagement is indispensable in the critical first days of disaster response. Spontaneous local initiatives and organized volunteer networks acted faster and more flexibly than formal systems, ensuring lifesaving assistance reached even hard-to-access communities. However, volunteer response has limits—heavy equipment and large-scale shelter provision required state and international capacity. The key lesson: integrating and supporting community responders with training, resources, and inclusive participation, particularly of women and marginalized groups, can transform grassroots solidarity into a sustained, coordinated force in national disaster management.

6.5 Challenges and Limitations of Community Engagement in Conflict Settings

While the above cases highlight success stories, it is important to critically examine the limitations and challenges of community engagement in fragile and conflict-affected situations. Implementing community-led approaches in these environments is not a panacea and comes with its own set of difficulties in ethical, logistical, and operational dimensions.

Perhaps the most serious challenge is the physical danger to community volunteers and participants. In active conflict zones, local responders often operate at great personal risk. Unlike international staff who might be evacuated when violence escalates, community volunteers are on the front lines with no such refuge. The tragic reality is that many have been injured or killed while trying to help others. For instance, in Gaza during the 2023 conflict, several Palestinian Red Crescent volunteers were killed in the line of duty, including ambulance crew members struck by munitions. In Syria, the White Helmets lost hundreds of volunteer rescuers over the years to airstrikes, often being targeted while saving lives. This raises a profound ethical dilemma: Relying on community members to respond can mean asking them to take life-threatening risks, especially if warring parties do not respect humanitarian norms. Ensuring volunteer safety is incredibly challenging; local actors typically lack protective equipment or security guarantees that international organizations might negotiate. They can also become targets of suspicion (e.g., seen as aligning with one side). Ethically, organizations must consider how to best protect and support community responders—including through training in personal safety, providing insurance or compensation, and advocating with conflict parties to respect the neutrality of local aid volunteers. Community engagement initiatives must adhere to "Do No Harm" principles, but in polarized conflicts, this is easier said than done. Involving certain community leaders could inadvertently politicize an effort or expose those individuals to retribution. For example, a community committee working with international aid might be accused by one side of "collaborating" with the enemy or diverting aid to fighters. Aid can become a pawn in conflict dynamics. To navigate this, practitioners emphasize conflict sensitivity—thoroughly analyzing how interventions might fuel tensions or endanger participants and adapting accordingly. In Afghanistan, NGOs have used conflict analysis tools to adapt community DRR projects, ensuring they do not exacerbate local conflicts and applying *do-no-harm* approaches. This kind of careful planning is necessary but time-consuming and complex.

Community groups in FCAS often operate with very limited resources and support. They typically do not have stockpiles of relief goods, backup funding, or advanced equipment. This can hinder their effectiveness or sustainability. For instance, in Sudan, the resistance committees' biggest challenge was lack of supplies—*"What we lack right now is supplies—from drugs to first aid kits, to gauze,"* one organizer said. In Gaza, local responders faced severe shortages of fuel, medicine, and power, which at times forced critical community-run services (like ambulance stations) to halt. Unlike international agencies that can appeal for global funds or bring in relief shipments, communities often must rely on what is immediately at hand or on local fundraising. This means their capacity can be quickly overcome in a major disaster. A community may heroically respond to an earthquake for a few days, but without external resupply, they cannot feed people or rebuild infrastructure alone. Additionally, conflicts pose access issues even for community responders. Frontlines, checkpoints, or ongoing fighting can prevent movement within a country. For example, in Yemen or Syria, community volunteers from one area might not be able to reach a neighboring area due to frontlines or siege tactics. Thus, community action tends to be localized, which may leave some pockets without sufficient help if those communities were particularly devastated or lack able-bodied responders. Coordination across communities is tough when communications are down (as often happens in war when the internet or phones are cut down). Community networks like Sudan's committees overcame some of this via WhatsApp and radio, but if the network goes out, each locale might become isolated.

A related operational challenge is that community engagement efforts can be fragmented and uneven. By nature, they are locally focused; this can lead to variability in response quality and coverage. Some communities may be very organized and capable, while others (perhaps those already marginalized) might not have a functioning committee or any resources to leverage. For example, within Palestine, some refugee camps have active community emergency committees, but others might not, meaning the level of preparedness differs. Because community initiatives are often informal, they might not cover all at-risk groups. People who are socially isolated or minority groups could be left out if the community leadership does not represent them. In times of chaos, there is a risk that aid stays within certain networks and inadvertently excludes others, e.g., along clan or sectarian lines. Effective community engagement requires inclusive representation—involving women, minorities, and all segments—which can be hard to achieve in conservative or divided societies. The Morocco example highlighted the importance of including women in decisions so that women and children's needs are addressed. If community leadership is homogenous (say, only elder males), certain voices might not be heard, limiting the effectiveness and fairness of the response.

Another challenge is integrating community-led efforts with formal humanitarian or government response structures. When communities act on their own, there can be duplication of efforts or clashes with official plans. Governments may be wary or dismissive of informal initiatives, perhaps seeing them as undermining authority. In Lebanon 2020, while civil society did nearly everything initially, eventually the army was tasked with coordinating aid—there were some frictions as the ad hoc volunteer system handed over to a more bureaucratic approach. In Palestine, OCHA notes that because of administrative fragmentation and access restrictions, *"the majority of communities are not part of an established... response plan,"* relying on local structures instead. This indicates a gap: Ideally, community mechanisms should link into a broader strategy, but in conflict settings that broader strategy may be absent or in disarray. Without coordination, you may get situations where some communities are flooded with multiple aid groups (often easier-to-reach areas) while remote ones get little—something observed in the Morocco quake, where villagers complained that some places received plenty of help and others almost none in the first week. Effective community engagement needs a way to scale up information to higher levels so that external actors know where to support, and likewise a way to scale down external resources to community level in an organized manner. Achieving this in a conflict is challenging because normal governance or disaster management hierarchies are broken.

Moreover, community responders are themselves victims/survivors of the crisis—they are often working while also dealing with personal losses. This can lead to physical and emotional burnout. They might not have the training to cope with traumatic situations over a long period. Unlike professional aid workers who rotate out, community volunteers must carry on and then also live in the aftermath with everyone else. The mental health toll can be heavy. For example, many Beirut explosion volunteers were youths who had never seen such devastation; months after the event, some experienced trauma symptoms from the things they saw and the pressure they felt. In Gaza, many PRCS volunteers continued working *despite losing family members*—an incredible testament to dedication, but also a potential recipe for long-term psychological distress if not addressed. Therefore, a challenge is providing psychosocial support and care for community caregivers. In conflict settings, there may be little to no resources for this. Similarly, sustaining engagement over protracted crises is hard. People eventually must return to making a living; volunteer zeal can wane if the emergency becomes chronic.

Additionally, community responders are frequently not formally recognized or trained by any authority. They may lack legal protection (e.g., insurance, access to compensation if injured) and may not have formal training in things like first aid, safe demolition, or managing hazardous debris. This can lead to avoidable accidents or inefficient methods. For instance, well-meaning volunteers might harm themselves lifting rubble without proper tools, or they might mishandle bodies and risk disease because they were not trained in proper procedures. Expanding training programs like Community Emergency Response Team (CERT) models in conflict areas is tough (security issues for trainers, etc.), so this gap persists.

In summary, while community engagement is a powerful approach in FCAS, it is not without challenges. Security and ethical issues, resource and access constraints, fragmentation, coordination difficulties, volunteer burnout, risk of politicization, and lack of training all complicate the picture. These challenges do not negate the value of community engagement—rather, they highlight areas where supporting and strengthening community mechanisms is necessary. Addressing these challenges requires thoughtful approaches, some of which inform the recommendations in the next section. It is vital to mitigate these issues so that community engagement can realize its full potential without exploiting or endangering the very people it aims to empower.

6.6 Recommendations for Institutionalizing and Enhancing Community Engagement in MENA

Building on the above analysis and case study lessons, the following actionable recommendations are offered to institutionalize and improve community engagement mechanisms in disaster preparedness and response across conflict-affected countries in MENA. These recommendations are directed at a range of stakeholders—including governments in the region, international agencies, NGOs, and the communities themselves—and aim to create an enabling environment for effective community-led action.

1. **Integrate Conflict Contexts into Disaster Policy**—Update national DRM plans to include community engagement in fragile areas; create coordination cells for conflict-affected provinces; push global DRR frameworks to explicitly address conflict.
2. **Empower and Formalize Community Networks**—Train and equip local volunteer groups (CERTs, grassroots committees); register and integrate them into coordination systems; recognize them as legitimate responders to enable direct funding and support.
3. **Ensure Inclusive Participation**—Involve women, youth, elders, people with disabilities, and minorities; set representation quotas in community committees; support women's and youth groups; leverage trusted leaders to promote inclusion.
4. **Strengthen Two-Way Communication**—Establish trusted channels (radio, WhatsApp, SMS, meetings) for alerts and feedback; embed CEA teams in responses; use perception surveys and hotlines; ensure information flows both ways.
5. **Allocate Direct Funding to Local Actors**—Create donor funding windows for community-based organizations; pre-position community stockpiles; fund preparedness training; require INGOs to partner with and build capacity of local groups.
6. **Enhance Coordination with Formal Responders**—Involve community reps in cluster meetings/EOCs; create liaison roles; connect grassroots reporting (e.g., crowdsourcing) to official systems; maintain databases of community capacities.
7. **Protect and Support Volunteers**—Advocate for protection under IHL; provide safety training, gear, and radios; offer psychosocial support; set up insurance or solidarity funds; publicly recognize volunteer contributions.
8. **Localize Climate Adaptation and DRR**—Integrate resilience measures into humanitarian response; support climate-smart practices, early warning systems, and contingency planning in conflict zones; secure climate funding for FCAS communities.
9. **Foster Regional Knowledge Exchange**—Establish regional forums for sharing best practices; facilitate peer learning and training across countries; engage diaspora systematically as partners in community-led projects.

By implementing these recommendations, stakeholders can create an ecosystem where community engagement is not an ad hoc last resort, but a planned and well-resourced component of emergency management in MENA's conflict zones. Ultimately, institutionalizing community engagement means shifting mindsets: recognizing that local people are the primary agents of their survival, and our job in the humanitarian and disaster management sector is to support, equip, and amplify their efforts. As one volunteer from Lebanon said, after seeing what communities can do, the aim is not literally to eliminate formal structures, but to make those structures supportive of and accountable to the communities at risk. With the increasing overlap of conflict and disaster in MENA, the resilience of communities will be the determining factor in how catastrophic or manageable emergencies turn out to be. Investing in that resilience through engagement, even amid fragility, is not just wise—it is imperative.

6.7 Conclusion

Conflicts across the MENA region have created some of the world's most difficult environments for emergency response, where traditional top-down models often fail. In these fragile contexts, community engagement has proven to be an essential, not optional, approach. From Syrian crowdsourced earthquake mapping to Yemeni flood diversion, Gazan volunteer medics, and Lebanese youth-led rebuilding, the case studies show that empowered communities can act quickly, deliver culturally attuned solutions, and reach places outsiders cannot. These efforts build trust, save lives, and sustain response capacity even amid insecurity. However, community action alone cannot replace the roles of states and international actors. Its impact is greatest when supported with policies, funding, training, and integration into broader systems, positioning communities as equal partners in emergency planning and response.

Moving forward requires concrete measures: establishing local response teams in every district, ensuring inclusion of women and marginalized groups, maintaining trusted communication channels, securing resources for local initiatives, and advocating for the protection of community responders in conflict zones. MENA communities have shown extraordinary resilience and ingenuity—whether a Yemeni mother sleeping peacefully behind a flood barrier built by neighbors, a Moroccan woman training others to set up shelters, or a Sudanese volunteer coordinating hospital evacuation via WhatsApp. These examples illustrate that community engagement is both a moral imperative and a practical necessity. By institutionalizing it, governments, international actors, and communities can build a more effective, inclusive, and sustainable disaster response system—one that harnesses the strength of local solidarity to turn crises into opportunities for resilience.

References

Abbara, A., Marzouk, M., & Mkhallalati, H. (2021). Health system fragmentation and the Syrian conflict. In Bseiso J, Hofman, M., & Whittall, J., (Eds.). *Everybody's war: The politics of aid in the Syria crisis*. Oxford University Press.

Abbashar A. (2023). *Resistance committees and Sudan's political future. Peacerep: The peace and conflict resolution evidence platform*. University of Edinburgh. Report No.

Al, J. (2023). *Survivors question rescue efforts after Morocco's earthquake*. https://www.aljazeera.com/gallery/2023/9/12/photos-survivors-question-rescue-efforts-after-moroccos-earthquake

Alhaffar, M., & Janos, S. (2021). Public health consequences after ten years of the Syrian crisis: A literature review. *Globalization and Health, 17*(1), 111. https://doi.org/10.1186/s12992-021-00762-9

Alhaffar, M., Joury, E., & Eriksson, A. (2023). Community engagement and crowdsourcing for effective disaster response and rescue operations during the earthquake in Syria. *International*

Journal of Disaster Risk Reduction, 98, 104096. https://www.sciencedirect.com/science/article/pii/S2212420923005769

Al-Hajj, S., Ghamlouche, L., Nasser AlDeen, K., & El Sayed, M. (2023, Feb 15). Beirut blast: The experiences of acute care hospitals. *Disaster Medicine and Public Health Preparedness, 17*, e318.

Alokaily, F. (2024, Jan). War and health crisis in Gaza. *Saudi Medical Journal, 45*(1), 9.

Bashri, M. (2025). *Beyond the battlefield: The grassroots networks sustaining Sudan.* https://fuf.se/magasin/beyond-the-battlefield-the-grassroots-networks-sustaining-sudan/

Börzel, T. A., & Risse, T. (2010). Governance without a state: Can it work? *Regulation & Governance, 4*(2), 113–134. https://doi.org/10.1111/j.1748-5991.2010.01076.x.

Chehayeb, K. (2020). *Local groups step up to lead Beirut blast response.* https://www.thenewhumanitarian.org/news-feature/2020/08/18/Lebanon-Beirut-explosion-local-aid-response

Dal Zilio, L., & Ampuero, J. -P. (2023). Earthquake doublet in Turkey and Syria. *Communications Earth & Environment, 4*(1), 71. https://doi.org/10.1038/s43247-023-00747-z

Dureab, F., Hussain, T., Sheikh, R., Al-Dheeb, N., Al-Awlaqi, S., & Jahn, A. (2021). Forms of health system fragmentation during conflict: The case of Yemen. *Frontiers in Public Health, 9*, 659980.

Geneva Academy of International Humanitarian L, Human R. (2025). *Today's armed conflicts.* https://geneva-academy.ch/galleries/today-s-armed-conflicts

George, J, & Adelaja, A. (2022). Armed conflicts, forced displacement and food security in host communities. *World Development, 158*, 105991. https://www.sciencedirect.com/science/article/pii/S0305750X22001814.

Haar, R. J., Read, R., Fast, L., Blanchet, K., Rinaldi, S., Taithe, B., et al. (2021). Violence against healthcare in conflict: A systematic review of the literature and agenda for future research. *Conflict and Health, 15*(1), 37. https://doi.org/10.1186/s13031-021-00372-7

International Federation of Red C, Red Crescent S. (2023). *Israel/Palestine conflict: Our response so far*. Report No.

International Federation of Red C, Red Crescent S, Iraqi Red Crescent S. (2025a). *Enhancing emergency response effectiveness through CEA: The building trust project in Iraq.*

International Federation of Red C, Red Crescent S, Iraqi Red Crescent S (2025b). *Enhancing emergency response effectiveness through community engagement and accountability (CEA): The building trust project in Iraq*. IFRC & Iraqi Red Crescent Society. Report No.

Jabbour, S., Abbara, A., Ekzayez, A., Fouad, F. M., Katoub, M., & Nasser, R. (2023, Mar 11). The catastrophic response to the earthquake in Syria: The need for corrective actions and accountability. *Lancet, 401*(10379), 802–805.

Mat, N. (2023). *Sudan 'resistance' activists mobilise as crisis escalates.* https://www.aljazeera.com/news/2023/4/22/sudan-resistance-activists-mobilise-as-crisis-escalates

Mena, R. (2020). *Disaster response and DRR in high-intensity conflict settings: Research brief (December 2020).* International Institute of Social Studies (ISS), Erasmus University Rotterdam.

Newsfeed, W. (2023). *Morocco's earthquake crumbled much of this adobe village nestled in the mountain.* https://newsfeed.wtjx.org/2023-09-19/moroccos-earthquake-crumbled-much-of-this-adobe-village-nestled-in-the-mountain.

Overseas Development I. (2025). *When disasters and conflict collide.* https://odi.org/en/about/features/when-disasters-and-conflict-collide/

Riley, P. (2024). International Federation of Red C, Red Crescent S. *Morocco earthquake: One year later, a local volunteer breaks new ground as she helps her country rebuild.* https://www.ifrc.org/article/morocco-earthquake-one-year-later-local-volunteer-breaks-new-ground-she-helps-her-country

Rosvold, E. L. (2023). Disaster resilience in conflict-affected areas: A review of how armed conflicts impact disaster resilience. *Current Opinion in Environmental Sustainability, 65*, 101381. https://www.sciencedirect.com/science/article/pii/S1877343523001288

Ryan, B., Johnston, K. A., Taylor, M., & McAndrew, R. (2020). Community engagement for disaster preparedness: A systematic literature review. *International Journal of Disaster Risk Reduction*, 101655. https://www.sciencedirect.com/science/article/pii/S2212420919317893

Shirleyana, H. S., Sunindijo, R. Y., & Sanderson, D. (2023). The critical role of community networks in building everyday resilience—Insights from the urban villages of Surabaya. *International Journal of Disaster Risk Reduction, 98*, 104090. https://www.sciencedirect.com/science/article/pii/S2212420923005708

Spiegel, P. B. (2017). The humanitarian system is not just broke, but broken: Recommendations for future humanitarian action. *The Lancet*. https://doi.org/10.1016/S0140-6736(17)31278-3.

Stockholm International Peace Research I. (2024). *SIPRI yearbook 2024: Armed conflict instability and regions*. https://www.sipri.org/yearbook/2024/02

Sufri, S., Dwirahmadi, F., Phung, D., & Rutherford, S. (2020). A systematic review of community engagement (CE) in disaster early warning systems (EWSs). *Progress in Disaster Science, 5*, 100058. https://www.sciencedirect.com/science/article/pii/S2590061719300584

United Nations Development P, Public Aid O, Alliance of Iraqi M, Fight for H (2022). *Encouraging peace, social cohesion, and community trust-building with local authorities in Iraq. UNDP Iraq*. Report No.

United Nations Office for Disaster Risk R. (2015). *Sendai framework for disaster risk reduction 2015–2030*. United Nations.

United Nations Office for the Coordination of Humanitarian A. (2018). *Yemen: The world's worst humanitarian crisis enters another year*. Report No.

United Nations Office for the Coordination of Humanitarian A. (2025). *Sudan conflict: What's really going on*. Report No.

United Nations Office for the Coordination of Humanitarian Affairs oPt. (2017). *Emergency response preparedness (ERP) and disaster risk management (DRM)*.. https://www.ochaopt.org/content/emergency-response-preparedness-erp-and-disaster-risk-management-drm

Van Niekerk, D., Nemakonde, L. D., Kruger, L., & Forbes-Genade, K. (2018). Community-based disaster risk management. In H. Rodríguez, W. Donner, & J. E. Trainor (Eds.), *Handbook of Disaster Research* (pp. 411–429). Springer International Publishing.

WHO. (2025). *Risk communication and community engagement*. https://www.who.int/emergencies/risk-communications

Yang, X., Yao, Y., Tian, K., Jiang, W., Xing, Q., & Yang, J., et al. (2023). Disaster response strategies of governments and social organizations: From the perspective of infrastructure damage and asymmetric resource dependence. *Heliyon, 9*(10), e20432. https://www.sciencedirect.com/science/article/pii/S2405844023076405

Yore, R., Kelman, I., & Tofa, M. (2018). *Community-based disaster risk management: Research-into-action brief*. Report No.

Chapter 7
Syria's Disaster Risk Profile and Opportunities for Disaster Governance: From Crisis to Capacity

Rodrigo Mena, Tom Ansell, Ahmed Ekzayez, Mohamad Diab, and David MacLeod

Abstract North-western Syria presents a complex disaster risk landscape shaped by overlapping climatic, socio economic, conflict related and environmental hazards. This convergence of risks has produced chronic vulnerability and limited institutional capacity to prevent, respond to, or recover from disasters. Focusing on north-western Syria, this chapter offers an overview of Syria's evolving risk profile and disaster governance in a context of political transition. It draws on regional climate and hydrological evidence, community survey findings, interviews, field observations and humanitarian reporting, and presents an integrated reading of these sources rather than detailed models or datasets. The chapter traces the historical trajectory of hazards and disaster management and shows how recurrent floods, heat extremes and water stress intersect with displacement, damaged infrastructure and fragmented authority. It highlights key gaps that constrain effective analysis and planning, including sparse and inaccessible observations, weak data sharing and limited attention to intersectional vulnerabilities. The chapter argues for more integrated, context sensitive and community-based approaches to disaster risk reduction and management, and identifies opportunities to strengthen institutions, rebuild monitoring and warning systems, and connect community knowledge with national structures and the wider MENA region.

Keywords Disaster risk governance · North-west Syria · Climate change adaptation · Community-based disaster risk reduction · Conflict and fragility · Anticipatory action

---

R. Mena (✉) · T. Ansell
The Hague Humanitarian Studies Centre, International Institute of Social Studies, The Hague, Netherlands
e-mail: mena@iss.nl

A. Ekzayez · M. Diab
Syria Civil Defense—The White Helmets, Sarmada, Syria

D. MacLeod
School of Earth and Environmental Sciences, Cardiff University, Cardiff, United Kingdom

E. Alam (ed.), *Disaster, Crisis and Emergency Management in the Middle East & North Africa*, https://doi.org/10.1007/978-3-032-24436-9_7

7.1 Introduction

Disaster risk in north-west Syria emerges from the convergence of climatic change, protracted conflict, displacement, and environmental degradation (OCHA, 2023). Floods, droughts, and heat extremes interact with fragile governance and eroded infrastructure to produce a chronic state of vulnerability (Abbara et al., 2022; Ali et al., 2022; Daoudy et al., 2023; Meininghaus, 2016; Schellens et al., 2024). This chapter examines how these risks are evolving and how they might be managed more effectively in the current moment of political transition (Clark-Ginsberg et al., n.d.; Coughlan de Perez et al., 2015). Although the empirical base and modelling focused on north-west Syria (NWS) and, due to their relevance and recurrence in the contexts we study, also on floods, we draw cautious inferences for other regions where the evidence allows. The underlying analysis was completed in 2024, before the fall of the Assad regime; its interpretation has been updated to reflect subsequent developments as of mid-2025, including the formation of a new government and early moves toward a dedicated ministry for disaster management.

The analysis draws on a mixed-methods approach and is the result of a project developed in collaboration between the White Helmets (WH) and The Hague Humanitarian Studies Centre, with support from Christian Aid. It combines climate modelling and global hydrological trends with interviews, community surveys, field observations, humanitarian situation reports, and a review of the regional and academic literature. This has been combined to draw regional findings for north-west Syria and cautiously situate them within the national picture. Particular attention is given to balance qualitative assessment with the practical value of probabilistic forecasts, including the Global Flood Awareness System and the ECMWF Ensemble Prediction System, for early warning in a data-scarce and conflict-affected setting (Tall et al., 2012; WMO, 2022).

The study found an overlapping and interlocking risk profile for NW Syria, characterised by combined socio-economic and environmental vulnerabilities, significant climatic hazards, and compounded by a changing climate and fragility caused by over 10 years of high-intensity conflict. Additionally, large numbers of internally-displaced people, regional variation in terms of formal governance reach, and a lack of funding versus humanitarian need (Daoudy et al., 2023). This implies a difficult landscape for integrated risk-reduction work, somewhat mitigated by a large and active civil society sector, developed informal governance structures, and a high level of technical knowledge within local NGOs.

We argue that while hazard signals are uneven and data remains scarce, exposure is rising with demographic pressures and the scope for anticipatory action is considerable. Investment in observational networks, systematic forecast evaluation, community-led programming, and inclusive governance could yield disproportionate benefits.

The Syrian case thus illustrates not only the acute challenges of multi-hazard risk in a fragile state, but also the opportunities for rethinking disaster governance in the MENA region more broadly.

7.2 Research Methodology

This chapter adopts a mixed-methods design anchored in NWS, where the WH operated at the time of this project, and where primary data collection, training, and observation were carried out. While the empirical centre of gravity is NWS, we read our findings against Syria-wide and eastern Mediterranean evidence. Where possible, we have extrapolated these findings, flagging when we do so and with what degree of confidence. The study engages a multi-hazard landscape (heat, drought, storms, earthquakes), but with a soft, non-exclusive emphasis on floods and flood-adjacent risks, co-defined with WH. That emphasis reflects both their recurrence in the study area and the practical feasibility of anticipatory action given present data constraints.

Methodologically, the work combined four streams. First, a structured review of peer-reviewed literature and grey documentation (2019–2024) providing context on hazards, climate trends, governance arrangements and operational practice. This included cluster plans and situation updates, agency reports, and internal WH materials. The review informed the design of instruments and offered comparisons for primary findings.

Second, we conducted community-level fieldwork via semi-structured survey. It was administered to 251 respondents across ten flood-affected communities: Khirbet al-Joz, Atma, Sarmada, Batabu/Batabo, Ahtimlat, Jindires, Qabasin, Shamarin, Al-Kafr and Armanaz (see Fig. 7.1). They were selected using historical flood impacts and access considerations, and include both formal and informal settlements, including Internally Displaced People camps. WH volunteers were trained in research ethics, bias minimisation and interviewing techniques in conflict-affected settings, before they administered the instrument using a KOBO Toolbox in male–female teams to respect local norms. All interviewees were over 18 years old and, as it is representative in such settings, the achieved sample skewed male (64%) and towards ages 30–49; the perspectives of older adults and women are therefore underrepresented in the survey alone.

Third, we undertook 15 one-hour key informant interviews with Syrian NGOs, INGOs, cluster leads (shelter/NFI, WASH, early recovery, camp management) and technical advisors, complemented by anonymised consultations with donors and academics Field observations from WH site walks, camp visits and a Community Based DRR "train-the-trainers" workshop (40 participants) added grounded insight into *vernacular* (Kelman, Mercer & Gaillard, 2020) risk-reduction practices.

Fourth, we placed these social findings within a forward-looking physical risk frame. Using global climate models (CMIP5/6) under moderate and severe scenarios (RCP4.5 and RCP8.5) (Meinshausen et al., 2022; Thomson et al., 2011), we examined projected temperature and rainfall trends for Syria, using Idlib as a reference for the northwest. Key indicators included temperature, precipitation, soil moisture, and river discharge. To complement these projections, we drew on global datasets such as WRI Aqueduct and PCR-GLOBWB for indicative flood exposure (Kuzma et al.,

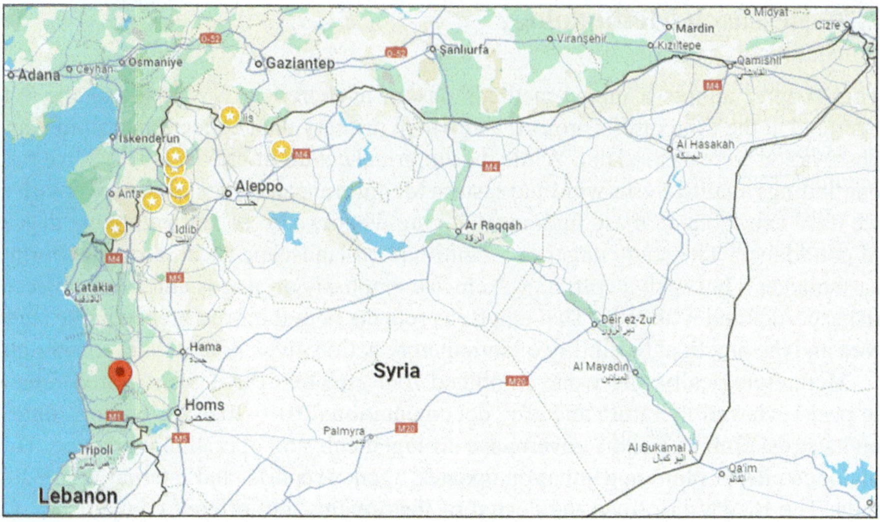

Fig. 7.1 Study area map showing the ten surveyed communities in northwestern Syria (Khirbet al-Joz, Atma, Sarmada, Batabu/Batabo, Ahtimlat, Jindires, Qabasin, Shamarin, Al-Kafr and Armanaz). Basemap: Google Maps, retrieved July 2025

2023).To connect long-term projections with recent experience, we reviewed EM-DAT flood records since 2019 and rainfall estimates from NASA's IMERG dataset (Delforge et al., 2023).

Analysis privileged coherence across streams, ensuring that different research methods complemented and balanced one another? The combined collected data was analysed using a thematic analysis. We summarised the model results by highlighting overall trends and the scale of possible change rather than exact figures. This reflects the uncertainties of semi-arid systems and the role of human interventions, such as irrigation and reservoirs as well as conflict dynamic, that affect how rainfall translates into river flow and exacerbate the general humanitarian crises of the country (OCHA, 2023).

Ethics and data protection followed GDPR[1] principles of data minimisation, secure storage and controlled access. All participants gave informed consent; interview recordings were made only with permission, and potentially sensitive details (e.g., exact camp locations) were generalised to avoid harm. The study adopted do-no-harm and conflict-sensitive practices throughout.

Two limitations shape interpretation. Data are conflict-conditioned and so sparse, whilst survey sampling delivered rich situational insight, it was not statistically representative. By utilising multiple forms of data for this study, we hope to build a composite and convergent 'picture' with each source complementing the others.

[1] General Data Protection Regulation (GDPR) (EU) 2016/679, is the European Union's legal framework for personal data protection and privacy.

Finally, although the analysis is centred on NWS, several elements (national climate signals, patterns of displacement and poverty, DRR/DRM coordination challenges, and the operational value of probabilistic forecasts) have wider salience. Where we generalise to other Syrian regions we do so cautiously, noting differences in hydrological systems, governance and exposure. We utilised triangulation and made uncertainty explicit, whilst translating probabilistic, multi-source evidence into choices that communities and humanitarian actors can act on now, and in the more connected governance environment that may follow Syria's political transition.

7.3 Disaster History and Risk Profile

7.3.1 Hazard History and Observed Change in a Levantine Context

Across MENA, disaster risk emerges at the intersection of hydro-climatic variability, rapid urbanisation and environmental degradation. In several settings, protracted conflict and displacement exacerbate disaster risk (Caso et al., 2023a, 2023b; Peters, 2022; Rosvold, 2023). In NW Syria, multiple hazards co-exist and interact: heat stress, episodic drought, pluvial and fluvial flooding, dust storms, landslides and, most recently, cascading earthquake impacts. At the same time whole of society vulnerability has risen with displacement, settlement densification, livelihood loss and damaged infrastructure (Daoudy et al., 2023; Schellens et al., 2024).

Although global catalogues undercount smaller or politically complex events, they nevertheless record recurrent floods alongside droughts and winter storms (Fig. 7.2); UNOCHA reporting similarly documents high-impact seasons in 2018/19, 2021 and 2023 centred on Idlib and Aleppo (Table 7.1). Our survey echoes this picture: three quarters of respondents reported flooding at least annually, a majority characterised onsets as rapid (often <24 h), and the most common local amplifiers were poor drainage (34%) and low-lying siting (30%). Our interviews verify these impressions, describing repeated rainwater and drainage overflows in dense camps and peri-urban basins, compounded by ad-hoc site selection for hosting IDPs.

Winter rainfall is commonly delivered by Cyprus Lows; summers are dominated by the Persian Trough, a regime historically associated with roughly four-month summers and winters in the Levant (Alpert et al., 2002). Observations indicate lengthening and intensifying summers, shifting hazard windows and antecedent soil-moisture states that condition flood generation due to climate change (Hochman et al., 2018a; OCHA, 2023). Twentieth-century rainfall declines across the Mediterranean have been linked to fewer rainy days under more persistent subtropical anticyclones; detection over the Levant is complicated by strong spatial variability and high inter-annual noise (Alpert et al., 2002; Trigo et al., 2000; Zappa et al., 2015; Zittis, 2018; Zittis et al., 2021). Warming is unequivocal: the intensity, duration and frequency of heatwaves have increased since the 1960s, and dangerous heat-stress days have

Fig. 7.2 A timeline of natural hazards leading to significant impacts in Syria (based on EM-DAT records)

Table 7.1 For EM-DAT flooding events shown in Fig.7.1, locations, onset date and number of people affected recorded in the database

Location	Onset date	Number affected
Al-Hasakah	29 October 2006	No data
Idlib and Al-Hasakah	30 March 2019	235,000
Idlib and Aleppo	14 January 2021	142,000
Idlib and Aleppo	19th March 2023	2,850
Idlib	17th January 2024	9,700

risen in parts of the region, with local humidity dynamics (including those linked to irrigation) compounding the signal (Krakauer et al., 2020; Kuglitsch et al., 2010).

Topography and human modification further shape risk in north-west Syria. The Orontes and its tributaries (the Afrin and Karasu/Saju) drain transboundary basins whose courses and storage have been altered by twentieth-century schemes (e.g. the Ghab drainage) and later dam development. Ephemeral channels, sediment dynamics and focused recharge complicate runoff; upstream reservoir operations, irrigation withdrawals and ad-hoc bank breaching to water livestock reshape hydrographs - factors which are entirely unrepresented in most hydrological and climate models. Additionally, both interviews and literature highlighted a strong cross-boundary governance issue for damming on the Orontes river, with (uncommunicated) spilling from the Afrin dam by Turkish authorities creating flash-flooding risk downstream, in Syrian territory, on land settled by IDPs.

Interviews and field observations also account for the impact that conflict dynamics and displacement have on the use of rivers and the land along their banks. Critically, displacement has placed large populations in exposed locations, often on marginal land with inadequate drainage, transforming heavy rain from a seasonal nuisance into a recurrent humanitarian emergency.

7.3.2 Projected Drivers: Climate, Hydrology, Inundation and Seasonality

Regional assessments point to consistent warming across Syria and a general tendency to drying (Fig. 7.3). In the north-west, the signal is clearest in the steady rise of hot nights, which erode nocturnal recovery and push up health risks, energy demand and pressure on outdoor labour and livestock (Zittis et al., 2022). Interviews echoed these concerns, noting fatigue during long heat spells, strained cooling, and growing difficulties, especially for camp residents.

Rainfall changes are more uneven, yet models agree that soils will be drier on average and that means runoff and river discharge are likely to fall (Jin et al., 2010; Mariotti et al., 2008; Zittis et al., 2022). A warmer atmosphere can still deliver short, intense downpours that overwhelm urban and camp drainage (Allen & Ingram, 2002). Local hydrology is also shaped by human action: reservoir operations for power and irrigation, ad hoc bank breaches to water livestock, unregulated boring for agricultural

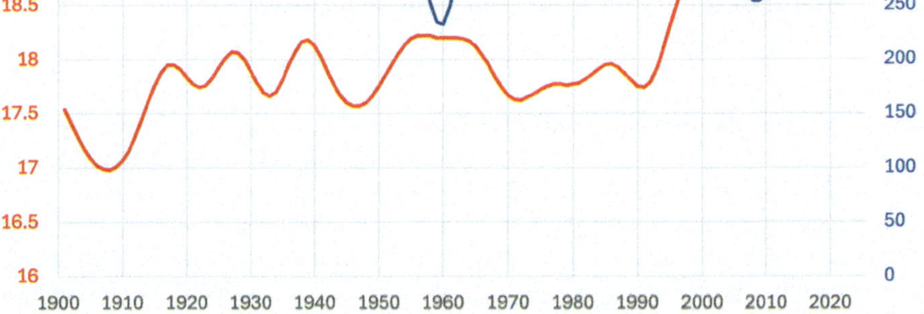

Fig. 7.3 Five-year running mean of annual average temperature and total precipitation over Syria. Based on CRU dataset (Harris et al., 2020), extracted from the World Bank Climate Data Portal

use and degraded infrastructure can sharpen or mute peaks in ways global models do not capture. For these reasons, the figures offered here should be read as directional, not as site-precise design values.

Indicative inundation modelling suggests that, for a one-in-fifty-year river event, the deepest water in the north-west tends to occur along the Orontes, at about 0.7 m on average grids, with shallower depths on the Afrin and Quweiq. Many surveyed localities lie outside the modelled 50-year floodplain, a situation that warrants caution given coarse hydraulics and absent local calibration. When these footprints are combined with demographic change, exposure grows: the number of people affected by one-in-five year floods rises from roughly 200,000 to about 300,000 per year by late century, even as the share of the national population exposed declines (Kuzma et al., 2023).

Heat is still more decisive, with much longer summers and shorter winters compressing the cool-season window for recharge and agriculture (Hochman et al., 2018b; Jin et al., 2010). Although the north-west sits inland of Syria's main marine exposures, expected sea-level rise on the eastern Mediterranean coast adds to national DRR burdens that will compete for attention and resources (Galassi & Spada, 2014; Satta et al., 2017). This, combined with decades-long agricultural policy orienting NW Syria's growing regions towards water-intensive crops for export including wheat, cotton, and olives, (Mohammed et al., 2020) and incomplete land reform, have led to inconsistencies within the region, as well as severely compromised food systems.

All of this sits on a thin empirical base. Gauges are sparse, many records are inaccessible, and event catalogues do not record sub-threshold but operationally important and numerous floods, which limits calibration and verification. Even so, ensemble rainfall and river-flow forecasts sometimes provide useful heads-up information several days in advance. The balance of evidence supports low-regret anticipatory action: clear but conservative trigger levels, procedures that can act on modest probabilities, reliable communication to at-risk communities, and steady investment in observational networks and impact logging so that forecast performance and local thresholds improve over time.

7.3.3 Anticipatory Potential and Conflict-Conditioned Vulnerability

The operational question is not only how hazards may change, but whether forecasts provide usable lead time and whether institutions can act under uncertainty. For several EM-DAT-listed floods since 2019, GloFAS indicated elevated probabilities of high flows on the Orontes, days in advance; in other cases, signals were weak or appeared only near onset. Forecasts sometimes pointed to heavy rainfall several days ahead, but probabilities were generally low and occasionally inconsistent as forecasts were updated in days approaching the event. In conflict-affected areas like northwest Syria, limited data and damaged monitoring systems make it difficult to verify or calibrate such predictions, or provide precise evaluation of forecast skill metrics which support development of anticipatory action (Coughlan de Perez et al., 2015; Jaime et al., 2022). In all cases, as substantiated by our survey and interviews undertaken, action was taken by WH and (i)NGO partners only after the flooding had begun—though annual 'winterisation' work to gravel roads, sandbag, rehabilitate or improve drainage systems, etc., did take place.

What matters most in these contexts is not statistical precision but the capacity to interpret and act on early signals. Forecasts become meaningful only when local actors can understand low-probability information, communicate it effectively, and coordinate responses. Community networks play a vital role: most people reported receiving flood warnings through social media or SMS, often reinforced by word of mouth and camp management structures. These informal systems show how local capacities can partly compensate for institutional gaps and technical uncertainty.

Conflict challenges resilience building (Rosvold, 2023), as much as it multiplies risk through multiple channels (Caso et al., 2023a, 2023b; Mena, 2020). Monitoring infrastructure is damaged or inaccessible, governance is fragmented, and human interventions (such as upstream releases, irrigation abstractions, or ad hoc bank breaching) decouple rainfall from discharge in ways global models cannot capture. Displacement concentrates vulnerable populations in exposed locations, often in tents or light shelters on poorly drained ground. Funding gaps skew priorities toward response rather than prevention, while earthquake damage to roads and basic services further erodes already fragile systems. Specifically, within NW Syria, continuous conflict also severely damaged drainage and dam systems, as well as creating space for uncontrolled boring for agricultural uses, all combining to create conditions of high vulnerability to climactic shocks.

Taken together, north-west Syria's risk profile is defined by robust warming, likely drying, declining soil moisture and reduced mean runoff and discharge, offset by the persistence of short-duration heavy rainfall capable of generating damaging pluvial floods. Riverine flood magnitudes may decline on average, yet absolute exposure is likely to rise with demographic trends and settlement expansion.

Heat emerges as a near-certain, intensifying risk. Forecast systems offer partial, probabilistic early warning; their value depends on institutions capable of acting on uncertainty, and on linking formal products to trusted local channels.

7.4 Evolution and Institutional Frameworks

Disaster governance in Syria has never been static; it has moved with the political economy of the state, the shocks of armed conflict, and most recently, the reconstitution of national institutions. Each phase has altered, not only who acts and in which region, but what counts as knowledge, how risks are prioritised, and which instruments are available to manage them.

Before 2011, Syria's arrangements were emblematic of state-centred governance in the region: authority was concentrated in Damascus, responsibilities distributed through civil defence and security structures, and a dominant logic of emergency response rather than prevention (Meininghaus, 2016). Meteorological and hydrological observation existed but was under-resourced, fragmented across ministries, and rarely translated into open data for planning. Community engagement was peripheral; international DRR agendas were acknowledged more than enacted, as many interviews ascertain.

In short, the system could mobilise command-and-control responses but struggled to reduce risk ex ante. Frequently, government policy prioritised other areas than risk reduction or disaster prevention, including agricultural development policy focused on export-led crop choices, hydrological projects focused on increasing arable land, and urban development focused on speed and political considerations.

Since 2011, and the pulling back of government reach, a range of actors filled the vacuum- local councils with uneven mandates, international agencies working through the Cluster system, and a dense ecology of Syrian NGOs and civil society organisations. In the north-west, the WH became the operational backbone for many hazards: uniquely positioned with personnel, heavy equipment and community legitimacy, they were repeatedly identified in interviews as the first resort for flood clearance, road opening and camp support.

Cluster coordination enabled a sense of order but also revealed systemic frictions: partners varied widely in capacity and geography; funding shortfalls pushed the system toward response over prevention; and responsibilities for technically adjacent tasks (e.g. camp drainage versus access roads) were split across clusters, slowing action during rapidly unfolding events. Survey results confirm that early warning and public communication were often mediated through social media and SMS and then amplified by camp managers and word of mouth; networks that worked because they were trusted and immediate, not because they were formally institutionalised.

This fragmented period also created space for innovation. With sparse instruments and inaccessible national data, humanitarian actors coupled global products (e.g. EPS rainfall probabilities, GloFAS discharge outlooks) with ground reports to build "good-enough" anticipatory practices: pre-positioning pumps, clearing culverts ahead of forecast rain, moving tents upslope before expected overland flow.

Several organisations piloted low-cost drainage rehabilitation and separated stormwater/sewage lines in formalised sites; others advanced "dignified shelter" models and phased upgrades of camp infrastructure. These initiatives were modest by engineering standards, but they mattered operationally and crucially were owned by communities who helped design and maintain them.

The political transition underway since late 2024 changes the opportunity structure. Early signals point to an effort to recentralise disaster functions, including discussion of a dedicated ministry and moves toward national standard-setting. Consolidation could bring badly needed coherence: clearer legal mandates, budget lines for observation and maintenance, integration with Sendai reporting, and the ability to negotiate transboundary water operations with Turkey and Lebanon in the Orontes–Afrin–Karasu system.

Comparative experience in MENA suggests that Syria's most credible path is neither romantic localism nor blunt re-centralisation, but a polycentric regime that mixes ~~couples~~ national authority with local capability. In practical terms, that implies: codifying roles for civil society (including all national NGO) within national SOPs for floods and heat, and with national/regional coordination centres; converting ad hoc winterisation and drainage fixes into funded, multi-year maintenance programmes; building an open, shared data system that blends national observations with community sensors and incident logs; and institutionalising anticipatory action through calibrated triggers that link global forecasts to pre-agreed measures and budgets. It also implies reworking Cluster–LNGO interfaces so that the speed and social reach of local actors is not lost as formal systems mature and ensuring that DRR investments in camps and informal settlements, where exposure is concentrated, are treated as public goods rather than temporary humanitarian add-ons.

Seen this way, the "institutional framework" is less of a hierarchy than a negotiated field. Formal ministries, standards and treaties matter; so do the informal practices that move sandbags before a storm, the WhatsApp groups that carry a warning faster than a bulletin, and the camp committees that keep culverts clear. The task ahead is to design a governance architecture that recognises those complementarities, aligning national strategy, municipal implementation, and community initiative into a coherent system capable of acting under uncertainty and constraint.

7.5 Challenges in Disaster Management

Disaster management in Syria is constrained by weaknesses that are institutional as much as they are technical. The hazard mix facing the north-west (heat, drought, floods, dust storms, landslides, and episodic seismic crises) is broadly typical of the Levant; what is distinctive is the intensity of co-occurrence and the way these hazards intersect with protracted displacement, poverty, and degraded infrastructure. Multi-hazard sequences are routine rather than exceptional: heavy winter rain follows earthquake damage to embankments; flash floods arrive in camps where drainage has failed after a summer of extreme heat; road closures delay access to health care and reinforce cycles of vulnerability (Djalante, 2012; Federici, 2022).

Institutional fragmentation remains the primary constraint. In the north-west, the WH effectively served as a key actor for flood clearance, road opening and site support, precisely because they combined equipment, reach and local legitimacy. Fragmentation brought agility, but also duplication, uneven coverage and blurred mandates, illustrated by the routine separation of responsibilities for camp drainage (Shelter/CCCM) and access roads (Early Recovery), which slows action during fast-moving events. Funding scarcity has compounded the problem: 2024/25 winterisation plans faced an aggregate shortfall approaching US$100 million, with, for example, CCCM receiving c. US$1.1 million against a requirement of US$10.3 million for critical site works, and Early Recovery receiving c. US$0.8 million of US$4.3 million requested for pre-emptive road and drainage rehabilitation. Unsurprisingly, the system defaults to response over prevention given these constraints.

Data poverty is a second systemic weakness. Hydrometric and meteorological observations are sparse, discontinuous and often inaccessible; GRDC archives yield only short, dated records, and domestic series are difficult to obtain. In this vacuum, global products stand in for national systems: EM-DAT captures only large events (four NW Syria floods since 2019, though our survey revealed many smaller flood events in this period), while many damaging, sub-threshold episodes are absent; IMERG fills gaps for rainfall, but its spatial and process limitations at catchment scale constrain local planning. The absence of verified discharge and impact data prevents bias correction and formal verification of forecast systems, undermining both risk assessment and the design of calibrated anticipatory triggers: compounded by dormant, or expired, data-sharing agreements between Turkiye and Syria for the Orontes river basin.

Hydrological complexity and human intervention further erode predictability. In semi-arid basins, ephemeral channels, focused recharge and sediment dynamics produce unstable rainfall–runoff relationships. In NW Syria upstream dam operations for irrigation and power, ad hoc bank breaches to water livestock, and decaying conveyance infrastructure alter hydrographs in ways that global models cannot capture. Projections therefore have directionality but limited precision on local flood magnitudes. The paradox is operationally important: even if median return-period flows decrease, short-duration, high-intensity rainfall over impermeable surfaces and clogged drains will continue to generate damaging pluvial floods in camps and towns (Allen & Ingram, 2002; Kuzma et al., 2023).

Forecast performance mirrors these constraints. GloFAS occasionally raised two- to five-year exceedance probabilities on the Orontes several days before EM-DAT onsets; in other cases, signals were weak or emerged only near the event. EPS rainfall forecasts sometimes lifted the probability of exceeding a 25 mm/day threshold at week-ahead lead times, but peak probabilities were modest (rarely >20%), volatile across updated forecasts, and accompanied by misses and false alarms. In practice, weak signals are valuable only when institutions are configured to act on low-to-moderate probabilities without exhausting scarce resources or eroding trust; precisely the calibration that requires long, local verification records that Syria lacks.

Exposure trends run counter to any decline in median hazard. Population growth and settlement expansion in marginal locations (low-lying ground, wadis, peri-urban basins) are increasing the absolute numbers at risk, with Aqueduct-based estimates suggesting exposure to five-year floods rising from roughly 200,000–300,000 people annually by 2080, even as the national share exposed falls.

Camps and informal settlements are structurally vulnerable: siting on poorly drained soils, improvised layouts, thin-section shelters, and constrained options for relocation due to tenure insecurity, livelihoods, and the politics of return. Administrative hurdles, like absent landowners, or fragmented permissions routinely delay drainage upgrades in informal sites. Heat risk amplifies these inequities: projected increases in tropical nights reduce nocturnal recovery and elevate health risks for older adults, children and outdoor workers—precisely those concentrated in settlements with the weakest access to cooling, water and care (Zittis et al., 2022).

A further weakness is the system's reliance on humanitarian actors to perform public functions without the mandate or financing horizons of the state. Anticipatory initiatives (e.g. Forecast-based Financing) demonstrate promise, but the combination of short funding cycles, risk-averse donor criteria, and prohibitions on "permanent" works constrains investment in durable risk reduction (e.g. covered stormwater networks, raised platforms, paved access) (Tall et al., 2012). In addition, investment is constrained by the current (2024–2025) global funding squeeze, including the suspension of USAID operations and cuts in funding across many major donors (Lambert, 2025; United Nations, 2025).

The survey indicates that early warnings and hazard information are most effectively transmitted through social media, SMS and word of mouth, often relayed by camp managers and the WH. These channels are fast and trusted but largely extra-institutional: coverage is uneven; rumours and misinformation can spread quickly; and two-way feedback, crucial for ground-truthing, depends on volunteer labour. Women and older adults were under-represented in the survey sample (64% male; low 65+ participation), signalling potential blind spots in both information reach and programme design unless deliberately corrected.

Finally, normative frameworks and transboundary realities complicate system design. Alignment with the Sendai Framework offers reputational and practical benefits (access to platforms, finance, and standards) but presupposes stability, data and multi-year investment. Yet, while usually seen as a disaster-related framework and less often considered in humanitarian action (Mena, 2025), the Sendai Framework can provide valuable support to crisis-related action in places like Syria (Patel et al., 2021). Moreover, risk-sensitive land-use planning, building codes and routine asset maintenance (core to Sendai) require legal authority and fiscal space that are only now being (re)constructed. At the same time, the Orontes–Afrin–Karasu system is embedded in cross-border governance: upstream operations and bilateral agreements shape downstream risk yet remain outside the control of local actors.

In sum, Syria's disaster-management challenges and weaknesses are not primarily a deficit of hazard knowledge; they are the product of fragmentation, thin data, hydrological and geopolitical contingencies, short funding horizons, and exposure patterns rooted in displacement, inequality, and poverty. Addressing them requires polycentric governance that protects community innovations while building national capability; an open, shared evidence base that blends observations, incident logs and satellite products; financing that moves from seasonal "winterisation" to multi-year maintenance and upgrade cycles; and anticipatory protocols calibrated to probabilistic signals. Without ~~Absent~~ these shifts, the system will continue to respond, sometimes heroically, but struggle to reduce risk at scale.

7.6 Research Gaps in DRR and DRM

If disaster management in Syria is to move beyond improvisation, it must be grounded in robust and context-specific knowledge. Yet the evidence base for risk reduction remains thin, fragmented, and unevenly distributed across hazards and regions at the moment of this research. The deficits are not only technical (missing data, uncalibrated models) but institutional (limited mechanisms for producing, sharing, and using knowledge). In a setting where hazards interact with displacement, poverty, and degraded infrastructure, these gaps cause systematically biased decisions toward short-term response and away from prevention, preparedness, and anticipatory action.

The most immediate deficit is observational. Over a decade of conflict has damaged or rendered inaccessible, hydro meteorological networks. Attempts to source river-gauge time series yielded fragmentary records; station rainfall, soil moisture, groundwater levels and urban drainage measurements are largely absent or siloed. In consequence, global systems (e.g., GloFAS, EPS) cannot be locally calibrated and estimates of their true forecast performance are fuzzy, hydrological projections remain "directional" rather than diagnostic, and flood mapping relies on coarse global hydraulics rather than city-scale conveyance and blockage dynamics. In semi-arid basins where ephemeral channels, sediment loads, informal embankment breaches and reservoir operations shape flows, the absence of in-situ data makes it impossible to establish stable rainfall–runoff relationships or to separate natural climate signals from human intervention.

Additionally, EM-DAT captures only a handful of large flood events in northwest Syria since 2019, but requires a certain threshold of event impact for inclusion. Yet the lived disaster landscape is dominated by sub-threshold episodes, such as camp inundations after cloudbursts, sewer backflow in dense urban fringes, road washouts that cut access to clinics. The level of information for other regions of the country is even less present, especially in the North-East. These events are visible in humanitarian sitreps, local logs, and remote sensing, yet there is no consolidated, geocoded registry of hazard, exposure, and impact with common metrics (people affected, shelter damage, service disruption, lead times, warnings issued). Without such a cataloguing, model verification and trigger calibration couched in high uncertainty and can only proceed with caution; lessons learned remain anecdotal; and vulnerability profiling cannot progress beyond generalities (Delforge et al., 2023).

Forecast evaluation is likewise under-developed. The study shows that ensemble river flow and rainfall forecasts sometimes provide actionable heads-up information, but signals are modest, inconsistent across runs, and unverified for lack of gauges. In anticipatory action terms, the missing elements are clear: ground truth data to support robust skill score estimation for relevant thresholds; false-alarm and miss rates by basin and season; and decision-analytic work that translates probabilistic signals into triggers that are credible, affordable, and acceptable to communities and responders. Until those pieces are in place, early action will remain contingent on expert judgement rather than institutionalised protocols.

The widening climate signal exposes a further gap: integration of physical projections with the social cartography of risk and vulnerability across the whole country. The survey and interviews illuminate where and how people are affected (poor drainage, low-lying sites, informal camps, fragile livelihoods), but systematic intersectional analysis is scarce. Urban hydrology and heat–health are especially under-researched for Idlib- and Aleppo-area settlements where impermeable surfaces, blocked culverts, and micro-topography dominate outcomes.

Institutional knowledge systems are also understandably thin. At present, there is no national platform that routinely combines humanitarian and municipal data, aligns standards, and returns actionable analytics to local actors. Universities and technical institutes are under-mobilised; data-sharing is episodic; and feedback loops from operations to policy are weak. As national structures are reconstituted, there is a risk that a centralised knowledge model will recentralise data without co-production, reproducing pre-2011 opacity. A credible system will require governance for open standards (metadata, geocoding, privacy), distributed data stewardship (including community sensors), and procedures that make analytics legible and useful to camp managers, municipalities, and frontline organisations, not only to a ministry.

Finally, research on compound and cascading risk remains embryonic. The Syrian record repeatedly shows multi-hazard chains, including earthquake damage to dams and drainage, followed by seasonal storms; heat-hardened soils amplifying pluvial runoff; conflict-driven settlement on marginal land that converts routine rainfall into disasters. Yet most analyses treat hazards in isolation, and very few couple hydrology with infrastructure fragility, public health, market access, or governance shocks in an integrated way. This is precisely the analytic move demanded by Sendai-aligned practice.

Taken together, these gaps define a pragmatic research agenda that is as much institutional as technical. In the near term, priorities are clear and feasible:

- Rebuild observations, bottom-up and top-down. Deploy low-cost rain and level sensors (including upstream of at-risk reaches), begin an open, quality-controlled national time series, and pair with selective restoration of reference-grade stations and urban drainage monitoring.
- Create community-based verification loops to observe and record the lived risk versus the data-led probability of a hazard occurring, tailored to power-minority population groups to ensure coverage by early-warning systems, risk reduction efforts, or (early) recovery programmes
- Establish a Syrian Disaster Impact Registry. A geocoded, privacy-aware event and impact catalogue that fuses humanitarian reports, remote sensing, and community submissions, using common indicators and enabling Sendai/DesInventar reporting.
- Verify forecasts and design triggers. Run rolling skill assessments for EPS/GloFAS against the registry; co-produce FbF-style trigger matrices with WH and camp managers, explicitly costed for low-probability signals.
- Study urban pluvial risk and heat–health. Map micro-catchments, culvert capacity, blockage points and shelter layouts in priority towns/camps; link heat metrics

(including tropical nights) to health utilisation and labour productivity to identify protective thresholds.
- Integrate social vulnerability at scale. Address sampling biases in existing surveys (gender, older adults, disability), and conduct longitudinal studies of displacement, tenure and livelihoods that can be joined to hazard layers.
- Build a co-production backbone. Create a standing DRR "risk lab" linking the prospective ministry, WH, municipal services, universities and CSOs, with open data standards, shared methods, and rapid translation from analysis to SOPs.

In Syria's transition, knowledge is not a technical afterthought but a component of institutional legitimacy and the precondition for moving from ad hoc response to anticipatory, risk-informed governance.

7.7 Recommendations and Future Strategies

Syria's transition creates an opening to convert humanitarian and non-governmental improvisation into a coherent, nationally owned and community-rooted risk management system. The north-west of the country offers some suggestions on ~~for~~ how to do this. Priority setting should be pragmatic, with 'good enough' solutions initially implemented as triage for huge needs. Infrastructure, shelters and operations should be stress-tested against the largest recent flood in the north-west (March 2019) so that preparedness meets a demonstrated impact scale. Heat must be treated as a first-order hazard: including adaptation of automated heat alerts based on forecast thresholds, pre-identifying cooling spaces in camps and dense settlements, and disseminating simple heat-health guidance through channels that residents already use, including camp management, WhatsApp and SMS.

Evidence systems require immediate attention, with a national network combining community rain gauges, low-cost river level loggers ~~on~~ in areas guided by historical flood events, and simple monitors at urban drainage pinch-points, and selectively rehabilitate reference meteorological stations. In parallel, a Syrian Disaster Impact Registry can collate geocoded impacts from humanitarian situation reports, municipal logs, community submissions and remote sensing under common metadata, open by default and suitable for routine planning as well as reporting to Sendai and DesInventar. These two assets will enable systematic verification, more reliable hotspot mapping and faster situational awareness.

Anticipatory action should be institutionalised. A small risk desk can monitor EPS rainfall and GloFAS river flow products, run rolling skill checks against the impact registry, and issue graded heads-up bulletins to municipalities, camp managers and the NGOs. Co-produce trigger matrices that link forecast probabilities to specific, pre-costed measures, for example pre-positioning pumps and gravel, temporary perimeter protection for the lowest camp rows, early evacuation of highly exposed clusters, or activating cooling centres ahead of forecast heatwaves. Triggers should reflect Syria's resource constraints and tolerance for false alarms, and should be recalibrated as evidence accrues.

Near-term risk reduction should focus on the failures that cause the greatest loss in frequent events, especially urban pluvial flooding. A rolling "winterisation plus" programme can clear and cover drains, widen undersized culverts, add swales and temporary berms, and, where feasible, separate stormwater from sewage in formalised sites while in informal camps, standardise surface conveyance that communities can maintain. These measures are low cost, rapidly deployable and yield visible benefits.

Governance should bridge national authority and local capability. If a ministry is created or reconstituted, its comparative advantage lies in orchestration rather than substitution. Three practical instruments can anchor this approach: a DRR steering group that convenes the ministry, the White Helmets, cluster leads, municipalities and selected civil society organisations to set seasonal priorities and share data; a risk lab that links government, universities and NGOs to maintain open data standards, verify forecasts and translate analysis into standard operating procedures and training; and a network of community committees embedded in camps and municipalities, connected to early warning channels and resourced for micro-interventions and local monitoring. All investments should be screened for conflict sensitivity and inclusion so that access to warnings, shelters and upgrades is equitable and transparent.

Implementation should proceed in steps. In the next 0 to 2 years, deploy the observation mesh, launch the impact registry, adopt heat protocols, undertake drainage fixes in priority sites, and pilot anticipatory frameworks on two basins and a cluster of high-risk camps. In years 2–5, scale triggers and standard operating procedures, integrate registry data into routine planning and cluster operations, expand dignified shelter programmes in flood-prone locations, and formalise the risk lab. Beyond 5 years, embed DRR standards in building codes and land-use plans, connect national systems to regional warning platforms, and consolidate a legal framework for data sharing and multi-level disaster risk management. Financing should blend climate adaptation windows for observations and heat preparedness, humanitarian lines for anticipatory actions and winterisation plus, development finance for drainage upgrades and dignified shelter, and research grants for the risk lab, with proposals framed around co-benefits for health, livelihoods and education continuity and aligned with Sendai and the Sustainable Development Goals.

7.8 Conclusion and Lessons for the Wider MENA Region

Syria's recent history shows how environmental stress, displacement, degraded infrastructure, and fragmented authority combine to amplify disaster risk.

North-west Syria concentrates these dynamics, with recurrent rain- and drainage-driven floods, chronic water stress, and rapidly intensifying heat. Global datasets and models illuminate broad directions of change, yet they miss many locally significant events and do not resolve the hydrological effects of damaged infrastructure, informal water management, or settlement expansion. The result is a risk landscape where mean flood magnitudes may decline, but damaging episodes persist, exposure grows in absolute terms, and heat becomes an unequivocal, first-order threat. In summary, this is a scenario in which climate change, disaster, and conflict compound risks and crises in Syria and across other areas of the MENA region (OCHA, 2023; Peters, 2022).

Against this backdrop, the operational lessons are clear. Community-based practice has delivered workable gains: targeted drainage improvements, relocation within camps to higher ground, use of multi-channel alerting, and routine consultation of probabilistic forecasts for heads-up decisions. Forecast skill for river flow and heavy rain remains uneven, but it is often sufficient to justify anticipatory actions when triggers are calibrated to low and moderate probabilities and coupled to simple, pre-agreed measures. Heat risk is more tractable still, since signals are stronger and actions are well known. These practice-level insights should anchor near-term investment choices.

The strategic opportunity lies in the current political transition. A national authority for disaster risk management can add coherence, align standards, and channel resources, provided it recognises and integrates the capabilities that emerged during state retrenchment. The immediate priorities are practical and measurable: rebuild an observation mesh for rainfall, river levels, and temperature; establish an open, national impact registry that captures sub-threshold events; institutionalise anticipatory action around explicit triggers; and mainstream heat preparedness across health, shelter, and urban services. Each of these steps depends on partnerships that connect ministries, municipalities, universities, civil society organisations, and international agencies through shared data, clear operating protocols, and international agendas (Djalante, 2012; Mena, 2025; Mena et al., 2022; Patel et al., 2021).

Finally, the Syrian case confirms that disaster governance is not only a technical project but also a political settlement about how knowledge is produced and used, and about who is authorised to act under uncertainty (Ahrens & Rudolph, 2006; Bankoff & Hilhorst, 2022; Blaikie et al., 1994). Credible evidence, transparent data systems, and inclusive coordination are as essential as pumps, culverts, and shelters. If the state can convene without crowding out, and if international support is aligned with a locally owned agenda, Syria can move from improvisation to a stable, learning-oriented system that reduces losses from frequent floods, protects people from escalating heat, and rebuilds the relationship between institutions and communities. This is the pathway to a risk governance architecture that serves in the near term while laying foundations for national resilience.

There are clear learning points that can be drawn from NW Syria, that can be applicable to the wider MENA region. The first of these is the creeping risk of extreme heat, with predicted maximum temperatures on the hottest summer days predicted to reach over 46 degrees Celsius by the middle of this century; based on a more conservative global warming rate than is currently predicted (Lelieveld et al., 2016). This hazard provides a different risk profile to fast-onset hazards such as flash flooding, and thus requires a wider range of anticipatory activities, from mapping to predict heat spots, pre-emptive health programming, to, over a longer period, reconsidering urban heat and effects on food systems.

The second applicable pull-out from the experience of NW Syria for the wider MENA region is the civil-society and non-governmental safety net that conflict, weak governance, and protracted crisis can activate. As demonstrated in NW Syria, a *vernacular* network of CSOs, working with international humanitarian organisations, has been able to provide a limited response capacity for fast-moving hazards (in the case of our study, pluvial flooding). We can draw from this that providing general support for CSOs and CSO networks to flourish is an important resilience-building technique. Utilising times of less intense conflict (or post-conflict) to build datasets, coordination mechanisms, and community-based risk reduction committees can help to mitigate a centralised state's limited capacities during crisis.

Thirdly, NW Syria demonstrates the importance of integrated multi-hazard planning and risk reduction efforts in contexts that have experienced highly centralised government with differing priorities in the last few decades, as a way to mitigate the risk of this centralised government losing its power and reach. The interaction of large infrastructure projects (for example the Afrin Dam, or Orontes basin drainage) completed in the last century, with a changing climate, can have wide-ranging effects for people's exposure and vulnerability to climatic shocks that are increasing in frequency. We would also advocate for a longer-term view around co-owned or cross-border governance of water and rivers, with developed mapping of downstream effects and agreed-upon protocols and information sharing.

It is only through the combination of multi-hazard preparedness, risk and vulnerability mapping, predictive models verified by lived experience, and a clear set of protocols and programming that integrates these learnings into pre-emptive and community-rooted action that wide scale risk reduction for all is likely to be achieved within the MENA region.

Acknowledgements We are deeply grateful to all those who shared their experiences and insights with us, and to the many humanitarian workers, community members, and local organisations who made this research possible. Their openness, commitment, and daily efforts to support people in need have shaped this work and continue to inspire it.

References

Abbara, A., Rayes, D., Ekzayez, A., Jabbour, S., Marzouk, M., Alnahhas, H., Basha, S., Katurji, Z., Sullivan, R., & Fouad, F. M. (2022). The health of internally displaced people in Syria: Are current systems fit for purpose? *Journal of Migration and Health, 6*, Article 100126. https://doi.org/10.1016/j.jmh.2022.100126

Ahrens, J., & Rudolph, P. M. (2006). The importance of governance in risk reduction and disaster management. *Journal of Contingencies and Crisis Management, 14*(4), 207–220. https://doi.org/10.1111/j.1468-5973.2006.00497.x

Ali, E., Cramer, W., Carnicer, J., Georgopoulou, E., Hilmi, N. J. M., Cozannet, G. L., & Lionello, P. (2022). Cross-Chapter Paper 4: Mediterranean Region. In *Climate Change 2022: Impacts,*

Adaptation and Vulnerability (1st ed.). Cambridge University Press. https://doi.org/10.1017/9781009325844
Allen, M. R., & Ingram, W. J. (2002). Constraints on future changes in climate and the hydrologic cycle. *Nature, 419*(6903), 224–232. https://doi.org/10.1038/nature01092
Alpert, P., Ben-Gai, T., Baharad, A., Benjamini, Y., Yekutieli, D., Colacino, M., Diodato, L., Ramis, C., Homar, V., Romero, R., Michaelides, S., & Manes, A. (2002). The paradoxical increase of Mediterranean extreme daily rainfall in spite of decrease in total values. *Geophysical Research Letters, 29*(11), 31–1–31–34. https://doi.org/10.1029/2001GL013554
Bankoff, G., & Hilhorst, D. (Eds.). (2022). *Why vulnerability still matters: The politics of disaster risk creation.* Routledge.
Blaikie, P. M., Cannon, T., Davis, I., & Wisner, B. (1994). *At risk: Natural hazards, people's vulnerability, and disasters.* Routledge.
Caso, N., Hilhorst, D., & Mena, R. (2023a). The contribution of armed conflict to vulnerability to disaster: Empirical evidence from 1989 to 2018. *International Journal of Disaster Risk Reduction, 95,* Article 103881. https://doi.org/10.1016/j.ijdrr.2023.103881
Caso, N., Hilhorst, D., Mena, R., & Papyrakis, E. (2023). Does disaster contribute to armed conflict? A quantitative analysis of disaster–conflict co-occurrence between 1990 and 2017. *International Journal of Development Issues, 23*(1). https://doi.org/10.1108/IJDI-01-2023-0015
Clark-Ginsberg, A., Blake, J. S., & Patel, K. V. (n.d.). Hybrid governance and disaster management in Freetown, Sierra Leone, Monrovia, Liberia, and Dar es Salaam, Tanzania. *Disasters, n/a*(n/a). https://doi.org/10.1111/disa.12466
Coughlan de Perez, E., van den Hurk, B., van Aalst, M. K., Jongman, B., Klose, T., & Suarez, P. (2015). Forecast-based financing: An approach for catalyzing humanitarian action based on extreme weather and climate forecasts. *Natural Hazards and Earth System Sciences, 15*(4), 895–904. https://doi.org/10.5194/nhess-15-895-2015
Daoudy, M., Duried, M., & Mayer-Rich, E. (2023). *Northwestern Syria in the Time of Cholera, Earthquakes, and Environmental Degradation* (Working Paper). Carnegie Endowment for International Peace. https://carnegieendowment.org/research/2023/07/northwestern-syria-in-the-time-of-cholera-earthquakes-and-environmental-degradation?lang=en
Delforge, D., Wathelet, V., Below, R., Sofia, C. L., Tonnelier, M., Loenhout, J. van, & Speybroeck, N. (2023). *EM-DAT: The Emergency Events Database.* Research Square. https://doi.org/10.21203/rs.3.rs-3807553/v1
Djalante, R. (2012). Review Article: Adaptive governance and resilience: The role of multi-stakeholder platforms in disaster risk reduction. *Natural Hazards and Earth System Sciences, 12*(9), 2923–2942. https://doi.org/10.5194/nhess-12-2923-2012
Federici, F. M. (2022). Translating hazards: Multilingual concerns in risk and emergency communication. *The Translator*. https://doi.org/10.1080/13556509.2023.2203998
Galassi, G., & Spada, G. (2014). Sea-level rise in the Mediterranean Sea by 2050: Roles of terrestrial ice melt, steric effects and glacial isostatic adjustment. *Global and Planetary Change, 123,* 55–66. https://doi.org/10.1016/j.gloplacha.2014.10.007
Harris, I., Osborn, T. J., Jones, P., & Lister, D. (2020). Version 4 of the CRU TS Monthly High-Resolution Gridded Multivariate Climate Dataset. *Scientific Data, 7,* 109. https://doi.org/10.1038/s41597-020-0453-3
Hochman, A., Harpaz, T., Saaroni, H., & Alpert, P. (2018a). Synoptic classification in 21st century CMIP5 predictions over the Eastern Mediterranean with focus on cyclones. *International Journal of Climatology, 38*(3), 1476–1483. https://doi.org/10.1002/joc.5260
Hochman, A., Harpaz, T., Saaroni, H., & Alpert, P. (2018b). The seasons' length in 21st century CMIP5 projections over the eastern Mediterranean. *International Journal of Climatology, 38*(6), 2627–2637. https://doi.org/10.1002/joc.5448
Jaime, C., Coughlan de Perez, E., van Aalst, M., Raju, E., & Sheaffer, A. (2022). What was known: Weather forecast availability and communication in conflict-affected countries. *International Journal of Disaster Risk Reduction,* 103421. https://doi.org/10.1016/j.ijdrr.2022.103421

Jin, F., Kitoh, A., & Alpert, P. (2010). Water cycle changes over the Mediterranean: A comparison study of a super-high-resolution global model with CMIP3. *Philosophical Transactions of the Royal Society a: Mathematical, Physical and Engineering Sciences, 368*(1931), 5137–5149. https://doi.org/10.1098/rsta.2010.0204

Kelman, I., Mercer, J., & Gaillard, J. (2012). Indigenous knowledge and disaster risk reduction. *Geography, 97*(1), 12–21. https://doi.org/10.1080/00167487.2012.12094332

Krakauer, N. Y., Cook, B. I., & Puma, M. J. (2020). Effect of irrigation on humid heat extremes. *Environmental Research Letters, 15*(9), Article 094010. https://doi.org/10.1088/1748-9326/ab9ecf

Kuglitsch, F. G., Toreti, A., Xoplaki, E., Della-Marta, P. M., Zerefos, C. S., Türkeş, M., & Luterbacher, J. (2010). Heat wave changes in the eastern Mediterranean since 1960. *Geophysical Research Letters, 37*(4). https://doi.org/10.1029/2009GL041841

Kuzma, S., Bierkens, M., Lakshman, S., Luo, T., Saccoccia, L., Sutanudjaja, E., & Van Beek, R. (2023). *Aqueduct 4.0: Updated decision-relevant global water risk indicators (Technical Note).* World Resources Institute. https://files.wri.org/d8/s3fs-public/2023-08/aqueduct-40-technical-note.pdf?VersionId=G_TxTR2LAnlgXGzy7xtdUP_5lmkXJY7d

Lambert, J. (2025, July). Study: 14 million lives could be lost due to Trump aid cuts. *NPR.* https://www.npr.org/sections/goats-and-soda/2025/07/01/nx-s1-5452513/trump-usaid-foreign-aid-deaths

Lelieveld, J., Proestos, Y., Hadjinicolaou, P., et al. (2016). Strongly increasing heat extremes in the Middle East and North Africa (MENA) in the 21st century. *Climatic Change, 137*, 245–260. https://doi.org/10.1007/s10584-016-1665-6

Mariotti, A., Zeng, N., Yoon, J.-H., Artale, V., Navarra, A., Alpert, P., & Li, L. Z. X. (2008). Mediterranean water cycle changes: Transition to drier 21st century conditions in observations and CMIP3 simulations. *Environmental Research Letters, 3*(4), Article 044001. https://doi.org/10.1088/1748-9326/3/4/044001

Meininghaus, E. (2016). Humanitarianism in intra-state conflict: Aid inequality and local governance in government- and opposition-controlled areas in the Syrian war. *Third World Quarterly*, 1–29. https://doi.org/10.1080/01436597.2016.1159509

Meinshausen, M., Lewis, J., McGlade, C., Gütschow, J., Nicholls, Z., Burdon, R., Cozzi, L., & Hackmann, B. (2022). Realization of Paris Agreement pledges may limit warming just below 2 °C. *Nature, 604*(7905), 304–309. https://doi.org/10.1038/s41586-022-04553-z

Mena, R. (2025). Humanitarianism and the Sendai framework: A 10-year review of converging and diverging paths. *International Journal of Disaster Risk Science.* https://doi.org/10.1007/s13753-024-00595-1

Mena, R., Brown, S., Peters, L. E. R., Kelman, I., & Ji, H. (2022). Connecting disasters and climate change to the humanitarian-development-peace nexus. *Journal of Peacebuilding & Development, 17*(3), 324–340. https://doi.org/10.1177/15423166221129633

Mena, R. (2020). The (im)possibilities of disaster risk reduction in the context of high-intensity conflict: The case of Afghanistan. *Submitted to Journal Natural Hazards.*

Mohammed, S. A., Alkerdi, A., Nagy, J., & Harsányi, E. (2020). Syrian crisis repercussions on the agricultural sector: Case study of wheat, cotton and olives. *Regional Science Policy & Practice, 12*(3), 519–537. https://doi.org/10.1111/rsp3.12222

OCHA. (2023, November). Syria: Facing the dual challenge of climate change and conflict. *United Nations Office for the Coordination of Humanitarian Affairs.* https://www.unocha.org/news/syria-facing-dual-challenge-climate-change-and-conflict

Patel, S. S., McCaul, B., Cáceres, G., Peters, L. E. R., Patel, R. B., & Clark-Ginsberg, A. (2021). Delivering the promise of Sendai framework for disaster risk reduction in fragile and conflict-affected contexts (FCAC): A case study of the NGO GOAL's response to the Syria conflict. *Progress in Disaster Science, 10*, Article 100172. https://doi.org/10.1016/j.pdisas.2021.100172

Peters, L. E. R. (2022). Disasters as ambivalent multipliers: Influencing the pathways from disaster to conflict risk and peace potential through disaster risk reduction. *Journal of Peacebuilding and Development*, *17*(2). Scopus. https://doi.org/10.1177/15423166221081516

Rosvold, E. L. (2023). Disaster resilience in conflict-affected areas: A review of how armed conflicts impact disaster resilience. *Current Opinion in Environmental Sustainability, 65*, Article 101381. https://doi.org/10.1016/j.cosust.2023.101381

Satta, A., Puddu, M., Venturini, S., & Giupponi, C. (2017). Assessment of coastal risks to climate change related impacts at the regional scale: The case of the Mediterranean region. *International Journal of Disaster Risk Reduction, 24*, 284–296. https://doi.org/10.1016/j.ijdrr.2017.06.018

Schellens, M., Ozdemir, A., Greenwood, M., Zwijnenburg, W., & Tammar, A. (2024). *Thirst for Peace: War and Water Security Challenges along the Orontes River in Syria.* PAX. https://paxforpeace.nl/wp-content/uploads/sites/2/2024/11/PAX_report_Thirst-for-Peace_2024.pdf

Tall, A., Mason, S. J., van Aalst, M., Suarez, P., Ait-Chellouche, Y., Diallo, A. A., & Braman, L. (2012). Using seasonal climate forecasts to guide disaster management: The red cross experience during the 2008 West Africa Floods. *International Journal of Geophysics, 2012*(1), Article 986016. https://doi.org/10.1155/2012/986016

Thomson, A. M., Calvin, K. V., Smith, S. J., Kyle, G. P., Volke, A., Patel, P., Delgado-Arias, S., Bond-Lamberty, B., Wise, M. A., Clarke, L. E., & Edmonds, J. A. (2011). RCP4.5: A pathway for stabilization of radiative forcing by 2100. *Climatic Change, 109*(1), 77. https://doi.org/10.1007/s10584-011-0151-4

Trigo, I. F., Davies, T. D., & Bigg, G. R. (2000). Decline in Mediterranean rainfall caused by weakening of Mediterranean cyclones. *Geophysical Research Letters, 27*(18), 2913–2916. https://doi.org/10.1029/2000GL011526

United Nations. (2025, September). Humanitarian aid: The most vulnerable already severely impacted by budget cuts. *United Nations Western Europe.* https://unric.org/en/humanitarian-aid-the-most-vulnerable-already-severely-impacted-by-budget-cuts/

WMO. (2022). *Early Warnings For All: Executive Action Plan 2023–2027* (p. 56). World Meteorological Organisation. https://library.wmo.int/idurl/4/58209

Zappa, G., Hawcroft, M. K., Shaffrey, L., Black, E., & Brayshaw, D. J. (2015). Extratropical cyclones and the projected decline of winter Mediterranean precipitation in the CMIP5 models. *Climate Dynamics, 45*(7), 1727–1738. https://doi.org/10.1007/s00382-014-2426-8

Zittis, G. (2018). Observed rainfall trends and precipitation uncertainty in the vicinity of the Mediterranean, Middle East and North Africa. *Theoretical and Applied Climatology, 134*(3), 1207–1230. https://doi.org/10.1007/s00704-017-2333-0

Zittis, G., Bruggeman, A., & Lelieveld, J. (2021). Revisiting future extreme precipitation trends in the Mediterranean. *Weather and Climate Extremes, 34*, Article 100380. https://doi.org/10.1016/j.wace.2021.100380

Zittis, G., Almazroui, M., Alpert, P., Ciais, P., Cramer, W., Dahdal, Y., Fnais, M., Francis, D., Hadjinicolaou, P., Howari, F., Jrrar, A., Kaskaoutis, D. G., Kulmala, M., Lazoglou, G., Mihalopoulos, N., Lin, X., Rudich, Y., Sciare, J., Stenchikov, G., Lelieveld, J. (2022). Climate change and weather extremes in the eastern Mediterranean and Middle East. *Reviews of Geophysics, 60*(3), e2021RG000762. https://doi.org/10.1029/2021RG000762

Chapter 8
Strengthening Disaster Management in the United Arab Emirates: A Review of Organizational and Institutional Capacity Development

Glenn Fernandez, Hamad Abushehab, Ghalya Almansoori, Mariam Almansoori, Shamma Almheiri, Raghad Alshamsi, Mahra Binhammad, and Edris Alam

Abstract This chapter examines the evolution of disaster management (DM) in the United Arab Emirates (UAE), with particular emphasis on the development of organizational and institutional capacities that underpin national resilience. Over the past decade, the UAE has established a comprehensive and well-coordinated DM system through the formulation of national policies, enactment of legal frameworks, creation of specialized institutions, and implementation of structured education and professional training programs. At the core of this system is the National Emergency, Crisis and Disaster Management Authority (NCEMA), which leads policy formulation, interagency coordination, and workforce capacity development across all seven emirates. The chapter explores how NCEMA's centralized governance model balances federal oversight with emirate-level adaptability, supported by key institutions such as the Abu Dhabi Emergency, Crisis and Disaster Management Centre (ADCMC) and Dubai's Supreme Committee of Emergency, Crisis, and Disaster Management (SCCDM). It also highlights the complementary roles of academic institutions, particularly Rabdan Academy and the American University in the Emirates, as well as non-governmental and volunteer-based organizations, including the Emirates Red Crescent, Jaheziya Program, and Sanid Program, in cultivating technical, managerial, and community capacities. Using official publications, case examples, and institutional reports, the chapter maps the UAE's multi-layered system of emergency governance and human resource development. It shows how the country's integration of policy, education, professional associations, and community participation contributes to disaster preparedness and effective response. The UAE's experience demonstrates the value of a centralized yet adaptive model for strengthening disaster management and provides actionable insights for other nations seeking

---

G. Fernandez (✉) · H. Abushehab · G. Almansoori · M. Almansoori · S. Almheiri · R. Alshamsi · M. Binhammad · E. Alam
Faculty of Resilience, Rabdan Academy, Abu Dhabi, United Arab Emirates
e-mail: gfernandez@ra.ac.ae

E. Alam (ed.), *Disaster, Crisis and Emergency Management in the Middle East & North Africa*, https://doi.org/10.1007/978-3-032-24436-9_8

to professionalize and institutionalize their disaster management capabilities within evolving risk environments.

Keywords Disaster management · Institutional capacity · Human resource development · United Arab Emirates · National Emergency Crisis and Disaster Management Authority (NCEMA)

8.1 Introduction

Effective disaster management (DM) depends not only on robust infrastructure and sound planning but also on the continuous development of skilled human resources capable of ensuring preparedness, response, and recovery during crises. Recognizing this, the United Arab Emirates (UAE) has prioritized the establishment of a comprehensive and adaptive emergency, crisis, and disaster management system (hereinafter referred to as disaster management for brevity) underpinned by strong institutional, legislative, and educational foundations.

Over the past decade, the UAE has significantly advanced its DM capacity through the enactment of national policies, creation of specialized institutions, and implementation of structured education and professional training programs. These initiatives reflect the country's commitment to developing a competent and agile workforce that can address the increasing complexity and frequency of disasters and emergencies. However, the description and analysis of human resource development in disaster management in the UAE remain fragmented across disparate sources. This chapter consolidates and critically examines these efforts within the country's disaster governance system.

This chapter examines how the UAE has approached the development of organizational and institutional capacity in DM, with particular emphasis on human resource development strategies. It analyzes key national policies, organizational mandates, and training systems, while also considering the complementary roles of governmental, non-governmental, and professional organizations in strengthening DM capabilities. By situating the UAE's experience within the broader context of disaster risk governance, the chapter highlights how the country's investments in human and institutional capacity contribute to building resilience and ensuring effective coordination across all levels of government.

8.2 Disaster Risk Profile and Recent Events

The UAE faces a diverse array of hazards, including biological, technological, and hydrometeorological threats. Its role as an international hub for tourism and trade increases exposure to biological risks such as pandemics, which have the potential to negatively affect the UAE's economy, including the travel sector and the labor market (Alalawi et al., 2022). Climate change has intensified hydrometeorological hazards, most notably extreme rainfall and flooding, which are exacerbated by limited drainage capacity in urban areas (Ramadan et al., 2025). In April 2024, the UAE experienced its heaviest rainfall leading to unprecedented flooding that overwhelmed the capacity of the drainage systems (Almheiri et al., 2024). The widespread flooding caused transport disruptions and significant damage to properties, particularly in Dubai.

The UAE's arid climate also exposes it to frequent heatwaves, which strain energy systems and increase the risk of power outages. Dust storms are common, impairing visibility, degrading air quality, and reducing economic productivity (Middleton et al., 2021). Flash floods present another recurring threat, particularly during and after heavy rainstorms (Yagoub & Al Yammahi, 2022). The July 2022 Fujairah flash flood, for instance, led to the evacuation of 4225 individuals and transport disruptions that halted economic activity (Terry et al., 2023).

According to the World Risk Report 2025, the UAE ranks 78th out of 193 countries in terms of having the highest disaster risk, with high score in exposure of its geographic location and population to natural hazards, such as floods, but very low score in vulnerability (BEH & IFHV, 2025). As stated in the Global Risk Report 2025, extreme weather events (floods, heatwaves, etc.) and cyber insecurity are two of the top 5 risks of the UAE (WEF, 2025).

With a population of approximately ten million, the UAE's rapid urbanization, especially in Dubai, Abu Dhabi, and Sharjah, has heightened exposure to climate-related and technological hazards. Urban flooding, sandstorms, cyclonic disturbances, and extreme heat events increasingly test the resilience of the nation's infrastructure and disaster management systems.

These evolving risks underscore the importance of developing robust human resource and institutional capacities to reduce vulnerability and enhance community resilience. A clear understanding of the UAE's hazard landscape remains fundamental to building an adaptive, evidence-informed disaster management system capable of addressing both existing and emerging threats.

8.3 National Policies and Frameworks

National disaster management policies provide the foundation for developing a capable workforce and ensuring coordinated disaster management. The UAE has established a comprehensive and standardized institutional framework led by the National Emergency, Crisis and Disaster Management Authority (NCEMA). Operating under the Supreme Council for National Security, NCEMA directs national policy formulation, emergency planning, interagency coordination, and capacity development across all emirates.

8.3.1 Legal and Institutional Framework

The UAE's disaster management system is anchored in Federal Decree-Law No. 2 of 2011, which defines the responsibilities of government entities in emergency, crisis, and disaster management (President of the UAE, 2011). Article 13 sets the foundational pre-event obligations (prevention and preparedness) of concerned parties, while Article 19 requires public and private sector organizations to maintain business continuity plans to ensure the uninterrupted delivery of critical services during emergencies. Complementing this, Federal Law No. 23 of 2006 concerning Civil Defense reinforces the need for integrated response mechanisms (President of the UAE, 2006), and Federal Decree-Law No. 6 of 2009 mandates radiological emergency training within the nuclear sector (President of the UAE, 2009).

The National Response Framework (NRF), developed by NCEMA (2013), adopts an all-hazards approach that unifies the efforts of ministries, local authorities, and private organizations to manage both natural and man-made risks. The NRF is used to coordinate emergency response from at the national, federal, and local levels (Alam, 2022a), as shown in Fig. 8.1. At the federal level, the lead federal agency (LFA) depends on the hazard (NCEMA, 2013). For example, for hydrometeorological hazards (storms and floods), geological hazards (earthquakes), and land transportation accidents, the designated LFA is the Ministry of Interior. For contagious diseases and epidemics affecting humans, it is the Ministry of Health and Prevention (MOHAP). For diseases affecting animals and plants, it is the Ministry of Climate Change and Environment (MOCCAE), formerly known as the Ministry of Environment and Water. Structured training and certification initiatives, along with simulation exercises and workshops, promote interagency coordination and ensure consistent competency standards across the nation.

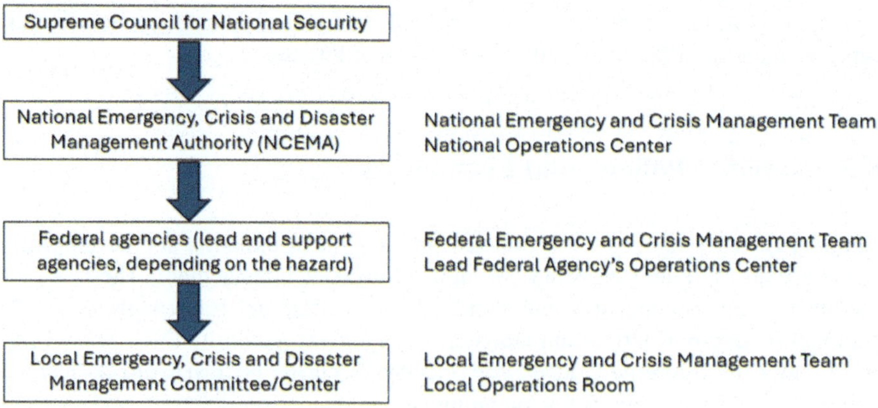

Fig. 8.1 UAE's national response system *Source* NCEMA, 2013

8.3.2 Complementary Emirate-Level Frameworks

While NCEMA oversees the centralized framework, several emirate-level laws ensure localized implementation and coordination. Law No. 22 of 2019 established the Emergency, Crisis, and Disaster Management Centre in Abu Dhabi (Ruler of Abu Dhabi, 2019), and Decree No. 4 of 2021 created the Supreme Committee of Crisis and Disaster Management (SCCDM) in Dubai (Ruler of Dubai, 2021). These bodies are aligned with federal priorities under NCEMA supervision. Furthermore, Federal Decree-Law No. 35 of 2024 reorganized the General Command of Civil Defence, reinforcing its alignment with national emergency planning objectives (President of the UAE, 2024).

8.3.3 NCEMA 7000:2021 and Standardization

A key policy instrument, NCEMA 7000:2021, strengthens disaster management by standardizing business continuity practices across sectors and enhancing organizational resilience (NCEMA, 2021). It complements the national disaster management framework overseen by NCEMA and promotes alignment among sectors such as health, environment, defense, and others under a unified business continuity approach. Section 8.6 of the standard requires organizations in the UAE to undertake planning activities, including risk assessments, business impact analyses, and setting continuity objectives, to enhance resilience within the national disaster management framework.

Although the UAE employs a centralized governance model, its disaster management system allows flexibility at the emirate level to tailor implementation to local needs and challenges. This balance between national coordination and local adaptation, reflected in the application of NCEMA 7000:2021, has strengthened the country's disaster management capacity by ensuring consistent national standards while accommodating regional differences in hazards and institutional arrangements.

Through its integrated legal, policy, and institutional framework, the UAE ensures that workforce development, education, and operational preparedness are systematically aligned with national resilience objectives, thereby strengthening the country's capacity to anticipate, withstand, and recover from disruptions.

8.3.4 Human Resource Development within Disaster Management Policies

Human resource (HR) development is a central pillar of the UAE's approach to disaster management. According to Article 13 of the Federal Decree-Law No. 2 of 2011, federal and local government entities concerned with disaster management should prepare programs and policies related to trainings and exercises for all individuals and groups concerned with the implementation of preparedness, response, and prevention measures (President of the UAE, 2011).

At the local level, in Abu Dhabi, Article 5 of Law No. 22 of 2019 states that the culture of disaster preparedness in the local entities can be enhanced through the development of training plans (Ruler of Abu Dhabi, 2019). In Dubai, Article 5 of Decree No. 4 of 2021 states that there should be commitment to implementing training based on disaster scenarios which may compromise public health and safety (Ruler of Dubai, 2021). This has been reinforced in Article 5 of Decree No. 48 of 2024, which states that training should be provided to representatives of concerned entities responsible for addressing disasters (Ruler of Dubai, 2024).

The UAE's framework for HR development in disaster management is operationalized primarily through the NCEMA. NCEMA oversees the Training Hub, which provides programs in risk management, crisis communication, disaster preparedness, and business continuity management. These programs are aligned with international standards and delivered in partnership with academic institutions and professional bodies to ensure alignment between education, practice, and policy.

Practical skill enhancement is reinforced through large-scale national simulation exercises such as "Shield of the Nation" and "UAE Barakah Exercise," with the latter centered on the Barakah Nuclear Energy Plant in Abu Dhabi. These exercises bring together the military, civil defense, health services, and other critical sectors under NCEMA's coordination to test response readiness and interagency coordination. Performance evaluations from these exercises are systematically analyzed to inform subsequent training curricula, policy updates, and workforce development strategies.

The UAE's approach demonstrates a mature and integrated HR development model, combining regulatory mandates, competency-based training, and cross-sectoral collaboration. By institutionalizing continuous learning and rigorous evidence-based evaluation, the UAE strengthens both individual competencies and organizational adaptability, two prerequisites for sustaining national resilience amid an increasingly complex risk environment.

8.4 Main Disaster Management Agencies and their Roles

8.4.1 National Leadership and Coordination

The UAE operates a centralized disaster management system led by NCEMA, functioning under the Supreme Council for National Security. NCEMA formulates national policies, coordinates disaster response, and maintains the National Response Framework, an all-hazards model supported by sector-specific policies and contingency plans, such as the National Policy for Combating Health Risks in the United Arab Emirates (MOHAP, 2025), Contingency Planning for Air Traffic Services (GCAA, 2014), Marine Pollution Response Plan (RAK Ports, 2024), and

Regulation for Emergency Preparedness and Response for Nuclear Facilities (FANR, 2021). According to the fourth planning principle of the National Response Framework, "response agencies should strengthen their capacity to ensure an effective response to all types of emergencies and crises, in order to mitigate their impact and ensure a fast recovery" (NCEMA, 2013).

According to the National Response Framework, one of NCEMA's responsibilities is to "supervise the development of response capabilities by proposing, coordinating, and updating programs involving stakeholders at both local and national levels" (NCEMA, 2013). NCEMA's mandate includes developing national standards, conducting training and exercises, and integrating efforts across ministries, agencies, and local governments. It coordinates closely with key ministries, such as the Ministry of Interior, Ministry of Health and Prevention, Ministry of Energy and Infrastructure, Ministry of Climate Change and Environment, and others, depending on the nature of the incident.

NCEMA has a Planning and Preparedness Department which organizes joint training sessions in collaboration with strategic partners to improve readiness to successfully deal with potential dangers and threats (Alzaabi, 2021). The department plays a central role in building workforce capabilities by designing a specialized training strategy informed by a training needs analysis of both the Authority's staff and its local and federal partners. This strategy is implemented through specialized training programs in disaster management, accredited by national and international institutions.

8.4.2 Emirate-Level Disaster Management

While NCEMA provides centralized leadership, each emirate maintains its own disaster management body aligned with national directives, allowing localized adaptation to context-specific risks.

8.4.3 Abu Dhabi

The Emergency, Crisis, and Disaster Management Centre—Abu Dhabi (ADCMC), established under Abu Dhabi Law No. 22 of 2019, oversees local preparedness, training, and coordination of government entities in the Emirate of Abu Dhabi (Government of Abu Dhabi, 2019). Some of its capacity-building activities include signing a Memorandum of Understanding (MOU) with the Mohamed bin Zayed University of Artificial Intelligence (MBZUAI) to strengthen cooperation and exchange expertise in technological innovation and artificial intelligence to strengthen the crisis management ecosystem and enhance data-driven decision-making (WAM, 2025a). ADCMC also organized a comprehensive field exercise at the Barakah Nuclear Energy Plant to address any potential radiological and nuclear emergencies related to its operations (Abu Dhabi Media Office, 2025). The exercise simulated a highly realistic and complex scenario and involved 22 local and federal entities, supported by seven field teams and 20 operation centers, highlighting shared commitment to enhancing the crisis response system.

8.4.4 Dubai

The Emirate of Dubai's Supreme Committee of Crisis and Disaster Management (SCCDM) was officially established by Decree No. 4 of 2021 (Ruler of Dubai, 2021). The decree states that the SCCDM replaces the former "Crisis and Disaster Management Team" and becomes the highest reference authority in the Emirate responsible for risks, emergencies, crises, and disasters. One of the duties of SCCDM is to coordinate and cooperate with NCEMA on preventative and preparedness measures. Another of its duties is to identify the needs of the Emirate for resources, including human resources, for managing disasters. SCCDM also supervises community awareness plans and programs related to disaster management. In 2024, the Dubai Resilience Centre (DRC) was established in 2024 to enhance the Emirate's capabilities in preventing, preparing for, responding to, and recovering from disasters (Ruler of Dubai, 2024). DRC has the duty to maintain and regularly update a register of representatives, including official spokespersons, coordinators, negotiators, and other officials, from the entities responsible for addressing disasters and to provide training to those representatives on addressing disasters. DRC is also in charge of preparing and qualifying Emirati cadres specialized in areas related to the work of the DRC, as well as forming volunteer teams for addressing disasters.

8.4.5 Other Emirates

The other five emirates of the UAE (Ajman, Fujairah, Ras Al Khaimah, Sharjah, Umm Al and Quwain) each have their own local emergency, crisis, and disaster management team. For example, the local disaster management team in Sharjah is led by the Chief of Sharjah Police and includes members from important local and federal government agencies, specifically those who hold senior positions equivalent to a Director General (Sharjah Executive Council, 2025). The local disaster management team in Ras Al Khaimah is headed by the Commander-in-Chief of the Ras Al Khaimah Police (WAM, 2021). Local disaster management teams organize their own capacity-building activities. For example, during the 2024 rainy season, the head of the local disaster management team of Ras Al Khaimah stressed the need for training and exercises on various rain and flood emergencies (RAK Police, 2024).

NCEMA has a Local Centers Department which coordinates the local disaster management centers across all seven Emirates. The Department is in charge of local and national coordination.

8.5 Cross-Sectoral and Community Collaboration

The UAE's disaster management architecture emphasizes both horizontal coordination among ministries and vertical coordination between national and emirate levels. This integrated system connects civil defense and humanitarian actors. NCEMA's Safety and Prevention Department provides support through training and through the management of the National Volunteers Program that aims to create a special volunteer base to handle disasters to support national disaster response efforts.

During the COVID-19 pandemic, for example, NCEMA coordinated the National Disinfection Program in partnership with the Ministry of Interior (MOI) and Ministry of Health and Prevention (MOHAP) (NCEMA, 2020). The program involved nationwide curfews, sterilization of public utilities and transportation, movement restrictions, and synchronized public messaging to enhance compliance (WAM, 2020). Media agencies, with inputs from the National Media Council, published COVID-19-related news to help disseminate accurate information to the public (Alam, 2022b).

8.6 Academic Programs in Disaster Management

Higher education institutions in the UAE play a pivotal role in cultivating the next generation of disaster management professionals. Recognizing that workforce competency is central to resilience building, the UAE has embedded disaster management education within its broader national strategy for capacity development.

Two leading academic institutions, Rabdan Academy and the American University in the Emirates (AUE), offer specialized disaster management-related undergraduate programs. These programs complement the UAE's policy commitment to developing a professional, knowledge-based disaster management workforce.

8.6.1 Rabdan Academy

The Faculty of Resilience at Rabdan Academy in Abu Dhabi offers a Bachelor of Science degree in Integrated Emergency Management (IEM) (Davies & Al Sharefeen, 2022). This program provides courses on "standard practices, procedures, and technologies to collect information, conduct risk assessment, and develop effective plans for emergency management, response, leadership, relief, and recovery. Through a combination of lectures, discussion, simulation exercises, and applied real-world scenarios, students learn team organization, professional ethics, interagency operations, and resilience through the core principles of contemporary relevance, integration, cooperation, multi-agency coordination, and interoperability" (Rabdan Academy, 2025a).

Compulsory core courses include Introduction to Risk Management; Introduction to Business Continuity Management; Introduction to Crisis Leadership; Creating Emergency Management Plans; Introduction to Emergency Response; Disaster Relief and Recovery; Tactical Emergency Management; Resilience Program Validation; Working with the Media in Emergencies; Strategic Integrated Emergency Management; Emergency Operation Centre Management; and Comparatives Approaches to IEM. The courses in the program are aligned with NCEMA frameworks and global best practices, preparing graduates for leadership roles in government agencies and humanitarian organizations. One course, SEC 301 Critical National Infrastructure Protection, instructs students on how to identify and prioritize existing important assets in the UAE and assess relevant risks, anticipating their involvement in protection programs once they graduate from the Academy (DelGrosso & Arlikatti, 2022).

Rabdan Academy also offers vocational and professional development training, leadership courses, and accredited international qualification through its Vocational Education Affairs Division (Rabdan Academy, 2025b). The Crisis and Disaster Management courses are categorized under the following thematic domains: Fundamentals of Crisis and Disaster Management; Assessment and Prevention; Planning and Preparedness; Response and Recovery; and Business Continuity and Organizational Resilience. Course levels vary from basic to advanced.

The Academy's close coordination with national stakeholders, such as NCEMA and the Ministry of Interior, ensures that its graduates possess both the academic foundation and the operational competencies required for contributing to the enhancement of the resilience of the nation.

8.6.2 American University in the Emirates

The College of Security and Global Studies of the American University in the Emirates (AUE) offers a Bachelor of Arts in Security and Strategic Studies with a specialization in Disaster Management, Emergency Management, Risk Management, and Crisis Management (AUE, 2024). Compulsory courses in the Disaster Management concentration include Disaster Management; Climate Change and Natural Disaster (sic; it is now increasingly accepted by the disaster risk management community that there are no "natural" disasters); Economics of Natural Disasters (sic); and Incident Response and Disaster Recovery.

8.6.3 Other Universities

While other universities do not have degree programs, they offer short courses or training for helping build capacity in disaster management. For example, the

University of Sharjah's Center for Continuing Education and Professional Development (CCEPD), in collaboration with NCEMA, ran the Business Continuity System Auditor program, which consisted of four intensive training sessions designed to sharpen the participants' skills in auditing business continuity systems and managing emergencies (University of Sharjah, 2025). The program, aligned with NCEMA's regulatory standard, focused on enhancing professional expertise in assessing institutional plans and strategies to ensure uninterrupted delivery of vital services during critical incidents.

The academic degree programs and short courses exemplify how the UAE integrates formal education with national policy objectives for disaster management. By aligning curricula with NCEMA's framework and standards and with international best practices, higher education institutions contribute to the professionalization of the disaster management workforce.

8.6.4 Non-Governmental and Volunteer-Based Training

Non-governmental organizations (NGOs) and volunteer-based initiatives play a crucial complementary role in the UAE's disaster management system. These organizations extend the reach of formal government programs by strengthening community preparedness, enhancing public awareness, and delivering specialized professional training that supports safety and resilience.

8.6.5 Emirates Red Crescent

The UAE Red Crescent Authority was established in 1983 (Krzymowski, 2022). Over time, the organization became commonly referred to in English as the Emirates Red Crescent (ERC). ERC is the UAE's primary humanitarian organization and a key partner in disaster preparedness and response. In collaboration with universities and the police, the ERC provides training in humanitarian operations, first aid, psychosocial support, and disaster management to help foster a culture of preparedness and volunteer work (AUS, n.d.; GMU, n.d.; WAM, 2015; Abu Dhabi Police, 2014).

8.6.6 Jaheziya Program

Another notable volunteer-based capacity-building platform is the Jaheziya Program (also known as the UAE National Programme for Medical Readiness and Response), established by the Sheikha Fatima bint Mubarak Volunteer Program and the Frontline Heroes Office "to provide UAE frontline professionals accredited emergency and disaster management training through a new comprehensive program that is recognized as the international gold standard for medical and emergency response" (Jaheziya Initiative, n.d.; WAM, 2025b). Jaheziya translates as "readiness" in Arabic. The Jaheziya Program provides advanced medical and response training for all types of potential situations, from infectious disease and airborne viruses to natural hazard-caused disasters, fires, and nuclear incidents. Participants must successfully complete examinations after each section of the program in order to earn their certifications.

8.6.7 Sanid Program

The Sanid Program is the national emergency response volunteer program of the UAE, launched in 2009 by the Emirates Foundation to recruit and train volunteers nationwide to equip them to respond to national and local disasters (Alzaabi, 2021). Sanid means "support" in Arabic (Emirates Foundation, n.d.). Over 37,000 individuals have enrolled as Sanid volunteers (UN Volunteers, n.d.). Sanid trains volunteers on first aid, cardio-pulmonary resuscitation (CPR), fire suppression, crowd management, search and rescue methods, and formation of emergency teams (NCEMA, 2010).

8.7 Conclusion

A resilient and adaptive disaster management system depends on the capacity, competence, and coordination of its workforce. The UAE has demonstrated steady progress in institutionalizing these dimensions through a centralized yet flexible framework led by NCEMA. Operating under the Supreme Council for National Security, NCEMA ensures unified national policies, standardized training, and well-defined coordination mechanisms that link federal and emirate-level agencies.

The UAE's approach integrates governance, education, and professional development into a coherent national strategy. Investments in undergraduate programs, such as Rabdan Academy's Bachelor of Science in Integrated Emergency Management. Meanwhile, volunteer-based and non-governmental initiatives, such as the Sanid Program and the Emirates Red Crescent, foster community engagement and expand local response capacity.

By combining centralized oversight with localized adaptability, the UAE has created a mature and integrated disaster management system that bridges national policies, institutional frameworks, academic programs, and public participation. Its standardized training, simulation exercises, and multi-sectoral coordination platforms, supported by legal and educational infrastructures, exemplify a model of governance-driven resilience.

Other nations, particularly within the Gulf Cooperation Council (GCC) and broader Global South contexts, can draw valuable lessons from the UAE's experience. The emphasis on human resource development, interagency collaboration, and education-based capacity building illustrates how workforce-oriented strategies can transform disaster management systems from reactive structures into proactive, learning-oriented institutions.

As emerging risks from climate change, technological disruption, and global health emergencies continue to evolve, continued collaboration across governmental, academic, and non-governmental sectors will be essential to sustaining national preparedness and resilience. The UAE's trajectory shows that investing in people, and in the systems that support their continuous development, is the foundation of effective disaster management.

Acknowledgments This chapter was developed in collaboration with some students from IEM403: Comparative Approaches to Integrated Emergency Management at Rabdan Academy during the Spring 2025 semester. Their research outputs and analyses contributed to the development of several sections of this work.

Declaration During the preparation of this work, the authors used ChatGPT (OpenAI, 2025) to assist with language editing and improvement of readability. The authors reviewed and edited the content as necessary and take full responsibility for the content of this publication. The use of this tool was limited to linguistic enhancement, as the authors are non-native English speakers, and did not influence the study's analyses, interpretations, or conclusions.

References

Abu Dhabi Media Office. (2025). *Emergencies, crises and disasters management centre—Abu Dhabi conducts comprehensive exercise at Barakah Nuclear Energy Plant*. https://www.mediaoffice.abudhabi/en/security/emergencies-crises-and-disasters-management-centre-abu-dhabi-conducts-comprehensive-exercise-at-barakah-nuclear-energy-plant/

Abu Dhabi Police. (2014). Abu Dhabi Police and red crescent sign MoU regarding 'Hifz Al Ni'am' project. https://www.adpolice.gov.ae/en/Media-Center/News/2021/06/19/Abu-Dhabi-Police-and-Red-Crescent-Sign-MoU-Regarding%2D%2DHifz-Al-Ni-am%2D%2DProject

Alalawi, S., Issa, S. T., Takshe, A. A., & ElBarazi, I. (2022). A review of the environmental implications of the COVID-19 pandemic in the United Arab Emirates. *Environmental Challenges, 8*, 100561. https://doi.org/10.1016/j.envc.2022.100561

Alam, E. (2022a). Cognition and collective action for COVID-19 prevention and response in The United Arab Emirates. In L. K. Comfort & M. L. Rhodes (Eds.), *Global risk management: The role of collective cognition in response to COVID-19*. Routledge. https://doi.org/10.4324/9781003253280-12

Alam, E. (2022b). Enhancing the policing response to COVID-19 with media engagement: The United Arab Emirates experience. *Policing, 16*(2), 311–324. https://doi.org/10.1093/police/paac012

Almheiri, K. B., Rustum, R., Wright, G., & Adeloye, A. J. (2024). The necessity of updating IDF curves for the Sharjah emirate, UAE: A comparative analysis of 2020 IDF values in light of recent urban flooding. *Water, 16*(18), 2621. https://doi.org/10.3390/w16182621

Alzaabi, E. H. (2021). *Developing a strategic framework to enhance readiness of UAE law enforcement agencies to effectively respond to road traffic emergencies (unpublished doctoral thesis)*. University of Salford. https://salford-repository.worktribe.com/OutputFile/1486203

AUE. (2024). *2024–2025 undergraduate catalog*. https://aue.ae/catalogs-handbooks/

AUS. (n.d.). *Volunteer training and development*. https://www.aus.edu/life-at-aus/student-life/community-involvement/volunteer-training-and-development

BEH & IFHV. (2025). World Risk Report 2025, Bündnis Entwicklung Hilft (BEH). https://www.preventionweb.net/media/110940/download

Davies, A., & Al Sharefeen, R. (2022). Enhancing academic integrity in a UAE. Safety, security defence emergency management academy—the Covid-19 response and beyond. *International Journal for Educational Integrity*, 18,17. https://doi.org/10.1007/s40979-022-00110-3.

DelGrosso, B., & Arlikatti, S. (2022). Teaching critical infrastructure protection and resilience using exercises and WebEOC: An examination of UAE undergraduate Students' after-action reports. *International Journal of Disaster Risk Reduction, 68*. https://doi.org/10.1016/j.ijdrr.2021.102700

Emirates Foundation. (n.d.). *Sanid Program*. https://www.emiratesfoundation.ae/en/programsandinitiatives/sanid

FANR. (2021). *Regulation for emergency preparedness and response for nuclear facilities*. https://www.fanr.gov.ae/en/Documents/FANR-REG-12%20V1%20ENG%20for%20website.pdf

GCAA. (2014). *Guidance Material No. 3: Contingency Planning for Air Traffic Services*. https://www.gcaa.gov.ae/en/epublication/EPublications/Standalone%20GM/GM%2003%20CONTINGENCY%20PLANNING%20FOR%20AIR%20TRAFFIC%20SERVICES.pdf

GMU. (n.d.). *Certificate in First Aid Training*. https://gmu.ac.ae/institutes/tihwdl/center-for-humanitarian-education/certificate-in-first-aid-training/

Jaheziya Initiative. (n.d.). *About Jaheziya Program*. https://www.jaheziyainitiative.org/

Krzymowski, A. (2022). Role and significance of The United Arab Emirates foreign aid for its soft power strategy and sustainable development goals. *Social Sciences, 11*, 48. https://doi.org/10.3390/socsci11020048

Middleton, N., Kashani, S. S., Attarchi, S., Rahnama, M., & Mosalman, S. T. (2021). Synoptic causes and socio-economic consequences of a severe dust storm in the Middle East. *Atmosphere, 12*(11), 1435. https://doi.org/10.3390/atmos12111435

MOHAP. (2025). *National policy for combating health risks in the United Arab Emirates*. https://uaelegislation.gov.ae/en/policy/details/national-policy-for-combating-health-risks-in-the-united-arab-emirates

NCEMA. (2010). *"SANID": UAE's new national emergency response programme is launched in Abu Dhabi*. https://www.ncema.gov.ae/en/media-center/news/12/1/2010/

NCEMA. (2013). *National response framework. Government of the United Arab Emirates*. https://www.ncema.gov.ae/dassets/download/9ef6a125/NRF-English.pdf.aspx

NCEMA. (2020). *MoHAP, MoI, NCEMA: National Disinfection Programme to start from 10 PM to 6 AM from Saturday*. https://www.ncema.gov.ae/en/media-center/news/29/5/2020/mohap-moi-ncema-national-disinfection-programme-to-start-from-10-pm-to-6-am-from-saturday.aspx#page=1

NCEMA. (2021). *The national standard for business continuity management system (AE/SCNS/NCEMA 7000:2021)*. https://www.ncema.gov.ae/vassets/48c1c32a/Publication-en.pdf.aspx

President of the UAE. (2006). *Federal Law No. 23 of 2006 Concerning the Civil Defense*. https://moi.gov.ae/DataFolder/leg/legl4e.pdf

President of the UAE. (2009). *Federal Law by Decree No. (6) of 2009 concerning the Peaceful Uses of Nuclear Energy*. https://fanr.gov.ae/ar/Documents/20101024_nuclear-law-scan-eng.pdf

President of the UAE. (2011). *Federal Decree-Law No. (2) of 2011 In Respect of the Establishment of the National Emergency, Crisis and Disasters Management Authority (NCEMA)*. https://www.ncema.gov.ae/vassets/11bfa4f2/Federal%20Law%20No.%202%20of%202011%20-%20NCEMA%20(2).pdf.aspx

President of the UAE. (2024). *Federal Decree-Law No. (35) of 2024 on the Reorganization of the Civil Defence Authority*. https://uaelegislation.gov.ae/en/legislations/2771/download

Rabdan Academy. (2025a). *Undergraduate Academic catalogue 2025–2026*. https://ra.ac.ae/en/useful-links

Rabdan Academy. (2025b). Crisis and disaster management courses. https://domaincatalog.ra.ac.ae/flip-book/169065/1875581

RAK Police. (2024). *Meeting of the local emergency, crisis, and disaster team in Ras Al Khaimah.* https://www.rakpolice.gov.ae/MediaCenter/Details/0?uid=2c73fc701edd4a7695786930c336ef9e

RAK Ports. (2024). *Marine pollution response plan.*, Ports. https://rakports.ae/wp-content/uploads/2024/12/marine-pollution-response-plan.pdf

Ramadan, M. S., Almurshidi, A. H., Razali, S. F., Ramadan, E., Tariq, A., Bridi, R. M., Rahman, M. A., Albedwawi, S., Alshamsi, M., Alshamisi, M., Alrashdi, S., Alnaqbi, S., Alhammadi, H., & Al Hosani, N. (2025). Spatial decision-making for urban flood vulnerability: A geomatics approach applied to Al-Ain City. *UAE. Urban Climate, 59*, 102297. https://doi.org/10.1016/j.uclim.2025.102297

Ruler of Abu Dhabi. (2019). *Law No. (22) of 2019 concerning the emergency, crisis, and disaster management Centre–Abu Dhabi.* https://www.adcmc.gov.ae/-/media/Project/ADCMC/ADCMC/Business-Continuity-Management-implementation/Law-22-of-2019-Eng.pdf

Ruler of Dubai. (2021). *Decree No. (4) Of 2021 concerning the supreme Committee of Emergency, crisis, and disaster Management in the Emirate of Dubai.* https://dlp.dubai.gov.ae/Legislation%20Reference/2021/Decree%20No.%20(4)%20of%202021%20Concerning%20the%20Supreme%20Committee%20of%20Emergency.pdf

Ruler of Dubai. (2024). *Decree No. (48) of 2024 Establishing the Dubai Resilience Centre.* https://dlp.dubai.gov.ae/Legislation%20Reference/2024/Decree%20No.%20(48)%20of%202024%20Establishing%20Resilience%20Centre.html

Sharjah Executive Council. (2025). *SEC forms new emergency and crisis management team.* https://ec.shj.ae/seceservicesapi/en/news/sec-forms-new-emergency-and-crisis-management-team/.

Terry, J. P., Ruheili, A. A., Almarzooqi, M. A., Almheiri, R. Y., & Alshehhi, A. K. (2023). The rain deluge and flash floods of summer 2022 in The United Arab Emirates: Causes, analysis and perspectives on flood-risk reduction. *Journal of Arid Environments, 215*, 105013. https://doi.org/10.1016/j.jaridenv.2023.105013

University of Sharjah. (2025). *University of Sharjah and NCEMA graduate new cohort of business continuity auditors.* https://www.sharjah.ac.ae/en/News/UOS-and-NCEMA-Graduate-New-Cohort-of-Business-Continuity-Auditors

UN Volunteers. (n.d.). *Embedding volunteers into national disaster risk reduction planning.* https://knowledge.unv.org/practices-and-experiences/embedding-volunteers-into-national-disaster-risk-reduction-planning

WAM. (2015). *Emirates red crescent organises summer programmes for university and high school students.* https://www.wam.ae/en/article/hszfsbd7-emirates-red-crescent-organises-summer-programmes

WAM. (2020). *MoHAP & MoI to Conduct 'National Disinfection Programme' for all public utilities, public transport over weekend.* https://www.wam.ae/en/details/1395302833041

WAM. (2021). *RAK Crisis and emergency management team announces series of precautionary procedures.* https://www.wam.ae/en/article/hszrbzlo-rak-crisis-and-emergency-management-team-announces

WAM. (2025a). *ADCMC signs MoU with MBZUAI to advance cooperation.* https://www.wam.ae/en/article/bm90xia-adcmc-signs-mou-with-mbzuai-advance-cooperation

WAM. (2025b). *Jaheziya launches ofnational forum for preparing leaders in medical sector.* https://www.wam.ae/en/article/bm61rjz-jaheziya-launches-ofnational-forum-for-preparing

WEF. (2025). Global Risks Report 2025 (20th ed.). *World Economic Forum.* https://www.weforum.org/publications/global-risks-report-2025/

Yagoub, M. M., & Al Yammahi, A. A. (2022). Spatial distribution of natural hazards and their proximity to heritage sites: Case of The United Arab Emirates. *International Journal of Disaster Risk Reduction, 71*, 102827. https://doi.org/10.1016/j.ijdrr.2022.102827

Chapter 9
Disaster and Emergency Management in Oman: Institutional Transformation and Persistent Gaps

S. Al Zaabi

Abstract This chapter examines the evolution of Oman's emergency management system (EMS) amid global shifts from centralised, command-and-control models towards more collaborative, decentralised, and inclusive approaches. This chapter adopts a qualitative research design, using a single-case study approach. Secondary data including national laws and plans, legal frameworks, and interview material collected during prior doctoral research were analysed thematically to explore how institutional change has unfolded in response to past disaster experiences, national reform efforts, and broader international frameworks. The findings suggest that while Oman's EMS retains a predominantly hierarchical structure, there have been notable steps towards improved cross-sectoral coordination and integration of non-state actors. The research highlights the importance of localisation, inter-agency collaboration, and adaptive governance as key features of effective emergency management. At the same time, it recognises the persistence of certain institutional constraints that may limit deeper systemic reform. Oman's experience provides a valuable perspective on the opportunities and challenges of EMS transformation in centralised, rapidly developing states.

Keywords Emergency management · Crisis · Disaster · Cross-sectoral coordination · Inclusive approaches

9.1 Introduction

Emergency and disaster management in Oman has undergone significant development in recent decades, driven by the increasing frequency and complexity of natural and technological hazards. The system remains highly centralised, reflecting a traditional model of governance where authority and coordination are concentrated at the national level. While this structure has supported coherent decision-making and rapid

S. Al Zaabi (✉)
Muscat, Oman
e-mail: saqar.zaabi.16@alumni.ucl.ac.uk

E. Alam (ed.), *Disaster, Crisis and Emergency Management in the Middle East & North Africa*, https://doi.org/10.1007/978-3-032-24436-9_9

mobilisation during major events, it has also highlighted limitations in flexibility, information flow, and local-level participation. As Oman continues to modernise its institutions and strengthen national resilience, an important question emerges: how can a centralised emergency management system evolve to incorporate more decentralised, collaborative, and community-engaged approaches?

Globally, the evolution of emergency management systems (EMSs) reflects a broader transition from authoritarian, command-and-control models to more participatory and locally empowered approaches (Alexander, 2002a, 2007; Koehler et al., 2014; Quarantelli, 1998). While no single managerial model fits all national contexts, a growing consensus has formed around several foundational principles: inclusive participation, inter-organisational collaboration, and localisation of response capacities (Al Zaabi, 2023). These principles are grounded in the recognition that, while disaster triggers differ, response demands, such as rapid coordination, shared situational awareness, and efficient resource mobilisation, are largely universal (Quarantelli, 1997).

Although the conceptual value of collaborative and decentralised systems is well established, translating these principles into practice is often constrained by deeply rooted administrative systems, legacy practices, and bureaucratic inertia. Many EMSs remain anchored in command-and-control traditions that favour centralised decision-making, rigid hierarchies, and an over-reliance on governmental actors (Alexander, 2002b). Such systems frequently marginalise emergent community groups and informal responders (Neal & Phillips, 1995), despite decades of evidence underscoring their essential role in effective disaster response (Drabek & McEntire, 2003). Empirical studies have consistently shown that local communities and non-state actors are not passive recipients of aid but active participants who fill critical operational gaps when formal systems are overstretched or slow to adapt (Alexander, 2007; Nohrstedt & Bodin, 2014).

Recent scholarship highlights the importance of integrating non-state actors, including civil society, volunteers, and private entities, into formal disaster governance (Alexander, 2007; Drabek & McEntire, 2003). Effective EMSs also depend on empowering local levels of government, which are best positioned to manage crises that unfold within their jurisdictions (Lindell & Perry, 1992; Waugh & Streib, 2006). Moreover, the success of any national EMS increasingly depends on its ability to foster horizontal (inter-agency), vertical (multi-level), and cross-sectoral collaboration, particularly in the face of large-scale, complex emergencies (Nohrstedt & Bodin, 2014; Thomson & Perry, 2006). Studies of major disasters, from Hurricane Katrina to the Sewol ferry incident, underscore how bureaucratic rigidity, fragmented coordination, and failure to adapt to emergent needs can significantly undermine response efforts (Alexander, 2008; Jung et al., 2018).

There is growing recognition that rigid centralised systems often suppress local improvisation and adaptability, which are essential when standard procedures become ineffective. Participatory models, such as community-engaged, adaptive, or civil protection approaches, promote a more flexible EMS that accommodates emergent actors, enables information flow across all levels, and decentralises authority to where situational awareness is most acute (Comfort, 2007; Loewenson et al., 2021).

While these approaches are not without challenges, particularly in contexts where local governance structures remain underdeveloped, they offer a stronger alignment with the empirical realities of disaster response and recovery.

This chapter examines Oman's EMS in light of these global transitions, analysing how its institutional evolution has addressed, or struggled to address, calls for decentralisation, integration, and inclusive governance. Oman's experience provides valuable insight into how rapidly modernising, yet centralised states attempt to balance strong national coordination with the need for broader participation and local empowerment. By situating Oman's trajectory within a wider theoretical and comparative discourse, this chapter contributes to a deeper understanding of the conditions that enable meaningful transformation and identifies opportunities to enhance flexibility, inter-sectoral collaboration, and overall system resilience in national emergency management.

9.2 Research Methods

This study adopted a qualitative research approach to examine the development of Oman's EMS. Qualitative research facilitates in-depth, contextual understanding by providing detailed, rich, and extensive descriptions of the phenomenon under study (Thiel, 2014). The research design relied on secondary data sources, including the national emergency plan, the Civil Defence Law, Royal Decrees, policy reports, and institutional frameworks issued by relevant government bodies. These documents offered insight into the formal structure of the system, key reforms, and the legal and institutional shifts that have influenced its trajectory. In addition, the study reviewed academic literature and prior research addressing institutional change, governance challenges, and disaster risk management in Oman. This helped situate the analysis within the broader field of disaster studies and enabled a critical assessment of how specific political, social, and environmental factors have either supported or hindered reform.

To complement the review of official documents and literature, the study also drew on a set of semi-structured interviews previously conducted with key stakeholders as part of the author's doctoral research (Al Zaabi, 2023). These participants included representatives from government agencies and sector coordinators engaged in emergency planning and response. Although the interviews were originally collected for thesis purposes, relevant portions were re-analysed to identify recurring themes related to coordination, preparedness, institutional learning, and community engagement. These qualitative insights enriched the analysis by providing practitioner perspectives on the practical realities of how the EMS operates, the challenges it faces, and the institutional dynamics that influence its effectiveness.

Data were analysed thematically, with a focus on patterns related to system development, sectoral coordination, decentralisation, and barriers to institutional reform. Ethical considerations, including informed consent and participant confidentiality, were strictly observed during the original data collection, and no personally identifiable information is disclosed in this study.

This methodological approach enabled a multi-layered understanding of Oman's emergency management system, linking formal institutional arrangements with on-the-ground implementation experiences. By drawing on diverse data sources, the study was able to triangulate findings and enhance the credibility of the analysis.

9.3 Disaster History and Risk Profile

Oman's disaster history has been dominated by natural hazards, shaped by the country's geography, topography, and climate. However, as Oman continues to expand its industrial base, energy infrastructure, and transportation networks, the overall risk environment has become increasingly complex. The coexistence of natural and technological systems introduces the potential for interacting or cascading hazards, where natural events may trigger technological failures, resulting in multi-faceted disasters that are difficult to predict and manage (Pescaroli & Alexander, 2018; UNDRR, 2015). Such scenarios, though not yet experienced in Oman on a large scale, are becoming more plausible as industrialisation and urban development intensify. This section first examines the natural hazards that have historically defined Oman's risk landscape, before addressing the emerging technological and human-induced threats that are beginning to occupy a growing share of the country's overall risk profile.

Oman's exposure to natural hazards is shaped by a complex interplay of its geography, topography, demographic trends, and rapid urban development. Located along the Arabian Sea, Gulf of Oman, and the Arabian Gulf, the country's terrain comprises semi-arid to arid desert plains, steep mountain ranges, and densely populated coastal lowlands (Al-Housni et al., 2015; Al-Shaqsi, 2009). These physical attributes increase the country's vulnerability to environmental hazards such as tropical cyclones, coastal and flash flooding, extreme heat, desertification, and, to a lesser degree, seismic activity (Al Manji, 2018; Alruheili, 2017).

The Al Hajar mountain range, which runs through northern Oman, is characterised by steep rocky slopes and minimal vegetation. This geophysical configuration leads to rapid surface runoff during intense rainfall events, substantially increasing the risk of flash flooding (Membery, 2002). In addition, shallow coastal zones, especially along the Gulf of Oman, are vulnerable to storm surges during tropical cyclones (Fritz et al., 2010). These areas are also increasingly at risk due to sea-level rise, augmented further by climate change. Al-Buloshi et al. (2014) assessed the potential impacts of sea-level rise on Oman's coastal zones under scenarios ranging from 0.2 to 5 m using a GIS-based spatial model. Their results showed that even under the smallest scenario (0.2 m), nearly 400 km^2 of land could be inundated, indicating the high vulnerability of Oman's low-lying coastal areas to sea-level rise.

Historically, Oman has faced recurring storm and cyclonic events with devastating consequences. One of the earliest documented events, the Muscat Cyclone of June 1890, resulted in at least 727 fatalities and the destruction of several thousand palm trees, significantly affecting both lives and livelihoods along the coastal plain (Membery, 2002). Nearly a century later, the Masirah Cyclone of June 1977 brought winds exceeding 105 km/h, displacing approximately 20,000 people (Membery, 2002). The storm caused widespread devastation on Masirah Island, where all homes were reportedly destroyed, thousands of livestock perished, and extensive agricultural losses occurred, including the destruction of palm and lime trees (Membery, 2002). These historical cyclones highlight the deep-rooted and recurrent nature of tropical cyclone risk in Oman.

More recently, Cyclone Gonu in 2007 caused extensive damage to infrastructure and coastal communities, particularly in Muscat, Al Sharqiyah, and the Al Batinah regions. The destruction was exacerbated by Oman's topography, which funnels rainfall from steep mountainous areas towards low-lying coastal plains, placing infrastructure in the direct path of flooding and storm surge (Fritz et al., 2010). Cyclone Phet in 2010 and Cyclone Shaheen in 2021 further demonstrated the country's vulnerability to multi-hazard events, combining high winds, torrential rainfall, flash flooding, and coastal inundation. Each event resulted in loss of life, significant economic disruption, and long-term recovery challenges.

The growing scale of disaster impacts is also tied to rising social vulnerability. Oman's population has increased fivefold over four decades, from around 900,000 in 1977 to 4.6 million in 2019, bringing with it demographic and socioeconomic shifts (NCSI, 2020). The non-Omani population, many of whom are migrant workers in low wage sectors, now comprises nearly 45% of the total population. Language barriers, limited access to emergency communication systems, and lower income levels can constrain these communities' ability to respond to disasters effectively (Al Rasbi, 2019). Furthermore, Muscat Governorate, where much of this population is concentrated, has a population density of 355 persons per square kilometre, vastly exceeding the national average of 14.9 persons per square kilometre, further amplifying exposure to natural hazards (NCSI, 2020).

The expansion of the built environment, often into hazard-prone zones, has exacerbated vulnerability. Many residential and commercial buildings are located on coastal plains flanked by the sea on one side and mountains on the other (Al-Shaqsi, 2009). Approximately 80% of Oman's population resides in low-lying areas (Al-Buloshi et al., 2014), increasing exposure to cyclones, storm surge, and flash flooding. Moreover, the country's infrastructure, especially in older urban areas, still lacks widespread flood-resistant design (Alruheili, 2017). While Oman's topography includes approximately 82% desert, 15% mountainous terrain, and only 3% coastal land, most of its population and critical infrastructure are concentrated in the latter two zones, which are the most hazard-prone (2010; Alruheili, 2017; NCSI, 2020). This spatial concentration of exposure continues to challenge emergency response effectiveness, particularly when multiple hazards occur simultaneously or sequentially.

In addition to cyclonic hazard, Oman is increasingly facing flash flood events, especially in areas characterised by wadis, dry riverbeds that rapidly fill during heavy rain. Urban expansion into these areas has exacerbated physical vulnerability, as evidenced during recent floods in Muscat and Al Batinah (Al Hatrushi & Al Alawi, 2011). Although Oman lies outside major seismic fault lines, the northern region, particularly Musandam and parts of Al Batinah, is susceptible to moderate seismic activity due to its proximity to the active Zagros fold belt in Iran (Al-Shaqsi, 2009).

Tsunami risk is also a growing concern for Oman's coastal zones. Although infrequent, tsunamis generated by undersea earthquakes in the Indian Ocean pose a threat to low-lying areas, particularly along the Al Batinah coastal plain. A GIS-based spatial analysis by Hereher (2020) showed that critical infrastructure, including schools, hospitals, roads, fuel stations, and residential areas, fall within 1–10 m of elevation along the coastline, making these areas vulnerable to inundation under tsunami scenarios with wave heights of 8 and 10 m. Al Batinah was identified as the most at-risk region due to its flat terrain and high density of population and services. These findings highlight the importance of incorporating tsunami risk into spatial planning and coastal development policies.

Climate change has further complicated Oman's risk profile. Regional climate assessments suggest rising sea surface temperatures and changing rainfall patterns are contributing to more intense and frequent hydro-meteorological events (Al-Awadhi et al., 2019; IPCC, 2014). Inundation of coastal areas has become inevitable (Hereher, 2020). Other factors that contribute to increasing coastal flooding include land use, soil type, and proximity to natural features such as wadis particularly in Muscat governorate (Hereher, 2020). Additionally, prolonged droughts and rising temperatures are contributing to desertification in interior regions, while coastal zones face risks of saltwater intrusion and storm surge impacts (Alruheili, 2017; Al-Hatrushi et al., 2015). Table 9.1 synthesises Oman's major natural hazards based on their recurrence and potential consequences. This risk matrix was developed by the author based on data synthesised from national reports, peer-reviewed studies, and official disaster databases cited in this chapter.

While Oman's disaster history has been dominated by natural hazards, its evolving economic and industrial landscape has introduced a range of technological and man-induced risks that increasingly shape its national risk profile. The expansion of oil, gas, and petrochemical industries, particularly in Sohar, Duqm, and Fahud, has increased exposure to industrial accidents. Studies of industrial areas such as Sohar have documented deteriorating air quality and associated health impacts among residents living near heavy industrial zones (Alwahaibi & Zeka, 2016; Al-Khadouri et al., 2015), highlighting the interconnectedness between environmental and technological risks. The presence of large fuel storage depots, refineries, and pipelines across the country also creates the potential for cascading technological incidents, where an industrial accident could be triggered or worsened by natural events such as flooding or cyclones (Pescaroli & Alexander, 2018).

Oman's rapid urbanisation and transportation growth further present safety and technological challenges. Road traffic collisions remain one of the leading causes of accidental deaths, particularly among working-age males (Al Aamri et al., 2025). In

Table 9.1 Risk matrix for natural hazards in Oman

Hazard	Likelihood	Impact	Risk level	Relevant literature
Tropical storms and cyclones	High	Severe	High	Al Manji (2018) Al Zaabi (2023) Membery (1997, 2001, 2002)
Flash floods	High	Moderate	High	Al Barwani (2014) Al-Rawas (2009) Al-Awadhi et al. (2017)
Earthquakes	Low	Moderate	Medium	El-Hussain et al. (2012) Al-Kindi and Hird (2020) Hancilar et al. (2018)
Tsunamis	Low	Moderate	Medium	Mokhtari et al. (2008) Okal et al. (2006) Hereher (2020)
Drought	Medium	Moderate	Medium	Al-Buloshi et al. (2014) Al-Hatrushi et al. (2015)
Coastal flooding	Medium	Significant	Medium	Al-Buloshi et al. (2014) Hereher (2020) Al-Hinai and Abdalla (2021)

parallel, the expansion of ports, airports, and hazardous cargo movement increases risks of transport-related incidents and environmental contamination. Moreover, the growing reliance on digital systems across government operations introduces cybersecurity vulnerabilities that can disrupt essential services and infrastructure (Ahamed et al., 2024), a concern recognised by national authorities such as the Oman National CERT and the National Committee for Emergency Management.

While industrialisation and technological advancement have introduced new and complex risks, the country's primary large-scale disasters to date have been driven by natural phenomena, particularly coastal and flash flooding triggered by cyclonic systems. This reality reinforces the need for a balanced approach, one that sustains strong early warning, preparedness, and recovery mechanisms for natural hazards, while simultaneously developing capacity to anticipate and manage emerging technological and human-induced threats.

9.4 Evolution and Institutional Framework

Oman's EMS has evolved but did not follow a linear path. An interplay between institutional learning, governance traditions, and contextual pressures has shaped the nature and structure of the system. An analysis by Al Zaabi (2023) of the system's development reveals a persistent tension between forces pushing for reform and those preserving the status quo. On one side, experiences with major emergencies, particularly Cyclone Gonu in 2007, the growing influence of social media, and the gradual privatisation of essential service providers have catalysed a shift towards decentralisation, the inclusion of non-governmental actors, and the flattening of communication structures. On the other side, endogenous factors, such as the legacy of centralised control, institutional culture, societal norms, and shared perceptions among leadership, have favoured continuity and the preservation of a government-led, response-based model (Al Zaabi, 2023). This has resulted in institutional adaptation without full transformation. While Oman has made formal structural changes to modernise its emergency management framework, the governance logic remains largely centralised, hierarchical, and activated primarily during crisis periods.

Prior to the 1980s, emergency management responsibilities were largely undefined and handled informally, typically falling under the jurisdiction of one agency. The country's approach to EM was response-oriented, reactive and largely improvised, with minimal planning. The reactive approach proved inadequate in the face of increasing hazard exposure, particularly from tropical storms and cyclones. The 1977 cyclone that struck Masirah Island became a turning point in national awareness. With over 100 fatalities and 20,000 displaced, the event revealed the severe limitations of the country's fragmented response capacity (Membery, 2002; Watts, 1978). It highlighted that managing emergencies of such magnitude required multi-agency coordination and prior risk reduction and preparedness measures. However, sustained planning and preparedness did not follow (Al-Shaqsi, 2009).

Institutional reform began with the issuance of Royal Decree No. 32/1988, which established the National Committee for Civil Defence (NCCD), as the country's first strategic multi-agency coordinating body for large-scale emergencies (Al-Shaqsi, 2009). The NCCD functioned as an ad hoc central structure, comprising over 20 governmental bodies, and was mandated to develop a comprehensive national disaster plan and coordinate emergency response across agencies. Though its structure was relatively simple, it was considered advanced compared to other systems in the region at that time (Al Zaabi, 2023). However, it remained a high-level, temporary body with no permanent staff or dedicated facilities. Composed mainly of ministers and deputy ministers, the committee struggled to meet regularly outside of emergency contexts (Al-Shaqsi, 2009). The absence of a central coordinating office or department meant that the system was largely reactive and lacked strategic foresight. In 2018, the NCCD was renamed the National Committee for Emergency Management (NCEM), and it will be referred to as NCEM throughout this text.

The Gulf War in 1990 and Oman's economic crisis in 1991 acted as critical junctures that further motivated the institutionalisation of risk and disaster governance. The Civil Defence Law of 1991 marked Oman's first formal legislation on disaster management. It defined disaster scenarios broadly and included provisions for emergency planning, public awareness, and the integration of volunteers, even mandating civil defence education at all levels (Sultanate of Oman, 1991). The law was comprehensive in its structure, but its implementation lagged. Between 1996 and 1998, severe storms again struck the country, renewing attention to preparedness and planning needs. In response, the government expanded the NCEM's agencies. Still, the system remained exclusive to government actors, with no meaningful engagement from civil society or the private sector.

Due to insufficient inter-agency integration and limited progress in preparedness and planning efforts, the establishment of a permanent executive office for NCEM in 2002 became necessary. This office was intended to enhance strategic planning and intergovernmental coordination in emergency management. Yet, its limited staffing and resources at that time restricted its influence. The office remained a largely administrative body, known to few and unable to fully assume the responsibilities of a national disaster management agency. It was not until the catastrophic impact of Cyclone Gonu in 2007, widely described by as an unprecedented event, that the systemic gaps were fully exposed. The cyclone revealed significant coordination challenges, information silos, and the lack of shared situational awareness among agencies. In response, Royal Decree 27/2008 was issued to authorise the nomination of new NCEM members, decentralising authority and introducing greater flexibility into the system. This marked a significant departure from the long-standing centralised appointment model and reflected a recognition of the need for broader, more dynamic participation in crisis management (Al Zaabi, 2023).

The aftermath of Cyclone Phet in 2010 provided the impetus for the most profound institutional transformation to date. The NCEM proposed a shift from an agency-based to a function-based sectoral model (Alexander, 2002a, 2008), drawing inspiration from modular emergency management structures used internationally. The new system divided emergency operations into eight sectors (NCCD, 2018), each grouped by a function, and led by a central coordinating agency (Moynihan, 2009), see Table 9.2. This shift facilitated parallel operations and created semi-autonomous but centrally coordinated networks. Agencies with similar roles were grouped under one operational function, improving communication, joint planning, and shared decision-making.

The sectoral model introduced significant improvements in coordination and preparedness. A marked shift towards a culture of planning emerged, evidenced by the approval of Oman's first national emergency management plan in 2011 (Al Hajri, 2024). The country has increased investment in monitoring technologies, and the institutionalisation of scenario-based training and multi-agency drills. The establishment of the Multi-hazards National Early Warning Centre (EWC) played a transformative role. The centre became the scientific and operational hub of the EMS. Activation of emergency responses now relies heavily on its recommendations. Agencies across the system expanded their workforce and recruited new technical specialists, signalling a wider professionalisation of emergency management roles.

In 2020, the National Emergency Management Centre (NEMC) became a permanent structure (NCCD, 2018), which increased its visibility and authority. The focus has shifted from response to preparedness. The Centre adopted the incident command system (ICS) as its approach in responding to emergencies. The ICS focuses on coordinating necessary actions and making resources potentially available in a jurisdiction (Perry, 2003). Operations under this system are decentralised, while crisis management is centralised. The NCEM began implementing a comprehensive view of disaster management, dividing it into mitigation, preparedness, response

Table 9.2 Oman's EMS operational sectors

Sector	Lead agency	Main role
Monitoring and early warning	Multi-hazards early warning centre—meteorology authority	Responsible for assessing natural and technological hazards, evaluating risks and providing early warning against all hazards
Hazardous materials	Public authority for civil defence and ambulance	Manages incidents involving biological, chemical, and radiological hazards
Search and rescue	Public authority for civil defence and ambulance	Manages lifesaving operations and coordinates rescue efforts during emergencies and disasters
Medical response and public health	Ministry of health	Provides health and medical services
Shelter and relief	Ministry of social development	Responsible for organising operating shelters and provision of relief materials
Media and public awareness	Ministry of information	Manages public messaging and risk communication
Essential services (utilities)	Public authority for electricity and water	Ensuring continuity of essential services during and after crises
Disaster victim identification (DVI)	Forensic laboratory	Identifying victims and providing family assistance following mass casualty events

and recovery. This has resulted in an increased number and types of planning and preparedness initiatives.

9.5 Challenges in Disaster Management

Despite significant institutional advancements in Oman's EMS, including the shift to a modular, function-based structure, several persistent challenges continue to hinder its overall effectiveness. Many of these challenges are not unique to Oman but reflect broader structural, cultural, and operational issues common in centralised, state-led disaster governance systems (Alexander, 2007; Buck et al., 2006).

One of the main issues is the misalignment between sector leadership and technical expertise. In the current configuration, some agencies have been assigned leadership roles in operational sectors that fall outside their core mandates. This mismatch often results in uneven performance across sectors. In effect, while some sectors have become relatively well-developed, others continue to struggle to meet even the minimum operational requirements. Several sector coordinators have reported being unprepared for their assigned responsibilities, and overall sector effectiveness remains heavily dependent on the individual capacity of the coordinating agency and its appointed personnel.

The practical implications of these challenges have been observed in several major emergencies such as Cyclone Gonu (2007) and Cyclone Shaheen (2021). Post-event assessments highlighted recurring difficulties in sectoral coordination, delays in decision-making, and inconsistencies in operational leadership. Such issues reflect deeper institutional dynamics that cannot be addressed through procedural fixes alone. Instead, they require double-loop learning, a process that involves questioning the underlying assumptions, role assignments, and power structures that influence institutional behaviour (Argyris, 1977). In other words, improving sector performance is not only about better coordination, but also about rethinking who holds responsibility and why, and whether the current structure truly supports effective emergency management.

Another key challenge lies in the centralised nature of Oman's EMS, which limits the responsiveness of the system at the local level. Although decentralisation has been formally encouraged, most notably through the Sultan's directives to empower governorates with their own budgets, and through the establishment of regional NCCD subcommittees, actual practice remains highly centralised, with most decision-making power concentrated at the governorate centre if not at the national level. This centralised approach assumes accessibility of transportation networks, but Oman's topography often undermines this assumption. In many areas, rural villages are connected to main towns by roads that pass-through wadis, which become impassable during heavy rainfall or flooding. This isolation during emergencies reveals a critical vulnerability in crisis logistics and underscores the need for localised disaster management capabilities (Alexander, 2007; Schneider, 2005).

While the idea of regionalisation is gaining traction, true localisation, at the wilayat or community level, remains underdeveloped. Emergencies are inherently local in impact (Lindell & Perry, 1992), yet the capacity to act at the local level is constrained by a reliance on top-down support and delayed decision-making. Studies have increasingly emphasised the importance of a bottom-up approach, where community actors, local NGOs, and volunteers are integrated into the formal EMS, not merely as operational tools but as legitimate stakeholders in planning and preparedness (Kuipers et al., 2020). In Oman, however, community participation remains minimal and informal. Although community members and volunteers often mobilise spontaneously during crises, the system lacks flexible mechanisms to incorporate these emergent resources. The dominant command-and-control model used in Oman, and in many parts of the world, while useful for establishing order, is not conducive to managing spontaneous, decentralised volunteer efforts (Koehler et al., 2014; McGuire et al., 2009; Quarantelli, 1998).

An additional challenge concerns the scope and applicability of the national emergency management plan. While Oman has developed a comprehensive national plan that clearly outlines institutional roles at the central level, this document has limited relevance for operational needs at the community or wilayat level. An effective emergency plan must not only coordinate national efforts but also provide detailed, context-specific guidance for local actors who are often the first to respond (Alexander, 2002a). Localising emergency management must therefore include localising the emergency plan itself. This would involve developing area-specific plans that reflect local risks, capacities, and resources, and ensuring that these plans are synchronised with national and regional strategies. Without this level of integration, local authorities remain poorly equipped to act independently during emergencies, and the broader system remains vulnerable to delays.

A related challenge is the underutilisation of volunteerism as a strategic component of disaster management. Volunteerism represents a powerful force in disaster management in many parts of the world, offering manpower, local knowledge, and community trust. Numerous studies highlight how volunteers, both trained and spontaneous, can significantly enhance disaster response when properly integrated into formal systems (Whittaker et al., 2015). However, in Oman, the absence of a structured national volunteer framework limits the ability to organise, train, and deploy volunteers effectively (Al-Ani, 2017). While spontaneous volunteerism often emerges during major disasters, these efforts remain largely informal and disconnected from institutional response mechanisms. Without formal pathways for volunteer engagement, including registration systems, training programs, and legal protections, valuable human resources are overlooked, and opportunities for local empowerment are missed. This gap also contributes to the broader issue of limited community resilience and the over-reliance on central authorities during emergencies.

The issue is not merely structural but also operational. Emergency management remains largely response-oriented, with limited progress in the areas of mitigation and long-term disaster risk reduction. Although there has been an increased focus on preparedness, especially for storms and cyclones, this still falls short of the broader paradigm of disaster risk reduction (DRR), which emphasises prevention, mitigation, protection, and resilience-building (UNDRR, 2015). This planning deficit is compounded by growing physical vulnerabilities. Urban expansion, population growth, and the spread of infrastructure into hazard-prone areas have increased the exposure of both people and assets (Alruheili, 2017). As the built environment expands, so too does the complexity of managing disasters, especially in a system that has not fully integrated critical infrastructure interdependencies into its emergency management strategies (Al Zaabi, 2023; Boin, 2009).

The modular model adopted in 2010 was envisioned as a solution to many of these issues, offering a more collaborative framework through which resources, knowledge, and responsibilities could be shared across agencies. In principle, this structure should allow for unified vision, streamlined operations, and efficient use of institutional capacities. In practice, however, implementation challenges persist. Many sector coordinators are preoccupied with unrelated administrative tasks and lack the authority, time, or motivation to lead inter-agency collaboration, particularly during non-crisis periods dedicated to planning and preparedness. The system's effectiveness, therefore, continues to hinge on individuals rather than institutional design, limiting its scalability and consistency.

In summary, the challenges facing Oman's EMS are multifaceted. While institutional reforms have created the foundation for a more effective disaster management system, significant gaps remain in decentralisation, inclusivity, inter-agency coordination, and proactive risk reduction. Addressing these requires more than procedural adjustments, it demands a cultural shift towards collaboration, flexibility, and local empowerment. The persistence of centralised control, hierarchical communication, and rigid operational models constrains the ability of the system to adapt, learn, and evolve in response to increasingly complex and frequent hazards.

9.6 Research Gaps in Disaster Management

The field of disaster management has expanded considerably over recent decades. However, critical research gaps remain, both globally and within the specific context of Oman. Much of the literature has advanced in theory, focusing on disaster risk reduction, sustainable development, resilience and governance (e.g. Alexander, 2007; Graveline & Germain, 2022; Imperiale & Vanclay, 2021; Tierney, 2012), yet empirical applications remain uneven, particularly in countries with centralised systems or limited academic infrastructure for disaster studies. In Oman, there has been a rapid institutional progress in emergency management but has not been matched by a corresponding depth of research.

At present, disaster research in Oman is disproportionately focused on government institutions and the perspectives of officials. Numerous studies have explored the experiences and challenges faced by sector leaders, decision-makers, and emergency managers (e.g. Al Hajri, 2024; Al Zaabi, 2023). While this body of work has generated useful insights into institutional dynamics and strategic planning, it has largely excluded the perspectives of community members and the general public. This top-down research orientation has resulted in limited understanding of how ordinary citizens perceive, respond to, or prepare for disasters. Recent events, including the widespread community-led evacuation efforts during Cyclone Shaheen (2021) and the spontaneous volunteer mobilisation during the 2020–2022 flash flooding episodes, underscore the need to incorporate bottom-up approaches that engage local populations. Community-based research would shed light on local coping mechanisms, trust in institutions, barriers to preparedness, and the role of informal networks in disaster response (Kuipers et al., 2020; Twigg & Mosel, 2017).

The thematic focus of disaster studies in Oman is also relatively narrow. Most research has concentrated on weather-related hazards, especially tropical cyclones and flash floods, understandably so given their frequency and impact (e.g. Al Hatrushi & Al Alawi, 2011; Al-Housni et al., 2015). However, this has left other hazard categories significantly underexplored. The COVID-19 pandemic, the international rise of biosecurity concerns, and Oman's expanding industrial footprint highlight the need to examine biological, technological, and industrial hazards. Recent incidents involving industrial fires and hazardous material spills, as well as regional concerns regarding radiological risks, demonstrate that non-meteorological hazards should be integrated into Oman's research agenda. Expanding the scope of analysis is essential to support a multi-hazard emergency management system.

Another major gap lies in vulnerability analysis. While hazard mapping and risk identification have been well represented in the literature, there is a scarcity of research on vulnerability, both physical and social. Disasters are not merely the product of natural forces, but the result of the intersection between hazards, exposure, and vulnerability (Wisner et al., 2004). Yet Oman's vulnerability landscape remains poorly understood. For example, during Cyclone Mekunu (2018) and various recent flash floods, some rural communities became isolated due to limited road access and topographic constraints, while urban districts experienced disproportionate damage due to drainage limitations and informal building extensions. These events highlight how income levels, housing quality, geography, infrastructure access, and social marginalisation combine to create differential disaster risk across communities. Even fewer have evaluated how urban planning, environmental policy, or land-use regulations could be leveraged to reduce vulnerability over time. Filling this gap is essential for a more accurate national risk profile and for designing interventions that move beyond emergency response towards long-term disaster risk reduction.

In addition, the cultural dimensions of disaster risk remain largely unexplored. Culture shapes how risk is perceived, explained, and acted upon. In Oman, collective beliefs, religious interpretations, and traditional knowledge influence both individual and institutional responses to disasters. For instance, many communities rely on informal knowledge systems to judge flood risk, as observed during multiple recent wadi events, yet these systems have not been systematically studied or integrated into formal preparedness efforts. More research is needed, particularly using quantitative and mixed-methods approaches, to capture how both the public and officials perceive risk, disaster causation, and responsibility. This would help tailor communication strategies, risk education, and preparedness programmes to better align with the social context in which they are deployed (Wachinger et al., 2013).

A further underdeveloped research area is the role of the voluntary sector in disaster management. While volunteerism is often cited as a strength in crisis response, there has been little empirical investigation into the barriers that prevent the emergence of a reliable, well-integrated volunteer sector in Oman. Recent disasters—particularly Cyclone Shaheen—showed that volunteer groups mobilised rapidly and played significant roles in evacuation, debris removal, and relief distribution. However, these efforts remained largely informal and uncoordinated, reflecting a gap between community willingness and institutional mechanisms for engagement. Questions remain about why formal volunteer organisations are not active stakeholders, how cultural attitudes or institutional regulations affect volunteer engagement, and what mechanisms could support their involvement across all phases of the disaster cycle. More policy-relevant research is needed to identify practical steps for incorporating volunteer actors within the national system and across different tiers of government.

Methodologically, there has been over-reliance on qualitative interviews with government stakeholders, with few studies utilising participatory and quantitative methods. There is significant room for innovation in methodology, particularly in using household surveys, GIS-based vulnerability assessments, and scenario-based exercises, to build a more comprehensive and actionable evidence base. Strengthening the connection between academic research and government institutions through collaborative, interdisciplinary initiatives could help close this gap and improve the relevance of research to policy and practice.

In sum, Oman's EMS has advanced considerably in institutional terms, but its research foundations remain fragmented and top-heavy. Addressing the current gaps, especially in community engagement, multi-hazard risk analysis, vulnerability mapping, cultural understanding, and voluntary sector integration will be vital to building a more inclusive, data-driven, and resilient disaster management system.

9.7 Recommendations and Future Strategies

Oman has made important progress in building a more structured emergency management system, but several critical areas require further attention to improve the system's performance and effectiveness. One of the most pressing priorities is to ensure that leadership roles within the emergency management sectors are better aligned with technical expertise. In several cases, agencies have been tasked with coordinating sectors outside their core functions, resulting in uneven implementation and weak coordination. Revisiting these assignments to ensure they reflect institutional capabilities would help create a more balanced and effective system.

Another key area is the localisation of disaster management. While the current system includes regional and national structures, local-level leadership remains limited. Wilayat authorities should have a more active role in planning, preparedness, and response, supported by regional and national entities rather than dependent on them. Local emergency management plans should be developed through community consultation, tailored to local risks and resources, and tested through regular drills. These plans need to be operationally realistic and clearly understood by all relevant actors at the local level.

Participation is also essential. A more collaborative and people-centred approach would make preparedness efforts more relevant and create stronger links between communities and institutions. Local residents, volunteers, and non-governmental groups should be included not only in response but also in planning and preparedness activities. Encouraging this level of engagement will help shift the system from one that reacts to crises to one that anticipates and prepares for them. Volunteerism, in particular, is an untapped resource. Although people in Oman often come forward during emergencies, there is no national structure to organise, train, or support them. A framework should be developed to formalise volunteer roles, offer basic training, and establish legal protections. This would make it easier to integrate volunteers into formal emergency plans and improve overall capacity during large-scale events.

In addition to structural adjustments, the system needs to learn more effectively from past emergencies. This means going beyond procedural reviews to ask deeper questions about decision-making, role clarity, and the effectiveness of coordination mechanisms. Lessons learned should be documented systematically and shared across agencies to improve future performance. Without this kind of reflection, the same problems are likely to resurface.

Finally, there is a need to strengthen the relationship between research and policy. At present, much of the available research focuses on government perspectives and storm-related hazards, while other important issues, such as social vulnerability, public risk perception, and non-weather-related threats, receive far less attention. Promoting more applied research, including studies that involve communities, will help inform practical decision-making. Partnerships between academic institutions and government agencies could help fill these gaps and support more informed planning and preparedness.

These recommendations point to the need for a more locally grounded, participatory, and learning-oriented emergency management system. Oman has already laid important foundations. The next step is to build on them in ways that are inclusive, practical, and responsive to the realities on the ground.

9.8 Conclusions

Oman's EMS has developed significantly in recent decades. It has shifted from a reactive, agency-based structure to a more organised, sector-based model, improving coordination and planning at the national level. While Cyclone Gonu in 2007 was a major catalyst for reform, more recent events, such as Cyclone Mekunu in 2018, Cyclone Shaheen in 2021, and the recurring flash floods of 2020–2023, have further exposed the limitations of the centralised command-and-control approach and triggered important reforms. Since then, the country has invested in national planning frameworks, early warning systems, and a more professionalised emergency management workforce.

Despite these advancements, several challenges continue to limit the system's overall effectiveness. The misalignment between sector leadership and technical expertise, the limited operational authority at the local level, and the underutilisation of community participation and volunteerism point to the need for a more flexible and inclusive model. Moreover, Oman's risk landscape is evolving. Industrial expansion, technological hazards, and public health threats such as pandemics require an emergency management system capable of addressing a wider range of hazards beyond hydro-meteorological events alone. Some of the remaining limitations are not simply operational but reflect deeper structural and cultural patterns that influence decision-making, stakeholder engagement, and responsibility-sharing.

Oman is committed to building a more resilient EMS that does not only focus on technical tools or formal procedures but also built on collaboration and learning from experience. This includes engaging local actors and communities as active partners and ensuring that lessons learned from past events translate into meaningful changes in policy and practice. Oman's experience mirrors broader challenges faced by Gulf countries and other rapidly developing regions, where growing hazard exposure intersects with centralised governance and rapid urban expansion. The country's efforts, achievements, and remaining gaps offer useful lessons for others working to strengthen disaster management in similar contexts. The task ahead is to continue building a system that is not only ready to respond but is also equipped to reduce risk, adapt to new threats, and support communities before disaster strikes.

AI Disclosure Statement The author used OpenAI's ChatGPT to support the preparation of this paper. The tool was used solely to enhance the clarity, coherence, and language of text originally written by the author. It was also used to assist in rephrasing complex ideas, aligning tone, and integrating literature content developed earlier by the author. No original content or ideas were generated by the AI. All substantive writing, argumentation, and conceptual development were carried out by the author. The final content has been thoroughly reviewed and verified by the author to ensure academic integrity.

References

Ahamed, J. N., Mohammed, A. A., Al Ghaithi, H. K., & Al Salty, R. K. (2024). A study on the importance of cybersecurity and its significance in Omani society. In *2024 2nd international conference on computing and data analytics (ICCDA)* (pp. 1–6). IEEE. https://doi.org/10.1109/ICCDA64887.2024.10867272

Alexander, D. (2002a). *Principles of emergency planning and management*. Oxford University Press.

Alexander, D. (2002b). From civil defence to civil protection—And back again. *Disaster Prevention and Management: An International Journal, 11*, 209–213.

Alexander, D. (2007). Disaster management: From theory to implementation. *Journal of Seismology and Earthquake Engineering, 9*, 49–59.

Alexander, D. (2008). Emergency command systems and major earthquake disasters. *Journal of Seismology and Earthquake Engineering, 10*, 137–146.

Alruheili, A. (2017). *An approach to resilient strategic planning in the face of climate change: A case study of Oman* (Doctoral dissertation, UC Berkeley). https://escholarship.org/uc/item/7s5379p2

Alwahaibi, A, & Zeka, A. (2016). Respiratory and allergic health effects in a young population in proximity of a major industrial park in Oman. *Journal of Epidemiology and Community Health, 70*(2), 174–180. https://doi.org/10.1136/jech-2015-205609

Argyris, C. (1977). Double loop learning in organisations. *Harvard Business Review, 55*, 115–125.

Al Aamri, A., Padmadas, S. S., Zhang, L.-C., & Al-Maniri, A. (2025). Improving the Efficiency and Quality of Road Crashes Data in the Sultanate of Oman: Evidence-Based Recommendations. *Journal of Road Safety, 36*(2), 63–74. https://doi.org/10.33492/JRS-D-25-1-2563701

Al Barwani, A. S. (2014). Effective flood management in the sultanate of Oman. In *Proceedings of the 2nd international conference on water resources (ICWR2012), Bayview Hotel, Langkawi, Malaysia*. https://doi.org/10.5334/oq.141#main-text-B3

Al Hajri, F. S. (2024). *Enhancing disaster risk reduction and response: A comparison of the complexities of inter-sectoral coordination in the Omani disaster management system 2010–2020* (Doctoral dissertation, Bournemouth University). http://eprints.bournemouth.ac.uk/39674/1/AL%20HAJRI%2C%20Faisal%20Salim_Ph.D._2023.pdf

Al Hatrushi, S., & Al Alawi, H. (2011). Evaluating the impact of flood hazard caused by tropical cyclones on land use using remote sensing and GIS in Wadi Uday: Sultanate of Oman. In *Paper presented at the 34th international symposium on remote sensing of environment—The GEOSS era: Towards operational environmental monitoring, Sydney, NSW, Australia*.

Al Manji, S. (2018). *Planning for resilience to extreme weather events in Oman, 2000–2015* (Doctoral dissertation, University of Leeds).

Al Rasbi, T. S. S. (2019). *Social vulnerability to tropical cyclones: A case study in Muscat Governorate, Oman.* (Doctoral dissertation, University of Leeds). https://etheses.whiterose.ac.uk/23288/

Al Zaabi, S. (2023). *Organisational learning and dynamics of institutional change following large-scale emergencies: Oman's emergency management system as a case study* (Doctoral dissertation, UCL).

Al-Ani, W. (2017). Methodological analyses and applications to determine and maximize the added value of volunteer work in Oman. *International Journal of Humanities and Social Sciences, 6*(2), 53–64.

Al-Awadhi, T., Charabi, Y., & Choudri, B. S. (2019). Pathways for building urban resilience to climate change in Oman. *Development in Practice, 29*(5), 594–605.

Al-Awadhi, T., Charabi, Y., Choudri, B. S., & Bani Oraba, Y. (2017). Flooding risk analysis: A case study of Muscat Governorate, sultanate of Oman. *Human and Ecological Risk Assessment: An International Journal, 24*(3), 667–678.

Al-Buloshi, A., Al-Hatrushi, S., & Charabi, Y. (2014). GIS-based framework for the simulation of the impacts of sea level rise and coastal flooding on Oman. *Journal of Earth Science & Climatic Change, 5*(10), 5–10. https://doi.org/10.4172/2157-7617.1000238

Al-Hatrushi, S., Ramadan, E., & Charabi, Y. (2015). Application of geo-processing model for a quantitative assessment of coastal exposure and sensitivity to sea level rise in the Sultanate of Oman. *American Journal of Climate Change, 4*(04), 379.

Al-Hinai, H., & Abdalla, R. (2021). Mapping coastal flood susceptible areas using Shannon's entropy model: The case of Muscat Governorate, Oman. *ISPRS International Journal of Geo-Information, 10*(4), 252.

Al-Housni, M., Gunawardhana, L., & Al-Rawas, G. (2015). Wadi flow simulation using tank model in Muscat, Oman. *International Journal of Students' Research in Technology & Management, 2*, 178–182.

Al-Khadouri, A., Al-Yahyai, S., & Charabi, Y. (2015). Contribution of atmospheric processes to the degradation of air quality: Case study (Sohar Industrial Area, Oman). *Arabian Journal of Geosciences, 8*, 1623–1633. https://doi.org/10.1007/s12517-014-1295-0

Al-Kindi, M. H., & Hird, R. (2020). A review of natural geological hazards in Oman. *Environmental Earth Sciences, 79*(16), 381.

Al-Rawas, G. A. A. (2009). *Flash flood modelling in Oman wadis* (Ph.D. thesis, University of Calgary).

Al-Shaqsi, S. (2009). Emergency management in the Arabian Peninsula: A case study from the Sultanate of Oman. In D. McEntire (Ed.), *Comparative emergency management: Understanding disaster policies, organisations, and initiatives from around the world*. FEMA.

Boin, A. (2009). The new world of crises and crisis management: Implications for policymaking and research. *Review of Policy Research, 26*, 367–377.

Buck, D. A., Trainor, J. E., & Aguirre, B. E. (2006). A critical evaluation of the incident command system and NIMS. *Journal of Homeland Security and Emergency Management, 3*(3).

Comfort, L. K. (2007). *Shared risk: Complex systems in seismic response.* Emerald Group Publishing.

Drabek, T. E., & McEntire, D. A. (2003). Emergent phenomena and the sociology of disaster: Lessons, trends and opportunities from the research literature. *Disaster Prevention and Management, 12*, 97–112.

El-Hussain, I., Deif, A., Al-Jabri, K., Toksoz, N., El-Hady, S., Al-Hashmi, S., Al-Toubi, K., Al-Shijbi, Y., Al-Saifi, M., & Kuleli, S. (2012). Probabilistic seismic hazard maps for the Sultanate of Oman. *Natural hazards, 64*, 173–210.

Fritz, H. M., Blount, C. D., Albusaidi, F. B., & Al-Harthy, A. H. M. (2010). Cyclone Gonu storm surge in Oman. *Estuarine, Coastal and Shelf Science, 86*(1), 102–106. https://doi.org/10.1016/j.ecss.2009.10.019

Graveline, M. H., & Germain, D. (2022). Disaster risk resilience: Conceptual evolution, key issues, and opportunities. *International Journal of Disaster Risk Science, 13*(3), 330–341. https://doi.org/10.1007/s13753-022-00419-0

Hancilar, U., El-Hussain, I., Sesetyan, K., Deif, A., Cakti, E., Al-Rawas, G., Safak, E., & Al-Jabri, K. (2018). Earthquake risk assessment for the building inventory of Muscat, Sultanate of Oman. *Natural Hazards, 93*, 1419–1434.

Hereher, M. E. (2020). Assessment of infrastructure vulnerability to tsunamis upon the coastal zone of Oman using GIS. *Geosciences, 10*(5), 175.

Imperiale, A. J., & Vanclay, F. (2021). Conceptualizing community resilience and the social dimensions of risk to overcome barriers to disaster risk reduction and sustainable development. *Sustainable Development, 29*(5), 891–905. https://doi.org/10.1002/sd.2182

IPCC. (2014). Climate change (2014): Impacts, adaptation, and vulnerability. In *Contribution of working group II to the fifth assessment report of the intergovernmental panel on climate change*. Cambridge University Press. Retrieved May 2025, from https://www.ipcc.ch/report/ar5/wg2/

Jung, K., Song, M., & Park, H. (2018). Filling the gap between bureaucratic and adaptive approaches to crisis management: Lessons from the Sewol Ferry sinking in South Korea. *Quality & Quantity, 52*, 277–294.

Koehler, G. A., Kress, G. G., & Miller, R. L. (2014). What disaster response management can learn from chaos theory. In A. Farazmand (Ed.), *Crisis and emergency management: Theory and practice* (2nd ed., pp. 111–130). CRC Press/Taylor & Francis Group.

Kuipers, E. H. C., Desportes, I., & Hordijk, M. (2020). Of locals and insiders: A "localised" humanitarian response to the 2017 mudslide in Mocoa, Colombia? *Disaster Prevention and Management, 29*, 352–364.

Lindell, M. K., & Perry, R. W. (1992). *Behavioural foundations of community emergency planning*. Hemisphere Publishing Corp.

Loewenson, R., Colvin, C. J., Szabzon, F., Das, S., Khanna, R., Coelho, V. S. P., Gansane, Z., Yao, S., Asibu, W. D., & Rome, N. (2021). Beyond command and control: A rapid review of meaningful community-engaged responses to COVID-19. *Global Public Health, 16*, 1439–1453.

McGuire, M., Brudney, J. L., & Gazley, B. (2009). The new emergency management: Applying the lessons of collaborative governance to 21st century emergency planning. In R. O'Leary, D. Van Slyke, & S. H. Kim (Eds.), *The future of public administration, public management, and public service around the world: The Minnowbrook perspective*. Georgetown University Press.

Membery, D. (1997). A compilation of all known information relating to the severe cyclone that affected Muscat and the Batinah coast June 1890, Special investigations series 7. *Weather, 1*(5).

Membery, D. (2001). Monsoon tropical cyclones: Part 1. *Weather, 56*(12), 431–438.

Membery, D. (2002). Monsoon tropical cyclones: Part 2. *Weather, 57*, 246–255.

Mokhtari, M., Abdollahie Fard, I., & Hessami, K. (2008). Structural elements of the Makran region, Oman sea and their potential relevance to tsunamigenisis. *Natural Hazards, 47*, 185–199.

Moynihan, D. P. (2009). The network governance of crisis response: Case studies of incident command systems. *Journal of Public Administration Research and Theory, 19*, 895–915.

National Centre for Statistics and Information. (2020). *Statistical year book*. NCSI. https://www.ncsi.gov.om/Elibrary/LibraryContentDoc/bar_Statistical%20Year%20Book%202020%20%20Issue%2048_be9e58d8-e44a-41d8-ae4c-77e2d0d34bc6.pdf

National Committee for Civil Defence (NCCD). (2018). *National emergency management plan* [Unpublished internal document].

Neal, D. M., & Phillips, B. D. (1995). Effective emergency management: Reconsidering the bureaucratic approach. *Disasters, 19*, 327–337.

Nohrstedt, D., & Bodin, Ö. (2014). Evolutionary dynamics of crisis preparedness collaboration: Resources, turbulence and network change in Swedish municipalities. *Risk, Hazards & Crisis in Public Policy, 5*, 134–155.

Okal, E. A., Fritz, H. M., Raad, P. E., Synolakis, C., Al-Shijbi, Y., & Al-Saifi, M. (2006). Oman field survey after the December 2004 Indian Ocean tsunami. *Earthquake Spectra, 22*(S3), 203–218.

Perry, R. W. (2003). Incident management systems in disaster management. *Disaster Prevention and Management, 12*, 405–412.
Pescaroli, G., & Alexander, D. (2018). Understanding compound, interconnected, interacting, and cascading risks: A holistic framework. *Risk Analysis, 38*(11), 2245–2257.
Quarantelli, E. L. (1997). Ten criteria for evaluating the management of community disasters. *Disasters, 21*, 39–56.
Quarantelli, E. L. (1998). *Major criteria for judging disaster planning and managing their applicability in developing countries.*
Schneider, S. K. (2005). Administrative breakdowns in the governmental response to Hurricane Katrina. *Public Administration Review, 65*(5), 515–516.
Sultanate of Oman. (1991). *Royal Decree No. 76/91 issuing the Civil Defence Law*. Official Gazette.
Sultanate of Oman. (2008). *Royal Decree No. 27/2008 amending the Civil Defence*. Official Gazette.
Thiel, S. v. (2014). *Research methods in public administration and public management: An introduction*. Routledge.
Thomson, A. M., & Perry, J. L. (2006). Collaboration processes: Inside the black box. *Public Administration Review, 66*, 20–32.
Tierney, K. (2012). Disaster governance: Social, political, and economic dimensions. *Annual Review of Environment and Resources, 37*, 341–363.
Twigg, J., & Mosel, I. (2017). Emergent groups and spontaneous volunteers in urban disaster response. *Environment and Urbanization, 29*, 443–458.
UNDRR. (2015). *Sendai framework for disaster risk reduction 2015–2030*. United Nations Office for Disaster Risk Reduction. https://www.undrr.org/publication/sendai-framework-disaster-risk-reduction-2015-2030#downloads
Wachinger, G., Renn, O., Begg, C., & Kuhlicke, C. (2013). The risk perception paradox—implications for governance and communication of natural hazards. *Risk analysis, 33*(6), 1049–1065.
Watts, D. (1978). Severe cyclone in the Arabian Gulf—June 1977. *Weather, 33*, 95–97.
Waugh, W. L., & Streib, G. (2006). Collaboration and leadership for effective emergency management. *Public Administration Review, 66*, 131–140.
Whittaker, J., McLennan, B., & Handmer, J. (2015). A review of informal volunteerism in emergencies and disasters: Definition, opportunities and challenges. *International Journal of Disaster Risk Reduction, 13*, 358–368.
Wisner, B., Blaikie, P., Cannon, T., & Davis, I. (2004).*At Risk: Natural Hazards, People's Vulnerability and Disasters*. (2 ed.) Routledge, London.

S. Al Zaabi holds a Ph.D. in Disaster Risk Reduction. He is an independent researcher focusing on institutional development, emergency management systems, and disaster risk governance in the Gulf region. The author has no institutional affiliation to declare for this publication.

Chapter 10
The Imperative of Disaster Management in a Small Archipelagic State: The Case of the Kingdom of Bahrain

Major Hamad Swar

Abstract The Kingdom of Bahrain, a strategic archipelago in the Arabian Gulf, faces complex disaster management challenges due to its limited land area, high population density, dependence on coastal desalination and energy facilities, and its location within a critical global energy transit zone. This chapter examines the evolution of Bahrain's national disaster management capacity in addressing these vulnerabilities and developing a model that may guide other small MENA states. It highlights three pillars of Bahrain's success: a robust governance framework, effective inter-institutional coordination, and the rapid integration of advanced technologies across preparedness, response and recovery phases. The discussion further illustrates how Bahrain's disaster risk reduction strategy aligns with its national Vision 2030, the Sendai Framework for Disaster Risk Reduction, and the Sustainable Development Goals (SDGs). The chapter analyses Bahrain's engagement with international partners and the Gulf Cooperation Council (GCC), which facilitates capacity-building and strengthens regional resilience. Through selected case studies, it provides empirical evidence of the system's effectiveness, identifying key lessons and areas for improvement. The chapter concludes by addressing future risks and strategic pathways for sustaining Bahrain's disaster resilience amid evolving global and regional threats.

Keywords Disaster risk reduction · Technological solutions · GCC co-operation · International partnerships · Sustainable development · Bahrain

---

M. H. Swar (✉)
Disaster Management, Manama, Kingdom of Bahrain
e-mail: ncdm.info@interior.gov.bh; swar.bh.cd@gmail.com

E. Alam (ed.), *Disaster, Crisis and Emergency Management in the Middle East & North Africa*, https://doi.org/10.1007/978-3-032-24436-9_10

10.1 Introduction

The United National Development Programme has long recognised that Small Island Developing States (SIDS) face unique challenges due to their small size, isolation and vulnerability to environmental and economic disruptions.[1] Bahrain's position as a strategic, small archipelago state in the Arabian Gulf means that it confronts many of these challenges. This chapter aims to describe the model used in Bahrain, which may be of value to other small nations on how integrating risk and disaster management can support sustainable national development. It introduces Bahrain's system to demonstrate how building digital resilience and adaptive governance may be a useful framework for regions with similar levels of inherent vulnerabilities but also restricted financial capabilities.

The chapter assesses the governance system, which Bahrain has developed to oversee delivery of its all-hazards disaster management strategy, and examines how predictive analytics, early warning systems and digital communication platforms support the development of operational readiness and efficiency in risk and emergency management. Finally, it investigates how GCC membership, together with UN agency partnerships, helps Bahrain develop better resilience through international collaboration in handling cross-border threats.

10.2 The Unique Vulnerability of Bahrain

National resilience requires disaster, crisis and emergency management to function as fundamental non-negotiable elements.[2] The functions of a state become vital for Small Archipelagic States like the Kingdom of Bahrain, especially with its three main structural elements of limited geography, high population density and complex infrastructure systems creating a critical situation.[3,4] The small size of Bahrain makes it vulnerable to various risks,[5] including high-frequency risks such as:

- Natural hazards—Storms and coastal flooding, which are worsening due to sea level rise, extended hot weather events which endanger public health and infrastructure, and droughts which threaten food and water availability,[6] requiring long-term climate adaptation and infrastructure hardening.

---

[1] United Nations Development Programme (2018).

[2] Gulf Co-operation Council Secretariat General (2020). See https://gcc-sg.org/en/pages/default.aspx for more information on the GCC. Note: The Strategy itself is a Restricted document.

[3] United Nations Office for Disaster Risk Reduction (2015).

[4] World Bank (2020).

[5] Bahrain Ministry of Interior (2021). See https://www.interior.gov.bh/en/ for more information on the Bahrain Ministry of Interior. Note: The Framework itself is a Restricted document.

[6] World Bank (2020, p. 4).

- Technological failures—The industrial nature of Bahrain means that technological hazards represent a high proportion of all risks in its National Risk Register. The region faces multiple severe industrial risks including major petrochemical plant accidents and major maritime oil spills which occur frequently in Gulf shipping routes, extensive infrastructure breakdowns of power grids and communication networks, and advanced cyber-attacks on government and energy systems.[7,8]
- Emerging and hybrid threats—These include non-traditional security risks and systemic threats which combine physical and digital elements and result in terrorism and major disruptions to critical infrastructure because of geopolitical instability.[9,10]

In particular, the country's very existence as a high-density urban environment, heavily dependent on coastal desalination plants for its potable water and power generation facilities, creates a massive concentration of risk. Any disruption that takes place in a particular coastal area will create both immediate and long-term impacts throughout the entire nation. A national crisis could emerge from the loss of any major element of critical infrastructure such as a large power plant or major desalination facility. Such a loss would lead to power outages and water shortages, which threaten public health, economic operations and social order.

10.3 The Role of the Disaster Management System

In light of its risk profile, the disaster management system of Bahrain has to fulfil multiple objectives, which extend beyond the basic needs of emergency response. Resilience and preparedness are vital elements of national security that support both social stability and enable sustainable development, in line with Bahrain's Vision 2030.[11] The lack of proper hazard management would cause disruptions to essential public services including water desalination, energy distribution and healthcare delivery, which would create extensive economic and social impacts.[12,13] The well-established disaster management system thus functions as an economic protection system and social bonding mechanism that protects past development accomplishments while safeguarding future growth and development from increasing global instability.

---

[7] International Maritime Organization (2020).

[8] Al-Kuwari and Al-Sulaiti (2021).

[9] International Federation of Red Cross and Red Crescent Societies (2019).

[10] United Nations Environment Programme (2020).

[11] Al-Sulaiti and Al-Kuwari (2022).

[12] World Health Organization (2021).

[13] International Federation of Red Cross and Red Crescent Societies (2019, p. 9).

Financial logic supports this approach: proactive investments in resilience and preparedness through advanced planning, early warning systems and resilient infrastructure have been shown to be more affordable over the longer-term than the expenses incurred in the response to and recovery from disasters.[14] For a small state with limited fiscal reserves, this principle is an absolute necessity.

10.3.1 The National Governance Framework for Disaster Management

The development and implementation of Bahrain's disaster management framework are overseen by the National Disaster Management Committee (NDMC),[15] which operates as the main governance structure for disaster management in Bahrain. The NDMC serves as the central authority which holds supreme power to co-ordinate all hazard-related policy and operational interventions throughout the Kingdom.[16,17] The Committee derives its authority from its direct operational and reporting ties to the Civil Defence Council, chaired by the Ministry of Interior, which grants both operational and command-and-control capabilities. The collaborative partnership between the two entities enables the creation of strategic plans and the execution of policies that result in operational readiness for national disaster response throughout all disaster lifecycle stages.[18]

In line with global standards, the NDMC operates an all-hazards approach that requires readiness for every possible risk instead of concentrating on a few most likely events. This philosophy enables flexible and scalable planning and resource allocation, which can handle both unexpected crises and multiple simultaneous emergencies.

10.3.2 Institutional Co-ordination and Operational Excellence

The NDMC executes multi-agency planning and response through its technologically advanced National Civil Emergency Operations Centre (NCEOC). The NCEOC is the nerve centre during a crisis, where representatives from all key ministries (Health, Interior, Energy, Transportation, etc.) come together to establish situational awareness, collaborate on their interventions and oversee the efficient execution of the

[14] World Bank (2020, p. 4).

[15] The National Disaster Management Committee (NDMC) was renamed as the National Civil Emergency Management Committee (NCEMC) in 2023.

[16] Bahrain Ministry of Interior (2021, p. 5).

[17] National Committee for Civil Emergency Management (2023).

[18] Al-Sulaiti and Al-Kuwari (2022, p. 11).

response strategy. The NCEOC determines the most effective response measures and resource allocation by analysing historical incident records and current situation reports and running complex scenario-based models.[19]

Organisations implement proactive governance through strategic actions that move them beyond their current resource-dependent reactive systems. The Kingdom uses predictive tools to detect potential disruptions early so that it can seek to prevent them from becoming major disasters. Time is an especially limited resource in this small geographic area.

Bahrain has adopted systematic institutional methods for knowledge management to extract useful knowledge from experiences. Every major exercise and real-world event is followed by an exhaustive post-event analysis. This learning, combined with standard high-fidelity scenario training, helps policymakers and operational teams to improve their emergency response plans and procedures. The feedback loop functions as a fundamental element of adaptive governance. It allows Bahrain to create an operational governance framework which successfully handles emerging worldwide threats that challenge current systems.[20,21]

10.3.3 Adoption of Technological Solutions

The Kingdom of Bahrain deliberately devotes the most forward thinking—and one of the most important—parts of its disaster management strategy to technological developments.[22] It is vital that organisations develop digital resilience through technological progress and innovative solutions, helping them to identify and handle upcoming challenges. Furthermore, the nation also needs digital resilience because it operates at high density and has extensive connectivity, as conventional response mechanisms are not sufficient for such concentrated risk conditions.

The government has therefore committed substantial funding to the establishment of an advanced digital infrastructure system, providing real-time monitoring, predictive analytics and the immediate and targeted delivery of hazard information to those who need it. This includes a National Civil Protection Platform[23] to educate and support the building of capacity in the community.

The digital resilience framework consists of multiple technological elements, functioning as a unified system.

---

[19] Bahrain Government Portal (n.d.).

[20] National Committee for Civil Emergency Management (2023, p. 17).

[21] Bahrain Ministry of Interior (2021, p. 5).

[22] Al-Kuwari and Al-Sulaiti (2021, p. 8).

[23] Available from: https://www.ncpp.gov.bh/en/. Accessed 27 Oct 2025.

10.3.3.1 Geospatial Information Systems (GIS) and Remote Sensing

The development of shared operational understanding depends on GIS platforms, required by all emergency agencies for their operations. They are utilised extensively for:

- The process of vulnerability mapping, requiring the combination of demographic information with critical infrastructure sites and hazard zones including flood plains and industrial areas to determine the most hazardous locations.[24,25]
- Running simulations to predict disaster effects through capabilities such as storm surge modelling and chemical spill plume dispersal analysis.
- Using real-time traffic and damage assessment data to find the best emergency routes and locate emergency resources including field hospitals and relief supplies. The system operates with high precision and speed to deliver better results than are available from the use of static printed maps.

10.3.3.2 Wireless Emergency Alerts (WEA) and Rapid Communication

The government can use Wireless Emergency Alerts (WEA) to send alerts to the entire public without network congestion during emergencies. The mobile alert system enables immediate emergency alerts to reach the public in specific geographic areas.[26] The system needs to include links to regional meteorological services and global early warning networks to help Bahrain provide precise and immediate alerts for severe weather occurrences, maritime threats and public health crises.[27] The capabilities provided by the system, enable the conversion of unexpected events into controlled incidents. This is achieved through its ability to provide strategic early warning. Such early warning extends the time available for hazard avoidance and, if necessary, public protective measures.

10.3.3.3 Artificial Intelligence (AI) and Predictive Analytics

Bahrain is pioneering the use of AI-driven models and predictive analytics to shift the focus from merely reacting to actively anticipating crises. The models are developed using extensive datasets including historical climate records, infrastructure performance statistics, public health monitoring records and social media opinion data. The system can:

---

[24] World Health Organization (2021, p. 12).

[25] World Bank (2020, p. 4).

[26] Bahrain Ministry of Interior (2021, p. 5).

[27] International Maritime Organization (2020, p. 7).

- Run thousands of virtual simulations to evaluate the effectiveness of emergency plans against different compound threats such as power outages during hot weather conditions.
- Use forecasting to determine disaster trajectories, including both the expected movement and strength of hazards, such as tropical cyclones, and their specific effects on particular population groups and essential infrastructure.[28,29]
- Enable proactive resource deployment through predictive features, allowing strategic resource allocation of fire brigade and medical supplies before predicted events occur or reach their peak. Thus, reducing human casualties and property damage.

10.3.3.4 Social Media and Digital Public Engagement Platforms

Bahrain has created digital platforms that function as a permanent system to enhance public involvement and information management during crisis events. The BeAware app[30] represents one of many mobile applications that offer citizens real-time information, customised safety alerts and contact tracing capabilities. Furthermore, the app functions as the official source for verified information to combat rumours.[31] The public can obtain information through government social media platforms while officials answer public enquiries and establish readiness in their communities.

10.3.3.5 Smart Infrastructure and Internet of Things (IoT)

The Kingdom of Bahrain has initiated a programme to integrate Internet of Things (IoT) sensors into its fundamental critical infrastructure systems. The sensors are located at essential facilities, including desalination plants, major energy facilities and transportation centres. They monitor operational parameters including water quality, pressure, temperature and energy load. The systems use anomaly detection algorithms to identify potential risks before they become major problems, by sending alerts to the EOC when they detect unusual pipeline pressure drops or transformer temperature increases.[32]

Bahrain supports digital resilience through continuous major investments in this technology suitable to small, high-density urban areas. The Kingdom achieves

[28] Al-Kuwari and Al-Sulaiti (2021, p. 8).

[29] World Bank and United Nations Office for Disaster Risk Reduction (2019).

[30] Bahrain Government Portal (n.d., p. 19).

[31] Ibid.

[32] United Nations Environment Programme (2020, p. 10).

high and sustainable preparedness levels through the combined strength of predictive modelling, rapid communication and real-time monitoring systems. This helps minimise the unavoidable social and economic effects of hazards.[33,34]

10.4 Disaster Risk Reduction and Sustainable Development

The relationship between disaster risk reduction (DRR) and sustainable development stands as the fundamental basis of Bahrain's Vision 2030 programme. The basic principle is that disasters create an uncontrollable danger to economic development and social advancement. This makes resilience an absolute necessity for national economic and social planning.

Bahrain implements disaster management through a system that directly supports its national sustainable development goals. The combination of DRR with Vision 2030 and international frameworks, especially the Sendai Framework for Disaster Risk Reduction[35] and the Sustainable Development Goals (SDGs),[36] enables disaster risk reduction investments to become strategic integrated investments. This helps in achieving national goals and fulfil international commitments.[37]

10.4.1 Synergy Between the Sendai Framework and Vision 2030

Bahrain implements the Sendai Framework through its national policy framework of Vision 2030. Vision 2030 identifies resilience as a fundamental success factor because it works to create a lasting economic system that operates both efficiently and justly. The government achieves risk reduction through national economic planning, by conducting thorough risk assessments for all new development projects. This prevents the creation of new risks and systematically reduces current risks. The methodology integrates hazard mitigation into the existing national framework, so that it functions as a natural part of broader economic diversification and the achievement of social equity and environmental stewardship goals.[38,39] The operation of this system ensures that resilience functions as a strategic tool for sustainable development, beyond its traditional role of civil protection.

[33] International Federation of Red Cross and Red Crescent Societies (2019, p. 9).

[34] World Health Organization (2021, p. 12).

[35] United Nations Office for Disaster Risk Reduction (2015, p. 3).

[36] United Nations (2015).

[37] United Nations Development Programme (2018, p. 1).

[38] Ibid.

[39] International Federation of Red Cross and Red Crescent Societies (2019, p. 9).

10.4.2 Strategic Policy Integration

Bahrain also implements disaster risk reduction in support of sustainable development by establishing direct connections between resilience-building activities and specific measurable targets of the SDG framework. For example:

- SDG 3 (Good Health and Well-being) works to improve healthcare system readiness and emergency response capabilities through disease surveillance and mass casualty management.[40]
- SDG 9 (Industry, Innovation and Infrastructure) promotes infrastructure resilience through technological innovation to protect economic assets.[41,42]
- SDG 11 (Sustainable Cities and Communities) sets the goal for urban planning and infrastructure development to decrease flood risk and strengthen community resistance.[43]
- SDG 13 (Climate Action) recommends establishing climate adaptation plans which include coastal defence measures and heatwave emergency readiness and climate-sensitive urban development.[44]

In addition, the basic governance system of Bahrain requires all sectors to function as one through policy integration. The implementation of disaster management strategies exists within the core policies that direct health management, transportation systems and urban development and energy distribution and environmental protection.[45,46] The integrated system enables resilience goals to become fundamental elements of national development planning and capital investment decision-making processes.

This multi-sectoral approach brings two vital benefits to the community. The first is the identification of opportunities for joint mitigation work across sectors. For example, the risk of failure of power plants from flooding is identified through the Ministry of Energy's involvement in flood planning. This enables joint risk mitigation efforts through equipment elevation and coastal defence reinforcement. The second is by enabling the execution of mitigation strategies to fulfil multiple objectives, for example, urban planning. This will decrease flood danger while developing green areas for better community safety and quality of life.[47]

[40] World Health Organization (2021, p. 12).

[41] World Bank (2020, p. 4).

[42] Al-Kuwari and Al-Sulaiti (2021, p. 8).

[43] United Nations Environment Programme (2020, p. 10).

[44] World Bank and United Nations Office for Disaster Risk Reduction (2019, p. 29).

[45] United Nations Development Programme (2018, p. 1).

[46] World Bank and United Nations Office for Disaster Risk Reduction (2019, p. 29).

[47] Al-Sulaiti and Al-Kuwari (2022, p. 11).

10.4.3 GCC Co-operation

The GCC Framework[48] serves as a foundation for regional co-operation, enabling transboundary resilience.

As an Arabian Gulf nation, Bahrain is committed to regional co-operation as an essential security and resilience requirement because it shares maritime, industrial and geopolitical interests with neighbouring countries. Because numerous major hazards span across international borders, Bahrain actively joins Gulf Co-operation Council (GCC) disaster management programmes. The programmes serve as essential external components of its national disaster management strategy.

10.4.4 Mechanisms of GCC Co-operation

The GCC Framework enables Member States to work together to build their preparedness and response capabilities, thereby strengthening the overall resilience of all Member States, including Bahrain as a smaller nation with limited resources. Key mechanisms include:

- The GCC has established a system of joint exercises and simulations that entail complex scenarios. These include massive maritime oil spills in shared waters, cross-border disease outbreaks and industrial accidents near borders. The exercises verify that all equipment functions as one unit while following established protocols. They also confirm that command staff from a range of countries have knowledge of each other's operational procedures.[49,50]
- The GCC has established Mutual Aid Agreements (MAAs), enabling quick and standardised sharing of vital resources and specialised personnel and technical expertise for regional emergency response. The MAAs function as essential multipliers which benefit small nations during disasters which surpass their national response capabilities.[51,52]
- The GCC works to integrate monitoring systems for meteorological, environmental (air quality and sea temperature) and industrial hazards through co-ordinated early warning systems. The system provides a wider detection network, enabling the rapid distribution of warnings across the Gulf region for typical threats such as sandstorms and maritime traffic risks.[53]
- To improve cross-border health security, collaboration has been intensified in the areas of pandemic surveillance, co-ordination of regional vaccination strategies,

---

[48] Gulf Co-operation Council Secretariat General (2020, p. 2).

[49] International Maritime Organization (2020, p. 7).

[50] Gulf Co-operation Council Secretariat General (2020, p. 2).

[51] Ibid.

[52] International Federation of Red Cross and Red Crescent Societies (2019, p. 9).

[53] World Bank and United Nations Office for Disaster Risk Reduction (2019, p. 29).

and the joint execution of emergency medical response protocols. Hence, ensuring a harmonised public health front against infectious diseases.[54]

- The sharing of restricted information between regional partners, along with standardised operational procedures and strategic resource management, enables better resilience. Bahrain's central strategic position as a vital maritime hub in the Gulf, makes this collaboration essential for effectively mitigating the constant threat of cross-border marine environmental and security hazards.[55]

10.4.5 International Engagement: A Global Citizen in Resilience

Bahrain expands its disaster management activities into international disaster management through active participation in global resilience-building initiatives. The United Nations Office for Disaster Risk Reduction (UNDRR), World Health Organisation (WHO), International Maritime Organisation (IMO) and other major international organisations lead initiatives that achieve this goal. This is accomplished through active participation and firm adherence to key principles.[56,57]

Bahrain obtains a range of advantages from its strategic alliances. These advantages include being able to join international capacity-building programmes and specialised training curricula which use worldwide best practices for flood modelling. Furthermore, public health response and complex supply chain logistics enable Bahrain to share its distinctive digital resilience model. This model is built on the experience of high-density urban areas and contributes to the worldwide knowledge base on advanced cyber threats, local climate change effects and industrial risks.[58,59] At the same time, Bahrain benefits from standardisation and alignment through global forums which enable quick adoption of international benchmarks and standards such as ISO certifications for business continuity. This process helps maintain Bahrain's disaster risk management system as a modern, adaptive system that meets current global standards and regulatory requirements.[60]

[54] World Health Organization (2021, p. 12).

[55] International Maritime Organization (2020, p. 7).

[56] United Nations Office for Disaster Risk Reduction (2015, p. 3).

[57] United Nations Environment Programme (2020, p. 10).

[58] Al-Kuwari and Al-Sulaiti (2021, p. 8).

[59] World Bank and United Nations Office for Disaster Risk Reduction (2019, p. 29).

[60] United Nations Development Programme (2018, p. 1).

10.5 Case Studies and Lessons Learned: The Adaptive Capacity in Practice

The best validation of Bahrain's integrated disaster management model is achieved through its actual responses to real-world emergencies. These case studies serve as critical feedback tools that prove successful approaches while identifying areas that require immediate enhancement.

10.5.1 The COVID-19 Pandemic Response: Digital Governance and Adaptive Strategy

The COVID-19 pandemic became the biggest public health crisis since modern times in Bahrain, pushing its governmental and technological systems to their maximum capacity.

Response Highlights

The BeAware mobile application received fast digital deployment status after Bahrain became the first country to use it at a rapid pace. The platform served as a vital tool for contact tracing and real-time exposure alerts, as well as for the digital registration of testing and dissemination of health mandates. Bahrain was able to achieve its digital transformation goals in this case because of its previous investments in core digital infrastructure.[61]

The Ministry of Interior converted the NDMC structure described above into a unified national pandemic management centre. The centralised decision-making system reduced bureaucratic delays, enabling fast resource distribution and rapid implementation of a flexible national strategy.

The Kingdom undertook extensive proactive testing and vaccination programmes which proved the effectiveness of its emergency logistics and public health systems.

Lessons Identified

The first stage of the pandemic revealed two main problems: the worldwide instability of medical supply chains for PPE, and for specialised equipment. Identification of these issues triggered the development a national strategic plan to find new suppliers and build up domestic reserves of vital medical supplies. The extended duration of the crisis revealed that healthcare staff and emergency responders require enhanced safety protocols, extending beyond standard measures to include access to mental health support and measures to prevent fatigue. The BeAware public information initiative succeeded in its mission but needed to grow its reach, so that it offered technology access and digital literacy education to all population segments. These include migrant workers to establish equality in the provision of health information

[61] World Health Organization (2021, p. 12).

and services. Bahrain learned from the COVID-19 experience to enhance its planning for both epidemic and pandemic situations, as well as local foodborne and vector-borne disease outbreaks. Active monitoring of disease outbreaks continues.[62,63]

10.5.2 Maritime Oil Spills: Inter-governmental and Regional Co-ordination

Bahrain faces continuous threats from maritime environmental disasters because its position as a coastal state in a major global energy-shipping route, puts its vital desalination plants and delicate fishery resources at risk from oil spills.

Response Highlights

The Ministry of Interior (Civil Defence), the Ministry of Energy and the environmental authority have needed to establish immediate, complex co-ordination arrangements for recent incidents. These have involved specialised units using containment booms and skimmers to contain spills while redirecting desalination plant intake operations. GCC mutual aid agreements proved successful in regional mobilisation which enabled the fast deployment of specialised oil spill response vessels and technical experts from neighbouring countries.[64,65] Response strategies evolved from immediate containment to the establishment of specialised teams. Both identified risks swiftly and conducted specific restoration operations in damaged coastal areas, while keeping a permanent watch on marine ecosystems.

Lessons Identified

The incidents demonstrated the importance of the assessment of potential risks before accidents occur. Such assessments provided inputs to models that track spill paths, hence enabling the protection of critical infrastructure facilities such as desalination plants and power stations. The incidents also showed that national teams need continuous specialised training to learn specific environmental response techniques, vital for executing extensive multi-day coastal habitat clean-up operations.

---

[62] Abuzeyad et al. (2018).

[63] World Health Organization (2021, p. 12).

[64] International Maritime Organization (2020, p. 7).

[65] International Federation of Red Cross and Red Crescent Societies (2019, p. 9).

10.5.3 Urban Flooding Events: Infrastructure Resilience and Community Preparedness

Localised but disruptive urban flooding occurs frequently in Bahrain, caused by heavy rainstorms. Serious flooding is becoming more common because of fast urban growth and changing climate patterns.

Response Highlights

In recent flooding incidents, response teams have built community-based preparedness through prior awareness campaigns, using digital channels (WEA and social media) to distribute safety directions. This includes road closure information and shelter-in-place instructions to residents in the affected areas. This approach built a better state of preparedness in affected communities.

The recovery phase used GIS platforms to rapidly create maps of areas with damage and to locate broken infrastructure, which enabled both humanitarian and engineering teams to focus on the most critical areas for swift recovery.

Lessons Identified

Post-event assessments led to targeted investments in drainage system enhancements focussing on the areas, which were most vulnerable to flooding. New infrastructure projects are required to meet elevated requirements for flood resistance in their construction. A key lesson was that the enforcement of flood-conscious zoning and land use policies require improvement. New construction needs to stay out of natural drainage areas and avoid overloading current storm water systems. Moreover, it is evident that the design of future infrastructure needs to use climate change projections for rainfall intensity to embed proactive adaptation measures in urban planning.

10.6 Challenges and the Future Direction: Building for Tomorrow

The disaster management framework of Bahrain has achieved numerous successes. However, it remains in a state of ongoing development while the Kingdom deals with multiple intricate issues which need persistent senior-level policy engagement.

10.6.1 Persistent Vulnerabilities and Emerging Threats

The fundamental weaknesses of archipelago states are becoming more severe because of major worldwide events.

The fast growth of urban areas creates two major problems. It puts excessive pressure on current infrastructure systems, and it leads to higher population density in hazardous areas. Coupled with severe scarcity of natural resources (especially water and arable land), the consequences of a major disruption are magnified.[66,67]

The low-lying coastal areas face severe threats from rising sea levels, and the whole Kingdom faces more frequent severe heat waves, which require long-term investments in expensive and complex climate adaptation of critical infrastructure.

Organisations also need to address security threats that seek to exploit the deployment of modern, interconnected technological systems. The national power grid and banking systems face an increased risk of cyber-attack, requiring sustained attention to cyber-security techniques alongside emergency response protocols.[68,69] This includes the adaptation of existing protocols or the creation of entirely new protocols for the response to a 'cyber-physical' incident.

10.6.2 Financial and Capacity Constraints

The main ongoing difficulty stems from the need for permanent and significant levels of financial support to maintain technological progress, train specialised personnel and develop community outreach programmes.[70,71] Small island states face financial limitations that make it difficult to sustain advanced high-tech infrastructure. Such infrastructure requires ongoing public–private partnerships and extended investment commitments to avoid system capacity decline.

10.6.3 The Future Strategic Roadmap

The future disaster management strategy of Bahrain will concentrate on five essential directions, brought together into a unified system.

First, integration across ministries needs to go further. Ministries need to work together on risk modelling and long-term planning through active multi-sectoral collaboration. This requires, for example, health organisations to work with urban planners and the transportation, energy and environmental sectors. Such co-ordination is essential to manage intricate risks which exceed the capabilities of individual agencies.[72]

Second, resources will need to continue to be focused on AI, big data analytics and high-resolution GIS systems that support the development of sophisticated risk

---

66 World Bank (2020, p. 4).

67 United Nations Environment Programme (2020, p. 10).

68 Al-Kuwari and Al-Sulaiti (2021, p. 8).

69 International Federation of Red Cross and Red Crescent Societies (2019, p. 9).

70 United Nations Development Programme (2018), p. 1).

71 World Bank and United Nations Office for Disaster Risk Reduction (2019, p. 29).

72 Al-Sulaiti and Al-Kuwari (2022, p. 11).

prediction tools and autonomous emergency response systems[73,74] which maximise operational effectiveness and efficiency.

Third, ongoing community resilience programmes are needed to boost public readiness through awareness programmes, mobile systems and recurring training activities. The primary objective will be to focus on converting warning receivers into community-based first responders who actively participate in emergency response activities.[75,76]

Fourth, GCC countries need to strengthen their regional and international co-operation through formal agreements. These agreements enhance their participation in joint exercises and cover their shared engagement with global organisations (e.g. IMO, WHO), to achieve maximum preparedness, knowledge-sharing and emergency response capabilities.[77,78,79]

Fifth, there is a continuing need to align sustainability with risk reduction, to create disaster risk management programmes, which support the fulfilment of relevant SDG targets. These include the implementation of climate adaptation programmes, building resilient infrastructure and improving social well-being. This will establish resilience as the key factor which enables countries to achieve lasting national success.[80,81]

Through these five steps, and through its comprehensive adoption of digital resilience and adaptive governance, Bahrain will not only increase its resilience and its effectiveness in managing present and future complex threats. Bahrain will but also achieve its goal of becoming a regional leader in disaster management, serving as a model for other small MENA states.[82,83]

10.7 Conclusion

The Kingdom of Bahrain proves that small island and archipelagic states need efficient, adaptive governance systems, comprehensive technological integration and robust collaborative networks to build national resilience. The Bahraini model achieves its success through three essential components which support each other: governance, technology and regional/international collaboration.

---

73 Al-Kuwari and Al-Sulaiti (2021, p. 8).

74 World Bank and United Nations Office for Disaster Risk Reduction (2019, p. 29).

75 World Health Organization (2021, p. 12).

76 Bahrain Government Portal (n.d., p. 19).

77 Gulf Co-operation Council Secretariat General (2020, p. 2).

78 International Maritime Organization (2020, p. 7).

79 United Nations Environment Programme (2020, p. 10).

80 United Nations Development Programme (2018, p. 1).

81 World Bank (2020, p. 4).

82 International Federation of Red Cross and Red Crescent Societies (2019, p. 9).

83 United Nations Office for Disaster Risk Reduction (2015, p. 3).

The adaptive governance framework functions as the core organisational structure and policy framework, which enables NDMC. It enables the organisation to perform vital multi-sector collaboration, undertake adaptive decision-making and pursue the proactive all-hazards planning which Bahrain needs as a high-risk compact state.

Enhancements in the pace and effectiveness of disaster risk management can be achieved through the integration of technological developments. Organisations can use GIS integration with WEA, AI, IoT and digital communication systems to deliver enhanced operational capabilities, which produce both better operational intelligence and risk prediction for crisis prevention.

Bahrain achieves enhanced national security through its dual strategy, combining strong regional alliances with international participation to access global resources and expertise. The national capacity of Gulf states is enhanced through the GCC Framework which enables vital knowledge-sharing and co-ordinated hazard management to protect the common environmental and economic region of the Arabian Gulf.

The Kingdom's strategic approach of connecting disaster risk reduction with national sustainable development plans in Vision 2030 has made resilience a fundamental strategic element in securing social stability, economic growth and environmental sustainability.

The benefits have been demonstrated in the response to the COVID-19 pandemic, maritime oil spills and urban flooding incidents. These events have shown the need for adaptive governance, technological integration and community participation to establish sustainable long-term resilience.

Bahrain demonstrates a highly effective approach which other MENA countries and small island developing states worldwide can duplicate. The research shows that small nations can convert their strategic weaknesses into national power by using effective leadership and strategic alliances and technological risk management solutions.

References

Abuzeyad, F., Alqasem, L., Al Farras, M. I., Al Jawder, S. S., Al Qasim, G., & Alghanem, S. (2018). Emergency medicine in the Kingdom of Bahrain. *International Journal of Emergency Medicine, 11*(4). Retrieved October 27, 2025, from https://doi.org/10.1186/s12245-018-0163-1

Al-Kuwari, H., & Al-Sulaiti, M. (2021). The GCC nations should use technology to improve their disaster management systems according to the lessons learned from small states. *Middle East Journal of Disaster Management, 7*(2), 33–50.

Al-Sulaiti, M., & Al-Kuwari, H. (2022). The Gulf region needs new disaster risk governance systems because its current management systems do not produce sufficient results. *Journal of Emergency Management, 20*(3), 45–59.

Bahrain Government Portal. (n.d.). *Crisis and disaster management—Digital resilience*. Business Continuity Management, BeAware App, National Civil Protection Platform, Digital Transformation in Crisis Management. Retrieved October 27, 2025, from https://bahrain.bh/wps/portal/en/BNP/BahrainAtAGlance/CrisisAndDisasterManagement

Bahrain Ministry of Interior. (2021). *National disaster management framework: Policies and practices*. Government of Bahrain.

Gulf Co-operation Council Secretariat General. (2020). *Regional disaster risk reduction strategy*. GCC Secretariat General.

International Federation of Red Cross and Red Crescent Societies. (2019). *Annual report 2019*. IFRC. Retrieved October 27, 2025, from https://www.ifrc.org/document/annual-report-2019

International Maritime Organization. (2020). *Oil spill contingency planning*. IMO. Retrieved October 27, 2025, from https://www.imo.org/en/ourwork/environment/pages/oil%20spill%20contingency%20planning.aspx

National Committee for Civil Emergency Management. (2023). *The Civil Defence Council and National Emergency Centre operate based on established policies which guide their operational activities*. Government of Bahrain. Retrieved October 27, 2025, from https://ncpp.gov.bh/en/about-pro/31116/

United Nations. (2015). *Sustainable development goals*. Retrieved October 27, 2025, from https://www.un.org/sustainabledevelopment/sustainable-development-goals/

United Nations Development Programme. (2018). *Rising up for SIDS: UNDP's strategy to support SIDS in the next decade*. UNDP. Retrieved October 27, 2025, from https://www.undp.org/publications/rising-sids-undps-strategy-support-sids-next-decade

United Nations Environment Programme. (2020). *UNEP programme performance report 2020*. UNEP. Retrieved October 27, 2025, from https://wedocs.unep.org/handle/20.500.11822/35910

United Nations Office for Disaster Risk Reduction. (2015). *Sendai framework for disaster risk reduction 2015–2030*. UNDRR. Retrieved October 27, 2025, from https://www.undrr.org/publication/sendai-framework-disaster-risk-reduction-2015-2030

World Bank. (2020). *Natural disasters in the Middle East and North Africa: A regional overview*. World Bank. Retrieved October 27, 2025, from https://www.worldbank.org/en/region/mena/publication/natural-disasters-in-the-middle-east-and-north-africa

World Bank and United Nations Office for Disaster Risk Reduction. (2019). *Words into action guidelines: Multi-hazard early warning systems*. World Bank. Retrieved October 27, 2025, from https://www.undrr.org/words-into-action/guide-multi-hazard-early-warning

World Health Organization. (2021). *Strengthening WHO preparedness for and response to health emergencies*. WHO. Retrieved October 27, 2025, from https://apps.who.int/gb/ebwha/pdf_files/WHA74/A74_R7-en.pdf

Chapter 11
Crisis and Disaster Management in Kuwait: Institutional Development and Future Challenges

Raed Al-Husain

Abstract This chapter offers a comprehensive assessment of disaster and crisis management in Kuwait, analysing its historical experiences, current institutional frameworks, operational challenges, and strategic gaps. Drawing on government reports, academic literature, policy documents, and recent empirical studies, the chapter evaluates Kuwait's vulnerability to both natural hazards and human-induced disasters, ranging from the 1990 Iraqi invasion and recurrent urban flooding to sand and dust storms and the COVID-19 pandemic. The analysis highlights the fragmented nature of Kuwait's disaster governance system, characterised by the absence of a centralised authority, weak inter-agency coordination, limited public engagement, and an underdeveloped data infrastructure. Through a synthesis of national case studies and survey-based research, the chapter identifies critical weaknesses in legislative planning, preparedness capacity, and risk communication. It further outlines key research gaps and proposes forward-looking recommendations aligned with global frameworks such as the Sendai Framework for Disaster Risk Reduction (DRR) and the Kuwait Declaration on DRR (2025). The chapter concludes that, while Kuwait demonstrates episodic resilience, sustained institutional reform and strategic investment are essential to enable a transition towards an integrated, risk-informed disaster governance model capable of serving as a blueprint for the Middle East and North Africa (MENA) region.

Keywords Disaster risk management · Kuwait · Institutional fragmentation · National resilience · Crisis governance

---

R. Al-Husain (✉)
Information Systems and Operations Management Department, College of Business Administration, Kuwait University, Kuwait City, Kuwait
e-mail: raed.husain@ku.edu.kw

E. Alam (ed.), *Disaster, Crisis and Emergency Management in the Middle East & North Africa*, https://doi.org/10.1007/978-3-032-24436-9_11

11.1 Introduction to Jurisdiction and Its Governing Structure and Aims

Kuwait's disaster risk profile is shaped by low-frequency, high-impact shocks and increasingly recurrent climate-related stresses. Since 1990–91, the country has faced armed conflict and the ignition of over 700 oil well fires, with long-lasting environmental and health consequences. It has also experienced recurrent urban flooding (notably 2018, 2020, 2022) exposing weaknesses in stormwater management and urban drainage systems. In 2015, the Imam Al-Sadiq Mosque bombing resulted in 27 fatalities, necessitating a mass-casualty response. The COVID-19 public-health emergency revealed gaps in strategic foresight and coordination. Furthermore, frequent sand and dust storms have disrupted mobility and operations, while extreme heatwaves have placed strain on power systems. These events produced substantial impacts—for example, the oil-well fires entailed over US $50 billion in clean-up and ecological recovery, the 2015 attack caused 27 fatalities and 227 injuries, and COVID-19 resulted in 2,570 deaths and more than 600,000 cases—underscoring systemic governance and coordination gaps that this chapter analyses.

Despite these recurrent and high-impact events, Kuwait continues to face a fundamental governance challenge: disaster risk management remains institutionally fragmented. The absence of a permanent, legally mandated central authority has resulted in overlapping mandates, ad hoc coordination, and uneven preparedness across ministries and sectors. This institutional fragmentation widens the gap between strategic intent and operational performance in prevention, preparedness, response, and recovery.

While several reports and descriptive accounts have documented Kuwait's disasters and institutional arrangements, few studies have critically examined how these fragmented structures shape the country's overall disaster governance capacity. Most available analyses focus on sector-specific challenges or individual hazard events, offering limited integration of institutional, operational, and strategic dimensions. This lack of a comprehensive analytical lens leaves an important gap in understanding how fragmented governance affects national disaster preparedness and resilience, and how lessons from the Kuwaiti case can inform broader regional approaches.

Accordingly, this chapter addresses these gaps through a structured and critical analysis of Kuwait's disaster governance. It does so by synthesising the country's major disasters and their operational implications, tracing the historical evolution of its institutional framework, and diagnosing why fragmentation persists despite multiple crises. The chapter further examines key weaknesses in coordination, legal frameworks, data systems, and community engagement, while advancing actionable recommendations for reform. In doing so, it aligns Kuwait's experience with international disaster risk reduction frameworks and offers insights relevant to both national and regional contexts.

By analysing a hydrocarbon-dependent, highly centralised state with decentralised disaster functions, the Kuwait case offers a transferable lens for other MENA jurisdictions facing similar governance fragmentation. The chapter distils transferable design choices—such as the sequencing of a national DRM law, the establishment of a centralised emergency operations centre, and the development of national loss databases—and identifies common implementation pitfalls, including ministerial turf conflicts and ad hoc committee structures. These insights provide value for policymakers, practitioners, and scholars seeking to strengthen disaster risk governance across the region.

11.2 Literature Review and Conceptual Positioning

Global disaster risk governance has evolved from response-centric models towards integrated, risk-informed systems under the United Nations Office for Disaster Risk Reduction's Sendai Framework for Disaster Risk Reduction (2015–2030). Sendai emphasises four priorities—understanding risk, strengthening governance, investing in resilience, and enhancing preparedness for effective response and 'Build Back Better'—and sets measurable targets for mortality, affected populations, and losses. This chapter engages these priorities by interrogating how institutional design—particularly the presence or absence of a central authority—enables or constrains governance performance (UNDRR, 2015).

A substantial body of research in disaster governance and crisis management stresses that fragmented, multi-actor landscapes require strong coordination mechanisms and clear lines of authority. Tierney (2012) frames disaster governance as polycentric and multi-level, with frequent integration gaps; Boin et al. (2005) highlight leadership under pressure and the institutional conditions for strategic coordination across government silos. This foundation is reinforced by classic works emphasising the role of governance and institutional capacity in shaping disaster outcomes (Kelman, 2011; Mileti, 1999). These insights motivate the focus on Kuwait's institutional fragmentation and its consequences for risk management capacity.

Within emergency management systems, collaboration and interagency networks are repeatedly identified as critical performance levers—yet they rarely emerge spontaneously. Waugh and Streib (2006) show that collaboration must be actively designed and led, particularly where agencies retain strong vertical accountabilities; Kapucu (2006) demonstrates how interorganisational communication networks shape information flow and decision speed during emergencies. Subsequent studies have expanded this perspective, linking network structure to performance and adaptability under crisis conditions (Kapucu & Garayev, 2016; Nowell & Steelman, 2015). These findings imply that, in the absence of a national disaster authority, durable network governance and pre-agreed coordination protocols (e.g. Emergency Operations Centres (EOCs)—centralised facilities for coordinating multi-agency disaster response—as well as data protocols and incident command systems) are essential substitutes.

Recent scholarship extends this perspective to network governance under conditions of complexity, arguing that fit-for-purpose structures must balance representation, voice, and adaptability. Evidence from complex wildfire incidents (Steelman et al., 2021) and post-disaster coordination studies (e.g. Comfort, 2007; Lassa, 2015) indicates that performance improves when decision rights, data-sharing rules, and boundary-spanning roles are formalised in advance of crises. This literature informs our recommendations on Kuwait's proposed central authority, EOC architecture, and data protocols.

Risk information and loss databases constitute a second pillar. Globally, EM-DAT, maintained by the Centre for Research on the Epidemiology of Disasters (CRED), provides a baseline for cross-country comparison of disaster occurrence and impacts and underscores the value of standardised, regularly updated national loss registries. The Sendai Framework likewise calls for national databases to inform planning and investment. Complementary global and policy-oriented studies further highlight the importance of robust loss databases and risk monitoring to support early warning and preparedness (OECD, 2018; UNDRR, 2019; World Bank, 2013). Kuwait's current data gaps—scattered across ministries—mirror well-documented challenges in many countries and motivate the proposal for a unified national loss database (UNDRR, 2015).

At the regional level, the Arab Strategy for Disaster Risk Reduction (LAS/UNDRR) and subsequent regional platforms translate global priorities to MENA contexts, emphasising legal frameworks, institutional roles, and multi-sector coordination. The Kuwait case contributes by detailing how a highly centralised political system can still produce decentralised disaster functions—and by specifying a practical sequence for reform (national DRM law → permanent EOC → interoperable loss database) consistent with regional roadmaps (LAS, 2010 & Updates).

Taken together, these studies and frameworks highlight the institutional factors that shape the effectiveness of national disaster governance systems. Within this broader scholarly context, Kuwait offers a theoretically informative and policy-relevant case: it illustrates how institutional fragmentation can persist despite strong sectoral capabilities, and how governance design choices influence performance across the disaster management cycle. By synthesising historical events with institutional analysis and translating network-governance insights into concrete design choices (authority, EOC, data registry, community integration), the chapter advances a transferable template for MENA jurisdictions facing similar constraints.

11.3 Research Methods

This chapter adopts a qualitative, document-based research methodology to investigate the state of crisis and disaster management in Kuwait. Given the complexity of the topic, which spans national policy, institutional frameworks, infrastructure vulnerability, and socio-political dynamics, an interpretive approach is employed to synthesise insights from a diverse set of secondary sources in both English and Arabic. These include academic publications, government reports, international agency assessments, parliamentary proposals, media articles, and local newspaper investigations. This methodology enables a comprehensive analysis of Kuwait's disaster governance evolution, encompassing its institutional, operational, and strategic dimensions.

To ensure analytical rigor, the study employs triangulation of sources, integrating peer-reviewed literature (e.g. Alfaras, 2020; Al-Qattan, 2021), doctoral dissertations, World Bank and UN reports, and sector-specific documents, such as the Kuwait National Adaptation Plan (2019–2030) and audit reports from national oversight agencies. These are supplemented by empirical case studies, legislative proposals, and situational analyses drawn from recent disaster events. These include the 1990 Iraqi invasion, the 2018 flash floods, the COVID-19 pandemic, and industrial accidents, and serve as focal points for contextual analysis. Each offers insights into different dimensions of Kuwait's risk landscape, institutional preparedness, and response capacity, including inter-agency coordination, public communication, and community resilience. The chapter also draws on recent national survey-based studies examining the organisational dynamics of disaster management in Kuwait, which provide empirical evidence on sectoral disparities in preparedness and resilience (Al-Husain, 2023a, 2023b).

Cases were selected based on their national impact, policy relevance, and the insights they provided into institutional strengths and weaknesses. The 1990 Iraqi invasion exemplifies military conflict, large-scale infrastructural collapse, and the subsequent recovery process following occupation. The 2018 flood disasters highlight infrastructure vulnerabilities in a natural hazard context, while the COVID-19 pandemic serves as a contemporary and globally relevant case that illustrates prolonged public health crisis management, revealing both strengths and deficiencies in crisis coordination and policy adaptability. These cases represent diverse hazard types and reveal different modes of institutional response.

A structured disaster timeline, compiled from both official sources and media archives, is used in Sect. 11.3 to categorise major disasters in Kuwait by type, frequency, and severity of impact. This timeline facilitates comparative analysis of event characteristics and institutional responses over time, with particular attention to events that triggered inter-agency coordination, legislative debates, or reform efforts.

The inclusion of Arabic-language sources utilised in this chapter is essential for capturing local perspectives, political discourse, and public sentiment, particularly in areas such as audit findings, newspaper investigations, and commentary on institutional failures. These sources are critically cross-referenced with formal government documents to mitigate bias and ensure balanced interpretation.

While no primary fieldwork or interviews are conducted, the research is informed by a reflexive lens grounded in the author's academic expertise in operations and risk management, as well as contextual familiarity with Kuwait's institutional landscape. This vantage point allows the analysis to extend beyond descriptive accounts towards interpretive evaluations of governance performance and reform opportunities.

This multi-layered methodological approach provides an evidence-based foundation for assessing Kuwait's disaster governance. It combines strategic case analysis with documentary synthesis to identify institutional gaps, assess coordination mechanisms, and inform actionable reform strategies that align with both global frameworks and Kuwait's unique national context.

11.4 Disaster History, Hazard Profile, and Impacts

Kuwait's disaster history reflects a complex interplay of rare but devastating crises alongside recurring environmental and technological hazards. Over the past four decades, the country has experienced events ranging from armed conflict and terrorism to flooding, pandemics, industrial accidents, and extreme climatic conditions. Each episode has revealed different dimensions of institutional capacity, coordination, and resilience gaps, thereby shaping the country's contemporary disaster governance landscape.

Unlike many countries where natural hazards dominate, Kuwait's risk profile is heavily influenced by geopolitical shocks, urban infrastructure vulnerabilities, and environmental stressors exacerbated by climate change and economic dependence on the hydrocarbon sector. Table 11.1 provides a consolidated hazard profile, combining historical event data with available information on human and economic impacts.

11.4.1 Hazard Typology and Frequency

Kuwait's hazard landscape is shaped by a mix of hydro-meteorological, technological, biological, and human-induced risks. Urban flooding is among the most recurrent and visible hazards, typically resulting from intense seasonal rainfall that overwhelms drainage systems in densely populated areas. Sand and dust storms (SDS) and extreme heat events occur annually, disrupting transport, energy demand, and public health, with SDS episodes often lasting several days and significantly reducing visibility (Al-Hemoud et al., 2019).

Technological hazards are closely linked to Kuwait's energy sector—including oil spills, refinery fires, and gas leaks—reflecting the country's heavy reliance on hydrocarbon infrastructure (KOC, 2021a, 2021b). Pandemics and epidemics, while less frequent, have had widespread impacts, as illustrated by the COVID-19 crisis (Alnassafi, 2022; WHO, 2024). Finally, low-frequency, high-impact events such as armed conflict and terrorism remain part of the national risk profile due to persistent regional geopolitical volatility. This hazard typology highlights both the recurrence of climate-related and technological hazards and the potential for cascading impacts when governance, infrastructure, and response systems are simultaneously stretched.

Complementing this classification, recent survey-based research indicates that many Kuwaiti institutions—especially in the public and non-profit sectors—exhibit low confidence in their ability to implement the full disaster management cycle. The preparedness and recovery phases are consistently rated as the weakest,

Table 11.1 Major hazard types and impacts in Kuwait (1980–2023)

Hazard type	Notable events	Frequency	Estimated impacts (deaths/affected/ losses)	Primary impact domains
Armed conflict and oil fires	1990–1991 Iraqi invasion; burning of 700+ oil wells	1	≈1,000 deaths; massive ecological damage; >US $50 billion clean-up	Infrastructure, environment, economy, health
Urban flooding	2018, 2020, 2022 major events	3+	Several fatalities; hundreds displaced; millions in damages	Housing, transport, public works
Epidemics and pandemics	COVID-19 (2020–2022), MERS scare	2+	2,570 deaths (COVID-19); >600,000 cases	Health system, economy, public administration
Environmental hazards	Sand and dust storms (SDS), heatwaves	Annual	Recurrent respiratory impacts; work stoppages; energy demand surges	Transport, energy, public health
Technological/ industrial	Oil spills, refinery fires (e.g. 2000, 2021)	Sporadic	Isolated casualties; millions in clean-up costs	Energy infrastructure, environment
Terrorism/ security events	2015 Imam Al-Sadiq Mosque bombing	Few	27 deaths; 227 injuries	Security, emergency response, health services

reinforcing the gap between growing hazard exposure and institutional readiness (Al-Husain, 2023b).

11.4.2 Event Narratives and Impacts

The 1990 Iraqi invasion anchors Kuwait's modern disaster history as the most severe crisis the nation has faced. Over seven months, the country experienced the collapse of civil governance, widespread infrastructure destruction, and the ignition of over 700 oil wells. These fires resulted in massive air pollution, ecological damage, and adverse health outcomes for residents and first responders (Behbehani et al., 1994; Lange et al., 2002). The post-war recovery period, although extensive, did not fully institutionalise a permanent disaster management authority, leaving future risk governance fragmented and reactive.

The 2018 urban floods marked another turning point in Kuwait's disaster history. Triggered by intense rainfall, these floods submerged major roadways, damaged public and private infrastructure, and exposed critical weaknesses in stormwater drainage systems (Alghais, 2019; GulfInsider, 2020). Similar flood events occurred in 2020 and 2022, amplifying public pressure for sustainable infrastructure and pre-emptive drainage solutions. Despite mitigation efforts—including the construction of stormwater reservoirs—underlying vulnerabilities remain unresolved (Al-Qattan, 2021).

In 2020, Kuwait faced the global COVID-19 pandemic, which tested the nation's health system and administrative coordination. Early response efforts included the repatriation of citizens, quarantine protocols, and the establishment of field hospitals, but the crisis also revealed significant gaps in strategic foresight, inter-agency planning, and economic resilience (Alnassafi, 2022). The Kuwaiti Audit Bureau documented the absence of a national crisis plan and inefficiencies in procurement and coordination across ministries.

Kuwait has also faced several other high-risk incidents. The 2015 Imam Al-Sadiq Mosque bombing stands as a stark example of domestic terrorism, resulting in 27 fatalities and prompting a coordinated emergency medical response (Al-Sabah et al., 2018). Sand and dust storms have become increasingly disruptive in recent years, affecting air quality, transportation, and sensitive equipment in industrial and energy sectors (Al-Hemoud et al., 2019, 2020). The country has also experienced oil sector accidents, including fires and chemical leaks, which pose persistent industrial risks due to Kuwait's reliance on hydrocarbon infrastructure (KOC, 2021a, 2021b).

Finally, extreme heatwaves present a growing challenge. Summer temperatures have increasingly strained the national electricity system, pushing peak demand to critical levels. Authorities have implemented scheduled, short-duration power outages—sometimes lasting up to two hours daily—across residential, industrial, and agricultural areas to stabilise the grid during peak load periods (Ata, 2025; Reuters, 2025). Beyond energy strain, these temperature extremes pose severe risks to outdoor workers and vulnerable populations, including older adults (Alwadi et al., 2024). Such conditions illustrate how climate extremes generate cascading challenges across energy systems, critical infrastructure, and public health.

11.5 Analytical Overview

Together, these events demonstrate Kuwait's multi-hazard and overlapping risk environment. They reveal patterns of reactive rather than anticipatory governance, fragmented institutional responsibilities, and underdeveloped data integration mechanisms. Armed conflict and terrorism, though infrequent, carry catastrophic systemic consequences. Meanwhile, flooding, pandemics, and environmental hazards such as SDS and heatwaves are recurrent and intensifying. Technological accidents in the energy sector further highlight systemic risk exposure. These patterns underscore the need for an all-hazards risk governance framework that integrates infrastructure resilience, public health preparedness, and coordinated emergency response—a subject examined further in subsequent sections.

11.5.1 Disaster Risk Governance in Kuwait

Kuwait's disaster risk governance operates within a highly centralised political system but remains fragmented across multiple ministries, agencies, and semi-governmental actors. This arrangement enables sector-specific specialisation—for example, in firefighting, health response, and infrastructure management—but also creates gaps in strategic coordination, unified planning, and resource allocation during emergencies. This fragmentation has been repeatedly observed during major disaster events, such as the 2018 floods and the COVID-19 pandemic, where responses were effective within individual agencies but lacked integrated cross-ministerial coordination.

To better understand the institutional structure of disaster governance in Kuwait, Table 11.2 provides a summary of the key entities involved in emergency management, their primary functions, and their areas of responsibility. This mapping illustrates how operational roles are distributed and where coordination challenges may arise in the absence of a central disaster management authority.

This institutional landscape reveals a governance system that relies heavily on sectoral mandates rather than on a centralised disaster management authority. While some entities—notably the Kuwait Fire Force and the Ministry of Interior—exercise clear operational leadership during emergencies, other critical functions are distributed across multiple ministries, such as Health, Public Works, and the Environment Public Authority. Non-state actors, such as the Kuwait Red Crescent Society, also play key roles in humanitarian response. However, the absence of a legally mandated, permanent national disaster management authority means that coordination is often ad hoc, reactive, and event-driven rather than strategic and planned.

These structural characteristics directly mirror the challenges highlighted in Sect. 11.4. Major disasters, such as the 2018 floods and the COVID-19 pandemic, revealed that while individual agencies performed effectively within their mandates, inter-agency coordination remained fragmented, with multiple information flows and overlapping responsibilities. The lack of unified command, common operating procedures, and shared data protocols reduces Kuwait's capacity to implement a comprehensive disaster management cycle, particularly in the preparedness and recovery phases (Alnassafi, 2022; Al-Husain, 2023b; UNDRR, 2025).

Addressing these governance gaps requires the establishment of an integrated national disaster management framework—one that consolidates authority, strengthens cross-sectoral coordination, and institutionalises shared planning and communication mechanisms.

Table 11.2 Key entities involved in disaster and emergency management in Kuwait

Entity/ministry	Primary role in disaster management	Hazard domain/ responsibility	Remarks
Kuwait Fire Force	Civil protection, firefighting, search, and rescue	Fire, floods, industrial accidents	Operational frontline agency
Ministry of Interior (Kuwait)	Security coordination, traffic and evacuation management	Security, terrorism, evacuation	Strong command structure
Ministry of Health (Kuwait)	Emergency medical services, epidemic control	Pandemics, mass casualty incidents	Key role in COVID-19 response
Ministry of Defence (Kuwait)	Military logistics, infrastructure protection	Armed conflict, terrorism	Support role during national emergencies
Ministry of Public Works (Kuwait)	Drainage, roads, and urban infrastructure management	Flooding, storm water management	Key actor in urban flood mitigation
Kuwait Petroleum Corporation	Industrial hazard prevention and response	Oil spills, refinery fires	Sector-specific emergency response
Kuwait Environment Public Authority	Environmental monitoring and impact assessment	Dust storms, pollution, climate hazards	Leads early warning and monitoring
Ministry of Information (Kuwait)	Risk communication and public information dissemination	Multi-hazard	Central to warning dissemination
Kuwait Red Crescent Society	Humanitarian response, relief distribution	Natural hazards and man-made disasters	Non-state operational partner

Note Various proposals have called for creating a centralised emergency/crisis authority reporting to the Council of Ministers (Alsharamri, 2024); as of August 2025, this body has not been established by law

11.5.2 Early Institutional Responses and Post-war ARecovery: A Catalyst for Disaster Management Evolution

The current fragmented structure outlined above has its roots in Kuwait's early post-war recovery period, when disaster response was improvised through existing ministerial channels rather than designed within an integrated national system.

The 1990–1991 Iraqi invasion stands as a pivotal moment in Kuwait's disaster management history, marking the nation's first encounter with a large-scale catastrophe that catalysed institutional mobilisation on an unprecedented scale. Prior to this, Kuwait's disaster preparedness and response mechanisms were largely nascent, relying primarily on ad hoc measures within existing governmental structures (Persian Gulf War, 2025). The invasion exposed these vulnerabilities, highlighting the urgent need for a more formalised and coordinated approach to national emergencies.

In the immediate aftermath of liberation, the Kuwaiti government, with substantial support from a broad spectrum of international actors, including the U.S. Army Corps of Engineers (USACE), spearheaded the Kuwait Emergency Recovery Programme (KERP). The USACE was authorised to assist Kuwait in restoring critical public infrastructure, including electricity, water, defence, and internal security systems. To manage these efforts, the Kuwait Emergency Recovery Office (KERO) was established, overseeing project management, emergency operations, engineering services, and contracting. Together, USACE and KERO played a pivotal role in rapidly mobilising resources for emergency aid and reconstruction (U.S. Army Corps of Engineers, 2002a). Their efforts included restoring power to critical facilities, repairing thousands of kilometres of transmission lines and roads, reopening ports, and conducting over 1,200 damage assessments. As part of a broader coordinated recovery programme, they also helped restore core government services such as reopening schools and hospitals, enabling ministries to resume operations, and facilitating the removal of landmines and the extinguishing of oil well fires (U.S. Army Corps of Engineers, 2002a, 2002b).

Although KERP was successful in enabling a rapid post-crisis recovery and in laying a foundational framework for inter-ministerial coordination during emergencies, it did not establish a permanent disaster management structure. The recovery process demonstrated a high degree of institutional collaboration; however, these efforts did not result in the establishment of a centralised authority dedicated to disaster management. Instead, the responsibilities of recovery were undertaken by existing government ministries, each operating within its respective domain.

The Ministry of Public Works (MPW) played a central role in the repair and reconstruction of infrastructure, leading efforts to restore vital public facilities and transportation networks (Barakat & Skelton, 2014). Simultaneously, the Ministry of Health (MOH) played a crucial role in reinstating healthcare services and addressing public health challenges in the aftermath of the crisis. The Ministry of Interior played a key role in re-establishing security and public order. It also managed the logistics of population return and movement, ensuring stability as displaced citizens returned to their communities.

This reliance on existing ministerial structures, while pragmatic in the immediate post-war context, highlighted a key challenge: the absence of a centralised authority specifically mandated for holistic disaster risk reduction, preparedness, response, and recovery planning beyond a specific crisis. The lessons learned from the Iraqi invasion and the KERP experience, while initially channelled through existing departmental mandates, contributed to subsequent considerations of a more integrated and dedicated disaster management architecture in Kuwait (Al-Damkhi et al., 2009). The scale of the 1990–1991 disaster served as an undeniable, albeit tragic, impetus for Kuwait to re-evaluate its national security and civil defence strategies, contributing to the gradual evolution of its disaster management framework in the following years.

11.5.3 *Institutional Diversification and Sectoral Response Capacity*

Following the immediate post-invasion recovery, Kuwait entered a new phase of disaster management evolution, characterised by a crucial, albeit fragmented, institutional diversification and the strengthening of sectoral response capacities. During this period, which broadly spanned the early 2000s, key sectors within the Kuwaiti government and critical national industries began to develop more formalised internal emergency plans and specialised operational procedures. This maturation represented a natural progression from the ad hoc coordination of the KERP era, reflecting a growing awareness of specific vulnerabilities and the need for dedicated departmental readiness.

The oil and energy sector, the backbone of Kuwait's economy, became a pioneer in this regard. Recognising its critical national importance and inherent risks, entities such as the Kuwait Oil Company (KOC), the primary national oil company responsible for upstream exploration and production of crude oil, invested significantly in robust disaster recovery protocols. These comprehensive plans addressed a spectrum of potential catastrophic events, ranging from large-scale industrial fires, which could devastate production facilities, to environmental threats posed by oil spills, and the increasingly risk of cybersecurity threats affecting operational integrity (Kuwait Oil Company, 2019, 2021a, 2021b, 2022, n.d.). These efforts included not only written protocols but also investments in specialised equipment, training for rapid response teams, and regular drills simulating various incident scenarios. Similarly, the Kuwait National Petroleum Company (KNPC) is responsible for the downstream refining and marketing of petroleum products. The Petrochemical Industries Company (PIC) is focused on the manufacturing and marketing of petrochemicals. Both companies have developed their own sophisticated emergency preparedness frameworks, with an emphasis on industrial safety, environmental protection, and business continuity planning, reflecting global best practices in critical infrastructure protection (KNPC, 2020; PIC, 2021).

Beyond the energy sector, critical public services also made significant strides. The Ministry of Health (MOH), drawing lessons from the mass casualty incidents of the invasion, developed and refined its emergency operating procedures. These included protocols for mass casualty management, enhancements in rapid medical response capacity, surge capacity planning for hospitals, and the establishment of prehospital triage systems to efficiently categorise and treat victims during large-scale emergencies (Abuzeyad et al., 2025; Ministry of Health, n.d.; Ridha & Orlin, 1996; WHO, 2024). Concurrently, the Kuwait Fire Force (KFF), a first-responder agency with a broad mandate, significantly augmented its capabilities. It formalised hazardous materials (HAZMAT) response procedures, including specialised training and equipment for chemical spills and industrial accidents, and improved coordination mechanisms for urban search and rescue operations (KFF, 2024, 2025). The Ministry of Interior (MOI) also strengthened its emergency response units, focusing on law enforcement aspects of disaster management, including maintaining public order, managing evacuations, and securing affected areas (KUNA, 2024; Kuwait Times, 2025; Ministry of Interior, n.d.).

However, despite these institutional advances, a significant challenge persisted: these specialised capabilities largely remained siloed. While developing sophisticated internal protocols, each agency operated primarily within its own domain. Coordination among ministries and between government agencies and critical industries remained limited to ad hoc, bilateral arrangements or the formation of crisis-specific task forces when required (Alenezi et al., 2014; KISR, n.d.; UNDRR, Prevention Web, 2022). This absence of an overarching framework meant that inter-agency communication, joint training exercises, and unified command structures were not systematically embedded in the national disaster management landscape. Crucially, Kuwait lacked, and to a significant extent continues to lack, a comprehensive national disaster law or a binding regulatory framework capable of integrating these disparate but crucial efforts under a unified command and control system. This institutional fragmentation constituted a significant barrier to a holistic and coordinated national disaster response, often resulting in redundancies, communication gaps, and inefficient resource allocation during complex, multi-sectoral emergencies. The journey from sectoral strength to integrated national resilience thus remained, and largely remains, an ongoing challenge.

While these developments significantly enhanced Kuwait's sectoral capacities, they also entrenched a siloed institutional culture, setting the stage for the coordination gaps observed during subsequent crises such as the 2018 floods and the COVID-19 pandemic.

11.5.4 Contemporary Frameworks and Policy Efforts: Towards Integrated Resilience

While the early 2000s saw significant strides in developing specialised, sector-specific response capabilities across Kuwait's critical sectors, a key challenge persisted: the lack of a unifying national framework to integrate these diverse efforts. Recognising this gap, particularly as the country grapples with mounting environmental and public health risks exacerbated by climate change and rapid development, Kuwait has increasingly focused on developing comprehensive policy instruments to enhance national preparedness and resilience. These contemporary efforts represent a shift towards more holistic approaches, even as the integration of various initiatives remains an ongoing endeavour.

Among these pivotal developments is the Kuwait National Adaptation Plan (NAP) 2019–2030. Developed in collaboration with the UN Environment Programme and the United Nations Development Programme (UNDP), the NAP represents a significant step towards a strategic, multi-sectoral approach to climate resilience. It identifies key vulnerabilities across critical areas, including water resources, public health, vulnerable coastal zones, and biodiversity (Kuwait EPA, 2019). Beyond identifying risks, the NAP outlines an ambitious agenda of 56 sectoral initiatives designed to strengthen Kuwait's adaptive capacity. These initiatives include establishing robust early warning systems, developing GIS-based risk mapping for more precise hazard assessment, and implementing integrated public health surveillance systems to track and respond to climate-sensitive health threats.

Complementing these efforts, the Kuwait Integrated Environmental Management System (KIEMS) provides a foundational platform for environmental oversight. KIEMS focuses on environmental monitoring and proactive pollution control, reflecting a commitment to safeguarding Kuwait's natural resources and public well-being from ecological threats (UNDP, 2015). However, its scope, while important, is inherently limited. KIEMS primarily addresses environmental hazards and, by design, has restricted authority regarding non-environmental emergencies or the broader inter-agency coordination required during broader national crises. Its strength lies in its specialised environmental mandate rather than in serving as a central command structure for all-hazards response.

In terms of digital infrastructure, the Kuwait Information Network (KIN), operating under the Communication and Information Technology Regulatory Authority (CITRA), serves as a critical backbone for digital interconnectivity across government ministries. In an increasingly digital environment, KIN plays an important role in ensuring the continuity of essential government operations, especially during periods of disruption or crisis. Its architecture facilitates information sharing and administrative functions, thereby supporting state functionality (Alenezi et al., 2014; UNESCWA, 2013). However, KIN is designed primarily for data exchange and operational continuity rather than as a dedicated, real-time emergency command and control system capable of coordinating multi-agency field responses. Its utility in disaster management is indirect, supporting administrative resilience rather than direct operational leadership.

Finally, while not yet a formalised government policy, academic contributions continue to inform the discourse on national preparedness. The Disaster Preparedness Framework for Sand and Dust Storms (SDS), notably proposed by Al-Qattan (2021), exemplifies this. This framework offers a model integrating public risk perception, incorporating institutional feedback mechanisms, and proposing hazard-specific readiness metrics tailored to one of Kuwait's most prevalent natural hazards. Although its adoption as formal policy remains pending, such research highlights the growing understanding of specific environmental challenges and contributes valuable insights for future policy development.

These contemporary frameworks and initiatives collectively demonstrate Kuwait's evolving commitment to building a more resilient national system. They highlight a growing recognition of vulnerabilities, from climate change to digital security. Yet, as reflected in Table 11.2, these efforts remain distributed across multiple agencies without a unifying legal framework or central coordinating authority. This fragmentation continues to constrain Kuwait's capacity to develop an integrated, all-hazards disaster management system.

11.5.5 Towards a Unified National Command: Legislative Aspirations and Enduring Obstacles

The increasing recognition of Kuwait's fragmented disaster management landscape, underscored by the experiences of sectoral responses and contemporary policy initiatives, has strengthened calls for a more centralised and integrated national authority. This fragmentation—long evident in practice and now widely acknowledged by policymakers—has driven recent legislative debates and proposals for establishing a central disaster management authority. This drive has become particularly pronounced within the Kuwaiti Parliament, where interest in establishing a unified crisis management body has increased in recent years. It gained significant traction following public criticism of the nation's COVID-19 response, which, despite individual ministerial efforts, revealed critical gaps in overarching coordination and rapid decision-making (OECD, The Covid-19 Crisis: A catalyst for government transformation?, 2020). The pandemic served as a stark reminder of the costs of institutional fragmentation.

Responding to this need and a persistent public demand for greater accountability and efficiency in national emergencies, members of the National Assembly have pursued legislative solutions. Notably, in 2017 and again with renewed vigour in 2022, parliamentary proposals emerged advocating the creation of a Central Crisis and Disaster Management Authority. These legislative drafts envisioned a centralised body reporting directly to the Prime Minister's Office, underscoring its proposed strategic importance and direct linkage to the highest levels of government (Council News, 2017; Times, 2022). The core tenets of these ambitious proposals called for a restructuring of Kuwait's disaster preparedness and response architecture. They aimed to integrate existing response entities—from the military and civil defence to specialised ministries and critical industries—within a single, cohesive legal and operational framework. Such an authority would be mandated with overarching responsibilities, including the oversight of national risk assessments, the development of preparedness strategies, the conduct of emergency drills, and the coordinated management of post-disaster recovery efforts. This mandate was intended to bridge the 'siloed' approach discussed earlier, replacing bilateral arrangements with a truly unified command structure.

Despite this parliamentary interest and apparent political support for enhanced national resilience, these legislative proposals have not materialised into law. The journey from parliamentary initiative to enacted legislation has faced significant internal challenges. Evidence from stakeholder interviews and extensive press coverage points to a complex interplay of factors hindering progress (Al-Rai, 2020; Al-Romi, 2020; Ezzahi, 2020). Foremost among these are bureaucratic inertia and entrenched ministerial turf protection. Existing ministries, having developed their own competencies and mandates, often resist ceding authority or resources to a new, overarching body. This reluctance reflects concerns regarding operational autonomy, budgetary control, and established internal procedures. Furthermore, limited legislative prioritisation has contributed to the delay; disaster management legislation has competed with other pressing economic, social, and political issues on the parliamentary agenda. The 2024 suspension of the National Assembly further disrupted legislative momentum, effectively halting progress on disaster governance reform. This continuing legislative impasse underscores the challenge of transitioning from a decentralised, sector-specific approach to a truly integrated and centrally commanded national disaster management system in Kuwait.

To summarise the trajectory of institutional development and highlight both the progress achieved and the persistent gaps hindering a fully integrated national disaster management system, Table 11.3 provides a chronological overview of key milestones in Kuwait's disaster risk management evolution from 1990 to 2025.

Despite these persistent institutional challenges and legislative roadblocks, Kuwait's recent engagement at the regional level signals a growing political willingness to advance disaster risk governance. The Kuwait Declaration on Disaster Risk Reduction, adopted during the 2025 Arab Regional Platform, explicitly calls for stronger national DRR strategies, enhanced governance structures, investment in

Table 11.3 Key institutional milestones in Kuwait's disaster risk management evolution (1990–2025)

Year	Milestone/event	Description	Source
1991	Kuwait emergency recovery programme (KERP)	Post-Gulf War recovery effort supported by international partners; restored infrastructure and public services, but no central disaster authority was established	U.S. Army Corps of Engineers (2002a)
2000s	KOC and sectoral agencies develop internal emergency protocols	Kuwait Oil Company, MOH, and KFF develop mostly siloed industrial and health emergency procedures	KOC (2021a, 2021b), KFF (2024), Abuzeyad et al. (2025)
2018	National criticism after flash floods	Urban flooding reveals major infrastructure weaknesses and lacks a coordinated disaster response strategy	Alghais (2019), GulfInsider (2020)
2019	Kuwait National Adaptation Plan (NAP) 2019–2030	Comprehensive climate resilience roadmap, identifying water, health, and infrastructure vulnerabilities	Kuwait EPA (2019)
2020	COVID-19 pandemic response	Multi-ministry effort exposed coordination gaps; the Audit Bureau cited an absence of a national crisis plan	Alarabiya (2021), Alnassafi (2022)
2022	Parliamentary proposal for central crisis and disaster management authority	A draft law was introduced to create a national coordination body but was not passed	Times (2022)
2025	Hosting of the sixth Arab regional platform for DRR and Kuwait declaration	Kuwait hosts a regional DRR summit, reaffirming its commitment to the Sendai framework and regional coordination	UNDRR (2025)

science and technology, and the development of comprehensive risk data systems (UNDRR, 2025). While these commitments remain aspirational, they reflect a growing political and institutional recognition of the need for systemic reform. This is further supported by recent empirical research showing significant variation in perceived disaster management competence across organisations in Kuwait, suggesting uneven readiness and increasing awareness of resilience deficits across sectors and regions (Al-Husain, 2023a). The transition towards a unified national command structure remains critical to Kuwait's future disaster resilience.

The persistence of these institutional gaps and stalled legislative reforms underscores why Kuwait continues to face significant challenges in disaster risk management. These challenges are examined in detail in the following section.

11.6 Challenges and Weaknesses in Disaster Management

While the preceding sections have outlined Kuwait's strides in developing sector-specific emergency response capacities and legislative aspirations for a centralised command, these advancements remain constrained by persistent systemic vulnerabilities. Despite episodic improvements in preparedness across various ministries and critical industries, Kuwait's national disaster management system continues to face fundamental weaknesses that undermine its ability to anticipate, coordinate, and recover effectively from major crises. These challenges extend beyond structural and legislative limitations, encompassing embedded cultural, operational, and strategic shortcomings. Building on the institutional evolution analysed in Sect. 11.5, the following discussion translates those governance patterns into tangible weaknesses that continue to undermine Kuwait's disaster risk reduction capacity. This section identifies and analyses the most salient of these deficiencies, demonstrating how they collectively impede effective disaster risk governance in Kuwait.

11.6.1 Structural and Legislative Deficiencies: The Core of Fragmentation

At the heart of Kuwait's ongoing challenges in disaster management lies the same institutional fragmentation traced in Sect. 11.5—further exacerbated by the absence of a permanent, centralised authority empowered with a comprehensive legal mandate—hereafter referenced to as the proposed Crisis and Disaster Management Authority. This structural gap means that while individual agencies, such as the Ministry of Interior, Kuwait Fire Force, and the Ministry of Health, possess strong technical capabilities and specialised expertise, their collective actions remain largely reactive and siloed. Each entity typically operates within its domain, often with limited joint planning or pre-established cross-ministerial coordination mechanisms. As a result, the national response relies heavily on temporary committees or ministerial task forces formed on an ad hoc basis, lacking the consistent oversight and rapid decision-making capacity of a dedicated central body.

The aspiration for such a unified command has been evident in repeated legislative attempts, with proposals to establish a Central Crisis and Disaster Management Authority introduced in the National Assembly on multiple occasions, as summarised in Sect. 11.5.4. However, to date, these legislative efforts have not materialised into binding law, leaving the fundamental leadership gap unaddressed.

This lack of a central coordinating body is intrinsically linked to Kuwait's broader siloed governance structure, where multiple ministries and agencies often operate in isolation, leading to overlapping mandates and coordination failures. For instance, during complex urban flood events—a recurring hazard in Kuwait—responsibilities are distributed across several entities: the Ministry of Public Works for stormwater drainage, the Environment Public Authority for environmental restoration, and the Kuwait Municipality for urban sanitation (Alfaras, 2020). Despite the importance of their respective roles, these entities often lack joint operational protocols and unified incident command systems. The consequence is often duplicated efforts, critical communication delays, and the inefficient deployment of resources, undermining the overall effectiveness of crisis response (Alfaras, 2020; Al-Qattan, 2021).

Compounding these structural and institutional issues is the absence of comprehensive disaster legislation and regulatory frameworks. Kuwait currently operates without a foundational Disaster Management Act that clearly defines roles, responsibilities, and enforcement mechanisms across all sectors. In the absence of such a legal framework, agencies cannot be held systematically accountable for lapses in preparedness, nor can national-level planning or mandatory drills be uniformly enforced across governmental and private sectors. Furthermore, there is no binding requirement for municipalities or private-sector entities to develop or regularly update emergency preparedness plans. This absence of enforceable regulation contributes to inconsistent risk assessments, unclear chains of command during emergencies, and insufficient proactive investment in resilience-building measures. Collectively, these structural and legislative deficiencies constitute the primary impediment to Kuwait's transition towards an integrated and proactive national disaster management system.

11.6.2 Operational and Communication Gaps: Consequences of a Disjointed System

The structural and legislative deficiencies discussed above manifest directly as operational inefficiencies and communication gaps within Kuwait's disaster management framework. A primary concern is the nation's underdeveloped early warning and risk communication systems, which frequently fail to provide timely and actionable information to the public and relevant stakeholders. The 2018 urban floods (see Sect. 11.4) exposed gaps in early warning, joint protocols, and public messaging, consistent with findings on fragmented inter-agency operations (Alghais, 2019; Al-Qattan, 2021). Similarly, research examining sand and dust storms (SDSs), a recurrent hazard in Kuwait, highlights inadequate public alert mechanisms, limited dissemination channels for critical information, and a lack of integration between meteorological forecasting data and civil defence operational protocols (Al-Qattan, 2021). These examples point to a persistent disconnect between hazard detection and effective public notification.

Compounding deficiencies in early warning is a broader problem of inconsistent and sometimes contradictory public communication channels during crises. The COVID-19 pandemic provided a clear example, as inconsistent announcements regarding travel restrictions, quarantine requirements, and school closures generated public uncertainty and frustration (Ezzahi, 2020). This communication breakdown reflected deeper underlying issues in cross-ministerial coordination, the absence of a unified public information command, and underdeveloped crisis communication strategies.

Independent assessments further corroborate these operational and communication deficiencies. A report issued by the Kuwaiti Audit Bureau in 2020 (State Audit Bureau, 2020), following its review of the government's COVID-19 response, identified systemic weaknesses. The audit specifically cited the absence of a national crisis plan, uncoordinated procurement protocols across government entities, and inconsistencies in emergency logistics and resource allocation. These findings reinforce the broader observation that limited inter-agency integration and weak oversight mechanisms continue to constrain Kuwait's national disaster response capacity.

More recent institutional reviews indicate that such gaps persist even within critical sectors such as the oil and energy industry. For instance, the 2022–2023 Audit Bureau report identified recurring compliance and resilience issues at the Kuwait Oil Tanker Company (KOTC), indicating limited integration of continuity planning into corporate governance structures (Times, 2024). Similarly, the KOC Annual Review 2023/2024 acknowledged that, while audit-driven reforms are underway, core systems related to business continuity and crisis automation remain in transition and are not yet fully institutionalised (KOC, Annual Review Report 2023/24, 2024).

Collectively, these shortcomings—spanning both governmental and sectoral levels—highlight the operational consequences of Kuwait's fragmented disaster management architecture and its direct impact on preparedness and response efficiency. These inefficiencies are not isolated incidents but symptoms of deeper systemic weaknesses in data management, knowledge utilisation, and citizen engagement—the focus of the next subsection.

11.6.3 Data, Knowledge, and Community Integration: Untapped Potential and Information Gaps

Effective disaster resilience relies not only on operational capacity but also on knowledge creation, data integration, and societal participation—dimensions in which Kuwait's progress has lagged behind institutional and infrastructural advances. In addition to the structural and operational challenges outlined above, Kuwait's disaster management system faces limitations in its ability to effectively leverage data and knowledge and to integrate the resources of its citizenry. A persistent gap is the absence of a centralised national disaster loss and risk data platform. Without reliable, accessible, and disaggregated data on past disasters—including detailed impacts, precise response times, accurate cost estimates, and comprehensive recovery timelines—government agencies, researchers, and policymakers lack the evidence base required to conduct accurate vulnerability assessments or to strategically prioritise mitigation investments. This deficiency creates a reactive rather than proactive environment, constraining informed decision-making for future preparedness. While the Kuwait National Adaptation Plan (NAP) 2019–2030 and the Kuwait Declaration on Disaster Risk Reduction (2025) emphasise the importance of data and science in risk management, a formal operational mechanism to achieve this crucial goal remains largely absent (Kuwait EPA, 2019; UNDRR, 2025). This disconnect between high-level policy aspirations and on-the-ground data operationalisation significantly impedes a data-driven approach to disaster risk reduction.

Simultaneously, there is a limited institutional framework for integrating community engagement and volunteer mobilisation despite a demonstrated history of civil society solidarity during national crises. Collective action was evident during the 1990 invasion and again following the 2015 mosque bombing, where citizens rallied spontaneously to support recovery efforts. However, this inherent community resilience remains largely untapped and uncoordinated by official bodies. Calls for the establishment of formalised neighbourhood-level disaster preparedness teams or structured volunteer integration strategies have frequently been raised in policy discussions and local media but have not been incorporated into a comprehensive national strategy (Al-Hamar, 2020; Al-Romi, 2020). The absence of a formal mechanism to channel this public participation limits the ability to coordinate citizen engagement, disseminate public awareness effectively, and build resilience at the community level. This reflects a broader shortfall in knowledge management, as lessons from past community-led efforts are not systematically captured or integrated into future planning. This disconnect between institutional mechanisms and civic capacity represents a missed opportunity to foster a culture of shared preparedness—a key principle aligned with the Sendai Framework's Priority 4 on 'Build Back Better'. These combined gaps in data management, knowledge integration, and community engagement constrain Kuwait's ability to maximise its collective capacity for disaster resilience.

11.6.4 *Underlying Impediments to Reform: Cultural Inertia and Political Will*

While the preceding sections have detailed Kuwait's structural deficiencies, operational shortcomings, and critical gaps in data and community integration, the persistence of these challenges can be attributed to deeper, systemic factors, namely, ingrained cultural inertia and bureaucratic and political resistance to reform. Numerous scholars and policy experts have identified these as fundamental barriers to the effective implementation of proposed improvements, irrespective of their technical soundness (Arezki, 2025; BTI, 2024; Thafer, 2024).

A prevalent tendency within Kuwait's administrative landscape is a reactive approach focused on crisis containment rather than proactive, long-term risk planning. This inclination is driven by a combination of factors, including political caution, a preference for maintaining the status quo, and the entrenched ministerial turf protection (Alfaras, 2020). Each ministry, having developed its own mandate and operational procedures, frequently exhibits reluctance to cede authority, share resources, or integrate under an overarching command structure. This resistance constitute a significant impediment to unified, cross sectoral planning and coordination, which are essential for effective disaster management.

Furthermore, reform momentum often falters due to a lack of sustained political leadership and administrative commitment. While parliamentary interest in a centralised authority has been evident, as discussed in Sect. 11.5.4, legislative efforts have repeatedly stalled. Without a decisive top-level commitment to prioritise and implement fundamental reforms, even technically sound proposals face delays, are deprioritised in practice, or become mired in inter-agency disagreements. This administrative inertia means that critical reforms designed to address the fragmented system, improve communication, enhance data utilisation, and integrate community efforts—all vital for strengthening national resilience—are either shelved indefinitely or implemented in a piecemeal manner. In summary, Kuwait's disaster management challenges are as much institutional and cultural as they are technical. Addressing theses challenges will require sustained political commitment, the enactment of enabling legislation, and the cultivation of a proactive governance culture that values prevention alongside response.

11.6.5 *Research Gaps and Knowledge Deficits in Kuwait's Disaster Management*

While the preceding sections have analysed the structural, legislative, and operational challenges affecting Kuwait's disaster management system, a deeper examination reveals a parallel and equally important set of limitations related to knowledge generation, evidence-based understanding, and strategic learning. Despite episodic improvements in sectoral preparedness, substantial gaps persist in the scholarly, institutional, and evaluative understanding of disaster risk governance in the Kuwaiti context. These research gaps impede evidence-based policymaking, limit cross-sectoral learning, and constrain Kuwait's capacity to design and implement forward-looking resilience strategies aligned with its specific vulnerabilities. Importantly, they also restrict Kuwait's ability to operationalise its commitments under international DRR frameworks, particularly the Sendai Framework for Disaster Risk Reduction. This section identifies and analyses the most salient areas where academic inquiry, institutional knowledge generation, and operational evaluation remain limited, affecting the country's adaptive capacity.

11.6.6 Overarching Knowledge and Programmatic Deficiencies: A Fragmented Research Landscape

A fundamental impediment to advancing disaster risk governance in Kuwait is the absence of a comprehensive national disaster research programme and a clear institutional mechanism to systematically study, synthesise, and apply insights related to disaster risk, resilience, and response performance. Academic investigations in these areas are typically siloed, project-based, and often initiated by individual researchers or isolated departments rather than coordinated through a national planning body or centralised research agenda (Alkandari & Mohamed, 2024; UNDRR, 2024). Unlike countries that have developed dedicated national disaster research centres or consortia, Kuwait lacks a permanent institutional platform to coordinate research efforts and translate findings into policy.

This decentralised and uncoordinated approach means that findings from sectors such as health, energy, environment, and urban planning are rarely synthesised into holistic national-level risk assessments or comprehensive preparedness models. The consequence of this fragmented research landscape is a disaggregated knowledge base and missed opportunities for integrated learning (Alfaras, 2020; Al-Husain, 2023a; Al-Qattan, 2021). Without a sustained national research programme, Kuwait faces challenges in building a cumulative body of evidence that can inform policy decisions, identify emerging threats, or evaluate the long-term effectiveness of its disaster management interventions. This also limits the country's ability to support decision-making through longitudinal studies, scenario modelling, or cross-sectoral knowledge transfer. This programmatic gap reinforces a reactive approach to crisis, as insights from individual events or sector are not systematically applied across the national framework, thereby constraining the development of proactive, evidence-based resilience strategies.

11.6.7 Data, Organisational, and Sustainability Research Gaps: Critical Knowledge Voids

As noted in Sect. 11.5.3, the absence of a national disaster loss database constrains research. This restricts longitudinal impact estimation, severity trend analysis, and the evaluation of policy effects. Essential disaster-related statistics—such as casualty figures, economic losses, and recovery durations—are either unavailable or inconsistently reported across ministries and agencies (Al-Hemoud et al., 2023). While entities such as the Kuwait Fire Force may record data for specific incidents, including fires and transportation accidents, there is no cohesive national disaster loss database. This absence of an integrated disaster information management system contrasts with global best practices in disaster risk reduction and monitoring.

Furthermore, organisational and institutional dimensions remain underexplored in systematic research. While survey-based studies, such as Al-Husain (2023b), examine how organisational characteristics—including sector, size, and location—affect preparedness, their findings are often underutilised in national policy development. There is limited follow-up research to validate these findings over time or to apply them to critical infrastructure sectors. More broadly, internal institutional dynamics—including communication cultures, decision hierarchies, and levels of inter-agency trust—remain under-researched. Despite their documented impact on crisis coordination and response efficiency, these organisational factors receive limited academic attention necessary to inform targeted interventions and support more collaborative environments.

Finally, sustainability and continuity frameworks remain inefficiently addressed in local research. Although sustainability has been incorporated into broader public policy, the alignment of disaster preparedness with long-term organisational continuity and operational resilience remains underexplored. Al-Husain (2023a), for example, has highlighted the importance of this alignment. However, its practical application within national plans like the Kuwait National Adaptation Plan (NAP) 2019–2030 or existing sectoral emergency procedures is rarely examined. Equally limited are studies linking disaster resilience with enterprise risk management practices, including supply chain resilience, ESG integration, and financial risk modelling. These knowledge gaps constrain the development of a holistic and forward-looking approach to national resilience, leaving critical blind spots in preparedness.

11.6.8 Human Capital and Community Engagement Research Gaps: Uncharted Social Dimensions

Another persistent weakness in Kuwait's disaster knowledge landscape concerns the human and social dimensions of resilience. Research gaps persist in understanding and optimising human capital and community engagement dimensions in disaster management. A notable deficiency lies in the educational and capacity-building infrastructure for disaster studies within the country (UNDRR, 2025). Available evidence indicates that Kuwait lacks a dedicated higher education framework for disaster management, with no comprehensive degree programmes currently offered at national universities. Consequently, most emergency management personnel acquire their skills through short, discipline-specific courses tailored to their roles (e.g. firefighting, police science), rather than through integrated academic curricula. As Alfaras (2020) argues, this absence of formal academic training reinforces reliance on experiential learning rather than standardised, evidence-based practices, thereby limiting long-term institutional learning and professionalisation within the sector.

Furthermore, Community-Based Disaster Risk Reduction (CBDRR) remains inadequately explored in the Kuwaiti context. Community engagement is critical for fostering a culture of preparedness, particularly in relation to recurring urban floods, dust storms, and potential health crises. However, existing research rarely examines how local communities perceive risk, mobilise during disasters, or interpret government interventions. While studies such as Al-Qattan (2021) and Al-Hamar (2020) highlight public dissatisfaction with official communication and identify low levels of preparedness at the household and neighbourhood levels, these findings have not been incorporated into a sustained national research agenda. The limited research on community risk perception and response behaviours constrains the development of culturally appropriate early warning systems and risk communication strategies. The limited empirical understanding of community dynamics and resilience at the grassroots level constitutes a significant barrier to developing effective and widely adopted national preparedness strategies, thereby constraining the development of a resilient society.

11.6.9 Comparative Analysis and Benchmarking Gaps: Measuring Against Global Standards

Finally, despite Kuwait's increasing engagement at the international level and its alignment with global disaster risk reduction (DRR) frameworks, a significant gap persists in comparative analysis and benchmarking (UNDRR, 2023). Kuwait has demonstrated commitment to international DRR initiatives, as reflected in its adoption of frameworks such as the Sendai Framework for Disaster Risk Reduction and the Arab Strategy for Disaster Risk Reduction 2030. The country's role as host of the Sixth Arab Regional Platform for DRR and the subsequent adoption of the Kuwait Declaration for Disaster Risk Reduction 2025 further reinforce this commitment (UNDRR, 2025).

However, this high-level alignment has not been accompanied by systematic academic literature or policy research benchmarking Kuwait's disaster management practices against international performance indicators. There is a lack of

Table 11.4 Major research gaps in Kuwait's disaster management system

Thematic area	Observed gap	Key sources
Disaster data and impact documentation	No centralised national disaster database	Alfaras (2020), Alkandari and Mohamed (2024), State Audit Bureau (2020), UNDRR (2024)
Organisational dynamics	Understudied influence of sector, size, and geography	(Al-Husain, 2023a)
Sustainability and continuity	Weak integration of resilience into operations and supply chain planning	Al-Hemoud et al. (2023), Al-Husain (2023b)
Academic and professional training	Absence of university-level disaster management programmes	Alfaras (2020), Al-Qattan (2021)
Community-level risk understanding	Lack of research on public perception, local action, and preparedness culture	Al-Hamar (2020), Al-Qattan (2021)
Comparative benchmarking	Limited analysis against Sendai, Hyogo, or global DRR metrics	Times (2022), UNDRR (2023, 2025)

studies comparing Kuwait's institutional effectiveness, operational capabilities, and resilience outcomes with those of other countries or against the targets set by global frameworks such as Sendai and Hyogo. This scarcity of comparative analysis limts Kuwait's evidence-based understanding of its relative strengths and weaknesses in the global context (Times, 2022). Without such benchmarking, opportunities to learn from international best practices, adapt successful strategies, and position Kuwait within the global DRR landscape are constrained. This gap hinders the country's ability to objectively assess its progress and advance towards comprehensive disaster resilience in line with international standards.

To consolidate the main findings of this section, Table 11.4 summarises the key research gaps in Kuwait's disaster management landscape, organised by thematic area, observed deficiency, and relevant supporting sources.

11.7 Recommendations and Future Strategies: Forging an Integrated and Resilient Kuwait

Moving beyond the challenges and knowledge gaps indentified, strengthening Kuwait's disaster risk governance requires a decisive paradigm shift. This transition involves moving from a reactive, fragmented crisis response towards an integrated, proactive, and sustainability-oriented approach to disaster risk reduction. Drawing on the analysis presented in previous sections, this part of the chapter outlines actionable recommendations. These recommendations are aligned with the four priority areas of the Sendai Framework for Disaster Risk Reduction—understanding disaster risk, strengthening disaster risk governance, investing in DRR, and enhancing disaster preparedness and resilience for recovery. They are grouped into four strategic domains: governance and legal reform, institutional capacity and operations, data-driven strategies and applied research, and public and inclusive resilience. Importantly, these recommendations are designed to address Kuwait's specific vulnerabilities while aligning with international frameworks, particularly the Sendai Framework for Disaster Risk Reduction, and to support the operationalisation of national commitments outlined in the Kuwait Declaration on Disaster Risk Reduction 2025 (UNDRR, 2025).

11.7.1 Governance and Legal Reform: Laying the Foundational Pillars

To establish clear authority and ensure systematic learning from past events, Kuwait must implement several key reforms.

First, Kuwait should establish a permanent, legally mandated Central Crisis and Disaster Management Authority. This entity should be positioned under the Council of Ministers, granting it cross-sectoral jurisdiction over planning, coordination, capacity development, and post-disaster recovery. Such a body would function as a national command centre, integrating disparate efforts into a coherent strategy. Achieving this objective requires revisiting and refining legislative proposals from 2017 and 2022 to ensure political feasibility and secure inter-agency support (Times, 2022).

Second, Kuwait should enact a comprehensive National Disaster Risk Management Law. The absence of such a framework constitutes a significant vulnerability. A national law would clarify roles, establish accountability mechanisms for preparedness and response, and provide the legal basis for public and private sector compliance with disaster preparedness regulations. It would also mandate cross-sectoral planning, drills, and coordination, replacing the current fragmented approach with an enforceable and integrated system.

Third, to foster a continuous learning environment and prevent the recurrence of systemic failures, formal processes for institutionalising After-Action Reviews (AARs) and post-disaster evaluations are essential. Lessons from major incidents—such as the 2018 flash floods and the COVID-19 pandemic—should be documented, analysed, and integrated into planning cycles, training programs, and policy adjustments. Embedding these processes within the legal framework ensures that lessons are translated into practice. This process should be coordinated through the proposed Crisis and Disaster Management Authority and linked to a national knowledge repository to support dissemination and application across institutions.

11.7.2 Enhancing Institutional Capacity and Operations: Building Practical Readiness

Beyond governance and legal reforms, a robust national disaster management system requires enhanced operational capacity across institutions and personnel. These recommendations focus on developing specialised strategies, strengthening operational coordination, and improving professional training across government entities.

A key step involves mandating and supporting the development of comprehensive, sector-specific Disaster Risk Reduction (DRR) strategies within major ministry and state enterprises, including the Kuwait Oil Company (KOC), the Ministry of Health (MOH), the Kuwait Fire Force (KFF), and the Environment Public Authority (EPA). These internal DRR plans should be aligned with the national strategy and incorporate business continuity planning, emergency simulations, and clearly defined interagency coordination protocols (Alfaras, 2020; Al-Husain, 2023b). Regular joint exercises across ministries should be mandated to ensure that operational coordination evolves from ad hoc arrangements to institutionalised practice.

To facilitate real-time information exchange and unified operational command, Kuwait should establish a national Emergency Operations Centre (EOC). At present, the country lacks a unified platform capable of serving as a central coordination hub for crisis response. Such an EOC would integrate functions including Geographic Information Systems (GIS) for dynamic mapping, comprehensive logistics management, secure communications, and centralised command functions, thereby improving decision-making and resource deployment during emergencies. The EOC should also be digitally linked to sectoral operation centres to ensure interoperability and rapid information exchange.

Finally, sustained investment in disaster training across all levels of government is essential. Training initiatives should extend beyond tactical simulations to include strategic, policy-level understanding of complex disaster systems. Achieving this requires partnerships with academic institutions to develop professional certificate programmes and establish specialised disaster leadership academies. This holistic approach will support the development of a professional cadre capable of informed decision-making and effective leadership, reducing reliance on improvisational practices and strengthening long-term institutional learning.

11.7.3 Data-Driven Strategies and Applied Research: Cultivating an Evidence-Based Approach

Building national resilience requires a strong foundation of reliable data and context-specific knowledge. These recommendations address gaps in information management and research, enabling Kuwait to advance evidence-based decision-making in disaster risk reduction.

A key priority is the development of a national disaster loss database and risk observatory. Kuwait should invest in developing a central, open-access platform capable of aggregating disaster data across ministries, municipalities, and key sectors. Such a system would support vulnerability mapping, risk modelling, and the prioritisation of mitigation investments, in line with commitments under the Kuwait Declaration on Disaster Risk Reduction (UNDRR, 2025). This centralised data repository is essential for understanding historical impacts and anticipating future risks. It should also be interoperable with global databases, such as the Emergency Events Database (EM-DAT), to facilitate comparative analysis.

Concurrently, sustained investment in applied research on disaster management and resilience is required. Universities and research centres should be supported to generate context-specific knowledge on hazards, analyse institutional performance, assess critical infrastructure risks, and study public behaviour during crises. Government support should prioritise interdisciplinary and policy-relevant research that informs national strategies and operational improvements (Al-Husain, 2023a). This approach will strengthen a local knowledge base aligned with Kuwait's unique environmental and social context.

Finally, disaster risk considerations should be integrated into national development planning. New development initiatives—across sectors such as housing, infrastructure, healthcare, and education—should incorporate disaster risk assessments at the design stage and throughout approval processes. By embedding risk considerations in planning, Kuwait can promote resilient and sustainable development aligned with global DRR best practices and reduce the creation of future vulnerabilities.

11.7.4 Fostering Public and Inclusive Resilience: Empowering Communities and Vulnerable Populations

Beyond governmental and institutional reforms, a resilient nation requires active public participation and a commitment to inclusivity. These recommendations focus on strengthening community engagement, enhancing public awareness, and ensuring that all population groups are included in disaster preparedness.

A key step involves developing a national public education campaign on disaster risk. Public awareness and individual preparedness are essential components of national resilience. Such a multi-platform campaign should address practical aspects, including household preparedness, risks associated with Kuwait's recurrent hazards (e.g. urban floods and sand and dust storms), emergency procedures, and opportunities for citizen volunteerism (Al-Qattan, 2021). This campaign would promote a culture of preparedness and encourage citizens to participate in disaster risk reduction. In line with global best practice, it should target diverse demographic groups through tailored messaging, including youth, seniors, and migrant workers.

To leverage civil society engagement, Community-Based Disaster Risk Reduction (CBDRR) initiatives should be formalised. This involves developing neighbourhood-level preparedness programmes, particularly in flood-prone or underserved areas. These programmes should be implemented in partnership with local Non-Governmental Organisations (NGOs), religious institutions, and educational institutions. They should include basic training, local drills, and the development of hazard maps, enabling communities to strengthen their preparedness and response capabilities.

Finally, Kuwait should ensure the inclusion of vulnerable and marginalised populations in all aspects of disaster planning and response. This requires tailoring communication strategies and evacuation plans to meet the needs of diverse groups, including people with disabilities, the significant migrant worker population, and low-income communities, many of whom often face higher risks during emergencies. Prioritising accessibility and cultural sensitivity in preparedness measures is both a matter of equity and a strategic requirement for comprehensive national resilience. Embedding these measures within the proposed Disaster Risk management (DRM) law will ensure they are institutionalised rather than implemented on an ad hoc basis.

11.8 Conclusion: Charting Kuwait's Path to Integrated Disaster Resilience

Kuwait's experience in disaster and crisis management reflects a duality between demonstrated societal and institutional resilience and persistent structural vulnerabilities. While the country has shown a strong capacity to mobilise resources and public solidarity during major events—from the 1990 Iraqi invasion to the 2018 floods and the recent COVID-19 pandemic—these responses have also exposed weaknesses in the underlying disaster governance framework. The coexistence of operational capacity and systemic fragility reflects a system in transition from reactive response to risk-informed planning.

This chapter has examined these dynamics, identifying a system characterised by institutional fragmentation, the absence of a central coordinating authority, and underdeveloped legislative frameworks. These structural issues contribute to operational inefficiencies, communication gaps, and limited integration of data and research into policy. Despite strong technical capabilities within individual ministries and agencies, efforts remain fragmented and reactive, constraining Kuwait's ability to transition from mere emergency response to genuinely risk-informed development. In addition, limited integration of sustainability principles, research, and community engagement into planning further reinforces these challenges.

Nonetheless, a growing recognition of these challenges is evident. Recent developments such as the Kuwait Declaration for Disaster Risk Reduction 2025 and growing academic engagement indicate a shift in Kuwait's disaster governance trajectory. Kuwait's growing regional engagement, including hosting the Sixth Arab Regional

Platform for Disaster Risk Reduction, further demonstrates its willingness to align with global frameworks such as the Sendai Framework for Disaster Risk Reduction and the Arab Strategy for Disaster Risk Reduction 2030. Emerging research, including national surveys and doctoral studies, highlights sectoral disparities, organisational constraints, and the need for enhanced capacity-building and policy reform. Together, these developments present an opportunity to modernise Kuwait's disaster management system.

Strengthening disaster risk governance requires a structural shift towards an integrated and proactive approach. This includes establishing a central authority, enacting comprehensive legislation, investing in preparedness across sectors, and fostering a culture of resilience. Equally crucial is embedding disaster risk reduction principles into national development, climate adaptation strategies, educational curricula, and community engagement programmes—ensuring that resilience is sustained rather than reactive.

Given geographic, economic, and political position in the Gulf region, Kuwait remains both exposed to regional risks and positioned to influence disaster governance across the MENA region. Strategic investments in governance, data infrastructure, community engagement, and international cooperation could position Kuwait as a leader in operationalising DRR principles in a resource-rich, hazard-exposed environment. With sustained policy commitment and institutional reform, Kuwait has the potential to safeguard its population and infrastructure while contributing to regional approaches to disaster risk management.

AI Disclosure Statement The author declares that generative AI tools (ChatGPT by OpenAI) were used solely to support language refinement, formatting suggestions, and typesetting assistance during this chapter preparation process. All scientific content, analysis, and conceptual contributions were entirely developed by the author. The use of AI did not influence or contribute to the originality, integrity, or intellectual contributions of the research. The author takes full responsibility for the accuracy, validity, and integrity of all content presented in the chapter.

References

Abuzeyad, F. H., Alburaidi, A., AlRazzuqi, H. H., Alkandari, F. M., Alqasem, L., & Qabandy, S. A. (2025). The status of emergency medicine in the state of Kuwait. *International Journal of Emergency Medicine,18*(28). https://doi.org/10.1186/s12245-025-00833-3

Alarabiya. (2021, May 12). *The Kuwaiti Audit Bureau reveals the absence of a government plan to manage crises.* Alarabiya. https://www.alarabiya.net/aswaq/economy/2021/05/12/%D8%AF%D9%8A%D9%88%D8%A7%D9%86-%D8%A7%D9%84%D9%85%D8%AD%D8%A7%D8%B3%D8%A8%D8%A9-%D8%A7%D9%84%D9%83%D9%88%D9%8A%D8%AA%D9%8A-%D9%8A%D9%83%D8%B4%D9%81-%D8%BA%D9%8A%D8%A7%D8%A8-%D8%AE%D8%B7%D8%A9-%D8%AD%D9%83

Alenezi, K. A., Al-Shaikhli, I. F., & AlAhmad, M. A. (2014). Kuwait government entities e-business continuity and disaster recovery plan. *Journal of Advanced Computer Science and Technology Research, 4*(3), 62–75. https://scispace.com/pdf/kuwait-government-entities-e-business-continuity-and-53am4xljei.pdf

Alfaras, M. (2020, May). *Developing an improved disaster management framework to enhance resilience in Kuwait.* ProQuest dissertations and theses, University of Salford, United

Kingdom. https://www.proquest.com/docview/2440341052?pq-origsite=gscholar&fromopenview=true&sourcetype=Dissertations%20&%20Theses

Alghais, N. (2019). Evaluation of the impacts of lack of geoinformation data in crisis management during the 2018 Kuwait flood. In *Geoinformation for disaster management* (Vol. XLII-3/W8, pp. 17–24). The International Archives of the Photogrammetry, Remote Sensing and Spatial Information Sciences. https://doi.org/10.5194/isprs-archives-XLII-3-W8-17-2019

Alkandari, R., & Mohamed, I. (2024). Assessing community perceptions of infrastructure resilience to natural disasters in Kuwait. *Electronic Interdisciplinary Miscellaneous Journal, 10*(77), 21. https://www.eimj.org/uplode/images/photo/amal.megda2..pdf

Alnassafi, F. (2022). Intellectual capital and its role in crisis management during the COVID-19 pandemic: An empirical study in Kuwait. *Journal of Asian Finance, Economics and Business, 9*(3), 0113–0121. https://doi.org/10.13106/jafeb.2022.vol9.no3.0113

Alsharamri, F. (2024, October 7). The Cabinet is discussing it and is ready for approval. It has been allocated to agencies in a region plagued by wars, natural and environmental disasters, and the risks of disease, epidemics, and cyber attacks. The Emergency and Crisis Management Authori. *MOHAMOON-KW*. https://www.mohamoon-kw.com/default.aspx?Catid=46&action=DisplayNews&id=55834&type=1&utm_source=chatgpt.com

Alwadi, Y., Al-Hemoud, A., Khraishah, H., Al-Mulla, F., Koutrakis, P., Ali, H., Alahmad, B. (2024). Impact of extreme heat on cardiovascular health in Kuwait: Present and future projections. *Journal of Epidemiology and Global Health, 14*, 1711–1719. https://doi.org/10.1007/s44197-024-00330-5

Arezki, R. (2025, February 11). *Monopolization is stifling Kuwait's economy—It's time to rethink top-down policies*. Atlantic Council. https://www.atlanticcouncil.org/in-depth-research-reports/books/monopolization-is-stifling-kuwaits-economy/

Ata, H. (2025, May 22). *Kuwait announces power cuts in 51 areas as temperatures near 50 °C*. Gulf News. https://gulfnews.com/world/gulf/kuwait/kuwait-announces-power-cuts-in-51-areas-as-temperatures-near-50c-1.500136351

Al-Damkhi, A. M., Abdul-Wahab, S. A., & Al-Khulaifi, N. M. (2009). Kuwait's 1991 environmental tragedy: Lessons learned. *Disaster Prevention and Management, 18*(3), 233–248. https://squ.elsevierpure.com/en/publications/kuwaits-1991-environmental-tragedy-lessons-learned

Al-Hamar, J. (2020, March 5). Kuwait and crisis management. *Al-Anbaa Newspaper* (p. 14).

Al-Hemoud, A., Al-Dousari, A., Misak, R., Al-Sudairawi, M., Naseeb, A., Al-Dashti, H., & Al-Dousari, N. (2019). Economic impact and risk assessment of sand and dust storms (SDS) on the oil and gas industry in Kuwait. *Sustainability, 11*(1), 200. https://doi.org/10.3390/su11010200

Al-Hemoud, A., Al-Enezi, A., Al-Dashti, H., Petrov, P., Misak, R., AlSaraf, M., & Malek, M. (2023). Hazard assessment and hazard mapping for Kuwait. *International Journal of Disaster Risk Science,14*, 143–161. https://doi.org/10.1007/s13753-023-00473-2

Al-Hemoud, A., Al-Dousari, A., Al-Dashti, H., Petrov, P., Al-Saleh, A., Al-Khafaji, S., Behbehani, W., Li, J., & Koutrakis, P. (2020, March). Sand and dust storm trajectories from Iraq Mesopotamian flood plain to Kuwait. *Science of the Total Environment, 25*(710), 136291. https://doi.org/10.1016/j.scitotenv.2019.136291

Al-Husain, R. (2023a). Promoting sustainability in Kuwait: An exploratory study of disaster management preparedness and resilience in state organizations. *Sustainability, 15*(13), 10066. https://doi.org/10.3390/su151310066

Al-Husain, R. (2023b). Unveiling the dynamics of organisational characteristics in disaster management: Insights from Kuwait. *Sustainability, 15*(17), 12860. https://doi.org/10.3390/su151712860

Al-Qattan, M. (2021, February). *An emergency preparedness framework for sand and dust storms (SDSs) in Kuwait*. https://usir.salford.ac.uk/id/eprint/60911/1/AlQattan-SDSs-in-Kuwait.pdf

Al-Rai. (2020, Febreuary 27). Crisis and disaster management. *Al-Rai*. https://www.alraimedia.com/article/880100/

Al-Romi, M. (2020, March 8). *Crisis and volunteer management authority*. Aljarida (5047). https://www.aljarida.com/articles/1583597700286965800/

Al-Sabah, S., Alsharqawi, N., & Haddad, E. A. (2018). The aftermath of the Kuwait Mosque bombing:. *International Journal of Surgery, 56*, 15–20. https://www.researchgate.net/publication/325623251

Bank, W. (2013). *Building resilience: Integrating climate and disaster risk into development—The World Bank Group experience*. World Bank Group. https://documents.worldbank.org/en/publication/documents-reports/documentdetail/762871468148506173

Barakat, S., & Skelton, J. (2014). *The reconstruction of post-war Kuwait: A missed opportunity?* London School of Economics. https://eprints.lse.ac.uk/55337/1/Barakat_Skelton_2014.pdf

Behbehani, A., Hasaniya, N., Abu-Zidan, F., & Merei, J. (1994). War in juries during the Gulf War: Experience of a teaching hospital in Kuwait. *Annals of the Royal College of Surgeons of England,76*(6), 407–411. https://www.researchgate.net/publication/15497491_War_injuries_during_the_Gulf_War_Experience_of_a_teaching_hospital_in_Kuwait

Boin, A., Hart, P. T., Stern, E., Stern, E., & Sundelius, B. (2005). *The politics of crisis management*. Cambridge University Press. https://www.cambridge.org/core/books/politics-of-crisis-management/D91A08EC1089E0463A1BEDBF244176FA?utm_source=chatgpt.com

BTI. (2024). *Kuwait country report 2024*. BTI Transformation Index. https://bti-project.org/en/reports/country-report/KWT

Comfort, L. K. (2007). Crisis management in hindsight: Cognition, communication, coordination, and control. *Public Administration Review, 67*(S1), 189–197. https://doi.org/10.1111/j.1540-6210.2007.00827.x

Council News. (2017, December 13). National Assembly. http://www.kna.kw/clt-html5/news-details.asp?id=28955

Ezzahi, M. (2020, March 7). *Specialists to "Al-Qabas": For a clear strategy to deal with emergencies*. Al-Qabas. https://alqabas.com/article/5758224

GulfInsider. (2020, November 29). *Heavy rains flood streets, schools and hospitals in Kuwait*. GulfInsider. https://www.gulf-insider.com/heavy-rains-flood-streets-schools-and-hospitals-in-kuwait/

Kapucu, N. (2006). Interagency communication networks during emergencies: Boundary spanners in multiagency coordination. *The American Review of Public Administration, 36*(2), 207–225. https://doi.org/10.1177/0275074005280605

Kapucu, N., & Garayev, V. (2016). Structure and network performance: Horizontal and vertical networks in emergency management. *Administration & Society, 48*(8), 931–961. https://doi.org/10.1177/0095399714541270

Kelman, I. (2011). *Disaster diplomacy: How disasters affect peace and conflict*. Routledge. https://doi.org/10.4324/9780203806210

KFF, K. F. (2024, November 10). *Kuwait's Fire Force's reserve room to ensure continuous service during emergencies*. The Times. https://timeskuwait.com/kuwaits-fire-forces-reserve-room-to-ensure-continuous-service-during-emergencies/

KISR. (n.d.). *Kuwait Institute for Scientific Research*. Crisis Decision Support. https://www.kisr.edu.kw/en/program/13/

KNPC, K. N. (2020). *Non-technical summary: Clean fuels project environmental impact study*. https://www.knpc.com/KNPC/media/pdf/CFP-Environmental-Impact-Study-EIA.pdf

KOC. (2021a, March 29). *KOC conducts disaster management exercise*. Al-Kuwaiti. https://www.kockw.com/sites/EN/EMagazine/Pages/HSE/139.aspx

KOC. (2021b, March 29). *KOC conducts disaster management exercise*. Kuwait Oil Company. Retrieved 2023, from https://www.kockw.com/sites/EN/Pages/Media%20Centre/News%20And%20Events/139.aspx

KOC, K. O. (2022, April 24). *Environment cooperation appreciation: KOC conducts annual crisis and disaster management exercise*. KOC News (816). https://www.kockw.com/sites/EN/Periodicals/KOC%20News%20816%20English.pdf

KOC. (2024). *Annual review report 2023/24*. https://www.kockw.com/sites/EN/Other%20Publications/HSSE/HSE%20Annual%20Review%202023--2024.pdf?utm_source=chatgpt.com

KUNA. (2024, October 8). *Kuwait Cabinet reassures about readiness for any emergency.* Kuwait News Agency KUNA. https://www.kuna.net.kw/ArticleDetails.aspx?id=3185973&Language=en

Kuwait EPA, K. E. (2019). *Kuwait National Adaptation Plan 2019–2030: Enhanced climate resilience to improve community livelihood and achieve sustainability.* https://unfccc.int/sites/default/files/resource/Kuwait-NAP-2019-2030.pdf

Kuwait Fire Force. (2025, May 20). *KFF, Army sign protocol to boost joint operations.* Kuwait Times. https://kuwaittimes.com/article/28173/kuwait/other-news/kff-army-sign-protocol-to-boost-joint-operations/

Kuwait Oil Company. (2019, November 11). *Enterprise risk management communication gate.* Al-Kuwaiti. https://www.kockw.com/sites/EN/EMagazine/Pages/Participation/Enterprise-Risk-Management-Communication-Gate.aspx

Kuwait Oil Company. (n.d.). *Non-technical summary: KOC drilling operation.* https://www.kockw.com/sites/EN/External_Financing_Projects/KOC%20Drilling%20Operation.pdf

Kuwait Times. (2025, May 1). *Zain conducts evacuation drill at headquarters.* https://kuwaittimes.com/article/27185/kuwait/other-news/zain-conducts-evacuation-drill-at-headquarters/

Lange, J. L., Schwartz, D. A., Doebbeling, B. N., Heller, J. M., & Thorne, P. S. (2002). Exposures to the Kuwait oil fires and their association with asthma and bronchitis among Gulf War veterans. *Environmental Health Perspectives,110*(11), 1141–1146. https://doi.org/10.1289/ehp.021101141

LAS. (2010 & Updates). *The Arab strategy for disaster risk reduction.* The Council of Arab Ministers Responsible for the Environment. https://www.unisdr.org/files/17934_asdrrfinalenglishjanuary2011.pdf?utm_source=chatgpt.com

Lassa, J. A. (2015). Post disaster governance, complexity and network theory. *PLoS Curr, 7*(ecurrents.4f7972ecec1b6). https://doi.org/10.1371/4f7972ecec1b6

Mileti, D. (1999). *Disasters by design: A reassessment of natural hazards in the United States.* Joseph Henry Press.

Ministry of Health. (n.d.). *Critical care unit bed crisis management policy.* https://www.moh.gov.kw/UserRefs/Technical%20Dept/ICUTech.pdf

Ministry of Interior. (n.d.). *Safety guide for emergency: The civil defence.* General Civil Defence Department, State of Kuwait. https://www.moi.gov.kw/main/content/docs/civildefence/en/safety-guide-for-emergency.pdf

Nowell, B., & Steelman, T. (2015). Communication under fire: The role of embeddedness in the emergence and efficacy of disaster response communication networks. *Journal of Public Administration Research and Theory, 25*(3), 929–952. https://doi.org/10.1093/jopart/muu021

OECD. (2018). *Assessing global progress in the governance of critical risks.* OECD Publishing. https://doi.org/10.1787/9789264309272-en

OECD. (2020). *The Covid-19 crisis: A catalyst for government transformation?* OECD. https://oecd-opsi.org/wp-content/uploads/2022/03/covid-catalyst.pdf

Persian Gulf War. (2025). *Britannica.* https://www.britannica.com/event/Persian-Gulf-War

PIC, P. I. (2021). *Sustainability report 2020–21.* https://www.pic.com.kw/wp-content/uploads/2022/02/PIC_Sustainability_Report_2020-2021.pdf

Reuters. (2025, April 2). *Kuwait implements power cuts as demand outstrips restricted capacity.* Reuters. https://www.reuters.com/business/energy/kuwait-implements-power-cuts-demand-outstrips-restricted-capacity-2025-04-02/

Ridha, H., & Orlin, M. (1996). Dealing with disaster victims: A Kuwaiti model. *Journal of the Social Sciences, 24*(4), 236–258. https://journals.ku.edu.kw/jss/index.php/jss/article/download/1361/617/10049

State Audit Bureau. (2020). Covid-19 crises. *Alraqaba* (16). https://www.sab.gov.kw/sabweb/Files/SABMagazinePublishedIssue/%D8%A7%D9%84%D8%B3%D8%A7%D8%AF%D8%B3%20%D8%B9%D8%B4%D8%B1%20%D8%A5%D9%86%D8%AC%D9%84%D9%8A%D8%B2%D9%8A/files/assets/common/downloads/AlRaqaba%20English%2016.pdf

Steelman, T., Nowell, B., Velez, A.-L., & Scott, R. (2021). Pathways of representation in network governance: Evidence from multi-jurisdictional disasters. *Journal of Public Administration Research and Theory, 31*(4), 723–739. https://doi.org/10.1093/jopart/muab004

Thafer, D. (2024, June 2). *Interview: Unpacking Kuwait's parliament dissolution*. Middle East Council on Global Affairs. https://mecouncil.org/blog_posts/interview-unpacking-kuwaits-parliament-dissolution/

Tierney, K. (2012). Disaster governance: Social, political, and economic dimensions. *Annual Review of Environment and Resources,37*, 341–363. https://doi.org/10.1146/annurev-environ-020911-095618

Times, T. (2022, April 14). *Proposal to form Central Agency for crisis, disaster management approved*. The Times. Retrieved May 12, 2023, from https://timeskuwait.com/news/proposal-to-form-central-agency-for-crisis-disaster-management-approved/

Times, T. (2024, February 5). *Audit Bureau report highlights KOTC compliance, operational efficiency*. The Times. https://timeskuwait.com/audit-bureau-report-highlights-kotc-compliance-operational-efficiency/?utm_source=chatgpt.com

U.S. Army Corps of Engineers. (2002a, January). *U.S. Army Corps, Engineers—"Historical Vignette 042—The Corps Helped Kuwait Recover from the Gulf War"*. U.S. Army Corps of Engineers Headquarters. https://www.usace.army.mil/About/History/Historical-Vignettes/Military-Construction-Combat/042-Kuwait-Recover.aspx

U.S. Army Corps of Engineers. (2002b). *Historical Vignette 043—Reconstruct Kuwait Following the Persian Gulf War*. U.S. Army Corps. https://www.usace.army.mil/About/History/Historical-Vignettes/Military-Construction-Combat/043-Reconstruct-Kuwait.aspx

UNDP. (2015). *Kuwait integrated environmental management system (KIEMS): Evaluation report*. https://erc.undp.org/evaluation/documents/download/9312

UNDRR. (2015). *Sendai framework for disaster risk reduction 2015—2030*. United Nations Office for Disaster Risk Reduction. https://www.undrr.org

UNDRR. (2019). *Global assessment report on disaster risk reduction (GAR)*. UNDRR Publications. https://www.undrr.org/gar

UNDRR. (2022, May 27). *Prevention web*. Breaking silos to build resilience—Multi-hazard, multi-sectoral approaches to managing disaster risks. https://www.preventionweb.net/news/breaking-silos-build-resilience-multi-hazard-multi-sectoral-approaches-managing-disaster-risks

UNDRR. (2023). *Regional report: Midterm review of the implementation of the Sendai framework for disaster risk reduction 2015–2030 for Arab States*. United Nations Office for Disaster Risk Reduction. https://www.undrr.org/publication/regional-report-midterm-review-implementation-sendai-framework-disaster-risk-3

UNDRR. (2024, February 12). *Complex disaster risks call for urgent action in the Arab region*. Prevention Web. https://www.preventionweb.net/news/complex-disaster-risks-call-urgent-action-arab-region

UNDRR. (2025). *Kuwait declaration for disaster risk reduction*. United Nations Office for Disaster Risk Reduction. https://www.unisdr.org

UNESCWA. (2013). *National profile of the information society in Kuwait*. https://www.unescwa.org/sites/default/files/inline-files/Kuwait-2013-En.pdf

Waugh, W. L., & Streib, G. (2006). Collaboration and leadership for effective emergency management. *Public Administration Review, 66*(S1), 131–140. https://doi.org/10.1111/j.1540-6210.2006.00673.x

WHO. (2024). *Kuwait: A primary health care case study in the context of the COVID-19 pandemic*. World Health Organization. https://iris.who.int/bitstream/handle/10665/376640/9789240084711-eng.pdf

Chapter 12
Disaster, Crisis, and Emergency Management in the Republic of Iraq

Bassim M. Hashim

Abstract Iraq is among the countries most affected by climate change globally, suffering from severe climate crises such as water scarcity, drought, desertification, floods, dust storms, and the COVID-19 pandemic. These crises have led to climate-related population displacement, agricultural land degradation, and the drying up of marshlands, particularly in central and southern Iraq. Furthermore, decades of war and conflict have exacerbated the crisis and led to the emergence of extremist organisations such as ISIS, causing a large-scale internal displacement crisis, particularly in northern and western Iraq since 2014. The problem is further accentuated by weak infrastructure, financial and administrative corruption, and deteriorating services and living conditions in most Iraqi governorates. The main issue addressed by this chapter is the urgent need for effective and proactive strategies to manage risks, enhance scientific research, build national capacities, increase funding, and improve coordination and partnerships at all levels of government to address these multifaceted challenges. The findings show that a reactive approach ensures the continued accumulation of new risks, making a radical strategic shift from 'disaster management' to a proactive and preventive disaster risk management framework an imperative national necessity for achieving stability and resilience.

Keywords Iraq · Water crisis · Drought · Floods · Climate change · Disaster management

12.1 Introduction to the Country

The Republic of Iraq is located in the eastern region of the Middle East in Western Asia. Iraq is bordered by Kuwait and Saudi Arabia to the south, Jordan and Syria to the west, Turkey to the north, and Iran to the east (Ministry of Environment, 2024). Iraq is located between longitudes (38°47′, 48°35′E) and latitudes (29°02′, 37°23′N). Iraq's overall area is 437,072 km^2, and it includes 58 km of shoreline in

B. M. Hashim (✉)
Scientific Research Commission, Baghdad, Iraq
e-mail: bassim_hashim@src.edu.iq

E. Alam (ed.), *Disaster, Crisis and Emergency Management in the Middle East & North Africa*, https://doi.org/10.1007/978-3-032-24436-9_12

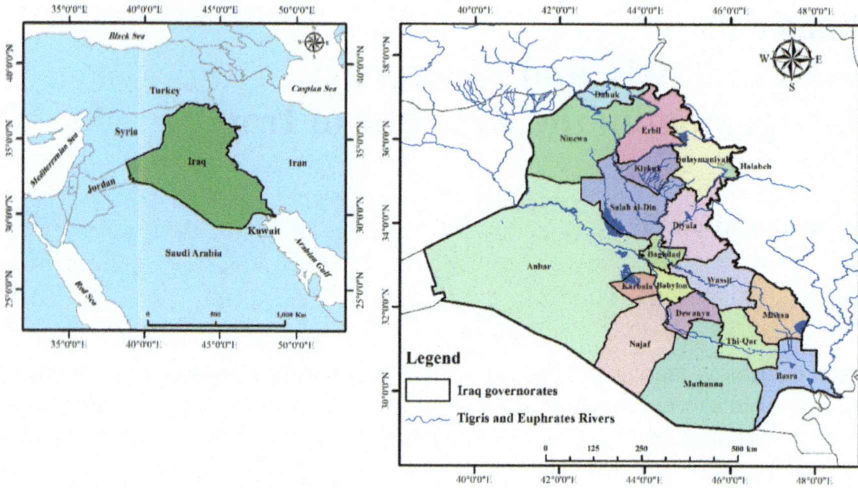

Fig. 12.1 Location of Iraq in the Middle East

Basra governorate, which is located along the Arabian Gulf's northern coast (Hashim et al., 2024). The fertile region of Iraq is divided between the Tigris and Euphrates, two significant rivers that flow from Turkey towards the Arabian Gulf, and Mesopotamia is the name given to this region (World Food Programme (WFP), 2024a). Figure 12.1 shows the location of Iraq within the Middle East.

Iraq's climate is semi-arid, continental, and subtropical in its centre and southern regions. The climate is Mediterranean in the north, with temperatures ranging from hot to extremely hot (World Food Programme (WFP), 2024b). Iraq's climate is characterised by severe temperatures and significant daily and annual thermal variations due to the lack of bodies of water to potentially reduce the country's winter cold and summer heat (Hameed et al., 2018). The 2024 comprehensive population census reported that there were over 46 million people living in Iraq. Of Iraq's population, 29.83% reside in rural areas while 70.17% live in cities. Additionally, the education, health, housing, and youth sectors are under heightened pressure due to the rapid population growth (Central Statistical Organization (CSO), 2024). The backbone of the Iraqi economy of Iraq is the oil and gas sector. It is the biggest industry and accounts for more than 90% of all government income, almost 60% of gross domestic product (GDP), and 95% of revenue from exports (Energy Information Administration (EIA), 2021). With 5% of GDP, the agriculture sector is one of the biggest non-oil sectors of the Iraqi economy and provides most jobs about 30% (CARBON LIMITS, 2022). The Republic of Iraq was established in 1921 and is a single, independent, and totally sovereign federal state. The federal system in Iraq is composed of a capital, Baghdad, as well as decentralised regions and provinces, as well as municipal administrations. The chapter examines the accumulated and complex disaster and crisis management challenges faced by Iraq. There are several significant research gaps in disaster management in Iraq that require attention from academic and research communities, including:

1. Studies assessing disaster risks such as drought, desertification, floods, earthquakes, environmental, and health risks, and the development of appropriate early warning and response systems.
2. Studies examining how to integrate disaster risk management into recovery and reconstruction plans in conflict-affected areas.
3. Absence of a unified and comprehensive national risk database in Iraq, which hinders planning and prevention efforts.
4. Lack of research on the long-term effects of displacement and unsustainable return processes.
5. The impact of disasters and crises on the most vulnerable groups (women, children, and people with disabilities) has not been adequately studied.

The chapter primarily employed the descriptive analytical approach as the main methodological framework for understanding and analysing the reality of disaster management in Iraq. It also relied on a quantitative approach to analysing statistical and numerical data related to crises and disasters, using a wide range of quantitative data, such as demographic and economic statistics, losses, and individuals affected by various disasters.

The significance of this chapter lies in its contribution to building a comprehensive conceptual framework for disaster risk management in Iraq, and in providing accurate data and in-depth analysis on the types of disasters Iraq faces. The objectives of the chapter are:

1. Highlighting the disasters, crises, and emergencies that Iraq has faced in recent and current years.
2. The role of Iraqi and international governmental institutions in responding to, mitigating, and recovering from risks and crises.
3. Identifying the strengths and weaknesses of the institutional and legislative frameworks in Iraq to address various crises and risks.

12.2 Research Methods

12.2.1 Research Approaches

This chapter employs an integrated descriptive and analytical approach. In the descriptive aspect, data related to the reality of disasters and crises in Iraq were collected, classified, and documented. The chronology of events was presented, the types of prevailing risks were identified, and human and material losses were estimated using official statistics and reports from international organisations. In the analytical aspect, the analysis moved beyond mere description to evaluate the effectiveness of existing institutional and legislative frameworks for managing these disasters. This included analysing the content of government documents and policies, comparing them with international frameworks such as the Sendai Framework, identifying strengths and weaknesses, and elaborating the root causes behind the repeated failures or limited successes in response and prevention efforts.

Recognising the notable scarcity of peer-reviewed academic literature specifically addressing disaster risk management within the contemporary Iraqi context, this chapter deliberately employs a mixed-source approach. Reports from international and national organisations are not treated merely as secondary sources but as primary data providers, offering essential, up-to-date empirical evidence and field-level analysis that is otherwise unavailable. This approach, while acknowledging the limitations of non-peer-reviewed data, is methodologically necessary to construct an accurate and timely profile of Iraq's risk landscape.

12.2.2 Data Sources

The chapter's database was primarily built upon Iraqi official sources, which drew from a vast array of documents, reports, and statistics published by different Iraqi government organisations. United Nations agencies provided an objective external perspective and crucial comparative data among the international and regional data sources used. In contrast, there is a notable scarcity of peer-reviewed academic sources in Iraq that address disaster risk assessment and management.

Used Landsat images acquired from the United State Geological Survey (USGS) by earth explorer website (https://earthexplorer.usgs.gov/) to study the southern Iraqi marshlands for five years from 2020 to 2025. The 18 images acquired in July based on the Landsat 8 and 9 OLI during 2020–2025 were utilised to monitor environmental changes and the effects of drought in marshlands. Three satellite images were downloaded for each year, which cover the marshlands in southern Iraq. Then, the images were combined using a mosaic to determine the changes in the marshland's water body. The ArcGIS 10.8 software was used to determine the marshland region and changes in water area. Modern events were also documented, and current information on crises and disasters in Iraq was gathered from trustworthy media and journalistic sources.

12.2.3 Data Analysis

To achieve a profound, multidimensional understanding, the analysis followed a sequential interpretive design within a mixed-method research methodology. In the first (quantitative) phase, statistical and numerical data (such as numbers of displaced persons, economic losses, and disaster frequency) were analysed to determine the magnitude and trends of crises in Iraq and map the most affected areas. In the second (qualitative) phase, the results of the quantitative analysis were used to guide the qualitative analysis. For example, the content of government reports, laws, and policies was analysed to understand and explain the reasons behind observed statistical trends, such as the reasons for the inadequacy of response to floods in a particular governorate despite their recurrence. Finally, geospatial data from Landsat satellite imagery was integrated to provide a visual and spatial dimension to the analysis, linking observed changes in the marshlands' water cover (spatial data) with population displacement statistics (quantitative data) and water policy shortcomings (qualitative data), allowing for more robust and integrated conclusions.

12.3 Disaster History and Risk Profile

Historically, Iraq was known as the 'Land of Blackness' due to the Tigris and Euphrates rivers' abundance of crops and palm palms. Iraq experienced many crises and disasters during the twentieth century, which profoundly impacted its history. These events, which included wars and political conflicts to disasters, presented serious difficulties for the government and society. Iraq had numerous natural hazards during the twentieth century, from the threat of terrible floods in the first half of the century to the severe drought and desertification in the second half. Due to its diverse topography, Iraq is exposed to drought, flash floods, sandstorms, and desertification (International Organization for Migration (IOM), 2023). On the other hand, rising temperatures and decreasing precipitation are two major negative effects of climate change that Iraq is currently experiencing, which are causing drought, desertification, and water scarcity (CARBON LIMITS, 2022). Water scarcity, pollution, and salinisation of the land are already causing a humanitarian crisis and internal displacement in southern Iraq. Drought-related displacement is also reported in historical records (International Organization for Migration (IOM), 2012). On the other hand, Iraq's location, climate variability, sociopolitical instability, emergency system fragility, and inadequate water management system all contribute to the country's ongoing environmental and humanitarian issues, particularly flash floods (Gleick, 2019). In addition to reducing water supplies, dust storms that hit Iraq continuously, especially during dry seasons, exacerbate desertification, and soil salinity. Both people's livelihoods and health are severely impacted by this extreme event (International Organization for Migration (IOM), 2022).

12.3.1 Types of Hazards, Crises, and Disasters Are Most Prevalent in Iraq

According to the United Nations Environmental Programme's (UNEP), based on the Global Environment Outlook 6 (GEO-6) report, Iraq is the fifth most vulnerable nation in the world to the consequences of climate change, which include drought, desertification, low precipitation, rising temperatures, and water scarcity (United Nations Environment Programme (UNEP), 2016). Due to several internal and external issues, such as inadequate management of water resources, internal tensions, lack of local policies, climate change, and fragile relations with neighbouring countries, Iraq is currently at heightened risk of water shortages (Al-Muqdadi et al., 2016). Many decades of conflict have severely damaged Iraq's water infrastructure, which remains unprepared for the challenges of the present and the future. Due to a severe water deficit, almost 70% of Iraqi farmers still employ flood irrigation (Norwegian Refugee Council (NRC), 2024). Local tension between tribes is already largely caused by droughts and a shortage of water in southern governorates, such as Missan and Dhi Qar (Export Working Group, 2018).

Recently, flash floods have increased in frequency, with hotspots found in many Iraqi governorates, particularly in the country's east on the Iranian border (International Trade Centre, 2021). Flooding is frequently caused by both excessive rainfall and the significantly diminished ability of soil to absorb precipitation due to land degradation. In addition to being a health risk, heavy rainfall and flooding may increase the spread of waterborne illnesses (International Organization for Migration (IOM), 2022). Earthquakes are one of the natural hazards that have affected Iraq over the past ten years. Iraq's impact from these earthquakes is relatively small, except for the border strip with Iran, particularly Wassit province, as well as the governorates of Sulaymaniyah, Erbil, Dahuk, and Ninewa in northern Iraq (Jassim & Goff, 2006). In the last ten years, 1042 earthquakes of magnitude four or higher have occurred within 300 km of Iraq. On November 12, 2017, the most powerful earthquake in the last ten years happened close to Iraq. It struck 87 km southeast of Sulaymaniyah at a depth of 19 km s and had a magnitude of 7.3 on the Richter Scale (https://earthquakelist.org/iraq/). Figure 12.2 shows the intensity of drought, floods, and earthquakes in Iraq during 2003–2024.

One of the biggest internal displacement crises in the modern era occurred in Iraq because of the Islamic State in Iraq and Syria (ISIS) after 2014. More than six million Iraqis, or approximately 15% of the nation's overall population, have been displaced by ISIS. This catastrophe included the genocide of thousands of Yazidis in Ninewa province by ISIS and the displacement of the remaining population inside Iraq and other countries (International Organization for Migration (IOM), 2018). This crisis has been concentrated in specific areas of Iraq, particularly those under ISIS control or under military operations to retake them. The governorates of Ninewa, Anbar, and Salah al-Din are the most affected (Displacements Tracking Matrix (DTM) IRAQ, 2025). The humanitarian situation in Iraq remains characterised by instability, due to prolonged internal displacement and inadequate shelter and accommodation, even after the fight with ISIL ended in 2017 (International Organization for Migration (IOM), 2023).

The COVID-19 pandemic is one of the largest health crises Iraq has faced in recent decades. The first confirmed case of COVID-19 virus in Iraq was on February 24, 2020, in Najaf, southern Baghdad (Hashim et al., 2021). From March 1 to August 31, 2020, the Iraqi government announced several partial and total lockdown measures to prevent the spread of COVID-19 in the country. These measures included closing

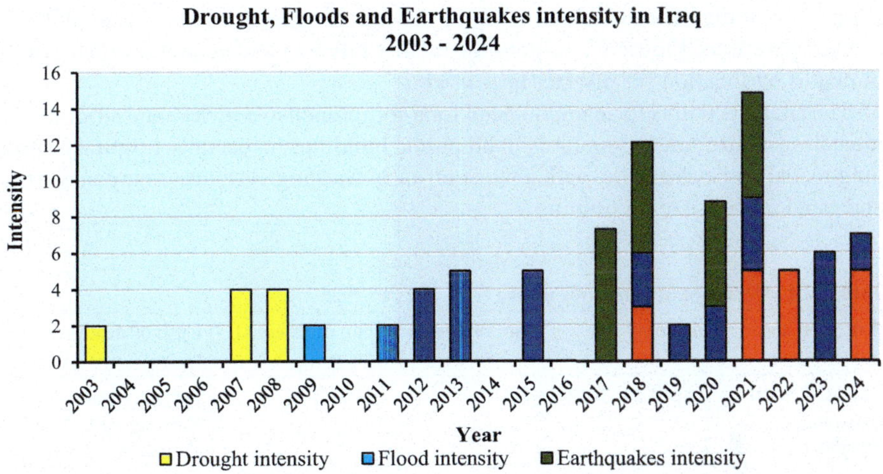

Fig. 12.2 Events of drought, floods, and earthquakes in Iraq from 2003 to 2024 (https://reliefweb.int/country/irq#disasters/; https://floodlist.com/tag/iraq; https://earthquakelist.org/iraq/)

universities and schools and limiting travel between governorates and restricting people's ability to move within them (Hashim et al., 2023). With the onset of the second wave of infections in May 2021, the COVID-19 Crisis Cell reimposed strict measures, such as a 10-day curfew during Eid al-Fitr. The measures also included road closures within Baghdad, the road to the Kurdistan Region, and the imposition of fines on violators. The total number of confirmed cases from February 2020 to April 2024 reached 2,465,545, while the death toll exceeded 25,375 (Our World in Data, 2025).

Iraq has experienced several catastrophic fires in recent years, claiming hundreds of lives and exposing fundamental weaknesses in safety systems and emergency plans. Table 12.1 outlines the location and timing of the four tragic fires from 2021 to 2025 (Al-Hussein Hospital, Nasiriyah, 2025; Al Kut Hypermarket, 2025; Ibn al-Khatib Hospital fire, 2025; Wedding hall in Al-Hamdaniya, Ninewa, 2025). The results revealed that the Al-Hamdaniya wedding hall accident was the deadliest in terms of the number of deaths (107 deaths), while the Al-Kut hypermarket accident recorded the largest number of injuries (200 injuries). The overall trend shows no clear decrease in the severity of these disasters over time, suggesting the persistence of their root causes. Pattern analysis revealed a clear pattern of targeting two types of locations: healthcare facilities, which are supposed to be the safest, and crowded public spaces with high traffic volumes. Hospitals accounted for 49.6% of the fatalities, both of which were directly linked to COVID-19 treatment and isolation units. The fatality rate (the percentage of deaths as a percentage of total casualties) varied significantly between incidents. Al-Hussein hospital recorded the highest fatality rate of 64.8%, meaning the probability of death was significantly higher than the probability of injury in that incident. In contrast, Al Kut hypermarket recorded the lowest fatality rate of 25.9%, despite recording the highest number of casualties, which may indicate a different nature of the fire or greater effectiveness of initial rescue operations. Figure 12.3 shows the fatality rate by incident and the distribution of deaths by location for disaster fires in Iraq.

The analysis points to an urgent need for a comprehensive review and strict implementation of fire safety standards in all public buildings, especially hospitals, shopping malls, and event halls, with a focus on early warning systems, emergency exits, and rapid response team training.

Table 12.1 Details of disaster fires in Iraq from 2021 to 2025

Location and governorate	Date	No. of deaths	No. of infected people	Case of deaths
Ibn Al-Khatib Hospital, Baghdad	April 24, 2021	82	110	A specialised centre for treating COVID 19 patients
Al-Hussein Hospital, Nasiriyah	July 12, 2021	92	50	COVID-19 isolation centre
Wedding hall in Al-Hamdaniya, Ninewa	Sep. 26, 2023	107	82	Guests at an Assyrian Christian wedding
Al-Kut Hypermarket	July 16, 2025	61–77	200	Most of the victims are women and children

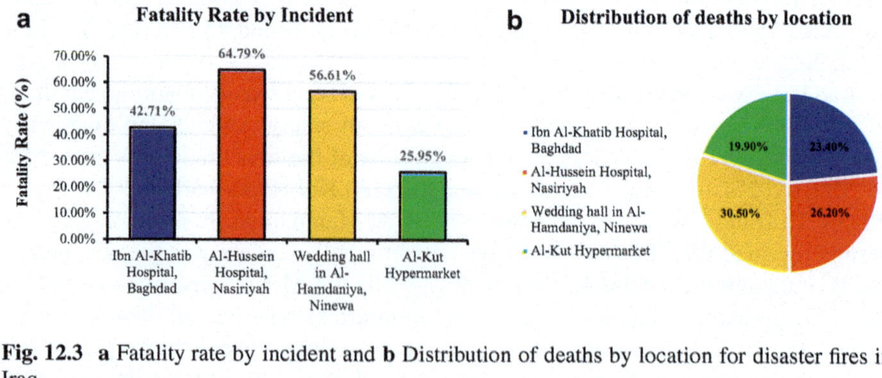

Fig. 12.3 **a** Fatality rate by incident and **b** Distribution of deaths by location for disaster fires in Iraq

12.3.2 Impact of Disasters on Communities, Infrastructure, and the Economy

Iraq has experienced several wars, authoritarian governance, the 2003 US-led invasion, sectarian violence, and confrontation with ISIS during the past four decades (World Food Programme (WFP), 2024c). These accumulated crises have led to a lack of infrastructure, widespread financial and administrative corruption, and deterioration in services and living conditions in most Iraqi governorates. In 2025, Iraq is facing the worst drought and water shortage since 1933 due to a lack of rainfall for several consecutive years, coupled with reduced flows from the Tigris and Euphrates rivers from Turkey and Iran. Accordingly, Water storage in reservoirs and dams has decreased to 8% of their storage capacity, a 57% decline from 2024 (Historic drought levels in Iraq and warnings of an impending disaster, 2025).

The Iraqi government has significantly reduced its winter and summer agricultural plans for the third consecutive year, banning the cultivation of water-intensive crops such as rice and corn in most governorates and allowing water to be used exclusively for drinking purposes. Basra province was most affected by the record-breaking salinity of the Shatt al-Arab, rendering the water unfit for drinking. Tribal and local protests escalated in the summer of 2025 over the deteriorating infrastructure and the lack of government's solution to the water crisis. Disputes between Iraqi governorates over water quotas also escalate from time to time, with downstream governorates (in the south) accusing upstream governorates (in the centre and north) of encroaching on their quotas (World Bank Group, 2022). Since 2021, the marshlands of southern Iraq have been experiencing one of the worst droughts in their recent history. Vast areas of the marshlands, particularly the Hawiza and Chebayish marshes, have dried up and become barren. Declining water levels and increased evaporation have raised the salt concentration in the remaining water to lethal levels, resulting in mass fish die-offs. Drought, a shortage of green fodder (reeds and papyrus), and the salinity of the water have led to the deaths of large numbers of buffaloes, the economic lifeblood of the marshlands, forcing thousands of families to migrate and sell their livestock. Consequently, migratory birds have disappeared, wildlife has been severely impacted, and dust storms have spread from the dry lands. Although the marshes of southern Iraq were listed as a World Heritage Site in 2016, this has not prevented the severe extinction of them for many years due to water scarcity (Norwegian Institute of International Affairs, 2023). Figure 12.4 illustrates the conditions of water bodies in marshlands from 2020 to 2025, which has been alarmingly diminished due to drought and mismanagement of the marshland's restoration issue.

Iraq's flash floods have caused deaths, worsened living conditions for those already displaced by the conflict, and damaged agricultural infrastructure (International Organization for Migration (IOM), 2025). Wassit and Missan governorates in the south are particularly prone to flooding due to flat terrain and the weak infrastructure for draining water coming from the Iranian highlands along the border strip (Rasn et al., 2021). More than 7,000 individuals are impacted by the floods in Erbil city, which killed 12 people in late 2021 and early 2022 and damaged about USD 14

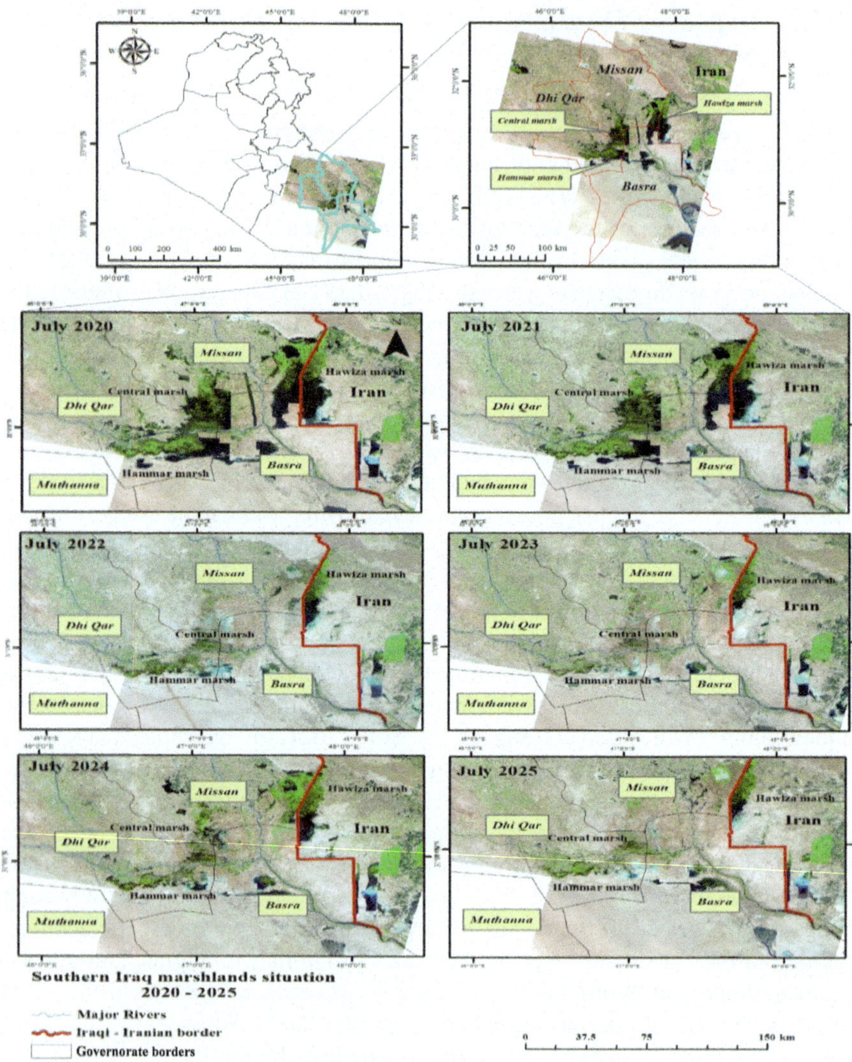

Fig. 12.4 Changes in water body in Iraqi southern marshlands from 2020 to 2025

million (International Federation of Red Cross (IFRC), 2021). In the same context, steep slopes amplify runoff due to flash floods in Erbil, Sulaymaniyah and Dohuk (Mohammad et al., 2024). Every year, 5% of utilised lands are lost due to salinisation, and Iraq loses over 2,000 ha of agricultural producing land due to salinity, costing the country nearly USD 300 million yearly (World Food Programme (WFP), 2024b). Sandstorms in the summer of 2022 cost approximately USD 7 million in damages to several sectors, especially agriculture and health (particularly respiratory diseases). During a mid-April sandstorm in 2022, over 10,000 people were hospitalised due to

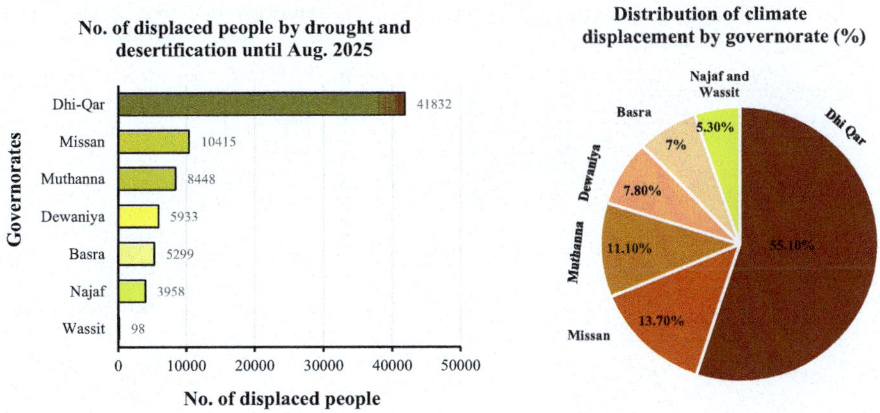

Fig. 12.5 Number of climates displaced by drought and desertification from 2014 to Aug 2025

respiratory issues. Iraq temporarily shut down airports, ceased movement in public buildings, and ordered a halt to all work in state run institutions, except for security and health, during the 2022 sandstorms (World Food Programme (WFP), 2024b). The 2017 earthquake was the worst in Iraq's modern history. Sulaymaniyah, Erbil, Dohuk, and Baghdad were all hit by an earthquake with a magnitude of 7.3, which left hundreds injured and nine dead. The western Iranian province of Kermanshah was the epicentre of the earthquake. Many residences were also damaged by the earthquake, especially in remote mountain settlements (https://earthquakelist.org/iraq/).

The internal displacement crisis, whether due to climate change or ISIS, has placed significant pressure on infrastructure and impacted social and security stability in the areas to which people have been displaced, whether in camps or cities (World Food Programme (WFP), 2024c). Figure 12.5 illustrates the number of climates displaced by drought and desertification until August 2025, especially in middle and southern Iraq. While Fig. 12.6 shows the number of people who were displaced due to ISIS related to Iraqi governorates (Ministry of Migration Displaced, 2025).

12.4 Evolution and Institutional Framework

Prior to 2003, little effort was made to comprehensively address disaster risks in Iraq. Existing legislation focused primarily on reactive disaster response without considering prevention, preparedness, and mitigation (Al-Shamsi & Al-Shamsi, 2019). There are key laws, Table 12.2, dealing with disaster and crisis management that were passed and are still in service, as follows (United Nations Development Programme (UNDP), 2014).

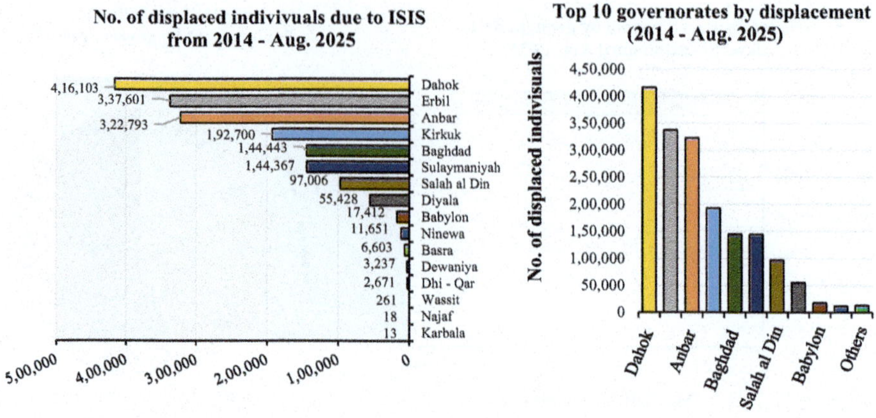

Fig. 12.6 Number of displaced individuals and the top 10 governorates by displacement due to ISIS from 2014 until Aug 2025

Table 12.2 Key laws dealing with disaster and crisis management in Iraq

Key laws	Year	Target
The Iraqi Red Crescent Society (IRCS)	1932	Assist groups affected by natural hazards and man-made disasters
Emergency Law 1961	1961	Attempt to include rules for disaster response in Iraq
The Civil Defence Law 1978	1978	Essential for disaster risk reduction (DRR) in Iraq
The Social Care Law	1980	The law mandates that the state of Iraq protect and aid victims
The Public Health Law	1981	Explains the roles and responsibilities that must be fulfilled in the case of an epidemic
The Environment Protection Act	2009	Significant part in developing the National Plan OF DRR
Immigration Law	2008	Addresses the challenges of internally displaced people
Civil Defence Law	2013	Regulate the work of institutions specialising in civil defence and disaster management
The Iraqi National Security Service Law	2021	Significant contribution to the legal framework for crisis management

Despite the significant challenges it faces, Iraq demonstrates a growing commitment to participating in regional and international efforts to reduce disaster risks. Iraq's participation in the Arab Regional Platform for DRR and its collaboration with the United Nations Office for Disaster Risk Reduction (UNDRR) reflect a growing recognition of the importance of regional cooperation in addressing common challenges (Ministry of Environment & Iraq, 2025).

Iraq began working with UNDRR in 2024 to develop a national disaster risk reduction strategy for the country (Ministry of Environment, Iraq, 2024). Similarly, a Disaster Risk Reduction Division was established within the Climate Change Directorate of the Ministry of Environment in 2024 as part of ongoing institutional development. In September 2024, the National Strategy for Environmental Protection and Improvement in Iraq for the period 2024–2030 was launched. This strategy represents an important milestone in Iraq's efforts to address environmental challenges, as it includes topics related to disaster risk management and climate change adaptation (Goodyear, 2009).

12.4.1 The Current Institutional Framework for Disaster and Crisis Management in Iraq

Iraq started to participate more actively in DRR efforts after 2003 and formed the Disaster Risk Committee in 2007. In accordance with the Hyogo Framework for Action, it coordinates the response to both natural hazards and man-made disasters using a preparedness, preventive, and mitigation approach. Iraq joined the regional DRR programme for the first time. The Arabian Centre for Earthquake Hazard and Disaster Prevention initiative received Iraq's approval in 2008 (Al-Shamsi & Al-Shamsi, 2019). The law provides provincial governors with the authority to establish Governorate Emergency Cells (GECs) and conduct other actions in the event of a disaster. The Law of Financial Management, which was passed in 2004 to coordinate the distribution of financial resources to areas affected by disasters (Al-Shamsi & Al-Shamsi, 2019; United Nations Development Programme (UNDP), 2014).

In times of crisis, Iraqi's government and UN agencies coordinate to offer relief and support, which includes:

- Ministry of Interior: through the Civil Defence Directorate in accordance with Law No. 44 in 2013.
- The Ministry of Water Resources: coordinates responding to drought and water scarcity.
- The Ministry of the Environment is responsible for several DRR-related tasks.
- Ministry of Health: offering emergency health services during emergencies.
- Ministry of Migration and Displaced: attending to the needs of displaced people (Ministry of Migration Displaced, 2025).
- Iraqi Meteorological Organisation and Seismology (IMOS) plays a pivotal role in the disaster and crisis response system through monitoring and early warning, providing historical data used in research studies and strategic planning to reduce hazards (Iraqi Meteorological Organization and Seismology (IMOS), 2025).
- Civil Defence Directorate is responsible for search and rescue and firefighting.
- JCMC is ensuring an efficient response to emergencies by coordinating the actions of international organisations and different government authorities.

- IRCS provides training to its employees in humanitarian aid distribution, camp construction, and relief operations. It also strives to improve its early warning and quick response skills by utilising contemporary technologies (Iraqi Red Crescent Society (IRCS), 2025).
- The National Crisis and Disaster Management Centre in Iraq represents the latest entity, established in 2023 following the recommendation of the Ministerial Council for National Security. The centre aims to unify efforts among ministries and coordinate immediate responses to emergencies and crises in Iraq and is affiliated with the Ministry of Interior. Crisis cells have been formed within ministries and governorates for direct field coordination (The General Secretariat for council of Ministries, Iraq, 2025).
- Directorate of Rapid Response and Central Control was established in the Ministry of Interior, by National Security Council Resolution No. (1) of 2024, with the aim of unifying all 26 emergency numbers, with one number (911), with the aim of ensuring a rapid response to accidents and emergencies (Ministry of Interior, 2025).

The UN agencies and international organisations include:

- The United Nations Office for the Coordination of Humanitarian Affairs (OCHA) oversees uniting aid workers to guarantee a coordinated emergency response (The United Nations Office for the Coordination of Humanitarian Affairs (OCHA), 2025).
- The United Nations Development Programme (UNDP) assists Iraq in developing climate change and disaster resilience (United Nations Development Programme (UNDP), 2025).
- The International Organisation for Migration (IOM) aids communities and displaced people impacted by conflict and disasters (International Organization for Migration (IOM), 2018).
- United Nations High Commissioner for Refugees (UNHCR) supports internally displaced people and refugees by providing shelter and essential crisis assistance (United Nations High Commissioner for Refugees (UNHCR), 2025).
- The World Food Programme (WFP), the Food and Agriculture Organisation (FAO), and the World Health Organisation (WHO) are additionally included as active agencies. These organisations support public health and food security in times of crisis.

12.5 Challenges and Weaknesses in Disaster Management in Iraq

Iraq's disaster management system faces profound structural and operational challenges that limit its effectiveness in responding to successive crises. These challenges extend beyond a mere lack of resources; they also include political and governmental failures, weak institutional frameworks, and significant research and knowledge gaps.

These major obstacles, which prevent the transition from a reactive approach to a proactive risk management strategy, can be summarised as follows:

1. Absence of a systematic and effective national disaster risk management plan: Iraq lacks a comprehensive and unified national disaster risk management strategy, rendering government efforts fragmented and inconsistent. The need to develop a 'comprehensive conceptual framework for disaster risk management' emphasises that this problem is not merely an implementation gap, but rather a strategic one.
2. Insufficient cooperation and coordination among various agencies: Poor coordination is one of the most significant obstacles to effective disaster management. This deficiency manifests itself in a variety of ways, including the absence of a binding law to coordinate governmental and non-governmental efforts, weak information and data sharing among various agencies, overlapping powers and responsibilities, and the negative influence of political and bureaucratic factors.
3. Lack of dedicated national financing channels: The lack of sustainable, dedicated financing for risk management is a major obstacle, which directly hinders disaster risk reduction efforts before they occur.
4. Weak early warning systems: Despite their vital importance in saving lives and reducing losses, early warning systems in Iraq remain weak and underdeveloped. There is an urgent need for risk assessment studies and the development of appropriate early warning and response systems, as well as the design and construction of integrated and effective early warning systems that consider Iraq's local and climatic characteristics.
5. Insufficient Scientific and Technical Capacities: Disaster management in Iraq faces a severe shortage of scientific and technical capabilities, impacting all phases of the disaster cycle, from assessment and planning to response and recovery. The lack of accurate data and scientific analysis makes decision-making be based on estimates rather than evidence.
6. The ongoing displacement crisis and its impact on communities: Internal displacement, whether due to conflict or climate change, represents a chronic challenge that burdens already fragile disaster management systems. The research gap concerns the long-term impacts of displacement and unsustainable returns, and the need to integrate disaster risk management into recovery and reconstruction plans in conflict-affected areas.
7. The necessity of integrating risk management into national planning: The most fundamental challenge is the failure to systematically integrate the concept of risk management into all aspects of national and development planning, which requires political will and a radical change in the planning culture at the national level.

12.6 Key Takeaways from the Analysis

The systematic analysis of the data and information presented in this chapter reveals a set of interconnected findings that form a picture of the reality and magnitude of disaster management challenges in Iraq.

1. The Dominance of Dual Hazards: Conflict and Climate Change

Iraq is facing a double-headed displacement crisis, fuelled by various drivers in diverse geographic areas:

- Conflict-related displacement: The largest displacement crisis, affecting more than 6 million people since 2014, is concentrated primarily in the northern and western governorates of Iraq (Ninewa, Anbar, and Salah al-Din) as a direct result of the ISIS takeover and the subsequent military operations.
- Climate-related displacement: In contrast, a different displacement pattern is emerging in the central and southern governorates of Iraq, driven primarily by the impact of climate change. The drought and desertification crisis, which peaked in 2025, is the main driver of this displacement, causing agricultural land degradation, the drying up of vast areas of marshlands, and the death of livestock, forcing thousands of families to migrate.

2. The Predominance of a Reactive Approach

Iraq's disaster management system suffers from profound systemic shortcomings, operating primarily on a reactive rather than proactive approach. These shortcomings are manifested in several critical areas:

- Absence of a unified national strategy: The country lacks a comprehensive national disaster risk management plan, leading to fragmented and inconsistent efforts.
- Weak institutional coordination: There are significant gaps in coordination between governmental and non-governmental agencies, exacerbated by the absence of a binding law, overlapping mandates, and weak accountability mechanisms.
- Weaknesses in early warning systems: Current systems are unable to provide accurate and timely warnings, reducing preparedness opportunities and increasing human and material losses.
- Lack of dedicated funding: Reliance on emergency funding after disaster strikes hinders investment in long-term prevention and risk reduction measures.

3. Cascading Effects of Disasters on Infrastructure and Communities

The findings reveal that the impact of disasters in Iraq is not merely a once-off event, but rather a series of successive shocks that weaken the already fragile infrastructure and exacerbate communities' suffering. This is evident in:

- Catastrophic Fires: The tragic fire incidents in hospitals and public places (2021–2025) revealed clear weaknesses in safety standards and emergency systems in critical infrastructure.

- Deteriorating Water Resources: The water crisis, caused by drought and reduced flows from neighbouring countries, has reduced dam reserves to 8% of their capacity, curtailed agricultural plans, and increased land salinisation, threatening the country's food and water security.
- Economic Losses: The country is incurring massive economic losses, with the cost of damage from sandstorms alone estimated at approximately $7 million in the summer of 2022, in addition to the loss of more than 2,000 ha of agricultural land annually due to salinisation.

4. Knowledge and Research Gaps as a Barrier to Decision-Making

Significant knowledge and research gaps directly hinder decision-makers' ability to formulate evidence-based policies. The most prominent of these gaps are:

- The absence of a unified national risk database: This makes it impossible to conduct accurate risk assessments and scientifically determine priorities.
- The lack of specialised studies: There is an urgent need for risk assessment studies for various types of disasters (drought, floods, earthquakes), studies on the impact of disasters on the most vulnerable groups, and research on integrating risk management into reconstruction plans.

12.7 Conclusions

This chapter concludes that Iraq's disaster management system is not merely facing individual challenges but is trapped in a vicious cycle of systemic shortcomings. The prevalence of a reactive approach rather than a proactive strategy has turned it into a tool for managing crises after they occur rather than preventing them. This outcome is evident in two fundamental facts: First, the system's inability to simultaneously address the dual and geographically disparate risks of conflict-related displacement in the north and west, and escalating climate-related displacement in the centre and south. Second, this inability is not only the result of lack of resources but that of deepened structural weaknesses, including the absence of a unified legal framework, conflicting institutional mandates, inadequate early warning systems, and exacerbating knowledge and research gaps that prevent evidence-based decision-making. The most alarming conclusion is that the continuation of this reactive approach not only leads to recurring disasters but also guarantees the accumulation of new overlapping risks, making the country more vulnerable to future shocks. Therefore, transitioning from the concept of 'disaster management' to a culture of 'disaster risk management' is no longer just a recommendation, but an imperative strategic necessity to ensure Iraq's stability and resilience.

12.8 Recommendations and Future Strategies

In response to the profound challenges and structural weaknesses analysed in the Iraqi disaster management system, building national resilience requires a radical strategic shift, according to the following priorities:

1. Accelerating the enactment and implementation of a comprehensive and effective DRR law, designed based on the Sendai Framework. This law should be the cornerstone regulating roles and responsibilities.
2. A proactive strategy focused on prevention and risk reduction should replace the post-disaster response approach and become the primary focus of national policy.
3. Intensive investment in technology and scientific research should be made by building a unified national platform for data and early warning and directing investment in data and research to address identified knowledge gaps.
4. Strengthening operational and institutional capacities by enhancing the preparedness and response of the healthcare sector to disasters, particularly fires and floods.
5. Integrated water resources management and drought response should be implemented by focusing on modern irrigation technologies, particularly in the southern governorates. Strategic projects should be launched to build small dams and harvest rainwater, while activating water diplomacy with Turkey and Iran to ensure equitable water shares.
6. Community awareness and public participation by launching public awareness campaigns and community education to maintain safety during disasters and risks.

Acknowledgements The author is thankful to Scientific Research Commission, Baghdad, Iraq (https://src.edu.iq/) for their support in completing this chapter. The author also extends his sincere thanks to the Ministry of Migration and Displaced in Iraq (https://momd.gov.iq/) for providing data on those displaced in Iraq due to ISIS, as well as drought and desertification. Many thanks to Dr. Edris Alam for his kind invitation to contribute to writing a chapter on Iraq in this book.

AI Discloser Artificial intelligence tools were not used to create the maps or graphs presented in the chapter.

References

Al-Hussein Hospital, Nasiriyah. (2025). Retrieved July 30, 2025, from https://aawsat.com/home/article/3079001/

Al Kut Hypermarket. (2025). Retrieved July 31, 2025, from https://edition.cnn.com/2025/07/17/middleeast/iraq-kut-building-deadly-fire-intl-hnk

Al-Muqdadi, S. W., Omer, M. F., Abo, R., & Naghshineh, A. (2016). Dispute over water resource management—Iraq and Turkey. *Journal of Environmental Protection (Irvine, Calif), 07*, 1096–1103. https://doi.org/10.4236/jep.2016.78098

Al-Shamsi, M. T. M., & Al-Shamsi, M. (2019). Disaster risk reduction in Iraq. *Jàmbá-Journal of Disaster Risk Studies*. https://doi.org/10.4102/jamba

CARBON LIMITS. (2022). Iraq—Technology needs assessment report for mitigation and adaptation. Ministry of Environment, Baghdad, Iraq

Central Statistical Organization (CSO). (2024). Ministry of Planning, Baghdad. 2024 General Population Census Results. https://cosit.gov.iq/ar/

Displacements Tracking Matrix (DTM) IRAQ. (2025). Retrieved August 1, 2025, from https://iraqdtm.iom.int/

Drought in Iraq. Retrieved August 1, 2025, from https://reliefweb.int/country/irq#disasters/

Energy Information Administration (EIA). (2021). Iraq Country Profile. Independent Statistics and Analysis. U.S. Energy Information Administration. https://www.eia.gov/international/overview/country/IRQ

Export Working Group. (2018). Iraq: Climate-related security risk assessment. https://www.preventionweb.net/files/61579_iraqclimaterelatedsecurityriskasses.pdf

Gleick, P. H. (2019). Water as a weapon and casualty of conflict: Freshwater and international humanitarian law. *Water Resources Management, 33*, 1737–1751. https://doi.org/10.1007/s11269-019-02212-z

Goodyear, E. J. (2009). The state of disaster risk reduction in Iraq. https://www.humanitarianlibrary.org/sites/default/files/2013/05/unpan050289.pdf

Hameed, M., Ahmadalipour, A., & Moradkhani, H. (2018). Apprehensive drought characteristics over Iraq: Results of a Multidecadal spatiotemporal assessment. *Geosciences, 8*, 58. https://doi.org/10.3390/geosciences8020058

Hashim, B. M., Al-Naseri, S. K., Al Maliki, A., et al. (2021). On the investigation of COVID-19 lockdown influence on air pollution concentration: Regional investigation over eighteen provinces in Iraq. *Environmental Science and Pollution Research, 28*, 50344–50362. https://doi.org/10.1007/s11356-021-13812-x

Hashim, B. M., Al-Naseri, S. K., Hamadi, A. M., et al. (2023). Seasonal correlation of meteorological parameters and PM2.5 with the COVID-19 confirmed cases and deaths in Baghdad, Iraq. *International Journal of Disaster Risk Reduction, 94*, 103799. https://doi.org/10.1016/j.ijdrr.2023.103799

Hashim, B. M., Alnaemi, A. N. A., Hussain, B. A., et al. (2024). Statistical downscaling of future temperature and precipitation projections in Iraq under climate change scenarios. *Physics and Chemistry of the Earth, Parts A/B/C, 135*, Article 103647. https://doi.org/10.1016/j.pce.2024.103647

Historic drought levels in Iraq and warnings of an impending disaster. (2025). Retrieved July 28, 2025, from https://www.aljazeera.net/climate/2025/7/28/

Ibn al-Khatib Hospital fire. (2025). Retrieved July 30, 2025, from https://www.bbc.com/arabic/middleeast-56879383

International Federation of Red Cross (IFRC). (2021). Iraq flash floods. https://adore.ifrc.org/Download.aspx?FileId=480634

International Organization for Migration (IOM). (2012). Iraq Special Report: Water scarcity. https://iraq.iom.int/

International Organization for Migration (IOM). (2018). Retrieved August 1, 2025, from https://iraq.iom.int/news/iraq-displacement-figures-drop-below-two-million-first-time-2014-nearly-four-million-have-returned-home

International Organization for Migration (IOM). (2022). Migration, environment, and climate change in Iraq. https://iraq.iom.int/

International Organization for Migration (IOM). (2023). IOM 2023 annual report. https://iraq.iom.int/

International Organization for Migration (IOM). (2025). Responds to Flooding in Iraq. Retrieved July 28, 2025, from https://www.iom.int/news/iom-responds-flooding-iraq

International Trade Centre. (2021). Climate change risks and opportunities in Iraqi Agrifood value chains: Strengthening the Agriculture and Agrifood Value Chain and Improving Trade Policy in Iraq (SAAVI). https://www.intracen.org/

Iraq—Flood List. Retrieved August 1, 2025, from https://floodlist.com/tag/iraq

Iraq earthquake Report. Retrieved August 3, 2025, from https://earthquakelist.org/iraq/

Iraqi Meteorological Organization and Seismology (IMOS). (2025). Retrieved July 30, 2025, from https://www.meteoseism.gov.iq/

Iraqi Red Crescent Society (IRCS). (2025). Retrieved July 29, 2025, from https://ircs.org.iq/en/ircs-founding/

Jassim, S. Z., & Goff, J. C. (2006). Geology of Iraq. Dolin, Prauge and Moravian Museum.

Ministry of Environment, Iraq. (2024). National strategy for the protection and improvement of the environment in Iraq. https://moen.gov.iq/

Ministry of Environment. (2024). Iraq's Second National Communiqué and First Biennial Update Report (p. 242).

Ministry of Environment, Iraq. (2025). https://moen.gov.iq/

Ministry of Interior. (2025). Retrieved July 30, 2025, from https://moi.gov.iq/

Ministry of Migration Displaced. (2025). Retrieved July 29, 2025, from MOMD.gov.iq

Mohammad, J. K., Saleh, S. A., & Amin, D. A. (2024). Determining the surface drainage network causing flash floods in Central Erbil Sub-Basins in Northern Iraq Using GIS and Geospatial Technique. *Iraqi Geological Journal,* 191–208. https://doi.org/10.46717/igj.57.1F.15ms-2024-6-24

Norwegian Institute of International Affairs. (2023). Climate, peace and security fact sheet: Iraq. https://www.nupi.no/admin/content/download/24802/1067504?version=2

Norwegian Refugee Council (NRC). (2024). Cracked Earth, shrinking harvest: Drought impact on displaced and returnee Iraqis. https://www.nrc.no/resources/reports/

Our World in Data. (2025). Iraq-Coronavirus Pandemic Country Profile. Retrieved August 1, 2025, from https://ourworldindata.org/coronavirus

Rasn, K. H., Nsaif, Q. A., Al-Obaidi, M. A., & John, Y. M. (2021). Designation of flood risk zones using the geographic information system technique and remote sensing data in Wassit, Iraq. *Geomatics and Environmental Engineering, 15*, 129–140. https://doi.org/10.7494/geom.2021.15.3.129

The General Secretariat for council of Ministries, Iraq. (2025). Decisions. Retrieved July 29, 2025, from https://www.cabinet.iq/en/category/decisions

The United Nations Office for the Coordination of Humanitarian Affairs (OCHA). (2025). Iraq: OCHA. Retrieved July 30, 2025, from https://www.unocha.org/iraq

United Nations Development Programme (UNDP). (2014). Iraq: Country case study report how law and regulation support disaster risk reduction. www.undp.org

United Nations Development Programme (UNDP). (2025).

United Nations Environment Programme (UNEP). (2016). Global environment outlook: GEO-6: Regional assessment for West Asia. https://www.unep.org/resources/

United Nations High Commissioner for Refugees (UNHCR). (2025). Retrieved July 30, 2025, from https://www.unhcr.org/

Wedding hall in Al-Hamdaniya, Ninewa. (2025). Retrieved July 30, 2025, from https://www.nytimes.com/2023/09/27/world/middleeast/iraq-wedding-hall-fire.html

World Bank Group. (2022). Iraq Country climate and development report. www.worldbank.org
World Food Programme (WFP). (2024a). Remote sensing & machine learning based high-resolution land: Use land cover classification map for Iraq a step towards Improved Land and Water Management for Food Security. https://reliefweb.int/report/iraq/
World Food Programme (WFP). (2024b). Drought in Iraq: Part 1-drivers and aggravating factors of recurring droughts. https://reliefweb.int/report/iraq/drought-iraq-part-1-drivers-and-aggravating-factors-recurring-droughts-july-2024
World Food Programme (WFP). (2024c). Drought in Iraq: Part 2-impact of the current drought on livelihoods and food security. Retrieved June 30, 2025, from https://reliefweb.int/report/iraq/drought-iraq-part-2-impact-current-drought-livelihoods-and-food-security-july-2024

Chapter 13
Evaluating Disaster Management Frameworks in Algeria: Progress, Challenges, and Future Directions

Zohra Boutaraa

Abstract The Middle East and North Africa (MENA) region is home to a large population in Arab and Muslim countries facing substantial risks from natural hazards and man-made disasters. The most common natural and man-made disasters that pose major risks to these nations are earthquakes, floods, droughts, conflicts, and civil wars. This chapter discusses crisis and disaster risk management in Algeria, a country particularly exposed to natural hazards such as earthquakes, floods, heat waves, and forest fires, which have repeatedly caused widespread human and economic losses. Since 1980, the Algerian government has adopted a prevention-focused disaster risk reduction policy to address these challenges. In line with this approach, public authorities have established technical and legislative measures aimed at reducing vulnerabilities and limiting impacts on communities and material assets. The chapter outlines the main disaster risks affecting Algeria and highlights their social, financial, and economic implications for national development. It examines the evolution of the technical and legislative framework introduced to enhance disaster preparedness, response, and mitigation efforts while contributing to societal resilience. A critical analysis of the current disaster management system is provided to assess its effectiveness in reducing risk. The chapter concludes by identifying priority areas for improvement and offering recommendations to guide future strategies that strengthen disaster resilience in Algeria.

Keywords Disaster risk governance · Natural hazards · Man-made disasters · Technical and legislative frameworks · Resilience

Z. Boutaraa (✉)
Hassiba Benbouali University of Chlef, Chlef, Algeria
e-mail: z.boutaraa@univ-chlef.dz

E. Alam (ed.), *Disaster, Crisis and Emergency Management in the Middle East & North Africa*, https://doi.org/10.1007/978-3-032-24436-9_13

13.1 Introduction

Algeria is the largest country in Africa and in the Arab world. It covers more than two million square kilometers and has a population of around 47.4 million. The northern part of this country sits on the border of the African tectonic plate, which is in continuous relative movement with the Eurasian plate. Thus, earthquakes are the most threatening natural hazard in this country. Several major earthquakes have been recorded in the northern part of the country over the past century, including the event of October 10, 1980, that shook El Asnam (Chlef currently) and whose magnitude attained 7.3. The May 21, 2003, Boumerdès earthquake in Algeria occurred near Zemmouri/Boumerdès with a moment magnitude M_w of 6.9. Both earthquakes caused thousands of casualties and billions in economic losses.

The other types of hazards including landslides, forest fires, and floods are also crucial hazards in Algeria. Despite the efforts made by the government, various disasters cause the loss of millions of Algerian dinars every year. This is why the Algerian legal system is showing great interest in disaster management by establishing public administrations and research institutes devoted to this task. Regarding anthropogenic hazards, Algeria experienced terrorism for ten years, known as the "Black Decade," during which more than 100,000 people lost their lives. The nation suffered financial losses because of terrorist organizations' vandalism, in addition to the human cost. From 1992 to 2002, the Black Decade was a period of extreme violence and unrest. It was typified by attacks, murders, and disappearances committed by terrorists connected to opposition organizations. The national army fought these terrorists. Ultimately, the Algerian government's determination, the sacrifices of the National People's Army, the mobilization of the Algerian people, and the realization by many supporters of the terrorist movement that these acts were detrimental to the nation's social, political, and economic future led to the end of the fight against terrorism in Algeria in 2005. Amnesty laws, compensation programs for victims, and a charter for peace and national reconciliation were the means of how this was accomplished.

Algeria also saw a significant increase in disaster-related deaths because of the COVID-19 pandemic. According to data published by the National Delegation for Major Risks,[1] there were approximately 2,745 deaths in 2020, affecting 99,610 people. In 2021, the number of affected individuals increased to 118,822, but decreased to 47,838 by 2022. During these two years, the number of people who died increased to 4,335. The latest report published by Statista[2] in 2023 stated that 6,881 deaths occurred in Algeria due to the pandemic. The COVID-19 pandemic also had a significant impact on the Algerian economy, as lockdown measures reduced both supply and demand. A severe reduction in worldwide demand aggravated the situation, resulting in a major drop in Algerian exports and a 50% drop in oil prices (Benouar & Belmokhtar, 2025).

This chapter provides a comprehensive overview of the technical and legislative policies governing the management of natural and man-made hazards in Algeria. It

---

[1] https://www.undrr.org/national-platform/algeria-national-platform-drr.

[2] https://fr.statista.com/statistiques/1101324/morts-coronavirus-monde/.

examines the evolution of the country's disaster management system and highlights both its strengths and areas where further improvement is required.

13.2 Data Sources and Methods

The chapter examines the evolution of Algeria's disaster management system and identifies areas where further improvement is required. A documentary research method is applied to systematically collect and analyze legal resources related to disaster risk management, including legislative texts that define the judicial and preventive responsibilities of the Algerian state. This approach allows a critical assessment of how regulations have been enacted and operationalized over time. The technical policy dimension focuses on the formulation and enforcement of engineering standards and safety regulations for the built environment and critical infrastructure. Notable examples include the seismic code (RPA, 2024) and the regulation dedicated to road infrastructures (RPOA, 2008), which play key roles in reducing structural vulnerability to earthquakes. The legislative component centers on decrees, orders, and laws that establish national risk governance, clarify institutional mandates, and define accountability mechanisms.

The research relies on secondary data obtained from authoritative sources such as the World Bank, the International Monetary Fund (IMF), and national statistical agencies across the MENA region. It also incorporates recent scientific articles and policy studies to strengthen interpretation. Both descriptive-analytical and statistical methods are employed to assess the effectiveness of current frameworks. Overall, the chapter draws on a thorough review of documentary sources, official records, and academic discourses to evaluate the strengths and limitations of Algeria's disaster risk management policies.

13.3 Study Area

13.3.1 Geographic Location of Algeria

With a total area of 2,381,741 km^2, Algeria is the largest country in Africa, the Arab world, and the Mediterranean Basin. The country is bordered in the northeast by Tunisia, in the east by Libya, in the west by Morocco, in the southwest by Western Sahara, Mauritania, and Mali, in the southeast by Niger, and in the north by the Mediterranean Sea. Its southern part includes a significant portion of the Sahara. To the north, the Tell Atlas alongside the Saharan Atlas forms two parallel sets of reliefs in approaching the east, between which are vast plains and highlands. The vast mountain ranges of *Aures* and *Nememcha* occupy the entire northeastern Algeria. Most of the coastal area is hilly, sometimes even mountainous. The area from the

Fig. 13.1 Geographical location of Algeria. *Source* https://www.turkey-visit.com/cities-map-of-algeria

coast to the Tell Atlas is fertile. South of the Tell Atlas is a steppe landscape, which ends with the Saharan Atlas.

Northern Algeria's climate is temperate, with a mild Mediterranean climate. However, due to its topography, there are significant local differences in both prevailing temperatures and rainfall incidence. Climatological circumstances vary year after year. Eastern Algeria's average temperatures range from pleasant to frigid, particularly on the High Plateau steppes. Algeria's relatively tiny Sahara region is across the Tropic of Cancer in the Torrid Zone,[3] yet even in winter, the desert can be quite hot during the day. Finally, rainfall is very copious in Tell's coastal region, ranging from 400 to 670 mm per year, with precipitation increasing from west to east (Fig. 13.1).

[3] One of the three theoretical climatic zones, lying between the Tropic of Cancer and the Tropic of Capricorn (Oxford reference).

13.3.2 Jurisdiction and Its Governing Structure

The judicial system in Algeria is completely independent. The Supreme Court oversees the activities of tribunals and courts. The State Council supervises the actions of the administrative domains. From an administrative point of view, the Algerian administrative structure is decentralized and divided into 69 provinces (Wilayat), each of which covers several municipalities (*Baladiyat*), making a total number of 1.541 municipalities throughout the country. The two Algerian administrative structures are as follows:

- The Province (*El Wilaya*) Delegate or administrative districts, established by Executive Decree No. 15-141. This structure oversees the coordination between the representatives of the different sectors concerned by the evaluation of the actions undertaken at the province level. The *Wali*, who is an administrative official, appointed by the state, and the Province People's Assembly, which is elected by the population of the province, are the two legal entities governing the province. The Municipality (*El Baladia*) delegate oversees the coordination between the representatives of the different sectors concerned and the evaluation of the actions undertaken at the provincial level.
- The Municipal People's Assembly, which is elected by the citizens, is the legal structure governing the municipality. The inter-sectorial committee of the province, formed by the representatives of the executive sectors, establishes the inter-sectorial strategy of any jurisdictional action to be taken in the context of economic, social, and cultural developments. This inter-sectorial strategy must integrate the prevention dimension into any development action. An inter-sectorial structure at the municipal level equivalent to that of the province is also installed.

These two administrative structures seek to advance development, encourage business ventures, generate economic activity, and provide administrative services in a more accessible manner. In the end, this allows for administrative decentralization, balanced territorial development, and the satisfaction of citizens' needs for social and economic advancement. Additionally, Algeria has started enacting structural reforms by embracing more adaptable procedures and shifting toward local governance in response to several issues brought on by globalization. Pertinent political parties must take part in this shift to implement efficient local governance and activate the partnership principle (Sekfali, 2025).

13.3.3 Population and Urbanization Rate

The population of Algeria is reported with slight variations, depending on the source. According to a third-party website[4] offering demographic data based on information from the United Nations and other demographic agencies, such as the Populationof.net website,[5] Algeria's population was 45,753,000 in 2024. The following year, in 2025, the population increased by 555,000, reaching a total of 46,308,000. According to this reference, the number is expected to reach 47,360,000 in 2027. The population density is 19 people per square kilometer. A significant proportion of this demographic, specifically 75.6%, is classified as urban residents. However, according to Algeria's National Statistics Office, the resident population was estimated to be 46,700,000 as of July 1, 2024. Assuming the 2023 growth rate of 1.53% is maintained, the total resident population would reach 47,400,000 by January 1, 2025 (ONS, 2024).

In the context of urbanization, the rate has increased from 31% in 1960 to over 60% in 2025. The number of people living in crowded cities continues to grow daily. Uncontrolled urbanization increases disaster risks by driving communities into hazard-prone places like floodplains and steep slopes, as well as eliminating natural protective ecosystems like forests and wetlands. This uncontrolled growth puts more people and infrastructure at risk, while growing usage of impermeable surfaces like concrete exacerbates issues like flooding and landslides. Furthermore, informal settlements usually lack proper infrastructure and building regulations, making their population especially vulnerable to disasters. Unchecked and often illegal urbanization often results in deforestation, which makes the local population more vulnerable to risks.

13.4 Disaster History and Risk Profile in Algeria

Due to its geographical localization, its natural and physical characteristics, Algeria is a country highly exposed to natural hazards: earthquakes, climate and weather risks, such as heat waves and precipitations, forest fires, floods, and insect infestations. Northern Algeria and its coastal strip are primarily affected by earthquakes, floods, forest fires, landslides, storms, and strong winds. The south of Algeria and its high plateaus are affected by floods, insect infestations, strong winds, and sandstorms.

[4] This includes Statista, Worldometer, and Opendataforafrica.org.

[5] https://www.populationof.net/algeria/.

13.4.1 The Predominant Disasters in Algeria

Floods and earthquakes are the most frequently reported disasters in Algeria, resulting in significant casualties and property damage. Between 1954 and 2022, 83 disasters were recorded, 70 of which were floods or earthquakes. A detailed breakdown of deaths, affected persons, and economic losses caused by these disaster events is provided in Table 13.1. In addition, Fig. 13.2 displays the total number of natural disasters in Algeria between 1980 and 2024.

Following catastrophic disasters such as the severe floods in Bab El Oued (Algiers) on November 10, 2001, the devastating Boumerdes earthquake on May 21, 2003

Table 13.1 Disasters recorded in Algeria during 1954–2022 (after World Bank Group, 2023)

Disaster type	Recorded events	Deaths	Affected people	Economic loss (10^3 USD)[a]
Floods	51	1870	809,145	1,543,917
Earthquakes	19	6483	1,387,704	10,270,929
Storms	4	27	10,122	Not available
Extreme temperatures	3	48	250,025	Not available
Forest fires	4	164	49,271	Not available
Landslides	1	15	696	Not available
Drought (Sécheresse)	1	Not available	Not available	Not available
Total	83	8,607	2,506,963	11,814,846

[a] 1 USD is equivalent to 130, 23 Algerian Dinar

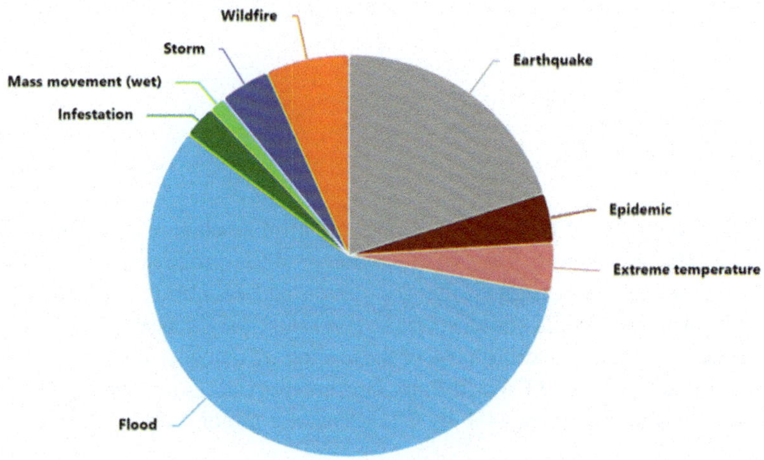

Fig. 13.2 Disasters in Algeria between 1980 and 2024 (World Bank Group, 2023)

($M_s = 6.8$), and the Ghardaia floods on October 1, 2008, the country's risk management capacity was strengthened further. These disasters have considerably improved Algeria's capabilities in disaster response and recovery. Continued political attention and commitment from the country's highest authorities have elevated disaster risk reduction to a national priority. Algeria has also demonstrated ingenuity in incorporating disaster risk reduction (DRR) into its sustainable development goals across many sectors. The governance of natural hazards in Algeria has progressed through several stages, often because of disaster response, and continues to do so. This improvement was particularly evident during the 2010s, with the strengthening of legislation and institutions (Benouar & Belmokhtar, 2025).

13.4.2 Seismic Monitoring and Instrumental Seismicity in Algeria

The Algerian seismic monitoring (so-called instrumental seismicity) began in the twentieth century. The first seismological station was installed in *Bouzaréah* in *Algiers* in 1910, followed by the installation of other stations at *Oued Fodda* in 1935, *Relizane* in 1955, and *Setif* in 1958. After the 1980 *El Asnam* earthquake, a telemetered seismological network was installed throughout Algeria. The Algerian seismic monitoring network is made up of the 35 stations of the telemetry network and the 45 digital stations acquired recently. Today, this network covers most of the northern part of the country, where the country's biggest activity is taking place, but also in the *Tamanrasset* region in the south of the country, where a station is located at the *Tamanrasset* Observatory (Boutaraa, 2018).

The Algerian Research Center in Astronomy, Astrophysics, and Geophysics (CRAAG) established a seismological database, which lists the most violent earthquakes that have struck the country since 1356. Table 13.2 summarizes these data.

13.4.3 Flood Risk

The combined effects of rapid urbanization and climate change are making Algeria more vulnerable to flooding. Most of the country's flood-prone areas are in the north, where a significant proportion of the population and economic activity is concentrated. According to official sources,[6] flood risk countrywide assessment reveals 865 locations vulnerable to flooding. 10% of these sites are categorized as high-risk locations. Additionally, the 2023 World Bank Group report states that exceptional storms in Algeria could increase by 41% by 2025, causing major damage, landslides, and flooding.

---

[6] The Ministry of Water Resources and the Directorate General of Civil Protection.

Table 13.2 Major earthquakes in Algeria recorded from 1365 to 2012

Locality	Date	Intensity/ magnitude	Consequences/damages
Algiers	January 03, 1365	High	Several victims
Algiers-Mitidja	February 03, 1716	X	20 thousand dead, the city of Algiers destroyed
Oran	October 09, 1790	IX–X	2 thousand dead
Blida	March 02, 1825	X	7 thousand dead, Blida destroyed
Jijel	August 22, 1856	X	Tidal waves from Algiers to Annaba
Gouraya	January 15, 1891	X	38 dead, important damages
Sor El Ghozlan	June 24, 1910	VIII	Important damage. Aftershocks felt until 1911
Cavaignac (El Asnam)	August 25, 1922	IX–X	Cavaignac completely destroyed. Vertical movement of 1 m
Carnot (El-Asnam)	September 07, 1934	IX	Damage to buildings, landslides, and 92 aftershocks recorded
M'SILA	February 12, 1946	5.6	264 dead, 1000 houses destroyed
Orléanville (El Asnam)	September 09, 1954	X-6.7	1,243 dead, 20,000 houses destroyed, landslides, liquefaction, vertical movement of 1.33 m and several aftershocks
M'sila	February 21, 1960	VIII-5.6	47 dead and 88 injured
M'sila	January 01, 1965	VIII-5.5	5 dead, 1,300 houses destroyed
Mansoura	November 24, 1973	VII-5.1	4 dead, 50 injured
El Asnam	October 10, 1980	IX-7.3	2,633 dead, 8,369 injured, 348 missing persons, 6,778,948 homeless, 70% of houses destroyed, and a strong aftershock recorded one hour after main shock ($M = 6.5$)
Constantine	October 27, 1985	VIII-5.9	10 dead, more than 300 injured, little damage, and surface break
El Affroun	October 31, 1988	VII-5.4	5 injured, many damages, and 500 homeless families
Chenoua	October 29, 1989	VIII-6.0	22 dead, a lot of damage to Sidi Moussa and little damage in Algiers. Inverse fault identified and several aftershocks recorded during 3 months
Mascara	August 18, 1994	VII-5.6	171 dead, important damages in Hacine and Shadlia

(continued)

Table 13.2 (continued)

Locality	Date	Intensity/ magnitude	Consequences/damages
Algiers	September 04, 1996	VII-5.7	A macroseismic radius of 140 km No damage recorded
Ain-Temouchent	December 22, 1999	VII-5.8	28 dead and several destroyed houses. A macroseismic radius ≥ 260 km
Béni-Ouartilane	November 10, 2000	VII-5.4	2 dead, cracking in masonry houses
ZEMMOURI (Boumerdes)	May 21, 2003	X-6.8	2,278 dead, important material damage, and thousands of homeless
Tadjna (Chlef)	December 16, 2006	5	No victims or damage
Chlef	November 01, 2009	4.8	No victims or damage
Chlef	April 25, 2012	5.5	Slight damage

Source Compiled by the author

To represent the most common hazards and disasters in Algeria, Fig. 13.3 shows the affected people and fatalities number by various types of disasters between the years 1980 and 2024. Note that two complementary datasets of disasters served as the foundation for these statistics: The Centre for Research on the Epidemiology of Disasters (CRED), which maintains the International Disaster Database (EM-DAT)[7] and provides freely available information on international mass disasters from 1900 to 2024. It standardizes the events, their effects on people, and the economic losses. The second one is the Climate Change Knowledge Portal, which is a World Bank Group website[8] that provides a wealth of information on disasters, including total deaths, damage, and affected people. These pie charts clearly show that, although floods are the most frequent hazard in Algeria, earthquakes are the most common and devastating disaster.

13.4.4 *Forest Fire Risk*

In addition to earthquakes, and floods, forest fires pose a major threat to northern Algeria. According to data released in 2021 by the White Book on Climate Change, published by the Algerian Ministry of Energy Transition and Renewable Energies, the vulnerability of forests to fires is expected to increase in the coming years due to the recurrence of long-term drought cycles and extreme temperatures. Indeed, the country has endured severe heat waves over the last decade, with weather bulletins placing areas of the country on orange or red alert, causing hospitals to declare a state of emergency. Heat-related illnesses kill an average of 262 individuals in Algeria each year (Hajat et al., 2023). A provisional Interior Ministry report states that forest

[7] EM-DAT, CRED/UCLouvain, Bruxelles, Belgique—www.emdat.be.

[8] https://climateknowledgeportal.worldbank.org/.

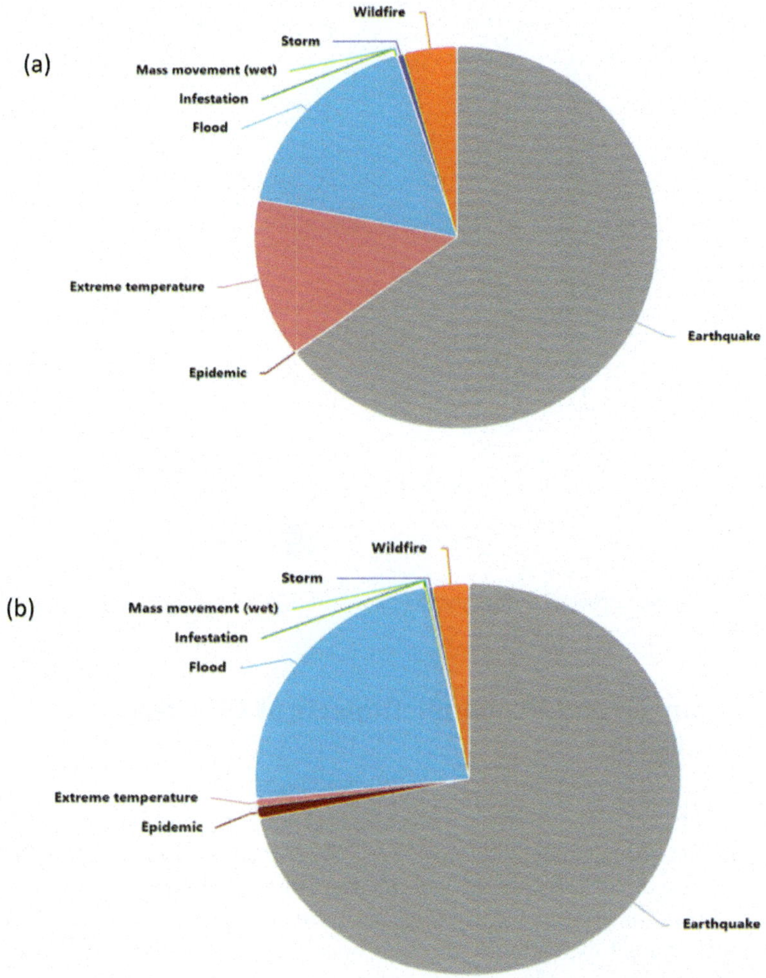

Fig. 13.3 Disaster effects in Algeria between 1980 and 2024: **a** affected people, and **b** number of deaths (World Bank Group, 2023)

fires in Béjaïa, Jijel, and Bouira resulted in 15 deaths and 36 injuries. According to official sources from the National Delegation for Major Risks, extreme temperatures and forest fires affected 125,025 people between 2019 and 2020 and killed 3,859 persons in 2021, rising to 6040 in 2022.

Consequently, over 99% of forested areas in Algeria are at medium to high risk of being impacted by climate change. According to the national plan for combating droughts established in 2019, the significant heat waves experienced by Algeria increases forest areas' vulnerability to forest fires.

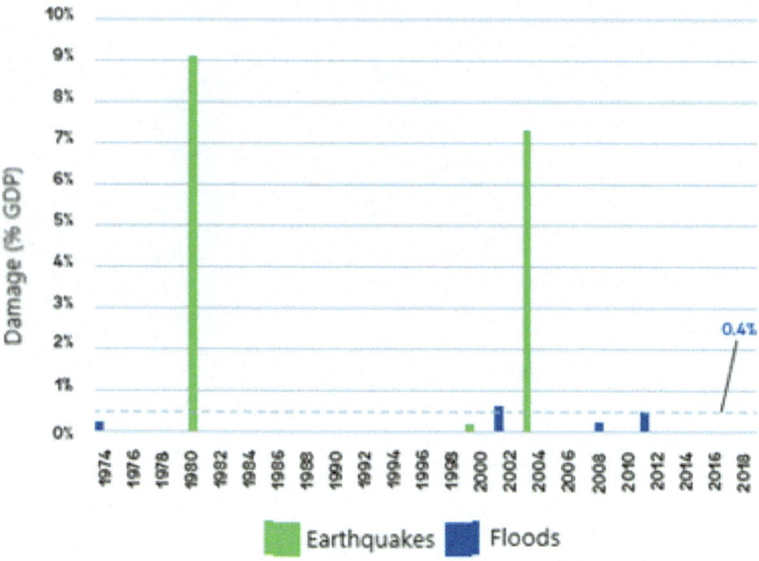

Fig. 13.4 Economic impact of earthquakes and floods in Algeria between 1974 and 2018 (World Bank Group, 2023)

13.5 Financial and Economic Impacts of Disasters in Algeria

Over the last 15 years, the Algerian government estimates that post-disaster interventions for floods, earthquakes, and forest fires have cost an average of $255 million (34 billion Algerian dinars). Within this period, more than $4 billion USD, or 545 billion Algerian dinars, have been set aside to deal with these three types of calamities. Furthermore, annual expenditure on the economic value of wood and cork, forest fire fighting operations, and vegetation restoration is estimated at $18.7 million (2.5 billion Algerian dinars). Between 1974 and 2020, earthquakes and floods in Algeria caused an annual loss of 0.4% of GDP[9] (Fig. 13.4). However, probabilistic risk evaluations show that this damage might total at least 0.7% of GDP per year. The latter statistic only accounts for the risk of earthquakes and river floods. The prospective losses calculated in this manner are about double the historical average and are heavily influenced by seismic risk. Note that disasters primarily have a local impact, only affecting the wider economy to a limited extent (WBG, 2023). Regarding forest fires, between 1985 and 2022, there were 75,375 forest fires, an average of 1,984 fires per year. During this period, 1,366,951 hectares were burned, an average of 18 ha per fire.

[9] Gross Domestic Product.

13.6 Evolution of the Institutional Framework

Prevention is given a lot of weight in Algeria's contemporary risk management regulatory framework. A 2004 framework law, regulatory texts pertaining to risk financing, disaster preparedness and response, and earthquake-resistant construction standards, as well as several sectorial laws and decrees pertaining to land use planning, water and forest management, and coastal development, are among the many texts that make up this framework. The National Drought Plan, the 2020–2030 National Climate Plan, and the National Strategy for Flood Risk Prevention and Management by 2030 are just a few of the many plans, policies, and initiatives that have been established. These seek direct initiatives to increase the nation's resilience to the dangers of disaster and climate change. A strong basis for developing disaster risk management (DRM) policy and enhancing resilience is provided by this collection of texts. In addition, Algeria is fully committed to all international framework agreements related to disaster risk management, including the Sendai Framework for Disaster Risk Reduction (2015) and the Hyogo Framework for Action (2005), as well as the Paris Climate Agreement (2015) and the New Urban Agenda Habitat III (2017).

In this regard, on December 25, 2004, the Parliament passed Law no. 04-20,[10] which is the most important law in the field of disaster risk reduction because, unlike other laws and regulations existing before 2004, which contain only a few elements on risk prevention and disaster, this law is solely dedicated to major risks and identifies ten major risks, namely:

- Seismic and geological risks, floods, climatic risks, forest fires, industrial and energy risks, radiological and nuclear risks, human health risks, risks to animal and plant health, air, terrestrial, marine, and aquatic pollution, and disasters caused by large gatherings of people.

Concerning the way by which the country's crisis and disaster management authorities evolved in Algeria, it is obvious that the national disaster prevention and management policy is implemented through institutions at the national, regional (58 *Wilayas*), and local (1541 *Baladias*) levels, addressing both risk prevention and the organization of interventions and relief efforts. In 2024, Algeria enacted the Major Risks Prevention and Disaster Management law (Law 24/04)[11] and the National Spatial Planning Scheme (SNAT, 2010), which was updated to include important prescriptions for land territory and urban planning. The implementation of national policy and programs is overseen by various institutions and organizations as follows: (1) at the central level, relevant ministerial departments are responsible; and (2) at the regional and local levels, the monitoring and enforcement of program implementation are entrusted to the regional authorities of the provinces (*Wilayas*) and the

[10] Law No. 04-20 of 13 Dhou El Kaada 1425 corresponding to December 25, 2004, on the prevention of major risks and disaster management in the context of sustainable development.

[11] Law no. 24-04 of 16 Chaâbane 145 corresponding to February 26, 2024, concerning the rules for prevention, intervention, and reduction of disaster risks in the context of sustainable development.

municipalities (*Baladiyats*), with technical support from decentralized ministerial departments (Benouar & Belmokhtar, 2025).

These measures include seismic micro-zoning, risk mapping, urban vulnerability assessment, risk teaching via the educational system, development of a national strategy for flood risk prevention, seismic monitoring networks, National Territorial Development Plan,[12] incorporating DRR counting climate change-related risks that includes requirements regarding building codes, land-use planning, and emergency response planning. The findings helped in establishing emergency plans for high-risk zones and the Emergency Response Plan (ORSEC).[13]

As to the legal and policy frameworks governing disaster management, Algeria has reviewed its national policies, plans, and laws related to civil protection, emergency management, and disaster management to align with the Sendai Framework for Disaster Risk Reduction (SFDRR) 2015–2030. The main texts of this legislative framework concern the organization and procedures of crisis management. These are Decrees Nos. 85-231 and 85-232, which were promulgated on August 19, 1985. The first defines the implementation of the ORSEC[14] plan, while the second standardizes the organization of major risk prevention at the national, provincial, and municipal levels. The most recent law regulating the risk management system in Algeria is Law No. 04-20, promulgated on December 25, 2004, concerning the prevention of major risks and disaster management in the context of sustainable development. This law will be analyzed in detail in Sect. 13.7.

Other executive decrees address the insurance aspect of the disaster risk management system in Algeria. These include Executive Decree No. 91-175 of May 28, 1991, which establishes the procedures for issuing planning certificates, subdivision permits, fragmentation certificates, building permits, certificates of conformity, and demolition permits; Executive Decree No. 04-268 of August 29, 2004, which identifies disasters to be covered by insurance; and Executive Decree No. 06-03 of January 7, 2006, which defines the general rules of planning and construction. The August 26, 2003, ordinance on the obligation to ensure disasters and compensate victims constitutes the latest legislative framework related to the insurance axis.

[12] SNAT (2010), updated in 2022.

[13] Executive Decree No. 19-59 of 26 Joumada El Oula 1440, corresponding to February 2, 2019, setting the procedures for developing and managing emergency organization plans (ORSEC).

[14] ORganisation des interventions et SECours (In French).

13.7 Agencies and Stakeholders Responsible for Disaster Response and Mitigation

13.7.1 State Agencies

Several organizations are involved in disaster response and mitigation. Belonging to multiple ministries, they are responsible for preparing and implementing preventive strategies. Table 13.3 gives a list grouping some of these organisms, their ministerial authority, and their allocated tasks.

13.7.2 Scientific Institutions

Research institutions supervised by the Higher Education and Scientific Research, or other ministries, such as the Ministry of the Interior and Local Authorities and Regional Planning, and the Ministry of Housing, Urban Planning, and City, contribute efficiently to improving disaster mitigation and resilience and promote scientific research for the prevention of disaster risk. In addition to those listed in Table 13.4, we can also cite the Directorate-General for Scientific Research and Technological Development (DGRSDT), the Center for the Development of Renewable Energy (CDRE), the Center for Research in Industrial Technologies (CRIT), and the Scientific Research Center on Arid Regions (CRSTRA). Moreover, national and international conferences are continually organized by Algerian universities to display the latest research findings in the field of disaster risk mitigation. Several research projects concerning cities' vulnerability to seismic risk, floods, and geological risks have been conducted for critical infrastructure in major Algerian cities.

13.8 Analysis of the Disaster Management System

The disaster management system adopted by Algeria has both strengths and weaknesses. Through a critical analysis, this section will expose the challenges faced by this system, its weaknesses, and shortcomings, as well as the aspects that make up the strength of this system, such as the contribution of Algerian scientific organizations to the improvement of this system, and the involvement of the civil community in awareness and preparedness.

Table 13.3 Agencies and stakeholders involved in risk management in Algeria

Organism	Ministerial authority	Assigned tasks
Algerian Space Agency (ASAL)	Prime Minister	The ASAL proposes space activity strategy aspects and ensures their design and execution. It also establishes a bilateral and multilateral cooperation policy tailored to national needs and monitors commitments resulting from the state's obligations under regional and international space agreements
National Institute of Cartography and Remote Sensing (INCT)	National Defense Ministry	The INCT's missions include creating national geographic databases, maintaining a basic geodetic and gravimetric framework, producing and updating topographic maps, collecting and storing aerial and satellite data, and preserving archives
National Council for Geographic Information (CNIG)		The CNIG proposes national policies on geographic information, as well as any measures that regulate the use, dissemination, and marketing of geographic data. It also ensures the integrity and security of this data while promoting scientific research in the field. Additionally, the SNIG develops common specifications, norms, and standards
General Directorate of Civil Protection (DGPC)	Ministry of the Interior and Local Authorities and Regional Planning (MICLAT)	The General Directorate of Civil Protection (DGPC) is responsible for ensuring the protection of people and property against accidents, disasters, and catastrophes resulting from anthropic or natural causes
National Delegation of Major Risks (DNRM)		The DNRM establishes safety procedures and standards for organizing, planning, and implementing emergency response measures. The DNRM also evaluates actions carried out within the national major risk prevention system and disseminates major risk prevention information to stakeholders and the public
National Statistics Office (ONS)	Ministry of Digitalization and Statistics (MNS)	The ONS is Algeria's official source for developing, providing, and disseminating reliable and regular information and statistical analyses tailored to the needs of economic and social stakeholders and the public
General Directorate of Forests (DGF)	Ministry of Agriculture and Rural Development	The DGF coordinates and monitors actions to prevent and control forest fires, diseases, and pest infestations. The DGF also ensures compliance with forestry laws and regulations, as well as overseeing the execution of forestry programs

13.8.1 Challenges of the Current Disaster Management System

The major challenges faced by disaster management authorities in dealing with past disasters are marked by insufficient capacity and coordination in reporting disaster

Table 13.4 Algerian scientific institutions involved in disaster resilience

Scientific institution	Mission
The Center of Research in Arrangement of the Territory (CRAT)	Responsible for conducting scientific research and technological development programs in the fields of land-use planning, focusing on the organization, management, and development of territories and cities in spatial, social, cultural, economic, and environmental terms
National Center of Studies and Research Integrated from the Building (CNERIB)	Responsible for undertaking all activities related to studies and realization, intended to facilitate the preparation of national policy on housing and construction, and to develop and implement national scientific research and technological development programs
National Center of Research Applied in Earthquake-Resistant Engineering (CGS)	Responsible for developing research in the fields of earthquake engineering, reducing seismic risk, contributing to the improvement of seismic regulation, and training and informing national technical managers on the practice of earthquake engineering and risk reduction
National Organism of Technical Construction Control (CTC)	Responsible for the standardization of construction risks as part of the prevention of technical hazards likely to be encountered in the construction of buildings and civil engineering works in accordance with the law on decennial insurance
The National Agency of Urban Planning (ANURB)	Constitutes the main tool of the State for the development and monitoring of development and urban planning plans
Research Center for Astronomy, Astrophysics, and Geophysics (CRAAG)	The CRAAG conducts research in astronomy, astrophysics, and geophysics. It monitors natural geophysical and astronomical phenomena to develop maps. It also expands the seismological network and ensures its operation and maintenance. The CRAAG contributes to seismic monitoring, identifies, and compiles old documents, including international ones

losses. When disasters occur, various ministries, organizations, and stakeholders record the damage, yet there is little collaboration among them to consolidate these reports. This fragmentation makes it difficult to form a clear and accurate understanding of the full extent of the damage. Conducting a comprehensive economic evaluation of disaster losses requires extensive research and coordination, a complex task for personnel working across different ministries, agencies, and organizations (Benouar & Belmokhtar, 2025).

Despite the existence of legal texts related to disaster management in Algeria, regulations and instruments are not being implemented. Indeed, implementing regulations is often slow to be adopted or, when they do exist, their application is problematic. Regulatory requirements pertaining to the consideration of risks in construction, land-use planning, and urban planning, for instance, have not stopped the proliferation of private buildings that do not comply with construction standards and/or are located in risk areas; only one of the 15 general risk prevention plans (PGPR) allowed in the framework law (2004) has been adopted; and the coverage rate for disaster insurance is still low despite being required.

However, better coordination across the various sectors and administrative levels is required to guarantee the coherence of this extensive collection of standards and plans as well as their efficient and well-coordinated implementation. The regulations provide several interdepartmental procedures in this regard. Nevertheless, there are differences in the operationalization of these different systems. When it comes to DRM inter-ministerial coordination, it remains largely ad hoc, focusing on crisis coordination rather than prevention or recovery. Implementing an integrated and coordinated national risk management policy is greatly aided by the DNRM's anchoring within the Ministry of the Interior, yet the DNRM currently lacks the resources required to carry out its duties.

Furthermore, given the close links between disaster risks and climate change, risk prevention initiatives could be more closely coordinated with those for climate change adaptation, under the leadership of the National Climate Committee. Climate change should also occupy a more prominent place in the national regulatory framework. An additional aspect that could enhance disaster management and recovery is planning for recovery before disasters occur. This involves implementing measures and procedures that would facilitate a quick return to normalcy, thereby achieving the concept of disaster resilience (Cao & Feng, 2025).

13.8.2 Weaknesses and Gaps in the Current Disaster Management System

The Algerian law, Law 24-04 needs to further clarify the concept of disaster and appropriately incorporate the concept of disaster preparedness (Siriwardhana et al., 2021). It should also clarify the criteria for declaring areas as disaster-affected to ensure greater ease in fulfilling responsibilities or obtaining rights for citizens, especially in relation to insurance and compensation. Among the risks that may be included are those arising from civil unrest, such as riots, acts of vandalism, major transportation accidents, and significant ideological conflicts. The law mentions marine, air, and water pollution, but does not refer to land pollution, which can be very dangerous in some cases and can cause both direct and indirect damage to the environment, human health, animals, and plants. It can also cause irreparable damage, so it is important to pay attention to this aspect as well.

According to Tout et al. (2025), although Law 24-04 clarifies and addresses many aspects overlooked in the previous law (Law no. 20-04) and despite adding value to the prevention policy and disaster management, it still neglects certain areas and lacks detailed guidelines on others. This may limit its effectiveness or impact in the short, medium, or long term. For instance, the law does not specify how local or international efforts and aid should be organized during disasters, nor does it establish criteria for preparing, evaluating, or revising prevention and intervention plans, or the timeframe for doing so. Instead, it refers these matters to regulatory provisions. Article 25 of the law states that the sectors responsible for drafting, amending, and implementing the general plans for disaster risk prevention will be designated by regulatory measures. The absence of regulatory texts that clarify these aspects raises questions about the importance of evaluating and reviewing prevention plans and the expected framework for organizing these evaluations and reviews.

Another gap in the current disaster risk management system is related to the coordination among governmental and non-governmental entities. This coordination may include inadequate communication, insufficient funding and resource allocation for both sectors, a lack of combined strategies and shared data across multiple agencies, and insufficient preparedness and response training and capacity building. These concerns can lead to duplicated actions, unsatisfied demands, and a less efficient disaster response overall.

While Algeria has made progress in reducing mortalities for some types of disasters, challenges remain in addressing others, particularly in the context of the pandemic. Despite having a strong and coherent legislative framework, its implementation among institutions and organizations is often lacking.

13.8.3 Strengths of the Current Disaster Management System

- **Community Involvement and Awareness**: The concept of participation mentioned in the law should evolve from merely informing and consulting about risks and preventive arrangements to actively encouraging the establishment of specific preventive measures. Civil society in Algeria has voluntarily become involved in risk management by participating in rescue and relief operations and providing material aid to affected populations during past disasters. The adoption of a local preventive culture is crucial, as it is based on real past experiences and should be promoted, especially in remote areas significantly exposed to risks. Conversely, current laws do not clarify the possibility of organizing training that simulates exposure to potential risks, which may prevent the actual framing of rescue operations and disaster management. Therefore, allowing this possibility may help prepare responders and citizens to better deal with emergency situations and avoid the disturbances that may accompany the process, ensuring a more effective intervention (Tout et al., 2025).
- **Alignment with International Frameworks**: Algeria's alignment with international frameworks and standards, such as the 2015 Sendai Framework for Disaster

Risk Reduction, strengthens its national preparedness and enhances its coordination with international partners during crises. By integrating these frameworks and standards, Algeria aims to improve its Disaster Risk Reduction and response ability. The updated regulations provide a clear roadmap for disaster risk reduction (DRR), stressing the importance of proactive measures, community engagement, and sustainable development (Benouar & Belmokhtar, 2025).

- **Risk Teaching in the Educational System**: The Algerian educational programs include teaching the 10 major risks at the different levels of national education. This includes raising awareness, providing information, and offering training in major risk and disaster risk reduction. Risk scenarios are also performed in elementary, middle, and high schools. These scenarios consist of simulating disaster events, such as earthquakes, and training both students and teachers on the proper behaviors to adopt during the evacuation procedures to prevent injuries. However, such actions must be monitored by experts, from the general directorate of civil protection to ensure their effectiveness (Matlakala, 2022) and to prevent practices that could lead to adverse outcomes.
- **The Use of Technology, Data, and Innovation in Enhancing Disaster Resilience**: The use of modern technologies in the prevention and management of major disasters has become very important and may be essential to avoiding compounding losses, especially during emergencies. The early warning system in Algeria, which is based on technology, data recording, and innovation, has shown notable and significant improvement over the last few years. This system covers seismic monitoring, epidemiological surveillance and alert, meteorological monitoring, floods, earthquake disasters, and forest fire monitoring and alert system. However, legislation suppleness is desirable to allow the use of certain technologies, such as drones, to provide better means of evacuation and intervention (Tout et al., 2025).

13.9 Recommendations and Future Strategies

A recent study suggests that the Algerian Law No. 24-04, which encompasses rules for prevention, intervention, and disaster risk reduction within the framework of sustainable development, requires revisions to include several additional aspects. The study proposes review and updating plans, incorporating into the legal text, a periodic review process, and an annual assessment. In addition, the study proposes the review of prevention plans based on reports from research institutions, conducting reviews based on major structural changes, including infrastructure and changes in human settlements, conducting a post-disaster review to ensure that the new conditions left by the disaster do not pose a threat of further disasters. Also, suggestions were made to review prevention plans based on reports from research institutions that include technological and technical changes in the fields of prevention, risk detection, forecasting, and warning. Plans should be reviewed based on legislative changes or outputs from international agreements, and expect reviews to follow community requests or consultations, and take into account, geographical, climatic, and potential vulnerability aspects during reviews, re-evaluate facilities after each exceptional incident, impose continuous monitoring procedures for protective structural facilities, and conduct assessment of preventive measures to identify potential imbalances (Tout et al., 2025).

Furthermore, measures intended to attain preventive actions are listed as follows:

- Establishing a comprehensive framework that integrates policies and coordinated action plans to increase flood and forest fires resilience.
- Controlling urbanization processes, which are considered essential and urgent (Chao & Hsu, 2018). It is insufficient to rely solely on prohibiting construction within specific areas of cities; there should be strong emphasis on managing urban expansion, especially in cities prone to risks, to avoid testing the regions' resilience and to reduce vulnerability factors (Bechara, 2020).
- Focusing on regulating evacuation operations, particularly large-scale evacuations that can lead to additional disruptions and chaos due to a lack of coordination. To protect public health, which is a central factor that disaster management aims to achieve overall, waste management and sanitation during relief and rescue operations should be a fundamental component in relief efforts to prevent negative outcomes (Evangelista et al., 2024).
- Improving storm water drainage, building retention basins, and strengthening dams in flood-prone locations, particularly metropolitan areas, are crucial initiatives. To accommodate significant rainfall occurrences, infrastructure should be modernized and maintenance performed on a regular basis (Hallouz & Meddi, 2025).

13.10 Conclusions

This chapter outlines the main natural and man-made hazards that Algeria, a country in the MENA region, faces. It presents the technical and legislative measures implemented by the Algerian government to mitigate the risks threatening the population and strengthen societal resilience. The chapter provides an overview of Algeria's jurisdiction and governance structure, as well as its geographic location, population, urbanization rate, economy, and the main natural risks threatening the country. By providing statistics collected from international and local sources, the chapter sheds light on three of the ten hazards identified by the Algerian law on the Prevention of Major Risks and Disaster Management in the context of Sustainable Development.

The chapter displays that the three hazards, which are earthquakes, floods, and forest fires, are the most dangerous in terms of loss of human life and economic damage. It summarizes all aspects related to disaster risk reduction, discussing the strengths and weaknesses of the system adopted for the prevention and management of major disasters in Algeria. The chapter also makes recommendations and suggests future strategies to improve the current system's efficiency. Ultimately, aligning with international frameworks and standards strengthens Algeria's national preparedness and enhances its coordination with international partners during crises. The updated regulations provide a clear roadmap for disaster risk reduction (DRR) that emphasizes the importance of proactive measures, community engagement, and sustainable development. Algeria's resilience increases as the country refines its disaster risk reduction (DRR) strategy. This commitment demonstrates Algeria's dedication to safeguarding its people and resources, as well as ensuring a safer future. Additionally, through critical analysis, the chapter makes recommendations and suggests strategies for improving the current system's effectiveness. It has been established that, aligning with international frameworks and standards strengthens the country's national preparedness and improves coordination with international partners in the event of a crisis. It is also recognized that the updated regulations provide a clear roadmap for disaster risk reduction (DRR), emphasizing the importance of proactive measures, community engagement, and sustainable development. Thus, Algeria's disaster management system is found to be based on proactive measures implemented before a risk occurs. Therefore, it can be concluded that as the country refines its DRR strategy, its resilience will increase. This commitment demonstrates Algeria's determination to protect its people and resources and ensure a safer future.

References

Bechara, S. (2020). Running the big risks: A reading of the Algerian experience (in Arabic). *Journal of Humanities and Social Sciences, 04*. https://www.asjp.cerist.dz/en/article/139750

Benouar, D., & Belmokhtar, A. (2025). Risk governance for the implementation of the Sendai framework. *International Journal of Disaster Risk Science, 16*, 58–70. https://doi.org/10.1007/s13753-025-00619-4

Boutaraa, Z. (2018). *Revelations of the 1980 El Asnam (Algeria) earthquake*. LAP LAMBERT Academic Publishing.

Cao, X. Y., & Feng, D. C. (2025). Probabilistic resilience assessment framework of structures considering the combined functionality-recovery-duration uncertainty. *Mechanical Systems and Signal Processing, 223*(July 2024). https://doi.org/10.1016/j.ymssp.2024.111856

Chao, J. C., Hsu, K. W. (2018). The impact of Taiwan's urban-renewal policies, laws and decrees on urban redevelopment and urban disaster prevention. *IOP Conference Series: Earth and Environmental Science, 164*(1). https://doi.org/10.1088/1755-1315/164/1/012034

Evangelista, S. S., Maturan, F., et al. (2024). Cumulative prospect theory under different types of input data for public health resilience assessment during natural disasters. *Expert Systems with Applications, 258*, 125172. https://doi.org/10.1016/j.eswa.2024.125172

Hajat, S., Proestos, Y., Araya-Lopez, J. L., Economou, T., & Lelieveld, J. (2023). Current and future trends in heat-related mortality in the MENA region: A health impact assessment with bias-adjusted statistically downscaled CMIP6 (SSP-based) data and Bayesian inference. *The Lancet Planetary Health, 7*(4), e282–e290.

Hallouz, F., & Meddi, M. (2025). Progress in multicriteria decision making models. A new paradigm to monitor hazards. In J. Das, & S. Halder (Eds.), *Progress in multicriteria decision making models, Advances in geographic information science*. https://doi.org/10.1007/978-3-031-89246-2_21

Matlakala, F. K. N. D. C. (2022). Traditional coping strategies among members of Runnymede, South Africa. *African Renaissance, 19*(1), 147–163. https://doi.org/10.31920/2516-5305/2022/19n1a7

ONS. (2024). Office National des Statistiques. "Démographie algérienne 2020–2023" No. 1030 (in French and Arabic). Retrieved November 03, 2025, from https://www.ons.dz/spip.php?article3002

RPA. (2024). *Règles Parasismiques Algériennes* (in French). Technical Regulatory Document D.T.R. B.C. 2.48. National Center for Applied Research in Earthquake Engineering & Ministry of Housing, Urban Planning.

RPOA. (2008). *Règles Parasismiques Applicables au Domaine des Ouvrages d'Art* (in French). Technical Regulatory Document D.T.R. Ministry of Public Works.

Sekfali, R. (2025). Local governance in Algeria: Reality and prospects. *Russian Law Journal, XIII*(202)(Issue 1).

Siriwardhana, S., Kulatunga, U., & Fernando, T. (2021). A systematic review to establish a definition for natural disaster preparedness. In *Proceedings of the international conference on industrial engineering and operations management*, Dekens 2007 (pp. 5386–5397). https://doi.org/10.46254/an11.20210903

SNAT. (2010). *Schéma National de l'Aménagement du Territoire* (in French). Ministry of Environment, Algeria & Ministry of Interior (Revised in 2022).

Tout, F., Rebouh, N., Dinar, H., Benzid, Y., & Zouak, Z. (2025). Evaluating Algeria's approach to major risk prevention and disaster response. *Journal of Geography, Politics and Society, 15*(1), 50–56.

World Bank Group. (2023). *Diagnostic on the management of climate and disaster risks in Algeria*. World Bank Group. Retrieved October 06, 2025, from https://documents.world-bank.org/curated/en/09912-10231-22035477/P1628-7902e-af5a0-1d0a1-940a7-2376b-420e5

Chapter 14
Hazards, Disaster, Crisis, and Emergency Management in Egypt: A Comprehensive Analysis Framework

Lobna Ahmed Elgamal and Amany Mounir Helal

Abstract Egypt faces increasing disaster risks from natural hazards, climate change, and urbanisation pressures, requiring a comprehensive evaluation of its disaster management capabilities. This chapter aims to conduct a comprehensive assessment of Egypt's disaster management framework, examining institutional evolution, current challenges, and future strategies for enhanced resilience. We employed mixed-methods systematic review following PRISMA guidelines, analysed literature from 2000 to 2024, including peer-reviewed journals ($n = 127$), government reports ($n = 45$), policy documents ($n = 47$), and international organisation reports ($n = 38$). Analysis employed thematic analysis, trend assessment, and comparative evaluation with inter-rater reliability of $\kappa = 0.85$. The findings suggest that Egypt's disaster management has evolved significantly since the 1992 Cairo earthquake, with the establishment of the National Centre for Crisis and Disaster Risk Reduction and expanding early warning systems. However, the system remains 70–80% reactive with limited community engagement and infrastructure vulnerabilities affecting 40% of buildings (95% GLF, 15% 4-5-4). Cumulative economic losses during 2000–2024 reached $23.1 billion. Pre-COVID-19 losses averaged $420 million annually, while post-COVID-19 losses (2020–2024) increased sixfold to $4.2 billion annually, reflecting heightened pandemic-related economic disruption. Egypt requires a transition from reactive to proactive disaster management through institutional modernisation, enhanced community engagement, infrastructure resilience improvements, and increased investment in prevention.

Keywords Disaster management · Egypt · Risk reduction · Institutional framework · Community resilience

L. A. Elgamal (✉)
MOHP, Cairo, Egypt
e-mail: lobna.icu@gmail.com

A. Mounir Helal
Egyptian Ambulance Organization (EAO), Giza, Egypt

E. Alam (ed.), *Disaster, Crisis and Emergency Management in the Middle East & North Africa*, https://doi.org/10.1007/978-3-032-24436-9_14

14.1 Introduction

This chapter aims to analyse Egypt's disaster management institutional evolution and current framework. This is achieved by addressing the following objectives: Evaluate hazard profiles and disaster impacts across different sectors; Systematically identify challenges, weaknesses, and gaps in current approaches; Assess best practices and innovations in Egyptian disaster management; and Provide evidence-based recommendations for enhanced resilience.

Egypt's 1,001,449 km^2 territory spans diverse hazard-prone regions, including the Mediterranean coasts vulnerable to sea-level rise, deserts with extreme temperatures, the flood/drought-susceptible Nile corridor, and the seismically active Sinai Peninsula at the African-Arabian tectonic plate intersection (El-Geziry & Radwan, 2022). With 104 million people (UN DESA, 2024), Egypt faces severe demographic pressures. The population density in the Nile Valley/Delta at 1,540 people/km^2, complicating disaster response. Critically, 12 million live in informal settlements with inadequate infrastructure and substandard buildings, creating compounded vulnerabilities. Only 43% are urbanised, but rural-to-urban migration intensifies infrastructure strain. A young population (60% under 30) presents both engagement opportunities and dependency challenges (United Nations Department of Economic and Social Affairs, Population Division. World Population Prospects, 2024).

Egypt's $469 billion economy (World Bank, 2024) faces critical vulnerabilities despite its growth potential. The service sector dominates (56% of GDP), industry 33%, and agriculture 11% (employing 25% of the workforce) (World Bank, 2024). Tourism peaked at 12% of GDP (2019) with $15.03 billion before collapsing to 3.4% (2020) during COVID-19, recovering to 10.7% by 2023, demonstrating extreme sensitivity to health emergencies (World Travel & Tourism Council, 2023). Suez Canal dependency creates significant risk: revenues reached $9.4 billion (2023), representing 1.5–2.0% of GDP and a major foreign currency source (Arab Finance, 2023). Climate-vulnerable agriculture threatens food security despite its small GDP share. The informal economy (40–60% of activity) complicates disaster assessment and response coordination (Food and Agriculture Organization of the United Nations, 2022).

Egypt's disaster management evolution demonstrates a clear alignment with global frameworks, including the Hyogo Framework for Action 2005–2015 and the Sendai Framework for Disaster Risk Reduction 2015–2030. The Sendai Framework emphasises four priority areas: understanding disaster risk, strengthening disaster risk governance, investing in disaster resilience, and enhancing disaster preparedness (United Nations Office for Disaster Risk Reduction, 2015). Following the transformative 1992 Cairo earthquake, Egypt progressively adopted international best practices while developing locally adapted approaches. The country's institutional development reflects broader regional patterns across the Middle East and North Africa while addressing unique geographic, demographic, and economic challenges (Abou-Korin & Al-Shihri, 2015). Egypt's adoption of the Sendai Framework in 2015 and establishment of the National Centre for Crisis and Disaster Risk Reduction in

2014 demonstrate a clear commitment to aligning with global standards. However, there is a lack of comprehensive research that examines Egypt's disaster management system evolution compared to international benchmarks (Arab Republic of Egypt, 2017).

The Arab region displays common challenges, including limited coordination mechanisms between government levels, inadequate early warning systems, and insufficient community engagement (Lassa, 2018). Regional cooperation initiatives, facilitated through the Arab League and Islamic Development Bank, provide frameworks for shared learning, though implementation effectiveness varies significantly (Lassa, 2018). Comparative analysis reveals Egypt demonstrates stronger institutional development than most Arab countries, benefiting from its larger economy, more developed educational institutions, and greater international engagement. However, challenges related to rapid urbanisation, infrastructure deficits, and resource constraints remain common across the region (Erian et al., 2021).

Despite significant evolution since 1992, substantial gaps persist in understanding system effectiveness, community resilience levels, and optimal resource allocation strategies. Existing research tends to focus on individual aspects without providing a holistic perspective necessary for understanding system-wide strengths, weaknesses, and improvement opportunities (Smith & Becker, 2023). It still lacks comprehensive analysis integrating institutional, technical, and community dimensions. The absence of systematic evaluation frameworks specifically adapted to Egypt's unique context suggests that policy makers lack robust evidence bases for prioritising investments, designing interventions, and measuring progress. Community-level resilience assessment remains underdeveloped despite growing recognition of its critical importance (El-Anwar et al., 2023).

14.2 Research Methods

This section outlines the comprehensive methodological approach employed to systematically analyse Egypt's disaster management system across multiple dimensions, following internationally recognised standards for systematic review methodology.

14.2.1 Study Design and Protocol

This mixed-methods approach integrates quantitative and qualitative evidence from multiple sources spanning 2000–2024, while incorporating historical context extends back to 1992. The protocol was developed and registered prior to data collection commencement. Two independent reviewers conducted literature screening, achieving inter-rater agreement of $\kappa = 0.85$ (95% CI: [0.78, 0.92]), indicating substantial agreement. Data extraction employed standardised forms with a 94%

agreement rate on key data points. Quality assessment utilised the Mixed Methods Appraisal Tool (MMAT) for mixed-methods studies and the Newcastle-Ottawa Scale for observational studies.

14.2.2 Search Strategy and Data Sources

The comprehensive search strategy employed multiple complementary approaches. Academic database searches utilised Web of Science, Scopus, PubMed, and Arab Impact Factor database. The complete Web of Science search string was: (("disaster management" OR "emergency management" OR "crisis management" OR "risk reduction" OR "disaster preparedness") AND ("Egypt" OR "Egyptian") AND ("policy" OR "institutional" OR "framework" OR "system")), resulting in initial identification of 1,847 potentially relevant records.

Government sources provided critical data through Egyptian Ministry of Interior reports ($n = 12$, 2010–2024), CAPMAS statistical yearbooks ($n = 15$, 2000–2024), governorate-level plans ($n = 27$), meteorological service reports ($n = 18$), and legislative records ($n = 5$). International organisation sources included UN system publications ($n = 22$), World Bank reports ($n = 8$), regional organisation documents ($n = 5$), and technical assistance reports ($n = 3$).

PRISMA Flow Summary

Initial database searches identified 1,847 records, with an additional 63 records from government sources and hand-searching, totalling 1,910 records. After removing 342 duplicates, 1,568 records were screened, with 1,311 excluded as clearly irrelevant. Following full-text review of 257 articles, 47 were excluded (insufficient quality $n = 18$, lack of Egypt-specific content $n = 21$, outside publication window $n = 8$), resulting in 210 studies included: 127 peer-reviewed articles, 45 government reports, 25 international organisation assessments, and 13 policy documents.

14.2.3 Inclusion and Exclusion Criteria

Systematic inclusion criteria ensured comprehensive coverage while maintaining quality standards. Included materials comprised studies published 2000–2024 with a specific focus on disaster management in Egypt, official government reports addressing the 2000–2024 period, international organisation assessments, and peer-reviewed publications. Language restrictions limited inclusion to English and Arabic publications.

Exclusion criteria eliminated studies published before 2000 (except for the 1992 earthquake historical context), non-English/Arabic publications without verified translations, opinion pieces without substantial evidence base, and duplicate publications.

14.2.4 Data Analysis Methods

The research employed a comprehensive mixed-methods approach that integrated quantitative and qualitative techniques through a convergent parallel design. Quantitative analysis utilised descriptive statistics to assess disaster frequency and impacts by calculating means, standard deviations, and confidence intervals, while linear regression models examined temporal patterns and GIS mapping identified spatial distribution patterns. Time series analysis with moving averages detected long-term trends in the disaster data. On the qualitative side, systematic thematic analysis was conducted using inductive coding by two independent coders who achieved a strong inter-rater reliability of $\kappa = 0.85$, and framework analysis aligned findings with Sendai Framework priorities and international best practices. Content analysis tracked the chronological evolution of policies, while pattern identification techniques sought convergent and divergent findings across multiple data sources. The integration of both approaches was achieved through triangulation techniques, which ensured a comprehensive understanding, with data merged during the interpretation phase using joint displays, narrative synthesis, and meta-inferences to produce cohesive findings.

14.2.5 Economic Loss Calculation Methodology

The economic loss data were compiled from multiple authoritative sources, including 25 CAPMAS economic impact reports spanning 2000–2024, the World Bank Egypt Economic Monitor series, Ministry of Finance disaster loss accounting records, and the CRED EM-DAT database. Economic losses were calculated using internationally recognised methodologies and categorised into direct components such as infrastructure damage, property destruction, and immediate response costs, as well as indirect components including business interruption, supply chain disruption, and productivity losses. Pandemic-related losses were calculated separately using the WHO health impact assessment methodology and World Bank economic modelling, with direct COVID-19 losses encompassing healthcare system costs and excess mortality impacts, while indirect losses included attributed GDP reduction, tourism sector collapse, and supply chain disruptions specifically resulting from pandemic-related restrictions. Statistical analysis involved aggregating annual losses by calendar year, with descriptive statistics calculated separately for pre-COVID-19 (2000–2019) and post-COVID-19 (2020–2024) periods. In addition, confidence intervals were calculated using the t-distribution at a 95% confidence level. The Shapiro–Wilk normality test for pre-COVID-19 data yielded $W = 0.94$, $p = 0.18$, indicating a normal distribution, and cross-validation was performed to assess reliability across the different data sources.

14.3 Results

This section presents comprehensive findings from the systematic analysis of Egypt's disaster management system, organised thematically to address key research questions and provide an evidence base for conclusions and recommendations.

14.3.1 Understanding Egypt's Disaster Landscape

Egypt's modern disaster management story truly begins with the 1992 Cairo earthquake. The magnitude 5.8 event on 12 October 1992, forever changed how the country approaches emergency preparedness. The earthquake claimed 561 lives, injured 6,512 people, and caused economic losses between $400 and 500 million when indirect impacts are included. More importantly, it exposed critical vulnerabilities in informal settlements and revealed significant gaps in emergency response coordination, building code enforcement, and public awareness that would shape reforms for decades to come (United Nations Office for Disaster Risk Reduction, 2015).

Since then, Egypt has faced a diverse array of challenges. Hydrological hazards have become a recurring threat, with notable flooding events in Sinai in 2010, the Red Sea coast in 2014, Alexandria in 2018 (causing 12 deaths and $100 million in losses), and Aswan in 2019. Heat waves have emerged as a growing concern, particularly the 2015 event that claimed 95 lives and the extended 2021 heat waves that tested public health systems. Transportation disasters have also left deep scars, including the devastating 2002 Beni Suef train fire that killed 383 people and the 2006 ferry disaster that claimed 1,034 lives. Most recently, the COVID-19 pandemic has proven to be the most economically impactful event in Egypt's recorded disaster history, with over 24,000 deaths and losses exceeding $20 billion (Abou-Korin & Al-Shihri, 2015).

When examining disaster patterns over the past three decades, several concerning trends emerge. The frequency of major events has doubled from 8 incidents during 1992–2001 to 16 events during 2012–2021, representing a statistically significant temporal trend. Even when excluding smaller events with losses below $10 million, the data reveals a 75% increase in significant disasters, confirming that this represents a genuine rise rather than improved reporting. Linear regression analysis reveals an upward trend with high statistical significance, and this pattern persists even when controlling for population growth, urbanisation rates, and GDP per capita (Arab Republic of Egypt, 2017).

As given in Table 14.1, Egypt's current hazard classification reveals the distribution of different disaster types across frequency, impact, and geographic patterns. This classification is based on multiple authoritative sources, including the Egyptian Meteorological Authority, National Research Institute of Astronomy and Geophysics, and international disaster databases, following ISO 31000:2018 risk management standards. The analysis covers 74.7 million people in formally assessed risk zones, representing 72% of Egypt's 104 million population.

The economic toll of disasters in Egypt tells a story that has been dramatically shaped by COVID-19. Between 2000 and 2024, cumulative losses reached $23.1 billion, with the pandemic alone accounting for $20 billion of this total. However, examining the pre-COVID-19 baseline reveals important patterns in typical disaster costs. During the 20 years before the pandemic, Egypt averaged $420 million in annual disaster losses with relatively stable patterns. As given in Table 14.2, the stark difference between the COVID-19 period and normal disaster patterns is evident when comparing mean annual losses, which jumped from $420 million to $4,200 million during the pandemic years. This tenfold increase demonstrates both the exceptional nature of health emergencies and the underlying stability of traditional disaster costs. When COVID-19 is excluded from the analysis, the baseline remains remarkably consistent at $430 million annually, providing a more accurate picture of Egypt's typical disaster risk profile.

Table 14.1 Egypt hazard classification matrix (1992–2024)

Risk category	Hazard type	Frequency	Impact level	Geographic distribution	Primary affected sectors	Return period
High frequency, high impact	Flash floods	Annual	High	Sinai, coastal areas	Infrastructure, settlements	1–2 years
	Extreme temperatures	Annual	High	Nationwide, urban heat islands	Public health, energy, agriculture	Every summer
	Sandstorms	Multiple/ year	High	Desert regions, urban areas	Transportation, health, aviation	10–15 events/ year
Moderate frequency, high impact	Earthquakes	Periodic	High	Southern Red Sea, Cairo	Buildings, infrastructure	5–10 years
	Coastal flooding	Episodic	High	Nile Delta, Mediterranean	Agriculture, settlements	3–7 years
	Drought	Cyclical	High	Agricultural regions, Upper Egypt	Agriculture, water, food security	7–12 years
Low frequency, very high impact	Major seismic svents (>6.0)	Rare	Very high	Fault zones, urban centres	Critical infrastructure, mass casualties	20–50 years
	Tsunami	Very rare	Very High	Red Sea, Mediterranean coasts	Coastal infrastructure, tourism, ports	50–100+ years
	Volcanic activity	Extremely rare	Very high	Red Sea region	Regional infrastructure, aviation	100+ years

Table 14.2 Economic loss analysis by period (2000–2024, values in 2024 USD)

Period	Years	Mean annual loss (M)	Standard deviation	95% Confidence interval	Median (M)	Range (M)
Pre-COVID-19 (2000–2019)	20	$420	±$180	[$340, $500 M]	$380	$150–$850
COVID-19 period (2020–2024)	5	$4,200	±$1,800	[$2,800, $5,600 M]	$4,000	$2,500–$7,000
Entire period (2000–2024)	25	$924	±$2,100	[$580, $1,268 M]	$400	$150–$7,000
Baseline (Excluding COVID-19)	23	$430	±$185	[$350, $510 M]	$390	$150–$900

Understanding how losses distribute across different disaster categories provides additional insight into Egypt's risk profile. As given in Table 14.3, the $23.1 billion in cumulative losses breaks down by category, revealing the overwhelming dominance of health emergencies due to COVID-19's exceptional impact. When this outlier is removed, the pattern shifts dramatically, with transportation disasters accounting for 58% of typical losses and natural hazards representing 26%. This distinction between pandemic-influenced and normal loss patterns is crucial for planning future risk reduction investments and insurance mechanisms (Lassa, 2018).

Geographic distribution of risk reveals stark disparities across Egypt's landscape. As given in Table 14.4, high-risk areas including Greater Cairo, Alexandria and coastal zones, and the Nile Delta collectively house 53.2 million people, just over half the national population. These regions have recorded 26 major events between 1992 and 2024 and face threats from seismic activity, heatwaves, flooding, sea-level rise, and infrastructure vulnerabilities, with a composite risk index averaging 7.6 out of 10. Medium-risk areas in Upper Egypt and Sinai contain an additional 21.5 million

Table 14.3 Cumulative economic loss distribution by category (2000–2024, total: $23.1B)

Category	Cumulative amount	Percentage of total (%)	Representative events
Health emergencies	$20.0B	87	COVID-19 pandemic (2020–2024)
Transportation disasters	$1.8B	8	Ferry disaster (2006), train accidents
Natural hazards	$0.8B	3	Floods, earthquakes, heat waves, sandstorms
Industrial accidents	$0.5B	2	Factory fires, chemical spills
Total	**$23.1B**	**100**	All disaster categories combined

people with a risk index of 4.8, primarily threatened by heat, drought, flash floods, and geographic isolation. In total, 74.7 million Egyptians live in formally assessed risk zones, while 29.3 million reside in remote or frontier areas that remain variably assessed, highlighting the disaster exposure concentration in Egypt's most densely populated regions (Erian et al., 2021).

Understanding what makes communities vulnerable requires looking beyond geography alone. As given in Table 14.5, correlation analysis across Egypt's 27 governorates reveals the statistical relationships between various vulnerability factors and disaster outcomes. Population density shows a strong positive correlation with disaster frequency at 0.73, meaning densely populated areas experience more frequent events. Even more striking, urban concentration exhibits a correlation of 0.81 with economic losses, revealing that urbanised regions suffer disproportionately severe financial impacts. Infrastructure age demonstrates a correlation of 0.68 with vulnerability scores, indicating that ageing infrastructure significantly increases disaster susceptibility. Perhaps most concerning from an equity perspective, income level shows a moderate negative correlation of -0.54 with resilience capacity,

Table 14.4 Geographic risk distribution and population exposure (2024)

Region	Population (Millions)	Major events (1992–2024)	Risk index (out of 10)	Risk level	Primary hazards
High-risk areas					
Greater Cairo Region	22.0	12	8.7	High	Seismic, heat waves, infrastructure
Alexandria and Coastal	6.2	8	7.2	High	Flooding, sea-level rise, storms
Nile Delta Governorates	25.0	6	6.8	High	Flooding, agricultural impacts
Subtotal high-risk	**53.2**	**26**	**Average: 7.6**		
Medium-risk areas					
Upper Egypt Governorates	20.0	4	4.5	Medium	Heat, drought, flooding
Sinai and Desert Regions	1.5	7	5.1	Medium	Flash floods, isolation
Subtotal medium-risk	**21.5**	**11**	**Average: 4.8**		
Total assessed areas	**74.7**	**37**	**Average: 6.6**		
Not formally assessed	29.3	N/A	N/A	Variable	Remote/ frontier areas
National total	**1040**				

Table 14.5 Correlation analysis of vulnerability factors ($n = 27$ governorates)

Correlation type	Correlation coefficient (r)	P-value	Significance	Interpretation
Population density versus disaster frequency	0.73	<0.05	Significant	Strong positive: higher density = more frequent disasters
Urban concentration versus economic losses	0.81	<0.01	Highly significant	Very strong positive: urbanised areas incur disproportionate damage
Infrastructure age versus vulnerability score	0.68	<0.05	Significant	Moderate-strong positive: older infrastructure = higher vulnerability
Income level versus resilience capacity	−0.54	<0.05	Significant	Moderate negative: lower income = reduced resilience

confirming that lower-income populations possess reduced adaptive capacity and recovery resources. These patterns inform where and how risk reduction strategies should be targeted for maximum effectiveness.

Infrastructure vulnerabilities present arguably the most concerning systemic weakness in Egypt's disaster risk profile. As given in Table 14.6, quantification of these gaps across six critical infrastructure categories reveals that only 60% of buildings have proper permits, and just 35% of critical infrastructure and hospitals meet seismic and disaster standards, respectively. These findings are based on comprehensive assessments by various government ministries examining thousands of structures and facilities nationwide. While schools demonstrate better preparedness with 78% having evacuation protocols in place, vulnerabilities in transportation infrastructure (58% lacking redundancy) and water systems (52% inadequately resilient) create cascading failure risks. The compliance rates represent current performance, while vulnerability rates indicate the proportion of infrastructure at risk during major disasters (Krichene et al., 2021).

14.3.2 *Building Institutions for Resilience*

Egypt's disaster management institutions have evolved through five distinct phases, each building on lessons learned from previous challenges. The pre-1992 period was characterised by limited formal frameworks and ad-hoc responses with minimal interagency coordination. The 1992 earthquake catalysed fundamental reforms during the Reform Catalyst Period of 1992–2000, including enhanced coordination mechanisms, scientific capacity building, and building code reforms. The Systematic

Table 14.6 Infrastructure vulnerability statistics (2024)

Infrastructure category	Compliance/ preparedness rate (%)	Vulnerability rate	Assessment method	Sample size
Buildings with proper permits	60	40% (±5% margin)	CAPMAS building survey	$n = 15{,}000$ structures
Critical infrastructure meeting seismic standards	35	65%	Ministry of Public Works structural assessment	$n = 2{,}300$ facilities
Hospitals meeting disaster standards	35	65%	Ministry of Health emergency preparedness audit	$n = 1{,}547$ facilities
Schools with evacuation protocols	78	22%	Ministry of Education safety programme	$n = 5{,}000+$ schools
Transportation infrastructure redundancy	42	58%	Ministry of Transportation network analysis	National assessment
Water/sanitation systems resilience	48	52%	Ministry of Water Resources evaluation	27 governorates

Development Period of 2000–2010 saw the development of Egypt's first National Disaster Risk Reduction Strategy in 2005, with the establishment of training centres, and equipment modernisation. Between 2010 and 2020, the modernisation period brought the adoption of the Sendai Framework, the establishment of the National Centre for Crisis and Disaster Risk Reduction in 2014, and technology integration initiatives. The current Contemporary Challenges Period, beginning in 2020, has been defined by COVID-19 response, emerging climate adaptation planning, and enhanced regional cooperation (Sakr et al., 2023).

Today's institutional structure reflects this evolution, creating a multi-level system with increasingly clear roles and responsibilities. At the national level, the Supreme Committee for Crisis and Disaster Management, chaired by the Prime Minister, provides the highest-level political leadership, while the National Centre for Crisis and Disaster Risk Reduction serves as the primary technical coordination body with 24/7 operational capacity. Various ministries maintain specialised responsibilities; Interior handles emergency evacuation and public safety through the Civil Defence Authority; Defence provides military engineering and specialised rescue operations; Health manages emergency medical services; and Public Works handle infrastructure assessment and emergency repairs. Scientific infrastructure has expanded substantially to support decision-making, with the National Research Institute of Astronomy and Geophysics now operating over 70 seismic monitoring stations (180% growth

Table 14.7 Institutional development score progression (1992–2024, 10-point composite index)

Year	Milestone/context	Score (out of 10)	Score category	Improvement from previous	Annual growth rate
1992	Post-earthquake baseline	1.2	Very low	–	–
2000	Initial frameworks established	3.1	Low	+1.9	+0.24/year
2005	Strategy development period	5.4	Moderate	+2.3	+0.46/year
2010	Capacity expansion phase	6.8	Moderate-high	+1.4	+0.28/year
2015	Sendai adoption completed	8.9	High	+2.1	+0.42/year
2024	Current assessment	9.2	Very high	+0.3	+0.03/year

from 1992) and the Egyptian Meteorological Authority maintaining over 200 weather monitoring stations (33% increase over three decades) (Ahmed et al., 2018).

Measuring institutional progress requires examining both qualitative developments and quantitative indicators. As given in Table 14.7, Egypt's institutional development score on a 10-point composite index has progressed from 1992 to 2024, showing remarkable improvement from 1.2 points immediately after the Cairo earthquake to 9.2 points currently, representing 667% growth from baseline. The table reveals that the largest single-period gain of 2.3 points occurred during 2000 and 2005, when the first national strategy was developed, while more recent periods show smaller incremental improvements as the system approaches maturity. Current performance at 92% of the maximum score suggests near-optimal performance given existing resource constraints, with strong statistical correlation between time and institutional capacity demonstrating that sustained investment in disaster management yields measurable results (Abdrabo et al., 2020).

14.3.3 Confronting System Weaknesses and Demonstrating Progress

Despite significant institutional progress, Egypt's disaster management system continues to face persistent coordination and communication challenges that undermine response effectiveness. The 2018 Alexandria floods starkly illustrated these problems when emergency response revealed systematic deficiencies across all phases. Initial warning generation took 45 min instead of the 30 min target, followed by cascading delays in transmission, activation, and escalation. Command fragmentation prevented a unified response, with inter-agency coordination consuming 5 h instead of 1.5 h, this is a gap that contributed to 12 deaths and $100 million in economic losses. These coordination problems manifest in measurable performance impacts, including multi-agency response delays, command clarity scores of only 4.2 out of 10, and resource duplication affecting 22% of equipment deployments. Early warning dissemination delays average 35–52 min from national detection to local activation, significantly reducing crucial flash flood lead time. Rural areas face particularly acute barriers, with language differences reducing warning comprehension by 45–55% and only 34% of rural communities have reliable SMS systems compared to 91% urban coverage (Baradei & Alsadeq, 2019).

Recognising these challenges, Egypt has implemented targeted innovations and best practices that demonstrate measurable progress. As given in Table 14.8, improvements in early warning system performance from 2010 to 2024 demonstrate how investing in monitoring infrastructure and forecasting technology has produced concrete results. Forecast accuracy improved from 68 to 85% correct predictions, representing a 25% performance enhancement credited with preventing an estimated 150–200 deaths between 2020 and 2024. Weather station coverage expanded by 33%, while warning lead times for floods increased by 50–100%, providing communities with more time to prepare and evacuate. The expansion of SMS population reach from 45 million to over 80 million users represents a 78% growth in early warning accessibility. These improvements demonstrate how systematic investment in monitoring capacity, forecasting technology, and communication infrastructure translates into enhanced public safety outcomes.

Table 14.8 Early warning system performance evolution (2010–2024)

Metric	Previous performance (2010)	Current performance (2024)	Absolute improvement	Percentage improvement	Time period
48–72 h forecast accuracy	68% correct	85% correct	+17 percentage points	+25% improvement	2010–2024
Weather station coverage	150 stations	200+ stations	+50 stations	+33% expansion	2010–2024
Warning lead time (floods)	24–48 h	48–72 h	+24 h average	50–100% increase	2015–2024
False positive rate	28% (2020 baseline)	24%	−4 percentage points	−15% reduction	2020–2024
Population reach (SMS)	45 M users (2015)	80 M+ users	+35 M users	+78% expansion	2015–2024
Earthquake detection threshold	Magnitude 2.5	Magnitude 1.8	−0.7 magnitude units	+40% sensitivity	2010–2024

Beyond technological improvements, community-based programmes demonstrate how engaging citizens directly improves preparedness and response capabilities. School safety programmes now operating in over 5,000 schools have shown 45% faster evacuation times, while training programmes have reached 250,000 people between 2015 and 2024. Household preparedness surveys reveal a 45% higher emergency supply kit rate in communities with active programmes compared to areas without such initiatives. The neighbourhood response teams established in 15 governorates provide critical local capacity during the first 30–60 min of an emergency when external help may be unavailable, with basic first aid training reaching approximately 50,000 community members. Public awareness campaigns now reach 85% of the population through multimedia channels, while community drills have increased by 150% since 2016, creating a culture of preparedness that complements institutional capacity. These community-level interventions represent a fundamental shift from purely top-down disaster management towards integrated systems that harness local knowledge and mobilise citizen participation in building resilience (Hallegatte et al., 2017).

14.4 Analysis and Discussion

This section synthesises findings from the systematic review to provide a comprehensive assessment of Egypt's disaster management system, examining strengths and achievements alongside critical gaps and weaknesses while contextualising performance within regional and international frameworks.

Egypt has demonstrated significant disaster management progress since 1992, establishing key institutions, including the National Committee for Crisis and Disaster Management and the National Centre for Crisis and Disaster Risk Reduction that provide clear governance aligned with international best practices. The country has achieved substantial capacity expansion with scientific monitoring capabilities growing through a 33% increase in meteorological stations and 180% growth in seismic networks, creating a robust technical infrastructure that supports disaster preparedness (Wanyama et al., 2025). Community development programmes have reached over 5,000 schools and established more than 500 neighbourhood response teams, delivering measurable improvements in evacuation times, household preparedness, and community resilience, while international cooperation has positioned Egypt as an emerging regional leader, facilitating knowledge exchange and resource mobilisation. Technological modernisation through GIS applications, mobile warning systems, and integrated data platforms has produced quantifiable improvements, including a 38% reduction in response times, 25% improvement in forecast accuracy, and enhanced situational awareness through real-time information sharing, supported by an evolving legal and regulatory framework that provides foundations for systematic disaster risk reduction (Red Cross Red Crescent Climate Centre, 2024).

However, critical gaps and weaknesses persist, particularly in resource allocation patterns that remain predominantly reactive, directing 70–80% of resources towards response and recovery rather than prevention and preparedness despite economic analyses suggesting that prevention investments yield benefit–cost ratios of 4:1 to 7:1. Community engagement remains limited beyond pilot programmes and urban centres, reducing resilience-building potential particularly among vulnerable populations in rural areas, informal settlements, and marginalised communities that face the highest disaster risks and preparedness needs. Infrastructure vulnerabilities are substantial, with 40% of buildings lacking proper permits, 65% of critical infrastructure failing to meet seismic standards, and 58% of transportation routes lacking redundancy, creating ongoing exposure despite policy improvements. Coordination and communication challenges include response delays averaging 35%, warning dissemination delays affecting 42% of available lead time, and communication reliability declining 31% during high-stress emergencies, while financial constraints from irregular funding cycles, limited disaster insurance coverage affecting less than 1% of the population, and competing budget priorities restrict capacity building efforts. Geographic disparities between high-capability urban centres and limited-capability remote areas create uneven risk landscapes, where disaster impacts disproportionately affect vulnerable populations (Moufti, 2021).

In comparative analysis, Egypt demonstrates regional leadership within the Middle East and North Africa through stronger institutional development, evidenced by a comprehensive Sendai Framework-aligned national strategy, extensive monitoring networks surpassing those of regional counterparts, and active international participation since 2014. Nevertheless, Egypt faces challenges similar to those of its regional neighbours, including limited community engagement, resource allocation prioritising response over prevention, enforcement gaps, and financial constraints. International benchmarking reveals mixed performance with opportunities for improvement by learning from Turkey's stronger disaster insurance schemes and stricter construction enforcement, Indonesia's expanded community engagement programmes, and Chile's successful transition from reactive to proactive approaches through strategic prioritisation and long-term commitment. The key insight from this analysis is that despite Egypt's institutional strengths, the allocation of 70–80% of resources to response rather than prevention suggests significant room for strategic reorientation towards the risk reduction approaches that have proven effective in comparable countries (Elboshy et al., 2019).

Three core hypotheses were tested to evaluate Egypt's disaster management system characteristics and effectiveness. Table 14.9 summarises the statistical analysis framework, including test types, test statistics, p-values, 95% confidence intervals, and conclusions for each hypothesis. The hypotheses examined: (1) the predominance of reactive resource allocation ($\geq$70% to response/recovery), (2) the correlation between institutional development and improved outcomes, and (3) the relationship between geographic concentration and vulnerability disparities. All tests were conducted at $\alpha = 0.05$ significance level, with results providing empirical evidence for systemic patterns within Egypt's disaster risk management approach, as shown in Table 14.9.

Table 14.9 Hypothesis testing results summary

Hypothesis	Test type	Test statistic	P-value	95% Confidence interval	Result	Conclusion
H1: Egypt's system remains predominantly reactive (≥70% resources to response/recovery)	One-sample t-test	$t = 4.2$, df $= 22$	$p < 0.001$	[68.2, 77.8%]	Significant	**Confirmed**—Reactive bias validated
H2: Institutional development correlates positively with improved outcomes	Pearson correlation	$r = 0.84$	$p < 0.001$	[0.72, 0.92]	Significant	**Confirmed**—Strong positive relationship
H3: Geographic concentration increases vulnerability disproportionately	One-way ANOVA	$F\ (4,18) = 12.8$	$p < 0.001$	Multiple comparisons significant	Significant	**Confirmed**—Regional differences exist

Rigorous hypothesis testing provides empirical support for major conclusions. Statistical validation confirms key propositions with high statistical significance (all $p < 0.001$), validation analyses demonstrate model robustness (82% cross-validation accuracy), and power analyses confirm adequate sample sizes (85% power, medium effect size detection).

14.5 Recommendations and Future Strategies

Based on a comprehensive analysis of Egypt's disaster management system strengths, weaknesses, opportunities, and threats, this section presents evidence-based recommendations organised by implementation timeline and strategic priority, providing actionable guidance for policy makers, practitioners, and international partners supporting disaster risk reduction efforts in Egypt.

The review recommends a phased approach that balances short-term reform with long-term structural transformation.

Immediate Priorities (1–2 years)

- Modernise the Civil Defence Law in line with Sendai principles
- Establish a unified national command system
- Standardise training across governorates
- Expand communication infrastructure, particularly in rural areas.

Medium-Term Goals (3–5 years)

- Decentralise disaster management authorities
- Expand community response programmes
- Retrofit critical infrastructure
- Strengthen regional cooperation.

Long-Term Transformation (6–10 years)

- Shift national resource allocation towards prevention
- Integrate AI-driven early warning systems
- Achieve nationwide resilience coverage
- Position Egypt as a regional leader in DRR innovation.

Investment estimates suggest that the required funding is approximately 0.03% of Egypt's GDP annually and is modest compared to potential savings, with benefit–cost ratios rising to 5:1 over two decades.

14.6 Conclusion

This systematic review provides a comprehensive, evidence-based assessment of Egypt's disaster management system, examining institutional evolution, current performance, challenges and gaps, and strategic priorities for continued improvement towards enhanced national resilience.

Egypt's disaster management system has evolved substantially since 1992, developing coordinated national frameworks with key achievements including enhanced monitoring capabilities such as 53% coverage of transportation routes and 180% growth in seismic networks, community programme development reaching over 5,000 schools and 500 neighbourhood teams, and modern technology integration. These efforts have produced measurable improvements, including a 38% reduction in mean emergency response time, 25% improvement in forecast accuracy, 40% reduction in school evacuation time, and institutional progression from 1.2 out of 10 in 1992 to 9.2 out of 10, representing a 667% improvement over 32 years. However, persistent challenges remain as the system continues to operate predominantly in a reactive mode, allocating 70–80% of resources to response and recovery rather than prevention and preparedness, while significant vulnerabilities persist including 40% of buildings lacking proper permits (95% CI: 35–45%), 65% of critical infrastructure failing seismic standards, 58% of transportation routes lacking redundancy, and approximately 1% of the population having disaster insurance. Coordination challenges include 35% response delays, 42% warning dissemination delays, and 31% declining communication reliability during high-stress emergencies, compounded by a changing risk profile driven by climate change, rapid urbanisation, ageing infrastructure, and emerging threats. Total cumulative economic losses from 2000 to 2024 reached \$23.1 billion, with COVID-19 representing an exceptional outlier at \$20.0 billion. Meanwhile, the baseline annual losses from natural hazards average \$420 million with increasing variance, establishing a reference point for disaster risk financing. The success of implementing recommendations depends on sustained political commitment, adequate resource mobilisation estimated at \$1.2–1.6 billion over 10 years, representing 0.025–0.034% of GDP investment, enhanced stakeholder coordination, and adaptive management enabling continuous learning and improvement.

The study acknowledges several limitations, including varying data quality across periods and sources, with older years having less comprehensive documentation, potential reporting biases in government statistics, where some agencies maintain detailed records while others prioritise successes, limited governance-level information that constrains spatial analysis and source triangulation, and rapidly changing conditions which means that the findings reflect the study period and require regular updating. The COVID-19 pandemic from 2020 to 2024 created exceptional conditions that complicated interpretation, with economic losses substantially exceeding the historical baseline by a factor of 10 and requiring separate analytical treatment. The findings focus specifically on Egypt's unique context and may not be directly applicable to other countries, while language limitations restricting inclusion to English and Arabic publications may have excluded relevant materials in other languages.

Future research priorities should include community resilience assessments examining local preparedness, social networks, and adaptive capacities, cost–benefit analyses comparing prevention versus response investments, evaluation of technology integration, including GIS, mobile warnings, and emerging tools, comparative studies with similar middle-income countries such as Turkey, Indonesia, Chile, and Morocco, and longitudinal tracking of the National DRR Strategy 2030 implementation. The analysis reveals common patterns in rapidly urbanising middle-income countries, including reactive resource allocation, infrastructure vulnerabilities, coordination gaps, and weak community engagement, suggesting that comparative research on reactive to proactive transitions in nations facing similar demographic, economic, and climate pressures could provide transferable insights for global disaster risk reduction frameworks and address governance challenges in developing economies confronting climate change and escalating disaster risk.

Acknowledgements We would like to express our sincere gratitude to all those who contributed to the successful completion of this research. We are particularly grateful to Dr Mohammed Rashad and Dr Ahmed Abdelsabour for providing the necessary facilities and resources that made this work possible.

We are grateful to the anonymous reviewers whose thoughtful comments and suggestions significantly improved the quality of this manuscript.

Author Contributions

Conceptualisation: All authors contributed to research design and methodology development.

Data Collection: Systematic literature search and screening conducted by the research team with independent verification.

Analysis: Quantitative analysis performed using SPSS and R statistical software; qualitative analysis employed NVivo for thematic coding.

Writing: Original draft prepared collaboratively with sections assigned based on expertise areas.

Review and Editing: All authors reviewed and approved the final manuscript.

Funding Statement This research received no specific grant from any funding agency in the public, commercial, or not-for-profit sectors. The work was conducted as part of institutional research programmes at participating universities and research centres.

Declaration of Competing Interests The authors declare no competing financial interests or personal relationships that could have influenced the work reported in this paper.

Data Availability Statement The data supporting the findings of this study are derived from publicly available sources, including government reports, international organisation publications, and peer-reviewed academic literature. Specific data sets compiled for this analysis are available from the corresponding author upon reasonable request, subject to any applicable restrictions on the sharing of government data.

References

Abdrabo, K. I., Kantoush, S. A., Saber, M., Sumi, T., Habiba, O. M., Elleithy, D., et al. (2020). Integrated methodology for urban flood risk mapping at the microscale in ungauged regions: A case study of Hurghada, Egypt. *Remote Sensing, 12*(21), 3548.

Abou-Korin, A. A., & Al-Shihri, F. S. (2015). Disaster management and response in Egypt: National and cultural considerations. *Disaster Prevention and Management, 24*(3), 358–371.

Ahmed, M., El-Raey, M., & Nasr, S. (2018). Review of earthquake demand in the building code of Egypt. *HBRC Journal, 14*(3), 356–363.

Arab Republic of Egypt. (2017). *National disaster risk reduction strategy 2030: Alignment with the Sendai Framework*. National Center for Crisis and Disaster Risk Reduction.

El Baradei, S., & Alsadeq, M. (2019). Governance of disaster risk reduction in Egypt: Lessons learned from a fire in a slum area. *International Journal of Disaster Risk Reduction, 33*, 228–231.

El-Anwar, H. O., Abdel-Kader, R. F., & Sherif, L. A. (2023). Disaster risk reduction in Egypt: Challenges and opportunities for building resilient communities. *International Journal of Disaster Risk Reduction, 95*, Article 103851.

Elboshy, B., Gamal, W., Ghoneim, E., & Haffa, A. (2019). An evaluation framework for disaster risk management in Egypt. *International Journal of Disaster Resilience in the Built Environment*. https://doi.org/10.1108/IJDRBE-05-2018-0024

El-Geziry, T. M., & Radwan, A. A. (2022). Coastal adaptation to sea level rise: An overview of Egypt's efforts. *Regional Studies in Marine Science, 49*, Article 102086.

Erian, W., Katlan, B., & Babah, O. (2021). Drought vulnerability in the Arab region: Case study—Drought in Syria. In W. Bank & U. Nations (Eds.), *Arab adaptation to climate change* (pp. 245–268). World Bank.

Finance, A. (2023). *The Suez Canal: A vital engine for the Egyptian economy*. Arab Finance.

Food and Agriculture Organization of the United Nations. (2022). *Towards climate-smart agriculture in Egypt: Scaling up sustainable agricultural practices and technologies*. FAO.

Hallegatte, S., Vogt-Schilb, A., Bangalore, M., & Rozenberg, J. (2017). *Unbreakable: Building the resilience of the poor in the face of natural disasters*. World Bank.

Krichene, H., Geiger, T., Frieler, K., Willner, S. N., Sauer, I., & Otto, C. (2021). Long-term impacts of tropical cyclones and fluvial floods on economic growth—Empirical evidence on transmission channels at different levels of development. *World Development, 144*, Article 105475.

Lassa, J. A. (2018). Disaster risk governance and vulnerability in the Arab region. *Disaster Prevention and Management, 27*(4), 456–472.

Moufti, A. (2021). A critical comparative review of emergency and disaster management in the Arab world. *Journal of Business and Socio-Economic Development, 1*(1), 24–35.

Red Cross Red Crescent Climate Centre. (2024). Egypt country profile: Climate change and disaster risk. Climate Centre.

Sakr, M., Shahin, H., & Elshazly, H. (2023). Evolution of disaster risk management in Egypt: From reactive response to proactive resilience. *International Journal of Disaster Resilience in the Built Environment, 14*(2), 215–234.

Smith, P., & Becker, P. (2023). Analysing disaster risk governance: The quest for shared responsibility. *Disaster Prevention and Management, 32*(4), 456–473.

United Nations Department of Economic and Social Affairs, Population Division. (2024). World population prospects 2024: Summary of results. United Nations.

United Nations Office for Disaster Risk Reduction. (2015). *Sendai framework for disaster risk reduction 2015–2030*. UNDRR.

Wanyama, J., Wesonga, J., & Odongo, W. (2025). Beyond institutional silos: Rethinking multi-level disaster risk governance in Africa a decade into the Sendai Framework implementation. *International Journal of Disaster Risk Science, 16*(3).

World Bank. (2024). Egypt overview: Development news, research, data. World Bank.

World Travel & Tourism Council. (2023). *Egypt 2023 economic impact research*. WTTC.

Chapter 15
Disaster, Crisis, and Emergency Management in Somalia: Risk Landscape, Institutional Evolution, and Future Resilience Strategies

Mohamed Mustaf Ahmed, Zhinya Kawa Othman, and Olalekan John Okesanya

Abstract Somalia's risk landscape is shaped by protracted conflict, recurrent droughts and floods, cyclones, epidemics, and environmental degradation, all of which are intensified by climate change and poverty. This chapter synthesises recent evidence on hazard trends and impacts, tracing the evolution of Somalia's disaster architecture from the post-1991 collapse to the creation and reform of the Somalia Disaster Management Agency (SoDMA), the 2020 National disaster risk management (DRM) Policy, and the launch of multi-hazard early warning capacities. We assess current arrangements across the federal, state, community, and international levels, including United Nations (UN) Office for the Coordination of Humanitarian Affairs (OCHA) clusters and Somalia's growing engagement with regional and global frameworks, to identify persistent gaps in financing, last-mile early warning, data systems, coordination, and local ownership. We highlight the need for research on disaster loss accounting, climate and conflict dynamics, coastal and urban risks, and the effectiveness of indigenous coping strategies. The chapter proposes practical, evidence-informed strategies: institutionalising whole-of-government DRM, linking early warning to anticipatory action, scaling layered risk finance, strengthening SoDMA–Federal Member States (FMS) coordination, investing in resilient infrastructure and services, and embedding inclusive, community-led preparedness to shift from reactive relief to durable resilience.

Keywords Somalia · Disaster risk management · Early warning systems · Drought · Institutional capacity · Resilience

---

M. M. Ahmed (✉)
School of Medicine and Health Sciences, SIMAD University, Mogadishu, Somalia
e-mail: momustafahmed@simad.edu.so

Z. K. Othman
Department of Pharmacy, Kurdistan Technical Institute, Sulaymaniyah, Kurdistan Region, Iraq

O. J. Okesanya
Department of Public Health and Maritime Transport, University of Thessaly, Volos, Greece

E. Alam (ed.), *Disaster, Crisis and Emergency Management in the Middle East & North Africa*, https://doi.org/10.1007/978-3-032-24436-9_15

15.1 Introduction

Somalia is a country in the Horn of Africa with the longest coastline on mainland Africa, bordered by the Indian Ocean to the east (UNEPGRID, 2025). It has an estimated population of approximately 16–17 million people (as of 2025) (World Population Review, 2025) and a small, fragile economy with a gross domestic product (GDP) of approximately $12 billion in 2025 and is heavily dependent on agriculture (especially livestock) (World Bank, 2025), remittances, and aid. Decades of civil war, state collapse (since 1991), and recurring humanitarian crises have severely undermined governance, making Somalia one of the world's most fragile states (Congress, 2020). In 2012, Somalia ended a transitional period by establishing a Federal Government under a provisional constitution, laying the foundation for a federal parliamentary republic (Resources, 2014). Somalia is now divided into seven Federal Member States: Puntland, Galmudug, Hirshabelle, South-West, Jubaland, North-East Somalia, and the self-declared Republic of Somaliland (Ministry of Planning, 2025). These states operate under a central Federal Government based in Mogadishu, although Somaliland remains internationally unrecognised, and its status is disputed (Ministry of Planning, 2025). However, governance remains weak and fragmented, with limited administrative capacity and ongoing insecurity due to the Al-Shabaab insurgency and clan conflict (Addan, 2022; Ahmed, 2024; BTI, 2024). These challenges have resulted in the poor delivery of basic services and have exacerbated the country's vulnerability to disasters and crises (Ahmed et al., 2024).

Given this context, this chapter aims to provide a comprehensive review of Somalia's disaster risk landscape, the evolution of its institutional framework for crisis and emergency management, and the future strategies required to build resilience. The study examines Somalia's major hazards and disaster history, assesses the evolution of disaster management authorities and policies over time, identifies key challenges and gaps (including governance and coordination weaknesses), highlights areas requiring further research, and recommends strategies to enhance disaster preparedness and reduce vulnerabilities. Ultimately, strengthening Somalia's capacity to manage disasters is critical for safeguarding its communities and supporting the country's broader goals of peace, stability, and development.

15.2 Research Methods

This chapter is based on a qualitative literature review and an analysis of secondary data sources. We surveyed academic studies, government documents, and reports from international organisations to gather relevant information on disasters and emergency management in Somalia. Key sources included scholarly articles and case studies on Somalia's vulnerability and resilience, official Somali government policies, United Nations reports on humanitarian situations, and assessments by development agencies. Data on disaster events and their impacts were obtained from historical disaster databases and previous studies. We also reviewed policy briefs and workshop reports that shed light on institutional and capacity issues.

All information was triangulated across multiple sources to ensure its accuracy and reliability. The analysis was organised thematically according to the study objectives. First, we synthesised data on Somalia's hazard profile and disaster history to characterise its risk landscape. Next, we examined documents on governance and institutional arrangements to trace the evolution of the disaster management framework in Somalia. We then identified common themes regarding the challenges and weaknesses from evaluations and after-action reports. Gaps in the literature and practice were noted to identify areas for further research recommendations. Finally, drawing on best practices in disaster risk reduction and context-specific recommendations from policy papers, we formulated strategies for resilience. Using a wide range of sources and applying critical analysis, this chapter provides a well-founded overview of disaster and emergency management in Somalia.

15.3 Disaster History and Risk Profile of Somalia

Somalia has a long history of natural-hazard-related disasters and human-induced crises that have devastated its people and economy (Ahmed et al., 2024). Foremost amongst these are droughts, which have periodically led to famine, notably the 1991–1992 famine during the civil war and the 2011 famine, each of which resulted in deaths of hundreds of thousands of people (CDC, 1992). The drought of 2010–2011 was the worst in 60 years, triggering an official famine declaration; an estimated 260,000 people died during the 2010–2012 famine period (about half of them were children) (Concern, 2022). More recently, an even longer drought from 2020 to 2022 pushed Somalia to the brink of another famine (Humdata, 2022; WHO, 2025a). Whilst large-scale aid in 2022 narrowly averted an official famine declaration, the effects were still severe: a scientific study estimated ~ 43,000 'excess' deaths in 2022 alone due to drought, and by mid-2024, the cumulative drought-related deaths were reported to have reached approximately 71,000, with young children accounting for approximately 40% of the fatalities (WHO, 2025a). In terms of those affected, the 2016–2017 drought left 6.2 million Somalis (approximately half the population) acutely food insecure, displacing at least 739,000 people by mid-2017 (SODMA, 2022). Table 15.1 summarises some of the major disasters that have impacted Somalia in recent decades, illustrating the scale of their human impact.

As shown in Table 15.1, drought-related crises are the deadliest disasters in Somalia's modern history. Multi-year droughts recur regularly (often linked to climate cycles such as El Niño/La Niña) and can escalate into famines when coupled with conflict and lack of aid (CRISIS GROUP, 2024; UNDRR, 2024a). The country experienced recurrent major drought emergencies in 1991–92 (famine), 2011–12 (famine), 2016–17, and 2021–22, driven by repeated rainfall failures and compounded by conflict and access constraints (Concern, 2022). Droughts devastate

Table 15.1 Major disaster events in Somalia–historical impact

Year(s)	Disasters	Impacts (deaths, displacement or affected)	Source(s)
1991–1992	Civil war and famine	12,255 deaths recorded Aug–Nov 1992 in Baidoa town alone (37% of town's pop). Overall, thousands died nationally in the early 1990s famine	CDC (1992), UN (1992)
2010–2012	Drought and famine	260,000 deaths (estimated total during famine period); half were children under five years. Approximately 13 million people were affected across Horn of Africa (including Somalia)	Concern (2022), GFDRR (2017)
2016–2017	Drought (no formal famine)	6.2 million Somalis acutely food insecure (at crisis or emergency levels) in early 2017; ~ 739,000 people were displaced by drought by May 2017	SODMA (2022), OCHA (2017)
May 2018	Cyclone *Sagar*	53 deaths (Somaliland/Puntland); 229,000 people were affected (heavy rains, flash floods). Massive loss of livestock and destruction of farms was reported	Xinhuanet (2018), OCHA (2018a), Reuters (2018)
Nov 2020	Cyclone *Gati*	Eight deaths; 42,000 people displaced in the Bari region (Puntland); caused millions of dollars in damage	OCHA (2020a)
2020–2022	Drought and food crisis	43,000 estimated excess deaths in 2022 alone (drought-induced); by Jan 2023, ~ 1.1 million people newly displaced due to drought (since 2021). In 2022, 7.8 million Somalis were affected by drought (nearly half the population)	WHO (2025a), UNICEF (2025) Reliefweb (2022), UNDRR (2024a)

pastoral and agro-pastoral livelihoods by killing livestock, collapsing food production, and triggering mass displacement (as people abandon rural areas in search of aid) (Ahmed et al., 2024; OCHA, 2025a). For instance, in the 2016–17 drought, high livestock losses were reported in the north (Somaliland and Puntland), and over one million people were internally displaced in Somalia (UN, 2017; UNDP, 2017a; UNHCR, 2019).

Floods are another frequent hazard: Somalia has both seasonal river floods (along the Shabelle and Juba rivers in the south) and flash floods in normally arid areas. In years, with above-normal rainfall, flooding can affect hundreds of thousands of people (FAO, 2020a; Reliefweb, 2025). For example, flooding in the 2018 Gu rains (April–June) affected over 700,000 people in southern Somalia, and in 2020, seasonal floods inundated parts of Beledweyne, displacing approximately 240,000 residents. Flood management is hindered by limited river embankment maintenance, weak early warning dissemination, and the absence of a coherent national flood management plan, despite regular flooding along the Shabelle and Juba rivers (OCHA, 2018b, 2020b). Coastal storms are less common but do occur; in recent years, Somalia was struck by two rare tropical cyclones (*Sagar* and *Gati*), which caused localised severe damage. These events exposed gaps in Somalia's disaster preparedness legislation and the limited operational capacity of the newly established Ministry of Humanitarian Affairs and Disaster Management (MoHADM) (OCHA, 2018a, 2020a; Reuters, 2018; Xinhuanet, 2018). Cyclone *Gati* in November 2020 was the strongest cyclone ever recorded in Somalia, dumping a year's worth of rain in two days on Puntland's coast (OCHA, 2020a).

In addition to natural hazards, Somalia faces complex emergencies driven by conflicts and diseases. The decades-long civil conflict and terrorism have killed tens of thousands and continually uprooted civilians from their homes. The protracted civil conflict has fragmented institutional disaster response, with NGOs and international agencies often filling gaps that should be managed by national disaster authorities (CFR, 2025; UN, 2023). Somalia has one of the world's largest internally displaced populations, estimated at 2.6 million people (IDPs) as of 2019, growing to around 2.9 million by 2021 (approximately 20% of the population) (OCHA, 2021; UN, 2019). These conflicts interact with natural hazards to exacerbate suffering; for example, conflict limits access to aid during droughts, turning food crises into famines (Warsame et al., 2023). Disease outbreaks are another hazard: cholera, measles, and other epidemics often spread extensively during the wake of floods or in overcrowded IDP camps (Reliefweb, 2025b; WHO, 2023a). In 2017, concurrent with drought, Somalia experienced a large cholera (over 78,000 cases) and measles outbreaks (Reliefweb, 2025b; WHO, 2017). The COVID-19 pandemic in 2020–2021 further strained Somalia's weak healthcare system amid ongoing locust infestation and floods (FAO, 2020b). Overall, Somalia's risk profile is characterised by recurrent climate-related hazards (droughts and floods), compounded by conflict, poverty, and weak infrastructure. Table 15.2 provides an overview of the major hazard types in Somalia and their characteristics.

In summary, Somalia's dominant natural hazards are droughts and floods, which together account for the majority of the disaster-related deaths and economic losses. A 2005 analysis by Columbia University ranked Somalia in the top decile globally for multi-hazard risk when weighted by mortality and impact on GDP (Columbia University, 2005). Drought stands out as the hazard posing the greatest risk to Somali lives and livelihoods, as evidenced by its role in repeated famines (GFDRR, 2017; Musei et al., 2021; UNDRR, 2024a; Wordlbank, 2023a). Floods, whilst less lethal, severely impact agriculture and infrastructure, perpetuating poverty cycles (Acaps, 2018; Reliefweb, 2020, 2025). The interconnected nature of these hazards is important to note; for example, a severe drought can weaken communities and force displacement, and subsequent poor rains or floods find populations extremely vulnerable (living in informal settlements prone to floods or disease) (UNDP, 2022a; UNDRR, 2024a).

Climate change is expected to further intensify these hazard trends, with models projecting rising temperatures (up to 3.2 °C by 2080) and more erratic rainfall, resulting in more frequent droughts and extreme flood events (Centre, 2021; UNDRR, 2024a; UNFCCC, 2013). Indeed, since 2011, Somalia has rarely experienced a normal rainy season, and drought conditions have almost become the 'new normal'

Table 15.2 Key hazards in Somalia and their characteristics

Hazard	Pattern and prevalence	Typical impacts	Source(s)
Drought	Most frequent and multi-season failures (often La Niña-linked); climate change intensifies severity	Highest mortality; livestock losses; crop failure; displacement; famine risk	GFDRR (2017), UNDRR (2024a), Wordlbank (2023a), Musei et al. (2021)
Flood (river & flash)	Seasonal along Shabelle/Juba; flash floods in arid zones; spikes in above-normal rainfall years	Hundreds of thousands affected in bad seasons; urban/rural inundation; WASH and health impacts	Reliefweb (2025), Reliefweb (2020), Acaps (2018)
Tropical cyclones	Rare but increasing salience (e.g. *Sagar* 2018; *Gati* 2020)	Localised wind/rain damage; coastal flooding; fishing/ livelihood losses	OCHA (2020a), NASA (2020), Reliefweb (2018), OCHA (2018c)
Conflict and insecurity	Persistent insurgency and clan conflicts	Access constraints; protection risks; sustained displacement (~2.6–2.9 M IDPs 2019–2021)	OCHA (2021), UN (2019)
Epidemics (cholera, measles)	Recurrent, often after floods and in crowded IDP sites	High caseloads (e.g. > 78,000 cholera cases in 2017); strain on weak health system	WHO (2017), Lubogo et al. (2023), UNICEF (2018)
Environmental degradation	Deforestation, overgrazing, desertification; coastal erosion	Amplifies drought/ flood impacts; loss of ecosystem services	Worldbank (2020a); UNEP (2022), UNFCCC (2022)

(Concern, 2022; UNDRR, 2024a; WMO, 2023). The increasing volatility of the climate, coupled with ongoing conflict and pervasive poverty, with roughly 70–74% of Somalis living below the international extreme-poverty line of US$2.15/day (2017 PPP), creates a complex landscape of compounded risks. Each disaster, whether hazard-related or rooted in conflict and other human factors, hits an already vulnerable population, often pushing communities beyond their ability to cope (Concern (Anaheim), 2023; World bank, 2025). The impacts are felt in all sectors: recurring crises have eroded household assets, undermined food security and health, and stalled economic growth (Somalia's per capita GDP has stagnated or declined in real terms, in part due to shock after shock). Notably, disasters in Somalia also have regional dimensions (Somalia National Bureau of Statistics, 2024; World bank, 2024). Droughts and floods in Somalia are usually part of wider Horn of Africa climate events, leading to transboundary crises (e.g. refugee flows to Kenya/Ethiopia or shared river flooding) (UNDRR, 2024a; UNHCR, 2023). Conversely, regional conflicts (such as the Yemen war) have sent refugees into Somalia during crises

(FMR, 2024). This underscores that Somalia's risk landscape is intertwined with that of the broader East Africa and Middle East and North Africa (MENA) regions.

15.4 Evolution of Disaster Management Institutions in Somalia

Somalia's ability to manage disasters has been shaped by its turbulent political history. Following the collapse of the Siad Barre government in 1991, formal state institutions disintegrated, and for over two decades, there was no central authority to coordinate disaster response (SODMA, 2022). During the 1990s and the early 2000s, when crises struck, the response came almost entirely from UN agencies, the Red Crescent, international NGOs, and local community groups, with minimal government involvement (SODMA, 2022). This ad hoc humanitarian system saved lives (for instance, NGOs and the International Committee of the Red Cross (ICRC) provided famine relief in 1992, and UN agencies led aid efforts in the 2011 famine), but it lacked overarching strategy or government ownership (Seal & Bailey, 2013). Some regions that achieved relative stability in the 1990s, notably Somaliland (northwest) and Puntland (northeast), established their own rudimentary disaster management or food security agencies. For example, the Somaliland administration formed a Disaster Preparedness and Food Reserve Authority (NADFOR) in the early 2000s, and Puntland established a Humanitarian Affairs and Disaster Management Agency (HADMA) (NADFOR, 2003; Somali JNA, 2005). However, elsewhere in Somalia, there was no unified framework; each drought or flood was dealt with on an emergency basis by fragmented actors.

The formation of the internationally recognised Federal Government of Somalia (FGS) in 2012 marked the beginning of rebuilding state capacity, including disaster management (Congress, 2020). A key milestone came in 2011 amid the devastating famine: the Transitional Federal Government established the Somali Disaster Management Agency (SoDMA) (SoDMA, 2011). SoDMA was tasked with coordinating humanitarian assistance and disaster response, effectively becoming the nascent national disaster management authority. Initially, SoDMA had very limited resources and authority reflecting a weak state, but it was a starting point for institutionalising disaster risk management (DRM) (SoDMA, 2011). In 2016, the Federal Parliament passed legislation formally authorising SoDMA as a government agency under the oversight of a new ministry. This led to the establishment of the Federal Ministry of Humanitarian Affairs and Disaster Management (MoHADM), a cabinet-level body overseeing disaster risk management and humanitarian coordination. By 2017, MoHADM was in place with a mandate to develop policies and lead coordination amongst federal ministries and international partners (Worldbank, 2025a). During this period, the Jubaland state formally established a Ministry of Humanitarian Affairs and Disaster Management, and several other Federal Member States

(Galmudug, Southwest, and Hirshabelle) began to improve coordination mechanisms for drought and emergency response (IOM, 2018). However, formal disaster management ministries remain nascent or absent in many states.

Since the late 2010s, Somalia's institutional framework for disaster management has become increasingly structured. In 2020, with support from international partners, the government launched its first National Disaster Risk Management Policy, which laid out a comprehensive framework for DRM governance at all levels (SODMA, 2022). The 2020 National Disaster Risk Management Policy named SoDMA (formerly under MoHADM) as the central agency for coordinating DRM in Somalia and emphasised the need to integrate disaster risk reduction into national development planning (SODMA, 2022). The policy aligns with regional frameworks, such as the Intergovernmental Authority on Development (IGAD)'s DRM strategy, and underscores the need for whole-of-government coordination, although publicly available documents do not clearly define the mechanism of a Prime Minister-chaired National Disaster Risk Management Council (SODMA, 2022). A significant development achieved in July 2020 was the opening of the National Multi-Hazard Early Warning Centre (NMHEWC) in Mogadishu. This centre, set up under MoHADM's Disaster Management Department, was intended to aggregate climate and hazard data and issue early warnings for droughts, floods, and other threats (OCHA, 2020c). The NMHEWC represents Somalia's commitment to moving from reactive crisis management to proactive risk monitoring.

A major institutional change occurred in 2022: the FGS reorganised, abolishing the separate MoHADM and transferring its responsibilities directly to SoDMA (SoDMA, 2011; Worldbank, 2025a). SoDMA was elevated to the central national authority for both disaster risk reduction and emergency response, reporting directly to the Office of the Prime Minister. (SoDMA, 2011; Worldbank, 2025a) This streamlining was partly to reduce bureaucracy and clarify leadership (previously, there was an overlap between the Ministry and SoDMA). As of 2023–2024, SoDMA, led by a commissioner, is the focal government body for disaster management, coordinating with line ministries (for example, Health for epidemics and Agriculture for locust control), state-level disaster management agencies, and international partners (SoDMA, 2023; WHO, 2023b).

15.4.1 Current Institutional Framework

15.4.1.1 Federal Government (FGS)

Through SoDMA, the FGS leads policy, coordination, and response to disasters (SoDMA, 2011). SoDMA hosts the National Emergency Operations Centre (NEOC) and the Early Warning Centre and works closely with the Prime Minister's Office for high-level decisions (SoDMA, 2025a). Other federal ministries play supporting roles in their respective sectors. For example, the Ministry of Health handles health emergencies, the Ministry of Agriculture and Environment deals with drought resilience and climate adaptation, and the Ministry of Finance manages disaster financing, often coordinating with donors on emergency funds (Reliefweb, 2025c; WHO, 2025b). There is also a Somalia Disaster Management Coordination Group (SDMCG) which unifies government and humanitarian partners. Importantly, Somalia's policies aim for a multi-sectoral approach, recognising that disaster risk reduction (DRR) must be integrated into all ministries' plans (IOM, 2021; SODMA, 2022).

15.4.1.2 Federal Member States (FMS)

Each FMS (and Somaliland) has some structure for disaster management, although its capacity varies. For example, Puntland's HADMA and Somaliland's NADFOR are relatively active, whilst newer states rely more on ad hoc committees (NADFOR, 2003; Somali JNA, 2005). The linkages between the federal and state levels are still developing, and the national policy calls for building institutional links such that states and districts take primary responsibility for local risks, supported by the federal framework (SoDMA, 2011, 2022). In practice, FMS disaster bodies coordinate operationally with local NGOs and UN partners; however, capacity gaps persist, and they often rely on SoDMA and international actors for technical and financial support. Several states are developing or using state-level disaster plans aligned to national policy, for example, Southwest State's 2020–2023 Disaster Response and Recovery Plan, whilst national and sub-national systems are being strengthened through United Nations Office for Disaster Risk Reduction's (UNDRR) Early Warnings for All (EW4All) work and related coordination initiatives. Evidence of improving but still uneven state-level coordination includes IOM-supported mechanisms in Galmudug, Jubaland, Hirshabelle, and Southwest, as well as SoDMA's own 2024–2029 strategy prioritising stronger SoDMA–FMS coordination (SoDMA, 2023; Somalian Consortium, 2023).

15.4.1.3 Community and District Level

At the grassroots level, traditional structures (clan elders and community committees) and local administrations are the first responders when crises hit remote areas (Care, 2025; Lwanga-Ntale & Owino, 2020). The formal system is attempting to integrate these; for instance, the Somali Red Crescent Society (SRCS) has volunteers in many districts, and the government intends to leverage SRCS's local presence in early warning dissemination and response (Hub, 2025; IFRC, 2025). The DRM Policy explicitly mandates engaging communities and leveraging traditional networks as a priority for building resilience at the village level. However, due to insecurity, many rural communities remain outside the effective reach of the government and rely on NGOs or their own coping mechanisms (SODMA, 2022; Somalia, 2022a; UNDP, 2024).

15.4.1.4 International Humanitarian Architecture

Given Somalia's limited capacity, the UN-led cluster system plays a central role in the crisis response. The United Nations (UN) Office for the Coordination of Humanitarian Affairs (OCHA) coordinates humanitarian partners through clusters (Food Security, Health, Water, Sanitation and Hygiene (WASH), Shelter, etc.), which conduct joint needs assessments and responses in emergencies (IOM, 2023; OCHA, 2020d, 2023a). There is a close collaboration between SoDMA and UN/OCHA in Somalia's humanitarian architecture. SoDMA and the UN have co-launched national response plans and co-chaired operational coordination forums, including monthly humanitarian coordination meetings and Humanitarian Country Team (HCT) sessions attended by OCHA leadership (SoDMA, 2025b; SONNA, 2025). Somalia is also part of regional bodies; it joined IGAD and, in 2024, became a member of the East African Community (EAC), which could open avenues for regional disaster risk reduction initiatives (EAC, 2024; IGAD, 2024). The country aligns with international frameworks: it is a signatory to the Sendai Framework for DRR (2015–2030) and the Paris Climate Agreement, indicating its commitment to global standards for resilience (SODMA, 2022; UN Climate Change, 2016). Table 15.3 provides a brief timeline of the key milestones in the evolution of Somalia's disaster management institutions and policies. The table below provides a brief timeline of the key milestones in the evolution of Somalia's disaster management institutions and policies.

As seen above, Somalia's institutional framework for disaster, crisis, and emergency management is still evolving and has made notable improvements over the past decade. The establishment of SoDMA and policy frameworks provides a foundation that did not exist during earlier crises (SoDMA, 2011). However, it is important to recognise that the implementation remains weak. Many of the structures outlined in the paper (e.g. the National DRM Council or consistent federal-member state coordination mechanisms) are not yet fully functional or effective. Somalia's public sector capacity and resources are extremely limited; for example, the Somali DRM institutions have little to no core funding from the national budget, relying heavily on external donor projects (SODMA, 2022, 2023). Additionally, customary and international systems still carry much of the weight: local clan leaders are often first responders, and UN-led humanitarian operations still mount major emergency responses, with the government in a supporting role (Care, 2025; Lwanga-Ntale & Owino, 2020). Nonetheless, the trend is towards greater government involvement. In recent emergencies, such as the 2022 drought, SoDMA played a visible role in public awareness and the coordination of aid appeals (Humdata, 2022). Somalia's participation in global initiatives, such as the Sendai Framework reporting and Climate Risk & Early Warning Systems (CREWS) (UNDRR, 2024b), indicates growing engagement in disaster risk reduction beyond immediate relief.

Table 15.3 Timeline of disaster management institutional development in Somalia

Year	Institutional milestone or policy	Description and significance	Source(s)
1990s	State collapse – no central DRM	After the government fell in 1991, formal disaster management functions ceased. Humanitarian response was led by NGOs, UN, and community groups. Some regions (e.g. Somaliland and Puntland) created their own emergency committees, but no national coordination existed	Congress (2020), NADFOR (2003), Somali JNA (2005), Amnesty (2021)
2000s	Regional agencies form in stable areas	Somaliland established NADFOR (Disaster Preparedness & Food Reserve Authority), and Puntland established HADMA for disaster/ humanitarian affairs. These were early attempts at structured DRM in the absence of a central government	NADFOR (2003), Somali JNA (2005)
2011	Somalia Disaster Management Agency (SoDMA) was established	During the 2010–11 famine, the Transitional Federal Government established SoDMA by decree as a focal agency for disaster response. This marked the re-emergence of a Somali government role in DRM after two decades. Initially under Office of PM	SoDMA (2011)
2012	Federal Government was established	The Federal Republic of Somalia was formed (post-transitional), with a Provisional Constitution envisioning a federal structure. This provided a mandate to build national institutions, including for disaster management	Resources (2014)
2016	Legal Act for SoDMA; Creation of MoHADM	Somali Federal Parliament passed legislation formally authorising SoDMA as the national disaster management agency under a new Ministry of Humanitarian Affairs & Disaster Management (MoHADM). MoHADM (headed by a Cabinet minister) was established to coordinate DRM across sectors and with donors	SoDMA (2011), Worldbank (2025a)

(continued)

Table 15.3 (continued)

Year	Institutional milestone or policy	Description and significance	Source(s)
2017	Operationalisation of MoHADM; State ministry counterparts	In 2017, the Federal Government commissioned the Drought Impact and Needs Assessment (DINA) and launched the Recovery and Resilience Framework (RRF) to link relief with recovery and resilience, whilst Federal Member States such as Galmudug, Jubaland, Hirshabelle, and Southwest strengthened state-level drought and emergency coordination mechanisms	UNDP (2017b), UNDP (2017c)
Jul 2020	National Multi-Hazard Early Warning Centre was launched	With support from UN/World Bank, Somalia opens an Early Warning Centre under MoHADM. Its purpose is to gather data (e.g. from meteorological services, FEWS NET) and issue alerts for droughts, floods, etc., enabling early action. This was a milestone in shifting from reactive to proactive risk management	SODMA (2022), Worldbank (2025a)
Dec 2020	National Disaster Risk Management Policy adopted	The FGS approved its first comprehensive National Disaster Risk Management Policy (revised edition) in December 2020. The policy outlines institutional roles for SoDMA, MoHADM, and Federal Member States and emphasises integrating disaster risk reduction (DRR) into development plans. It also aligns Somalia's framework with international standards such as the Sendai Framework and regional strategies under the African Union and IGAD	SODMA (2022)
2022	Recognition—MoHADM dissolved, SoDMA elevated	The MoHADM is abolished in a government reshuffle; SoDMA assumes all roles of the former ministry as the standalone national agency for disaster management. SoDMA now reports directly to the Prime Minister's office, simplifying the command structure	SoDMA (2011), Worldbank (2025a)

(continued)

Table 15.3 (continued)

Year	Institutional milestone or policy	Description and significance	Source(s)
2023	Early Warning for All (EW4All) Initiative; EWS Roadmap	Somalia joins the UN's EW4All initiative to ensure universal early warning coverage. In 2023, UNDRR and SoDMA developed a Road Map to strengthen Somalia's Multi-Hazard Early Warning System, identifying gaps in data, risk assessment, and communication networks. A national workshop in July 2024 (Nairobi) and exchanges with Italy's civil protection refined Somalia's EWS framework	UNDRR (2024b)
2024	Regional Integration and Partnerships	Somalia became the EAC's 8th member on March 4, 2024, opening avenues for regional cooperation on preparedness, early warning, and risk financing. It continued DRM capacity building with the World Bank/ GFDRR (e.g. *Strengthening Disaster Resilience in Somalia* 2023–2025; *Somalia Floods Rapid PDNA* 2–8/ 2024). Somalia also engaged the African Union's ARC mechanisms for drought risk financing, including Start Network/ARC Replica funding in 2023 and an ARC insurance payout to the Federal Government in March 2025	EAC (2024), GFDRR (2025), ARC (2023a)

15.5 Challenges and Weaknesses in Somalia's Disaster Management System

Despite the progress made in establishing agencies and policies, Somalia's disaster management capacity faces significant challenges and systemic weaknesses (Table 15.4). Past responses to disasters have often been only partially effective, hindered by various political, institutional, and resource-related problems (SoDMA, 2023; UNFPA, 2025; Worldbank, 2023).

15.5.1 Limited Government Capacity and Reliance on International Actors

The legacy of conflict and state fragility means that Somali institutions are still very weak. The government has rarely managed disaster response on its own; instead,

Table 15.4 Major challenges and weaknesses in Somalia's disaster management

Challenge	Description	Source(s)
Weak Government Capacity and Dependency	Limited staffing, skills, and resources in Somali institutions result in heavy reliance on UN/NGOs for disaster response. The Government rarely leads responses directly, hindering its learning and capacity building	UK HIH (2023), SoDMA (2024)
Funding Constraints	Almost no domestic funding is allocated for DRM; operations depend on donor aid. The lack of contingency funds or logistics implies slow, inadequate responses	UNFPA (2025), UK HIH (2023)
Reactive Approach (versus Preventive)	Emphasis on crisis relief rather than risk reduction. Little investment in mitigation (water infrastructure, etc.), leading to repeated crises. The early warning to early action link is weak	UNFPA (2025), UK HIH (2023)
Coordination Gaps (Govt–Humanitarian)	Poor integration of the government in humanitarian coordination. UN/NGOs coordinate in clusters largely without formal government leadership, causing parallel efforts and inefficiencies	UNFPA (2025), UK HIH (2023)
Federal–State Coordination Issues	Disaster management responsibilities between FGS and Federal Member States are not clearly delineated in practice. State agencies exist but have weak capacity; collaboration mechanisms are nascent	NAPGlobal Network (2022), SCRP (2024)
Insecurity and Access Constraints	Ongoing conflict in certain regions impedes aid delivery and assessments. This also poses risks to aid workers, delaying response	OCHA (2023c); OCHA (2024)
Information and Data Gaps	Until recently, Somalia lacked reliable disaster data and analytical capacity. Weak meteorological and statistical systems have resulted in poor risk analysis and planning. This continues to hamper evidence-based decision-making	ARC (2023b), PreventionWeb (2022)
Gender and Social Inclusion Gaps	Women, youth, and marginalised groups have had a limited voice in DRM institutions. Disaster policies have not been gender-sensitive, risking that responses overlook the specific needs of women/children	ARC (2023b), PreventionWeb (2022), ARC (2021)
Infrastructure Deficits	Poor transport, water, and health infrastructure magnify disaster impacts and complicate response logistics. Few paved roads, limited flood control, and scarce health facilities outside cities	UK HIH (2023), SCRP (2024)

international organisations take the lead, with the government playing a secondary role (OCHA, 2018b, 2020d; UNFPA, 2025). For example, forecasting and emergency response initiatives are 'predominantly funded and overseen by international actors outside Somalia's humanitarian responses', which in turn minimises the ability of local institutions to develop experience in these areas (HPF, 2019; UNDRR, 2024c). This external dominance creates a vicious cycle: it undermines government ownership and capacity building, perpetuating dependency on non-state actors. Even when SoDMA or line ministries have coordination roles, they often lack the staff, equipment, and expertise to provide leadership (SoDMA, 2023; Worldbank, 2025a). Government reports and diagnostics indicate that domestic contributions to disaster preparedness and response are minimal, whilst international donors fund most humanitarian operations (OCHA, 2023b; Reliefweb, 2025c).

15.5.2 Fragmented and Reactive Approach

Historically, Somalia's approach to disasters has been reactive, waiting for crises to hit and then mobilising emergency relief. Proactive risk reduction (mitigation and preparedness) was minimal (OXFAM Library, 2017; Lwanga-Ntale & Owino, 2020). Historically, DRM strategies in Somalia have been fragmented and reactive rather than unified or anticipatory. Many efforts have been project-based and led by external actors, with forecasting, early warning systems, and risk reduction often lagging the emergency response. The 2020 National DRM Policy itself recognises that past DRM practices 'do not necessarily reflect current best practices' (GFDRR, 2017; Lwanga-Ntale & Owino, 2020; SODMA, 2022). For instance, various NGOs might run separate drought resilience projects in specific regions without an overarching national risk reduction programme. On the government side, coordination amongst ministries has been weak, and disaster management has been seen as the job of the humanitarian sector, not integrated across development planning (Lwanga-Ntale & Owino, 2020; UNDP, 2022b). The government's 2020 DRM Policy acknowledges that Somalia had 'almost non-existent policies and protocols' for disasters as state institutions were just beginning to evolve (SODMA, 2022). Many measures in that policy (such as establishing disaster management committees at the district level or mainstreaming DRM into sector ministries) remain on paper only (SIDRA Institute, 2020). Early warning dissemination in Somalia has significant last-mile gaps. A 2023 roadmap identified weaknesses in data, risk assessment, and communication networks, which limited the systematic translation of warnings into early action. Since 2023, the EW4All effort via UNDRR/SoDMA workshops and capacity building in 2024 has focussed on fixing these dissemination and 'last-mile' issues (e.g. coordinated SMS/radio/community channels and loss-tracking skills) (OXFAM Library, 2017; UNDRR, 2024b).

15.5.3 Coordination Gaps (Internal and External)

One of the most cited weaknesses is the lack of effective coordination amongst the diverse actors, including government agencies, the UN, NGOs, and local groups. At the height of emergencies, coordination forums exist (e.g. the Humanitarian Country Team and cluster meetings), but the Somali government's role in them has not always been clear or consistent (UK HIH, 2023; UNDRR, 2024b; UNFPA, 2025). The 2020 DRM Policy critiques that UN and NGOs often 'mobilise separately' during emergencies and coordinate amongst themselves in a cluster system with little formal government participation (SODMA, 2022). Indeed, in cluster coordination meetings during recent crises, government representatives were either absent or had observer status, meaning that the government was not formally directing the response (DRC, 2021; Hih, 2023; ICVA, 2024). This was partly due to capacity; often, government personnel were simply not available or not trained to engage, and partly due to trust and procedural issues. The result is parallel systems: one led by humanitarian partners and the other by the government, leading to duplication or gaps (DRC, 2021; Hih, 2023; UNFPA, 2025). For example, during the 2017 drought, there were instances where the government announced a national drought committee and appeal, whilst in parallel, the UN issued its own international appeal; improved alignment emerged later under pressure from donors (HPF, 2019; Reliefweb, 2017). Coordination gaps also exist between the federal and state levels: some Federal Member States initially bypassed the federal government and worked directly with international NGOs (Hashi & Barasa, 2023; Said & Kicha, 2024). Whilst this is improving as SoDMA signs MOUs with states, there are still frictions, especially if political disputes arise between the Centre and a state.

15.5.4 Political Instability and Governance Issues

The political and security dynamics in Somalia significantly impede disaster response. Many parts of the country remain inaccessible or dangerous due to active conflicts and forcing humanitarian actors to operate with extreme caution or not at all in some areas. For instance, the 2024 Humanitarian Access Snapshot recorded 243 separate access incidents, with attacks, checkpoints, and threats frequently delaying or halting aid deliveries (UK HIH, 2023; OCHA, 2025b; Reliefweb, 2025d). During the 2011 famine, restrictions blocked many UN and international actors from entering vast stretches of southern Somalia, sharply reducing life-saving relief in those areas and contributing to elevated mortality rates (Seal & Bailey, 2013). More recently, in the 2022 drought, ongoing conflict and insecurity hindered humanitarian operations in several districts, making access unreliable or impossible in zones under contested controls (IOM, 2022; Menkhaus, 2012). Compounding this, Somalia's 18-month political deadlock during the 2020–2022 election standoff diverted critical attention and institutional capacity away from the early drought response. This prolonged instability undermined the continuity of governance and delayed coordinated action when urgency was highest (BTI, 2024). In 2022, when MoHADM was dissolved, and SoDMA inherited its mandate, this shift occurred amid active donor-sponsored development and humanitarian programmes, and SoDMA's internal plans explicitly acknowledged reliance on both government and donor funding, suggesting that external institutional backing may have assisted stabilise the transition and maintain continuity of disaster management functions (SoDMA, 2024; Worldbank, 2025a). However, to date, we have not found explicit independent confirmation of this attribution. Disaster governance in Somalia faces chronic integrity and regulatory weaknesses. In past relief operations, powerful actors from local gatekeepers, clan leaders, and security forces and officials have diverted aid (cash, food, or in-kind) through manipulative beneficiary lists, rent demands, or control of distribution routes (UK HIH, 2023; Integrity, 2011; Reuters, 2023). Political focus tends towards immediate crisis security, power-sharing arrangements, or conflict dynamics, leaving a limited appetite for proactive disaster risk reduction (DRR) (PeaceRep, 2024; The New Humanitarian, 2023). Meanwhile, institutional and legal voids inhibit effective risk reduction: the enforcement of land-use rules is minimal, settlement in floodplains remains unchecked, and deforestation continues as government supervision is weak (Somalia, 2022b). Together, these governance gaps undermine trust, discourage preventive investment, and amplify the vulnerability of communities that are already exposed to multiple shocks.

15.5.5 Resource Constraints and Infrastructure Weakness

Somalia's capacity to respond to disasters is constrained by extremely limited fiscal resources. The federal government's budget is only in the low-to-mid billions of USD, domestic revenue represents a small fraction of GDP, and a disproportionately large share is absorbed by security and administrative costs. For example, approximately 18 per cent of the 2022 budget was allocated to security, and in recent years, the government has spent 35–40 per cent of its budget on security functions alone (MOF, 2024; Nor & Yusof, 2025; SPA, 2022). Somalia's disaster management units generally lack stockpiles, emergency reserves, or logistics capacity and are thus heavily dependent on UN or cluster aviation and common logistics services when land routes fail. For example, in August 2025, the Somalia Logistics Cluster conducted air operations to reach areas where road access was severely constrained due to flooding (Cluster, 2025a). Infrastructure constraints further exacerbate these challenges. Heavy floods in late 2023 and 2024 damaged roads, bridges, and airstrips, often rendering them unusable or impassable, which severely impeded overland and fixed-wing air delivery. Only approximately 13% of Somalia's road network is paved, and many transport corridors are vulnerable to seasonal damage (Cluster, 2024, 2025b). On the risk monitoring and early warning side, Somalia has historically faced significant data and coordination deficits. The 2023 roadmap developed by the UNDRR identified major weaknesses in data collection, risk assessment, and communication networks. Recent UNDRR initiatives have aimed to strengthen disaster tracking, data management, and early warning systems, reflecting the need to close these gaps (UNDRR, 2024b). The country's early warning systems remain fragmented, with challenges in dissemination and reaching remote or pastoral communities. Broad global assessments (for example, MHEWS 2023) also confirm that many low-income and fragile states, including Somalia, still struggle with limited coverage, forecasting, and communication in EWS systems (PreventionWeb, 2023; UNDRR, 2024b).

15.5.6 Reactive Emergency Focus Versus Risk Reduction

Somalia faces a chronic imbalance between short-term emergency responses and long-term risk reduction. Because crises such as droughts, floods, and conflict are essentially continuous, most resources and institutional attention are funnelled into life-saving relief measures, with far too little investment in mitigation or preparedness. For example, investments in drought-resistant seeds, flood dikes, or environmental regulations remain minimal relative to food aid or emergency health care. The 2025 Somalia Disaster Risk Finance Diagnostic highlights this underinvestment, where risk finance instruments remain sparse and much funding is reactive rather than proactive (Reliefweb, 2025c). Donor and development programmes are beginning to support a shift; for example, the Somalia Crisis Recovery Project includes a component for long-term disaster risk management and preparedness beyond immediate recovery efforts (MOF, 2020). However, the concept of DRR is still relatively new in Somalia's policy space, gaining traction only in recent years through frameworks such as the National Disaster Risk Management Policy and renewed emphasis through partnerships with the UNDRR and early warning initiatives (SODMA, 2022; UNDRR, 2024b). Until the balance shifts more strongly towards prevention and resilience, each new disaster will continue to expose the unaddressed vulnerabilities.

15.5.7 Weaknesses in Local Engagement and Inclusion

Effective disaster management demands genuine community engagement and inclusivity; however, Somalia faces structural hurdles in this regard. Literacy and education levels remain low in many regions, and multiple languages or dialects complicate the task of communicating risk messages in a way that reaches diverse communities in the region. Trust in the government is often fragile after years of unfulfilled promises, implying that top-down directives may be ignored or resisted. The formal DRR/DRM system has not sufficiently leveraged traditional coping strategies or local social capital. Somali households routinely rely on clan networks, remittances, informal credit, and community reciprocity to respond to shocks; however, these systems often function outside or parallel to formal disaster structures (Lwanga-Ntale & Owino,

2020; NRC, 2022; SODMA, 2022). Some projects (e.g. BRCiS) have successfully engaged communities in decision-making, such as allowing communities to lead water procurement, thereby tapping social capital (NRC, 2022). However, such examples remain the exception rather than the norm; the institutional DRM system is still largely top-down and has limited mechanisms to institutionalise inclusive or community-led processes.

15.5.8 Coordination with Development and Climate Adaptation Efforts

There is a persistent gap between immediate disaster response and longer-term development and climate adaptation in Somalia. Although the Recovery and Resilience Framework (RRF) and the Drought Impact & Needs Assessment (DINA) were intended to embed 'build-back-better' principles, progress has been slow and fragmented (GFDRR, 2018; UNDRR, 2018). Analysts have noted that humanitarian and development programming often run on separate tracks, hampering coherence and sustainability (NRC, 2021). In practice, the World Bank-led Somalia Crisis Recovery Project (SCRP) now includes provisions to integrate climate-resilient rehabilitation and align with the RRF, which illustrates the recognition of the need to close this divide (Worldbank, 2020b). However, the scale of immediate needs and constraints on development funding mean that the transition from relief to resilience remains weak. Without scaling up climate adaptation investments (e.g. water infrastructure, reforestation, and livelihood diversification), Somalia risks remaining trapped in a reactive cycle of crisis and recovery.

15.6 Research Gaps in Disaster Management for Somalia

Effective disaster management relies on sound evidence and context-specific knowledge of the disaster. In Somalia, there are several areas where research is lacking or insufficient, hindering informed decision-making and the development of innovative solutions.

15.6.1 Disaster Data and Risk Analysis

Historically, one of the biggest voids has been the absence of comprehensive data on disasters (frequency, impacts, and losses). Until recently, Somalia has not maintained a national disaster loss database. This makes it difficult to analyse trends or plan based on evidence. For example, precise data on how each drought or flood affected communities were not systematically recorded by any government entity. This gap is addressed through the rollout of the DesInventar Sendai disaster loss database in Somalia, with national training and capacity building conducted in 2024 (UNDRR, 2024d). However, continuous research is needed to compile, validate, and use historical disaster statistics for risk modelling. Additionally, hazard mapping in Somalia remains rudimentary, and there is a need for updated maps of flood plains, drought risk zones, seismic hazard zones, etc., using modern techniques (GIS, remote sensing) (Osman & Das, 2023; UNDRR, 2024c). Climate projection research specific to Somalia is also limited; whilst global/regional models exist, downscaling them and studying localised climate impacts (such as changes in seasonal rain patterns in different Somali agro-climatic zones) would help tailor adaptation strategies (Risk, 2022).

15.6.2 Climate Change and Emerging Hazards

There is a gap in research on certain emerging or less understood threats. For example, Somalia, with its extensive coastline of over 3,000 km and numerous coastal communities, faces significant challenges due to sea level rise. Specifically, there is a notable 'lack of research on current and projected impacts of sea level rise' along Somalia's coast (IDC, 2024; Risk, 2022; Wordlbank, 2023b). Critical issues such as coastal erosion, saltwater intrusion into coastal aquifers, and the vulnerability of low-lying cities like Mogadishu remain insufficiently studied. Similarly, urban risks have not been studied sufficiently: Somalia's cities (Mogadishu, Hargeisa, etc.) have grown rapidly, often in unplanned ways, which increases disaster vulnerability (e.g. urban flooding, fire hazards, disease outbreaks in slums) (GFDRR, 2024; Risk, 2022; Worldbank, 2025b). One source noted 'a significant lack of research about the linkages between mass starvation and migration', which could be extended to understanding urban migration due to climate stress (Policy, 2018). Overall, climate-security intersections merit more study, for example, how do drought and resource scarcity contribute to conflict in Somalia (and vice versa)? Initial studies suggest strong links (drought can fuel communal violence), but deeper qualitative research could inform conflict-sensitive resilience programmes (Maystadt & Ecker, 2014). Groundwater and water resource research is another gap: in a drought-prone country, surprisingly little detailed hydro-geological mapping has been conducted to identify sustainable water sources for drought mitigation. Research in these emerging areas would provide an evidence base for long-term risk reduction in a changing environment (Worldbank, 2021).

15.6.3 Effectiveness of Traditional Coping Mechanisms and Indigenous Knowledge

Somali communities have lived with climate variability for centuries and developed their own coping strategies (such as pastoral mobility, water storage in *Berkad* cisterns, and social support networks like *zakat* charity) (Ahmed et al., 2023; Somalia, 2023; Worldbank, 2017). However, there is limited research on how these indigenous mechanisms function today and how they can be leveraged or strengthened. Studying local knowledge, such as traditional drought early warning signs that nomads observe, could complement scientific early warning systems. In addition, an anthropological look at how clans mobilise support during crises (e.g. urban relatives sending remittances to rural families in drought) would be valuable (Lwanga-Ntale & Owino, 2020). Such research can reveal social resilience factors that external aid may overlook. Simultaneously, research can identify where traditional systems are overwhelmed by modern challenges (e.g. repeated droughts may erode community coping capacity or change pastoral patterns) (Ahmed et al., 2023). However, there is a research gap in documenting these changes. Filling this gap would guide culturally appropriate disaster risk reduction efforts that work with (not against) community practices.

15.6.4 Humanitarian Response Evaluation and Lessons Learned

Whilst much has been written about Somalia's crises, there are relatively few systematic evaluations of past disaster responses that are publicly available. What worked and what did not in the 2011 famine response, the 2017 drought response, or the 2020 flood response. Some internal evaluations exist (by the UN or NGOs), but academic research analysing these responses in depth, especially from a coordination and local participation perspective, is scarce (OXFAM Library, 2012; Maxwell et al., 2016; Tufts University, 2014). For example, the 2017 drought response was considered more successful than that of 2011 (no famine declared), but a detailed study might examine the role of improved early action, use of cash transfers, etc., and identify gaps (perhaps certain regions were left out, or certain interventions were less effective) (Network, 2019; WFP, 2018a, 2018b). Similarly, learning from Somalia's experience with innovative approaches, such as forecast-based financing pilot projects or the use of mobile cash distribution at scale, can provide lessons for both Somalia and other countries (Hub, 2022; Network, 2021). Operational research that tests different approaches (e.g. comparing outcomes in villages that received resilience training versus those that did not) could strengthen the evidence on what builds resilience in the Somali context.

15.6.5 Linkages Between Disasters, Migration, and Urbanisation

Another understudied area is how recurrent disasters are reshaping Somalia's demographics, particularly the trends of internal displacement and urbanisation. Somalia now has millions of IDPs in camps around cities; drought is a key driver, but our understanding of these migration dynamics is incomplete (Gundel, 2002; Policy, 2018). Migration–disaster nexus research could explore questions such as: do most drought-displaced people return home when it rains, or do they become permanently urban poor? How do repeated shocks affect intentions to migrate abroad? One source pointed out that despite many famine studies, 'there is a remarkable lack of research about linkages between mass starvation and migration' (Migration Policy, 2018). Filling this gap is crucial for planning durable solutions, whether to invest more in rural resilience to prevent displacement or in urban services to support those who relocate.

15.6.6 Gender, Youth, and Social Dimensions of DRM

As noted earlier, gender issues in DRM have not been mainstreamed, and there is little research on the gender-differentiated impacts of disasters in Somalia (ARC, 2021, 2023b). Some qualitative studies shed light on increases in gender-based violence during crises or women's coping strategies. Likewise, research on youth engagement in DRR or the plight of persons with disabilities in emergencies would help ensure inclusive planning (Save the Children, 2012; UNDRR, 2024b). The ARC gender policy brief highlights gaps in disaggregated data; indeed, collecting and analysing data by gender, age, etc., is a research need (ARC, 2021, 2023b). Even the mental health impacts of repeated trauma and disaster on the population are an area with virtually no data in Somalia; given decades of crisis, one could hypothesise high levels of post-traumatic stress disorder (PTSD) or other issues, but little formal study exists (Salad et al., 2023).

15.6.7 Institutional and Governance Research in DRM

Finally, on the governance side, more research could be directed at which institutional models work best in fragile contexts, such as Somalia, for disaster management. Comparative research with other fragile states in the Middle Africa/Middle East (such as South Sudan, Yemen, and Afghanistan) could identify best practices in organising disaster management amidst conflict, which is directly relevant to Somalia (ICVA, 2024; SoDMA, 2011; UNFPA, 2025). Additionally, risk financing mechanisms (such as drought insurance via African Risk Capacity or contingent funds) are relatively new in Somalia and would benefit from research into their feasibility and design, for example what level of insurance payout Somalia could afford or how to structure cash-based safety nets that trigger before a famine (Somalia has piloted forecast-based cash in some areas, but research on effectiveness is needed) (UNDRR, 2024c).

Addressing these research gaps is not merely an academic exercise; it has practical implications for saving lives and livelihoods. If studies show that a certain region will suffer coastal flooding by 2030 due to sea level rise, the government and aid agencies can immediately begin working on developing preventive measures or relocation plans. If research finds that traditional pastoral early warning (such as reading stars or animal behaviour) is effective, it can be incorporated into formal early warning messaging to lend credibility. If evaluations reveal that cash aid was more effective than food aid during a certain drought, future responses can be adjusted accordingly. Moreover, highlighting the economic toll of disasters through data can help persuade decision-makers to fund risk reduction (shifting the narrative from 'disasters are humanitarian issues' to 'disasters are developmental and economic issues too') (SIDRA Institute, 2020; ODI, 2020).

Encouragingly, Somalia has begun to invest in knowledge. The formation of a Somali National University Disaster Management programme (hypothetically) or partnerships with local think tanks, such as the Heritage Institute or SIDRA, could foster home-grown research. Internationally, organisations such as FEWS NET, REACH, and academia (for example, LSHTM's mortality studies) are focussing more on Somalia, given the recurrent crises. It will be important for the Somali government and academia to set a research agenda aligned with these gaps and for donors to fund research as part of resilience programming (UK HIH, 2023; Impact, 2024, 2025; LSTHM, 2018). Ultimately, improved data and research will lead to better policies and interventions, creating a positive feedback loop for Somalia's disaster management and risk reduction capacity.

15.7 Recommendations and Future Strategies for Resilience

Somalia can transition from recurring crisis management to durable risk reduction by making DRM a standing state function, financing risk before shocks hit, acting on early warnings, empowering communities, and using better data to inform decision-making. First, resource SoDMA to lead coordination year-round staffing, systems, and a modest contingency fund with clear triggers whilst operationalising the National DRM Policy's governance architecture (a high-level council and federal–state SOPs) so that roles are unambiguous in emergencies (SoDMA, 2011, 2022). Simultaneously, Somalia's engagement with regional platforms (IGAD/EW4All; EAC) should be deepened to manage cross-border hazards and expand early warning coverage nationwide (EAC, 2024; IGAD, 2024; UNDRR, 2024b). Second, shift from ad hoc relief to layered risk finance: pair a domestic contingency fund and scalable social protection (*Baxnaano*) with sovereign risk transfer (ARC/Replica) so cash can flow rapidly and predictably when triggers are met; use the 2025 ARC payout to stress-test affordability, coverage, and delivery chains (ARC, 2025; Insurer, 2025; Wordlbank, 2022). Third, link early warning to early action as standard practice, finalise and drill Early Action Protocols so forecast triggers automatically release funds and mobilise predefined actions (livestock feed, mobile health, water) through existing cash systems (Hub, 2025; Network, 2021). Fourth, institutionalise 'One Somalia' coordination beyond crisis peak joint analyses, unified appeals, and a shared dashboard building on recent government–UN famine risk warnings to reduce duplication and align partners under national leadership (OCHA, 2025c).

Data and technology should underpin all of this: expand observation networks and remote sensing feeds, stand up robust loss accounting and Sendai reporting (DesInventar/UNDRR capacity building), and use the insights to prioritise mitigation in high-risk districts and to target last-mile communication in local languages (UNDRR, 2024b, 2024c; WMO, 2025). Coupling these system investments with practical risk reduction on the ground—such as flood protection and drainage in riverine towns, climate-smart agriculture and water storage in arid zones, and risk-informed siting/standards for schools and clinics—will ensure that development spending systematically lowers risk, not just restores it after each shock. Finally, anchor resilience in communities: support inclusive, locally led preparedness through Somali institutions and civil society, blending indigenous knowledge with scientific warnings (a core thrust of EW4All in Somalia) and using cash at scale to protect livelihoods before crises escalate (GWP, 2024; SODMA, 2022).

Acknowledgements None

AI Discloser The authors acknowledge the use of Paperpal AI (https://paperpal.com/) for its 'Language Edit' and 'Make Academic' features to improve clarity and readability. This assistance was limited to linguistic refinement; all analyses and interpretations are solely those of the authors.

References

Acaps. (2018). Somalia: Central and South Floods update II. Retrieved September 22, 2025, from https://www.acaps.org/fileadmin/Data_Product/Main_media/20180530_acaps_briefing_note_somalia_floods_update_ii.pdf

Addan, O. A. (2025) Challenges to national unity in Somalia: in-depth investigation from post-2012 to 2022. *Cogent Social Sciences, 11*. https://doi.org/10.1080/23311886.2025.2537787

Ahmed, M. M. (2024). Explosive violence in Somalia: A threat to health and development. *Medicine, Conflict, and Survival, 40*, 347–351. https://doi.org/10.1080/13623699.2024.2385337

Ahmed, M., Mohamed, M. D., Muhammad, F.: Pastoralists adaptation strategies, and resilience capacity to climate change in Somalia: A scoping review 2023. https://doi.org/10.21203/RS.3.RS-2535309/V1

Ahmed, M. M., Asowe, H. A., Dirie, N. I., Mohamud, A. K., Lucero-Prisno III, D. E., Okesanya, O. J., et al. (2024). The nexus of climate change, food insecurity, and conflict in Somalia:

a comprehensive analysis of multifaceted challenges and resilience strategies. *F1000Res, 13*, 913. https://doi.org/10.12688/f1000research.154400.1

Amnesty. (2021). Building human rights inthe disintegrated state. Retrieved September 23, 2025, from https://www.amnesty.org/ar/wp-content/uploads/2021/06/afr520031995en.pdf

Aniticipation Hub. (2022). Forecast-based financing in Somalia: New project kick-off—Anticipation Hub. Retrieved September 25, 2025, from https://www.anticipation-hub.org/news/forecast-based-financing-in-somalia-kicking-off-a-new-project-to-act-ahead-of-drought

Anticipation Hub. (2025). Somali Red Crescent Society activates its early action protocol for drought to support most vulnerable people—Anticipation Hub. Retrieved September 23, 2025, from https://www.anticipation-hub.org/news/somali-red-crescent-society-activates-its-early-action-protocol-for-drought

ARC. (2021). Gender Analysis on DRM Somalia. Retrieved September 24, 2025, from https://www.arc.int/gender-drmp/IMG/pdf/gender_analysis_on_drm_somalia.pdf

ARC. (2023). New funding streams activated after fifth successive failed rainy season in Somalia | African Risk Capacity Group. Retrieved September 23, 2025, from https://www.arc.int/news/new-funding-streams-activated-after-fifth-successive-failed-rainy-season-somalia

ARC. (2023). Key Gender issues & Priorities in DRM Somalia | Policy Brief—Plateforme AGGRC. Retrieved September 24, 2025, from https://www.arc.int/gender-platform/key-gender-issues-priorities-in-drm-somalia-policy-brief.html?lang=fr

ARC. (2025).The ARC Group makes climate insurance payouts to the Federal Republic of Somalia and the Start Network to assist people affected by drought.

BTI. (2024). BTI 2024 Somalia Country Report: BTI 2024. Retrieved September 22, 2025, from https://bti-project.org/en/reports/country-report/SOM

CALP Network. (2019). Reflections and recommendations from evaluations of the 2017 CVA Somalia Drought Response—The CALP Network. https://www.calpnetwork.org/publication/reflections-and-recommendations-from-evaluations-of-the-2017-cva-somalia-drought-response/

CALP Network. (2021). Somalia Shock Responsive Pilots Early Action Protocol Written by Barbara Leseni Somalia Social Transfers Technical Assistance Facility. Retrieved September 25, 2025, from https://www.calpnetwork.org/wp-content/uploads/2021/12/Somalia_Shock_Responsive_Early_Action_Protocol.pdf

Care. (2025). Resilience in action as early warning committees protect communities in Somalia—CARE Climate Change. Retrieved September 23, 2025, from https://careclimatechange.org/resilience-in-action-as-early-warning-committees-protect-communities-in-somalia/

CDC. (1992). Population-Based Mortality Assessment—Baidoa and Afgoi, Somalia, 1992. Retrieved September 22, 2025, from https://www.cdc.gov/Mmwr/preview/mmwrhtml/00018099.htm

CFR. (2025). Conflict with Al-Shabaab in Somalia | Global Conflict Tracker. Retrieved September 22, 2025, from https://www.cfr.org/global-conflict-tracker/conflict/al-shabab-somalia

Climate Centre. (2021). Somalia—Country Overview. Retrieved September 22, 2025, from https://www.climatecentre.org/wp-content/uploads/RCCC-ICRC-Country-profiles-Somalia.pdf

Columbia University. (2005). Somalia Natural Disaster Profile Natural Disaster Risk Hotspots: Droughts Floods Natural Disaster Risk Hotspots (Weighted by Mortality). Retrieved September 22, 2025, from https://www.ldeo.columbia.edu/chrr/research/profiles/pdfs/somalia_profile1.pdf

Concern (Anaheim). (2023). Understanding hunger and poverty in Somalia | Concern Worldwide. Retrieved September 22, 2025, from https://www.concern.org.uk/news/understanding-hunger-and-poverty-somalia

Concern USA. (2022). Timeline: Breaking down more than a decade of drought in Somalia. Retrieved September 22, 2025, from https://concernusa.org/news/somalia-drought-timeline/

Conciliation Resources. (2014). Somalia: legitimacy of the Provisional Constitution | Conciliation Resources. Retrieved September 22, 2025, from https://www.c-r.org/accord/legitimacy-and-peace-processes/constitutional-review-%C2%A0peace-processes-securing-local-0

Congress. (2020). Somalia—Rebuilding the State. Retrieved September 22, 2025, from https://www.congress.gov/crs_external_products/IF/PDF/IF10155/IF10155.14.pdf

CRISIS GROUP. (2024). Fighting Climate Change in Somalia's Conflict Zones | International Crisis Group. Retrieved September 22, 2025, from https://www.crisisgroup.org/africa/horn-africa/somalia/316-fighting-climate-change-somalias-conflict-zones

DRC. (2021). Diaspora Organizations and Their Humanitarian Response in Somalia. Retrieved September 24, 2025, from https://drc.ngo/media/iutlri0g/demac-report-somalia-final-rz-digital.pdf

EAC. (2024). Somalia finally joins EAC as the bloc's 8th Partner State. Retrieved September 23, 2025, from https://www.eac.int/press-releases/3049-somalia-finally-joins-eac-as-the-bloc-s-8th-partner-state

FAO. (2020). Somalia Facing Risk of Floods in the Juba and Shabelle Basins. Retrieved September 22, 2025, from https://faoswalim.org/article/somalia-facing-risk-floods-juba-and-shabelle-basins

FAO. (2020). Impact of COVID-19 on food security and agriculture. Retrieved September 22, 2025, from https://openknowledge.fao.org/server/api/core/bitstreams/1ef1299b-5d88-4df0-b57c-d16f780cdf78/content

FMR. (2024). Somalia-Yemen links: refugees and returnees - Forced Migration Review. Retrieved September 26, 2025, from https://www.fmreview.org/solutions/mohamud/

GFDRR. (2017). Somalie | GFDRR. Retrieved September 22, 2025, from https://www.gfdrr.org/fr/somalie

GFDRR. (2018). TOWARD DROUGHT RECOVERY AND RESILIENCE The Somalia Drought Impact & Needs Assessment and Recovery & Resilience Framework Background. Retrieved September 24, 2025, from https://www.gfdrr.org/sites/default/files/publication/somalia-dina-exec-summary.pdf

GFDRR. (2024). Somalia—Strengthening Climate Resilience in Urban Settlements. https://www.gfdrr.org/sites/default/files/2025-01/Project%20Fiche%20-%20Somalia%20%E2%80%93%20Strengthening%20Climate%20Resilience%20in%20Urban%20Settlements.pdf

GFDRR. (2025). Somalia. Retrieved September 23, 2025, from https://www.gfdrr.org/en/region/somalia

Gundel, J. (2002). The migration-development nexus: Somalia case study. *International Migration, 40*, 255–281. https://doi.org/10.1111/1468-2435.00219

GWP. (2024). Somalia Country Programme. Retrieved September 25, 2025, from https://www.gwp.org/globalassets/global/about-gwp/publications/gcf-outputs/somalia-country-programme---june-2024.pdf

Hashi, M. B., & Barasa, T. (2023). Enhancing federal inter-governmental relations and state building in Somalia. *Humanities and Social Sciences Communications, 10*, 1–10. https://doi.org/10.1057/S41599-023-02177-W;SUBJMETA

HPF. (2019). The 2017 Pre-famine response in Somalia Progress on reform? HPG Commissioned Report About the authors. Retrieved September 23, 2025, from https://media.odi.org/documents/12592.pdf

Humdata. (2022). Somalia: The 2022 Crisis in Four Charts—The Centre for Humanitarian Data. Retrieved September 22, 2025, from https://centre.humdata.org/somalia-the-2022-crisis-in-four-charts/

ICVA. (2024). Localisation in Humanitarian Leadership -Somalia. Retrieved September 24, 2025, from https://www.icvanetwork.org/uploads/2024/09/Localisation-in-Humanitarian-Leadership-Somalia.pdf

IDC. (2024). Country Private Sector Diagnostic Creating Markets In Somalia. Retrieved September 25, 2025, from https://www.ifc.org/content/dam/ifc/doc/2024/somalia-country-private-sector-diagnostic-en.pdf

IFRC. (2025). Somalia | Community Based Surveillance. Retrieved September 23, 2025, from https://cbs.ifrc.org/implementations/somalia-0

IGAD. (2024). SOMALIA | IGAD Resilience. Retrieved September 23, 2025, from https://www.resilience.igad.int/somalia/

Impact. (2024). Somalia. https://www.impact-initiatives.org/where-we-work/somalia/

Impact. (2025). REACH. Retrieved September 25, 2025, from https://www.impact-initiatives.org/what-we-do/reach/

IOM. (2018). Drought Response Coordination Improves in Four Federal Member States in Somalia. Retrieved September 23, 2025, from https://www.iom.int/news/drought-response-coordination-improves-four-federal-member-states-somalia

IOM. (2021). Bridging the Divide in Approaches to Conflict and Disaster Displacement. Retrieved September 23, 2025, from https://www.iom.int/sites/g/files/tmzbdl2616/files/inline-files/bridging-the-divide-in-approaches-to-conflict-and-disaster-displacement-english_0.pdf

IOM. (2022). Somalia Drought Response. Retrieved September 24, 2025, from https://www.iom.int/sites/g/files/tmzbdl486/files/situation_reports/file/IOM-Somalia-Drought-Response-November-2022.pdf

IOM. (2023). HUMANITARIAN NEEDS OVERVIEW—SOMALIA. Retrieved September 23, 2025, from https://www.crisisresponse.iom.int/sites/g/files/tmzbdl1481/files/appeal/documents/20230208_Somalia%20_HNO_2023.pdf

Logistics Cluster. (2024). Somalia Logistics Cluster Concept of Operations. Retrieved September 24, 2025, from https://www.s3.eu-west-1.amazonaws.com/logcluster-web-prod-files/public/2024-04/240403_Somalia%20Logistics%20Cluster_CONOPS%20%28v.final%29.pdf

Logistics Cluster. (2025). Somalia—Operation Overview—August 2025 | Logistics Cluster Website. Retrieved September 24, 2025, from https://www.logcluster.org/en/document/somalia-operation-overview-august-2025

Logistics Cluster. (2025). Concept of Operations, March 2025. Retrieved September 24, 2025, from https://s3.eu-west-1.amazonaws.com/logcluster-web-prod-files/public/2025-03/250318_Somalia%20Logistics%20Cluster_CONOPS.pdf

LSTHM. (2018). LSHTM-Crisis-attributable mortality in Somalia-Q&A Estimates of crisis-attributable mortality in Somalia. Retrieved September 25, 2025, from https://www.lshtm.ac.uk/media/40141

Lubogo, M., Evans, B., Abdinasir, A., Sherein, E., Muhammad, T., Mohamed, A., et al. (2023). Responding to cholera outbreaks in Somalia in 2017–2019. *East Mediterranean Health Journal, 29*, 734–41. https://doi.org/10.26719/EMHJ.23.096

Lwanga-Ntale, C., Owino, B. O. (2020) Understanding vulnerability and resilience in Somalia. *Jàmbá: Journal of Disaster Risk Studies, 12*, 856. https://doi.org/10.4102/JAMBA.V12I1.856

Maxwell, D., Majid, N., Adan, G., Abdirahman, K., & Kim, J. J. (2016). Facing famine: Somali experiences in the famine of 2011. *Food Policy, 65*, 63. https://doi.org/10.1016/J.FOODPOL.2016.11.001

Maystadt, J. F., & Ecker, O. (2014). Extreme weather and civil war: Does drought fuel conflict in Somalia through livestock price shocks? *American Journal of Agricultural Economics, 96*, 1157–1182. https://doi.org/10.1093/AJAE/AAU010

Menkhaus, K. (2012). No access: Critical bottlenecks in the 2011 Somali famine. *Global Food Security, 1*, 29–35. https://doi.org/10.1016/J.GFS.2012.07.004

Migration Policy. (2018). Article: The Emerging Crisis: Is Famine Returning. | migrationpolicy.org. Retrieved September 25, 2025, from https://www.migrationpolicy.org/article/emerging-crisis-famine-returning-major-driver-migration

Ministry of Planning. (2025). Federal Member States—Ministry of Planning. Retrieved September 22, 2025, from https://mop.gov.so/federal-member-states/

MOF. (2020). Somalia Crisis Recovery Project (WB) | Ministry of Finance - Somalia. Retrieved September 24, 2025, from https://www.mof.gov.so/the-ministry/programs/somalia-crisis-recovery-project

MOF. (2024). 2024 END-YEAR BUDGETPERFORMANCE REPORT. Retrieved September 24, 2025, from https://www.mof.gov.so/sites/default/files/Publications/2024%20END%20YEAR%20BUDGET%20PERFORMANCE%20REPORT_0.pdf

MOP Somalia. (2022). SOMALIA NATIONAL DEVELOPMENT PLAN 2020 to 2024. Retrieved September 23, 2025, from https://www.mop.gov.so/wp-content/uploads/2022/07/Somali-National-Development-Plan-9-2020-2024.pdf

Musei, S. K., Nyaga, J. M., & Dubow, A. Z. (2021). SPEI-based spatial and temporal evaluation of drought in Somalia. *Journal of Arid Environments, 184*, Article 104296. https://doi.org/10.1016/J.JARIDENV.2020.104296

NADFOR. (2003). Somaliland National Disaster Preparedness and Food Reserve Authority. Retrieved September 23, 2025, from https://www.nadfor.govsomaliland.org/article/about-nafor

NAPGlobal Network. (2022). Somalia's National Adaptation Plan (NAP) Framework. Retrieved September 24, 2025, from https://www.napglobalnetwork.org/wp-content/uploads/2022/11/napgn-en-2022-somalia-nap-framework.pdf

NASA. (2020). Gati Makes Historic Landfall in Somalia. Retrieved September 22, 2025, from https://www.earthobservatory.nasa.gov/images/147576/gati-makes-historic-landfall-in-somalia

Nor, B. A., & Yusof, Y. (2025). Government spending and economic growth dynamics in Somalia: Time series evidence from ARDL bounds testing. *Discover Sustainability, 6*, 1–21. https://doi.org/10.1007/S43621-025-01471-8/FIGURES/5

NRC. (2022). Building Resilient Communities in Somalia. Retrieved September 24, 2025, from https://www.nrc.no/globalassets/pdf/reports/brcis-phase-2/brcis-phase-2.pdf

NRC. (2021). Supporting longer term development in crises at the nexus Lessons from Somalia report. Retrieved September 24, 2025, from https://www.nrc.no/globalassets/pdf/reports/development-actors-and-the-nexus/supporting_longer_term_development_in_crises_at_the_nexus_lessons_from_somalia.pdf

OCHA. (2017). Somalia: Drought Response - Situation Report No. 1 (as of 24 March 2017) | OCHA. Retrieved September 22, 2025, from https://www.unocha.org/publications/report/somalia/somalia-drought-response-situation-report-no-1-24-march-2017

OCHA. (2018). OCHA Flash Update #2—Tropical Cyclone Sagar | 20 May 2018 | OCHA. Retrieved September 22, 2025, from https://www.unocha.org/publications/report/somalia/ocha-flash-update-2-tropical-cyclone-sagar-20-may-2018

OCHA. (2018). Somalia: Flood Response—May 2018 (as of 15 May 2018) | OCHA. Retrieved September 22, 2025, from https://www.unocha.org/publications/report/somalia/somalia-flood-response-may-2018-15-may-2018

OCHA. (2018). OCHA Flash Update #4 Tropical Cyclone Sagar | 14 June 2018 | OCHA. Retrieved September 22, 2025, from https://www.unocha.org/publications/report/somalia/ocha-flash-update-4-tropical-cyclone-sagar-14-june-2018

OCHA. (2020). OCHA Somalia—Tropical Cyclone Gati Update #4, As of 26 November 2020 [EN/SO] | OCHA. Retrieved September 22, 2025, from https://www.unocha.org/publications/report/somalia/ocha-somalia-tropical-cyclone-gati-update-4-26-november-2020-enso

OCHA. (2020). Somalia: Flood Response Plan (June 2020) | OCHA. Retrieved September 22, 2025, from https://www.unocha.org/publications/report/somalia/somalia-flood-response-plan-june-2020

OCHA. (2020). SOMALIA Situation Report.

OCHA. (2020). Joint Multi-Cluster Needs Assessment: Sool, Somalia - June-August 2020 | OCHA. Retrieved September 23, 2025, from https://www.unocha.org/publications/report/somalia/joint-multi-cluster-needs-assessment-sool-somalia-june-august-2020

OCHA. (2021). Somalia: Humanitarian Response Plan 2022 (December 2021) | OCHA. Retrieved September 22, 2025, from https://www.unocha.org/publications/report/somalia/somalia-humanitarian-response-plan-2022-december-2021

OCHA. (2023). We coordinate. Retrieved September 23, 2025, from https://www.unocha.org/we-coordinate

OCHA. (2023). Somalia 2023 | Financial Tracking Service. Retrieved September 23, 2025, from https://fts.unocha.org/countries/206/summary/2023

OCHA. (2023). Somalia: 2023 Annual Humanitarian Access Overview | OCHA. Retrieved September 24, 2025, from https://www.unocha.org/publications/report/somalia/somalia-2023-annual-humanitarian-access-overview

OCHA. (2024). Somalia: Humanitarian Access Snapshot (1 July to 30 September 2024)—As of 30 September 2024 | OCHA. Retrieved September 24, 2025, from https://www.unocha.org/publications/report/somalia/somalia-humanitarian-access-snapshot-1-july-30-september-2024-30-september-2024

OCHA. (2025). Monthly Humanitarian Update| August 2025 | OCHA. Retrieved September 22, 2025, from https://www.unocha.org/publications/report/somalia/monthly-humanitarian-update-august-2025

OCHA. (2025). Somalia: Humanitarian Access Snapshot (January–December 2024)—As of 22 January 2025 | OCHA. Retrieved September 24, 2025, from https://www.unocha.org/publications/report/somalia/somalia-humanitarian-access-snapshot-january-december-2024-22-january-2025

OCHA. (2025). Drought, conflict and high food prices risk pushing 4.4 million people into hunger, Federal Government of Somalia and UN warn | OCHA. Retrieved September 25, 2025, from https://www.unocha.org/publications/report/somalia/drought-conflict-and-high-food-prices-risk-pushing-44-million-people-hunger-federal-government-somalia-and-un-warn

ODI. (2020). Research in Somalia: Opportunities for Cooperation. https://www.media.odi.org/documents/research_in_somalia_opportunities_for_cooperation.pdf

Osman, S. A., & Das, J. (2023). GIS-based flood risk assessment using multi-criteria decision analysis of Shebelle River Basin in southern Somalia. *SN Applied Science, 5*, 1–17. https://doi.org/10.1007/S42452-023-05360-5/FIGURES/8

OXFAM Library. (2012). Somalia Drought Response 2011/2012. Retrieved September 25, 2025, from https://www.oxfamilibrary.openrepository.com/bitstream/handle/10546/247813/er-somalia-drought-effectiveness-review-081012-summ-en.pdf

OXFAM Library. (2017). From Early Warning to Early Action in Somalia: What can we learn to support early action to mitigate humanitarian crises? Retrieved September 23, 2025, from https://www.oxfamilibrary.openrepository.com/bitstream/handle/10546/620345/dp-early-warning-early-action-somalia-021017-en.pdf

PeaceRep. (2024). Fighting Corruption in Somalia: Continuing the Momentum, Building on the Technical 'Fixes'—PeaceRep. Retrieved September 24, 2025, from https://www.peacerep.org/2024/11/13/fighting-corruption-in-somalia-continuing-the-momentum-building-on-the-technical-fixes/

PreventionWeb. (2022). Toward Gender-responsive and Technology-oriented Disaster Management in Eastern Africa. Retrieved September 24, 2025, from https://www.preventionweb.net/media/79162/download?startDownload=20250924

PreventionWeb. (2023). Somalia: Improving early warning systems. Retrieved September 24, 2025, from https://www.preventionweb.net/news/improving-early-warning-systems-somalia

Public Integrity. (2011). Corruption, the war on terror hindering food aid to southern Somalia—Center for Public Integrity. Retrieved September 24, 2025, from https://www.publicintegrity.org/accountability/corruption-the-war-on-terror-hindering-food-aid-to-southern-somalia/

Reliefweb. (2017). Somalia President declares drought a national disaster—Somalia | ReliefWeb. Retrieved September 24, 2025, from https://www.reliefweb.int/report/somalia/somalia-president-declares-drought-national-disaster

Reliefweb. (2018). OCHA Flash Update #3—Tropical Cyclone Sagar | 23 May 2018—Somalia | ReliefWeb. Retrieved September 22, 2025, from https://www.reliefweb.int/report/somalia/ocha-flash-update-3-tropical-cyclone-sagar-23-may-2018

Reliefweb. (2020). Somalia: Flash Floods—Apr 2020 | ReliefWeb. Retrieved September 22, 2025, from https://www.reliefweb.int/disaster/ff-2020-000055-som

Reliefweb. (2022). Drought Displacement Monitoring Dashboard (August 2022)—Somalia | ReliefWeb. Retrieved September 22, 2025, from https://www.reliefweb.int/report/somalia/drought-displacement-monitoring-dashboard-august-2022

Reliefweb. (2025). Somalia: 2025 Gu (April to June) Seasonal Floods - Flash Update No. 2 (as of 30 April 2025)—Somalia | ReliefWeb. Retrieved September 22, 2025, from https://www.reliefweb.int/report/somalia/somalia-2025-gu-april-june-seasonal-floods-flash-update-no-2-30-april-2025

Reliefweb. (2025). Somalia 2025 Humanitarian Needs and Response Plan (HNRP)—Somalia | ReliefWeb. Retrieved September 22, 2025, from https://www.reliefweb.int/report/somalia/somalia-2025-humanitarian-needs-and-response-plan-hnrp

Reliefweb. (2025). Somalia Disaster Risk Finance Diagnostic - Somalia. Retrieved September 23, 2025, from https://www.reliefweb.int/report/somalia/somalia-disaster-risk-finance-diagnostic

Reliefweb. (2025). Somalia: Humanitarian Access Snapshot (January–March 2025)—As of 13 April 2025—Somalia | ReliefWeb. Retrieved September 24, 2025, from https://www.reliefweb.int/report/somalia/somalia-humanitarian-access-snapshot-january-march-2025-13-april-2025

Reuters. (2018). Storm in Somaliland kills dozens, wipes out farms, livestock | Reuters. Retrieved September 22, 2025, from https://www.reuters.com/article/business/environment/storm-in-somaliland-kills-dozens-wipes-out-farms-livestock-idUSKCN1IO1BI/

Reuters. (2023). Somalia aid theft—a daily reality for country's most vulnerable | Reuters. Retrieved September 24, 2025, from https://www.reuters.com/world/africa/somalia-aid-theft-daily-reality-countrys-most-vulnerable-2023-09-19/

Said, A. S., & Kicha, D. I. (2024). Implementing health system and the new federalism in Somalia: Challenges and opportunities. *Frontiers in Public Health, 12*, 1205327. https://doi.org/10.3389/FPUBH.2024.1205327/FULL

Salad, A. M., Malik, S. M. M. R., Ndithia, J. M., Noor, Z., Madeo, M., & Ibrahim, M. (2023). Prevalence of mental disorders and psychological trauma among conflict-affected population in Somalia: A cross-sectional study. *Frontiers in Public Health, 11*, 1219992. https://doi.org/10.3389/FPUBH.2023.1219992/BIBTEX

Save the Children. (2012). Country Risk Reduction Profile: Disaster Risk Reduction in Somalia—Save the Children's Resource Centre. Retrieved September 25, 2025, from https://www.resourcecentre.savethechildren.net/document/country-risk-reduction-profile-disaster-risk-reduction-somalia

SCRP. (2024). 23Somali Crisis Recovery Project (SCRP). Retrieved September 24, 2025, from https://www.scrp.gov.so/wp-content/uploads/2024/09/last-one-1.pdf

Seal, A., & Bailey, R. (2013). The 2011 Famine in Somalia: Lessons learnt from a failed response? *Conflict and Health, 7*, 22. https://doi.org/10.1186/1752-1505-7-22

SIDRA Institute. (2020). Community Survey and Mapping for District Disaster Risk Management. Retrieved September 23, 2025, from https://www.sidrainstitute.org/wp-content/uploads/2020/04/Disaster-Risk-Management-Study-Report.pdf

SoDMA. (2011). Somali Disaster Management Agency. Retrieved September 23, 2025, from https://www.sodma.gov.so/eng/about-us/

SODMA. (2022). NATIONAL DISASTER RISK MANAGEMENT POLICY. Retrieved September 22, 2025, from https://www.sodma.gov.so/eng/wp-content/uploads/2022/10/National-Disaster-Risk-Management-Policy.pdf

SoDMA. (2023). Strategic Plan Somalia Disaster Management Agency. Retrieved September 23, 2025, from https://www.sodma.gov.so/wp-content/uploads/2024/07/SoDMA-Stratigic-Plan-2024-2029-Book-Ok.pdf

SoDMA. (2024). Achievements of SoDMA for 2022–2024. Retrieved September 24, 2025, from https://www.sodma.gov.so/eng/wp-content/uploads/2025/06/SoDMA-Two-Year-Achievement-2022-2024-Report.pdf

SoDMA. (2025). National Emergency Operation Center (NEOC). Retrieved September 23, 2025, from https://www.sodma.gov.so/eng/national-emergency-operation-center-neoc/

SoDMA. (2025). Monthly Humanitarian update April 30-2025 | SODMA. Retrieved September 23, 2025, from https://www.sodma.gov.so/eng/sodma-monthly-humanitarian-update-april-30-2025/

Somali JNA. (2005). Humanitarian Affairs and Disaster Management Agency. Retrieved September 23, 2025, from https://www.somali-jna.org/downloads/Puntland%20Fact%20Sheet.pdf

Somalia. (2022). BTI 2022 Country Report—Somalia. Retrieved September 24, 2025, from https://www.bti-project.org/fileadmin/api/content/en/downloads/reports/country_report_2022_SOM.pdf

More Somalia. (2023). Rainwater Harvesting in Somali-Inhabited Regions of East Africa: A Solution to Drought and Water Scarcity—MoreSomalia. Retrieved September 25, 2025, from https://www.moresomalia.com/rainwater-harvesting/

Somalia National Bureau of Statistics. (2024). Somalia Household Shocks and Responses Report. Retrieved September 22, 2025, from https://www.nbs.gov.so/wp-content/uploads/2025/01/HH-shocks-Responses-Report-.pdf

Somalian Consortium. (2023). Southwest State's 2020–2023 Disaster Response and Recovery Plan. Retrieved September 23, 2025, from https://www.somaliangoconsortium.org/silo/files/southwest-disaster-response-and-recovery-plan-20202023.pdf

SONNA. (2025). SoDMA and UN Launch 2025 Humanitarian Needs Response Plan for Somalia—Somali National News Agency. Retrieved September 23, 2025, from https://www.sonna.so/en/sodma-and-un-launch-2025-humanitarian-needs-response-plan-for-somalia/

SPA. (2022). Reviewing the FederalGovernment of Somalia's 2022near billion-dollar budget: development priorities anddonor dependence. Retrieved September 24, 2025, from https://www.somalipublicagenda.org/wp-content/uploads/2022/07/SPA_Governance_Briefs_18_2022_ENGLISH1.pdf

The Insurer. (2025). Somalia drought triggers $2.2 million in payouts from African Risk Capacity | Parametric Insurer. Retrieved September 25, 2025, from https://www.theinsurer.com/parametric-insurer/news/somalia-drought-triggers-22-million-in-payouts-from-african-risk-capacity-2025-04-02

The New Humanitarian. (2023). Aid, corruption, and neighbours in need: A reporter reflects on Somalia's record drought. Retrieved September 24, 2025, from https://www.thenewhumanitarian.org/opinion/2023/03/06/aid-corruption-somalia-drought-reporter-reflection

Tufts University. (2014). Lessons Learned from the Somalia Famine and the Greater Horn of Africa Crisis 2011–2012—Tufts—Feinstein International Center. Retrieved September 25, 2025, from https://www.fic.tufts.edu/publication-item/famine-somalia-crisis-2011-2012/

UK HIH. (2023). Somali capacities to respond to crisis are changing; how are humanitarian actors responding? Retrieved September 24, 2025, from https://www.humanitarianoutcomes.org/sites/default/files/publications/ho_ukhih_somalia_1023_2.pdf

UN. (1992). UNITED NATIONS OPERATION IN SOMALIA I (UNOSOM I)—Background (Full text). Retrieved September 22, 2025, from https://www.peacekeeping.un.org/mission/past/unosom1backgr2.html

UN. (2017). FAO Rapid Results Drought Response Plan: Somalia 2016/17 | United Nations in Somalia. Retrieved September 22, 2025, from https://www.somalia.un.org/en/34550-fao-rapid-results-drought-response-plan-somalia-201617

UN. (2019). Somalia Humanitarian Response Plan 2019 | United Nations in Somalia. Retrieved September 22, 2025, from https://www.somalia.un.org/en/33920-somalia-humanitarian-response-plan-2019

UN. (2023). Amid Devastating Drought, Ongoing Rights Violations, Somalia Pushes Forward in Fight against Al-Shabaab Terror Group, Security Council Hears | Meetings Coverage and Press Releases. Retrieved September 22, 2025, from https://www.press.un.org/en/2023/sc15208.doc.htm

UN Climate Change. (2016). Somalia | UNFCCC. Retrieved September 23, 2025, from https://www.unfccc.int/node/61200

UNDP. (2017). Somalia Drought Impact & Needs Assessment VOLUME III Federal Member State and Administrative Region Reports. Retrieved September 22, 2025, from https://www.undp.org/sites/g/files/zskgke326/files/publications/GSURR_Somalia%20DINA%20Report_Volume%20III_180116_Lowres.pdf

UNDP. (2017). Somalia Drought Impact and Needs Assessment (DINA) Executive Summary | United Nations Development Programme. Retrieved September 23, 2025, from https://www.undp.org/somalia/publications/somalia-drought-impact-and-needs-assessment-dina-executive-summary

UNDP. (2017). Recovery and Response for Drought in Somalia Final Report. https://www.info.undp.org/docs/pdc/Documents/SOM/Progress%20Report%20EDRCR%20Funds%202017.pdf

UNDP. (2022). Resilience and Climate Change | United Nations Development Programme. Retrieved September 22, 2025, from https://www.undp.org/somalia/our-focus/resilience-and-climate-change

UNDP. (2022). Enhancing Climate Resilience of the Vulnerable Communities and Ecosystems in Somalia | UNDP Climate Change Adaptation. Retrieved September 23, 2025, from https://www.adaptation-undp.org/projects/enhancing-climate-resilience-vulnerable-communities-and-ecosystems-somalia

UNDP. (2024). Community Resilience in The Horn of Africa: Insights from Ethiopia. Retrieved September 23, 2025, from https://www.undp.org/sites/g/files/zskgke326/files/2024-08/groundlevel_research_draft_8_july.pdf

UNDRR. (2018). Somalia: Recovery and Resilience Framework 2018|PreventionWeb. Retrieved September 24, 2025, from https://www.preventionweb.net/publication/somalia-recovery-and-resilience-framework-2018

UNDRR. (2024). Horn of Africa floods and drought, 2020–2023 - Forensic analysis | UNDRR. Retrieved September 22, 2025, from https://www.undrr.org/resource/horn-africa-floods-and-drought-2020-2023-forensic-analysis

UNDRR. (2024). Somalia is advancing disaster preparedness through EW4All and beyond | UNDRR. https://www.undrr.org/news/focus-how-somalia-advancing-disaster-preparedness-through-ew4all-and-beyond

UNDRR. (2024). Enhancing risk management in Somalia | UNDRR. Retrieved September 23, 2025, from https://www.undrr.org/news/enhancing-risk-management-somalia-building-national-capacity-disaster-tracking-data-management

UNDRR. (2024). DesInventar—Profile. Retrieved September 25, 2025, from https://www.desinventar.net/DesInventar/profiletab.jsp?continue=y&countrycode=som

UNEP. (2022). Somalia attempts to revive lands blighted by deforestation. Retrieved September 22, 2025, from https://www.unep.org/news-and-stories/story/somalia-attempts-revive-lands-blighted-deforestation

UNEPGRID. (2025). Somalia | Interactive Country Fiches. Retrieved September 22, 2025, from https://www.dicf.unepgrid.ch/somalia

UNFCCC. (2013). National Adaptation Programme of Action (NAPA). Retrieved September 22, 2025, from https://www.unfccc.int/resource/docs/napa/som01.pdf

UNFCCC. (2022). Somalia's First Biennial Update Report. https://www.unfccc.int/sites/default/files/resource/Somalia%20First%20BUR%20report%202022.pdf (accessed September 22, 2025).

UNFPA. (2025). Inter-agency humanitarian evaluation of the response to the humanitarian crisis in Somalia evaluation report Management, Funding, and Implementation of the Evaluation. Retrieved September 23, 2025, from https://www.unfpa.org/sites/default/files/2025-04/IAHE%20of%20the%20Drought%20Response%20in%20Somalia.pdf

UNHCR. (2019). UNHCR warns of growing climate-related displacement in Somalia | UNHCR US. Retrieved September 22, 2025, from https://www.unhcr.org/us/news/briefing-notes/unhcr-warns-growing-climate-related-displacement-somalia

UNHCR. (2023). As the Horn of Africa drought enters a sixth failed rainy season, UNHCR calls for urgent assistance | UNHCR US. Retrieved September 22, 2025, from https://www.unhcr.org/us/news/briefing-notes/horn-africa-drought-enters-sixth-failed-rainy-season-unhcr-calls-urgent

UNICEF. (2018). Measles vaccination campaign reaches the most vulnerable children affected by drought | UNICEF Somalia. Retrieved September 22, 2025, from https://www.unicef.org/somalia/stories/measles-vaccination-campaign-reaches-most-vulnerable-children-affected-drought
UNICEF. (2025). New study estimates that 71,000 drought-related "excess" deaths may have occurred in Somalia between 2022–2024. Retrieved September 22, 2025, from https://www.unicef.org/somalia/press-releases/71000-drought-related-excess-deaths
Warsame, A., Frison, S., & Checchi, F. (2023). Drought, armed conflict and population mortality in Somalia, 2014–2018: A statistical analysis. *PLOS Global Public Health, 3*, Article e0001136. https://doi.org/10.1371/JOURNAL.PGPH.0001136
Weathereing Risk. (2022). Climate Risk Profile Somalia. Retrieved September 25, 2025, from https://www.weatheringrisk.org/sites/default/files/document/Climate_Risk_Profile_Somalia.pdf
WFP. (2018). Somalia, Humanitarian Cash-Based Response (2017): Joint Evaluation | World Food Programme. Retrieved September 25, 2025, from https://www.wfp.org/publications/somalia-humanitarian-cash-based-response-2017-joint-evaluation
WFP. (2018). Summary of evaluation evidence. Retrieved September 25, 2025, from https://www.executiveboard.wfp.org/document_download/WFP-0000069935
WHO. (2017). WHO EMRO—2017 Cholera outbreak. Retrieved September 22, 2025, from https://www.emro.who.int/somalia/information-resources/2017-cholera-outbreak.html
WHO. (2023). Somalia crisis. Retrieved September 22, 2025, from https://www.who.int/emergencies/situations/somalia-crisis
WHO. (2023). One Health National Bridging Workshop Roadmap of the Federal Republic of Somalia. Retrieved September 23, 2025, from https://www.extranet.who.int/sph/sites/default/files/document-library/document/NBW%20Somalia%20-%20Final%20Roadmap.pdf
WHO. (2025). WHO EMRO—New study estimates that 71 000 drought-related "excess" deaths may have occurred in Somalia between 2022 and 2024. Retrieved September 22, 2025, from https://www.emro.who.int/somalia/news/new-study-estimates-that-71-000-drought-related-excess-deaths-may-have-occurred-in-somalia-between-2022-and-2024.html
WHO. (2025). How Who is Driving Impact on the Ground. Retrieved September 23, 2025, from https://www.who.int/about/accountability/results/who-results-report-2020-2021/country-profile/2021/somalia
WMO. (2023). Another poor rainy season forecast for drought hit Horn of Africa. Retrieved September 22, 2025, from https://www.wmo.int/media/news/another-poor-rainy-season-forecast-drought-hit-horn-of-africa
WMO. (2025). EW4All multi-stakeholder accelerator in LDCs and SIDS. Retrieved September 25, 2025, from https://www.wmo.int/activities/projects/project-portfolio/ew4all-multi-stakeholder-accelerator-ldcs-and-sids-0
Wordlbank (2022). Developing a State-led Social Safety Net System to Boost Human Capital and Build Resilience in Somalia: The Baxnaano Program. Retrieved September 25, 2025, from https://www.worldbank.org/en/results/2022/10/10/developing-a-state-led-social-safety-net-system-to-boost-human-capital-and-build-resilience-in-somalia-the-baxnaano-prog
Wordlbank. (2023). Publication: Somalia Climate Risk Review. Retrieved September 22, 2025, from https://www.openknowledge.worldbank.org/entities/publication/8f51dc1a-e342-40a3-8fbe-08e9aa4a2058
Wordlbank. (2023). Somalia Climate Risk ReviewPublic Disclosure Authorized. Retrieved September 25, 2025, from https://documents1.worldbank.org/curated/en/099062923035034613/pdf/P17624603756190c409e570193ea2ae944d.pdf
World bank. (2024). Somalia Overview: Development news, research, data | World Bank. Retrieved September 22, 2025, from https://www.worldbank.org/en/country/somalia/overview
World Bank. (2025). Somalia. Retrieved September 22, 2025, from https://www.thedocs.worldbank.org/en/doc/bae48ff2fefc5a869546775b3f010735-0500062021/related/mpo-som.pdf
World Bank. (2025). Five Facts About Poverty in Somalia. Retrieved September 22, 2025, from https://www.blogs.worldbank.org/en/africacan/five-facts-about-poverty-in-somalia-afe-0425

World Population Review. (2025). Somalia Population 2025. Retrieved September 22, 2025, from https://www.worldpopulationreview.com/countries/somalia
Worldbank. (2017). Remittances and Vulnerability in Somalia Rift Valley Institute. Retrieved September 25, 2025, from https://documents1.worldbank.org/curated/en/633401530870281332/pdf/Remittances-and-Vulnerability-in-Somalia-Resubmission.pdf
Worldbank. (2020). Diagnostic study on trends and threats for environmental and natural resources challenges Somalia Country Environmental Analysis. Retrieved September 22, 2025, from https://documents1.worldbank.org/curated/en/742491594100313982/pdf/Somalia-Country-Environmental-Analysis-Diagnostic-Study-on-Trends-and-Threats-for-Environmental-and-Natural-Resources-Challenges.pdf
Worldbank. (2020). Project Appraisal Document (PAD). Retrieved September 24, 2025, from https://documents1.worldbank.org/curated/en/155201589207732331/pdf/Somalia-Crisis-Recovery-Project.pdf
Worldbank. (2021). Somalia: Groundwater Assessment. Retrieved September 25, 2025, from https://documents1.worldbank.org/curated/en/099615012012129914/pdf/P1749940c70d5d09008a3203ad9eeb97d1a.pdf
Worldbank. (2023). Somalia—Disaster Risk Finance Diagnostic. Retrieved September 23, 2025, from https://documents1.worldbank.org/curated/en/099618001282524663/pdf/IDU15a5837ca1bdbd1447118d951d68f2a5c7867.pdf
Worldbank. (2025). Somalia Contingent Emergency Response Project, P509224. Retrieved September 23, 2025, from https://documents1.worldbank.org/curated/en/099062625115597058/pdf/P509224-a682e00e-a4b3-44d7-9526-e139d57d968c.pdf
Worldbank. (2025). Mogadishu's Flood Dilemma: How a Rehabilitated Pond Became a Model for Urban Resilience. Retrieved September 25, 2025, from https://www.worldbank.org/en/news/feature/2025/06/16/mogadishu-afe-flood-dilemma-how-a-rehabilitated-pond-became-a-model-for-urban-resilience
Xinhuanet. (2018). UN says 53 people killed by tropical cyclone in Somalia—Xinhua | English.news.cn. Retrieved September 22, 2025, from http://www.xinhuanet.com/english/2018-06/15/c_137254349.htm

Chapter 16
Syria–Turkey 2023 Earthquake: Lessons Learned and Way Forward

Shubham Chaudhary, Som Nath, Edris Alam, and Rinku Mahanta

Abstract Earthquakes are highly destructive natural hazards, with their impacts being significantly influenced by tectonic settings, structural vulnerabilities, governance frameworks, and emergency preparedness. The Mw 7.8 and Mw 7.5 earthquake sequence that occurred in Türkiye and Syria on 6 February 2023 constitutes one of the most severe seismic disasters of the twenty-first century, resulting in significant loss of life, extensive infrastructure failure, and enduring humanitarian crises. This chapter provides a multidisciplinary synthesis of seismological, geological, engineering, and humanitarian data sets to analyse rupture dynamics, spatial damage distribution, emergency response performance, and socio-political factors that exacerbated disaster impacts, especially in conflict-affected areas of Syria. Parameters of earthquakes are sourced from reputable global catalogues, such as the USGS and NGDC, whereas statistics on loss and damage are obtained from international disaster databases, including CRED, UNDRR, and the World Bank. The analysis reveals systemic non-compliance with seismic building codes, deficiencies in risk-sensitive urban planning, and significant disparities in response capacity influenced by governance stability and conflict. The 2023 Türkiye–Syria earthquakes highlight the necessity for enhanced enforcement of seismic design regulations, increased investment in local search-and-rescue capabilities, and improved regional cooperation in seismic monitoring, preparedness, and post-disaster recovery efforts.

Keywords Earthquake · Seismic risk · Structural vulnerability · Disaster governance · Conflict-affected regions · Türkiye–Syria

S. Chaudhary (✉)
Academy of Scientific and Innovative Research (AcSIR), Ghaziabad, India
e-mail: shubhamgeo7@gmail.com

S. Nath · R. Mahanta
Indian Institute of Technology (IIT) Roorkee, Roorkee, Uttarakhand, India

E. Alam
Faculty of Resilience, Rabdan Academy, Abu Dhabi, United Arab Emirates

E. Alam (ed.), *Disaster, Crisis and Emergency Management in the Middle East & North Africa*, https://doi.org/10.1007/978-3-032-24436-9_16

16.1 Introduction

Earthquakes are among the most destructive natural hazards, frequently resulting in substantial loss of life, serious injuries, and extensive damage to buildings and critical infrastructure (Dong & Shan, 2013). Their impacts are strongly influenced by population exposure, structural vulnerability, and local geological conditions, making them a continuing focus of research on global disaster risk reduction. The series of earthquakes that struck Türkiye and Syria on 6 February 2023, rank among the most devastating seismic disasters of the twenty-first century, severely affecting a region already burdened by political unrest, unequal development, and ongoing humanitarian crises. The event featured two significant earthquakes occurring within around nine hours: a Mw 7.8 ruptures along the East Anatolian Fault Zone (EAFZ), succeeded by a Mw 7.5 event on the Sürgü–Çardak Fault system. The combined multi-segment strike-slip ruptures caused devastating ground shaking over southern Türkiye and northern Syria, leading to significant structural collapse and huge human and economic losses. The seismic characteristics and rupture dynamics of the 2023 Türkiye–Syria earthquakes have been widely documented using global datasets and remote sensing observations (United States Geological Survey, 2023; NASA Jet Propulsion Laboratory, 2023; Tan & Öcal, 2023; Tokçer et al., 2023).

The earthquakes impacted approximately 15–18 million individuals throughout 11 provinces in Türkiye and many governorates in Syria, resulting in tens of thousands of deaths and injuries, widespread structural failures, and significant interruption to essential infrastructure systems (AFAD, 2023; UNDRR, 2023). Several cities, including Kahramanmaraş, Hatay, Gaziantep, Adıyaman, Aleppo, and Idlib, endured unparalleled devastation. In Syria, the disaster's effects were exacerbated by almost a decade of conflict, which had already compromised infrastructure, diminished state capacity, and restricted humanitarian access. The earthquakes triggered one of the largest and most intricate emergency responses in the recent regional history. The 2023 earthquake sequence revealed significant deficiencies in construction methods, governance frameworks, societal resilience, and emergency preparedness in both nations, in addition to the immediate physical destruction. Failure to adhere to seismic construction norms, unregulated urban development, deteriorating infrastructure, and disjointed administrative frameworks severely exacerbated the extent of losses. The catastrophe underscored the discrepancies in humanitarian response capabilities between Türkiye, which activated substantial national and international aid, and Syria, whose geopolitical obstacles significantly hindered prompt rescue and relief efforts. The 2023 Syria–Türkiye earthquake, because of its magnitude, intricacy, and extensive consequences, serves as a significant case study for enhancing knowledge in seismology, structural engineering, disaster risk mitigation, and transboundary hazard management. The 2023 Syria–Türkiye earthquake sequence, due to its magnitude, multi-segment rupture behaviour, and significant transboundary effects, serves as an important case study for enhancing knowledge in seismology, earthquake engineering, disaster risk governance, and humanitarian response in conflict-affected areas. The event revealed significant deficiencies in seismic code enforcement, preparedness planning, and institutional coordination, particularly in contexts where political instability and armed conflict hindered risk reduction and emergency response efforts. This chapter integrates scientific, engineering, and humanitarian perspectives to analyse earthquake sequences and extract lessons for enhancing seismic risk governance and regional cooperation in a highly tectonically active area.

16.2 Literature Review

Earthquake faults are fracture zones, where the accumulated tectonic stress is released suddenly in the form of seismic energy. Tectonic earthquakes are the result of strike-slip, normal, reverse, or oblique fault systems, depending on the direction of the accumulated stress regime (Scholz, 1989, 2019; Sibson, 1989; Wu & Hu, 2024). Faults at active plate boundaries often rupture in complex and multi-segment patterns that produce intense ground shaking and cascading natural hazard effects. Recent studies have emphasized the complex fault interactions and cascading rupture mechanisms associated with the East Anatolian Fault system (Tan & Öcal, 2023; Tokçer et al., 2023).

The 2023 Türkiye–Syria earthquakes highlighted the complex tectonic behaviour of the region. The sequence began with a moment magnitude (Mw) 7.8 mainshock, followed by an Mw 7.5 event, and involved multiple rupture segments extending up to approximately 300 km along the East Anatolian Fault Zone (EAFZ). The multi-segment, high-energy rupture, amplified by local basin effects, generated intense shaking across southern Türkiye and northern Syria, highlighting the geometry of the fault, tectonic stress, and regional structural complexity together shape the severity of the earthquake impact on the active plate-boundary zones. The recent studies highlight that the region's geological and tectonic configuration, characterized by the interaction of the Anatolian, Arabian, and African plates, results in a significant seismic hazard exacerbated by various vulnerabilities, including prolonged armed conflict, unregulated urban expansion, antiquated infrastructure, and susceptibility to climate-induced secondary hazards such as landslides and flooding (Hussain et al., 2023; IRC, 2023; OCHA, 2023). The interplay of these physical and socio-political elements influenced the intensity of impacts throughout Türkiye and Syria.

From an engineering standpoint, reconnaissance missions executed by the Earthquake Engineering Research Institute (2023) and the Geotechnical Extreme Events Reconnaissance (2023) teams recorded systemic structural deficiencies that significantly exacerbated the magnitude of the disaster. This encompassed non-ductile reinforced-concrete frames, soft-story failures in mid-rise edifices, insufficient confinement reinforcement, and extensive collapse of deteriorating unreinforced masonry structures. Site-specific amplification, particularly in the Amik Basin and certain areas of Hatay Province, significantly contributed to the enhancement of ground shaking, as corroborated by geophysical and geotechnical assessments

(Cheloni et al., 2023). Comparisons with prior strike-slip disasters in the region underscore persistent trends of inadequate construction methods and weak enforcement of building regulations.

Humanitarian and socio-economic research further elucidates the discrepancies in disaster impact and emergency response skills between Türkiye and Syria. In northwest Syria, prolonged fighting has significantly undermined infrastructure, disturbed government frameworks, and restricted access to heavy rescue apparatus and medical services (OCHA, 2023; IRC, 2023). Community-based first responders, particularly the White Helmets, were crucial during the vital opening 48 h, offsetting inadequate institutional capacity (Whittaker et al., 2023). The extensive disaster-science literature has emphasized the cascading public health effects, such as trauma, displacement, and mental health problems, which reflect the repercussions of previous mega disasters in the region.

Notwithstanding thorough studies, major deficiencies persist. This encompasses inadequate integration of seismological, engineering, and socio-economic statistics; insufficient cross-border seismic surveillance; and an absence of coordinated long-term reconstruction strategies in conflict-affected regions. The lack of a cohesive regional strategy for seismic risk governance impedes early warning, preparedness, and swift response. This chapter rectifies these deficiencies by integrating diverse research and presenting a framework for improved regional resilience.

16.3 Objectives

The primary aim of this chapter is to develop a thorough and multidisciplinary analysis of the 2023 Syria–Türkiye earthquake sequence and its extensive repercussions. The chapter specifically aims to analyse the tectonic and anthropogenic factors that led to the catastrophic outcomes of the event, encompassing geological rupture dynamics, structural vulnerabilities, and socio-political conditions. It aims to analyse the spatial patterns of loss and damage, evaluate the efficacy and constraints of emergency response operations in both nations, and consolidate essential insights from engineering, governance, and humanitarian viewpoints. In doing so, it presents evidence-based frameworks for seismic risk reduction, resilient reconstruction, and improved regional cooperation, thus fostering a more coordinated and sustainable strategy for disaster preparedness and mitigation throughout the region.

16.3.1 Methods and Materials

This chapter adopts a multidisciplinary synthesis approach integrating seismological, geological, engineering, geospatial, and humanitarian data sets. Earthquake source parameters, rupture characteristics, and aftershock distributions are compiled from global authoritative catalogues, including the United States Geological Survey

(USGS) and the National Geophysical Data Center (NGDC). Satellite-derived geospatial data sets, including Sentinel-1 InSAR deformation products and UNOSAT building damage assessments, are used to evaluate surface rupture and spatial damage patterns. Post-earthquake engineering reconnaissance reports from EERI, GEER, and ATC are synthesized with humanitarian situation assessments from OCHA, UNICEF, and international non-governmental organizations. Loss and damage statistics are corroborated using international disaster databases, including CRED (EM-DAT) and World Bank post-disaster needs assessments.

16.3.2 Tectonic Faults in the Syria–Turkey Region, 2023

The 2023 Syria–Türkiye earthquake sequence transpired in a highly tectonically active and structurally intricate area of the Eastern Mediterranean, influenced by the movements of the Anatolian, Arabian, and African plates. The principal rupture occurred along the East Anatolian Fault Zone (EAFZ), a significant left-lateral strike-slip system facilitating the westward movement of the Anatolian Plate at a slip rate of roughly 10–12 mm/year (Akyüz & Meghraoui, 2022). The Mw 7.8 mainshock fractured several segments of the EAFZ, notably the Amanos, Pazarcık, and Erkenek sections, resulting in a continuous surface rupture beyond 300 km and causing severe ground shaking throughout southern Türkiye. The intricacy of the rupture was exacerbated by the activation of neighbouring splay faults, particularly the Sürgü–Çardak Fault system, which experienced the Mw 7.5 event that occurred hours later (Duman et al., 2023). The participation of these secondary structures highlights the multi-segmented and cascading characteristics of the earthquake sequence. The tectonic influence extended southward into Syria via the Dead Sea Transform Fault (DSTF), a significant right-lateral strike-slip boundary that connects with the EAFZ and contributes to the regional stress regime impacting seismicity in Syria, Lebanon, and Jordan (Gomez et al., 2021). Other fault systems, including as the Antakya Fault and the Malatya Fault Zone, significantly contributed to the distribution of deformation and the amplification of seismic effects in basin areas. The interconnected fault networks generated a complex rupture pattern, elucidating the extensive and intense ground motions recorded during the 2023 occurrences (Fialko et al., 2023).

16.4 Seismic Sequence and Historical Earthquake Activity

On 6 February 2023, a destructive earthquake sequence initiated with a moment magnitude (Mw) 7.8 mainshock at 01:17 UTC along the East Anatolian Fault Zone, marking the boundary between the Anatolian and Arabian plates (Fig. 16.1). According to USGS and NGDC data, the rupture propagated over approximately 200–300 km, producing peak ground accelerations exceeding 1.0 g in near-fault areas and generating intense shaking lasting more than one minute. Approximately nine hours later, a Mw 7.5 earthquake ruptured the Sürgü–Çardak Fault system, dynamically triggered within the evolving regional stress field and compounding damage across southern Türkiye and northern Syria. Figure 16.2 shows epicentres of the mainshock (Mw 7.8), the major aftershock (Mw 7.5), and subsequent aftershocks are shown as circles scaled by magnitude and coloured by focal depth (km). Major affected cities in southern Türkiye and northern Syria are labelled for reference. The concentration and alignment of seismicity delineate the East Anatolian Fault Zone and associated splay faults responsible for the 2023 earthquake sequence.

This region, once seen as less perilous than the Northern Anatolian Fault, has a historical record of devastating seismic activity demonstrated by prior earthquakes. The 2023 Mw 7.8 incident emitted around 2.5 times the energy of the Mw 7.4 Izmit earthquake of 1999, which resulted in over 17,000 fatalities, and it equalled the magnitude of the 1939 Mw 7.8 Erzincan earthquake, one of Türkiye's most lethal natural hazards, accountable for more than 32,000 deaths. In Syria, significant urban centres like Aleppo have seen recurrent destruction due to important historical occurrences, notably the disastrous earthquake of 1822, which is estimated to have resulted in over 20,000 fatalities. The historical precedents highlight that the 2023 Syria–Türkiye sequence is part of a persistent trend of significant earthquakes linked to the active strike-slip systems in the Eastern Mediterranean region (Table 16.1).

16.5 Analytical Examination of the 2023 Earthquake in Syria and Turkey

16.5.1 Analysis of Loss and Damage

The 2023 Syria–Türkiye earthquake sequence resulted in catastrophic human, structural, and economic losses. According to consolidated estimates from CRED (EM-DAT), UNDRR, and national disaster agencies, the event caused approximately 59,000 fatalities, with 55,000 in Türkiye and more than 4,500 in Syria, along with over 120,000 injuries and the displacement of more than 3 million people. In Türkiye alone, over 230,000 buildings collapsed or were rendered uninhabitable, predominantly reinforced-concrete residential structures that failed due to inadequate seismic detailing and non-compliance with building codes. In northwest Syria, at least 10,600 buildings were destroyed or severely damaged, with losses exacerbated by pre-existing infrastructure degradation linked to prolonged armed conflict. The humanitarian crisis was further highlighted by international agencies reporting widespread displacement, health emergencies, and disruption to essential services (United Nations High Commissioner for Refugees, 2023; UNICEF, 2023; World Health Organization, 2023).

Critical infrastructure, including roads, bridges, airports, hospitals, electrical networks, and communication systems, experienced significant disruption, resulting in extended socio-economic stagnation in the impacted areas. Impaired hospitals

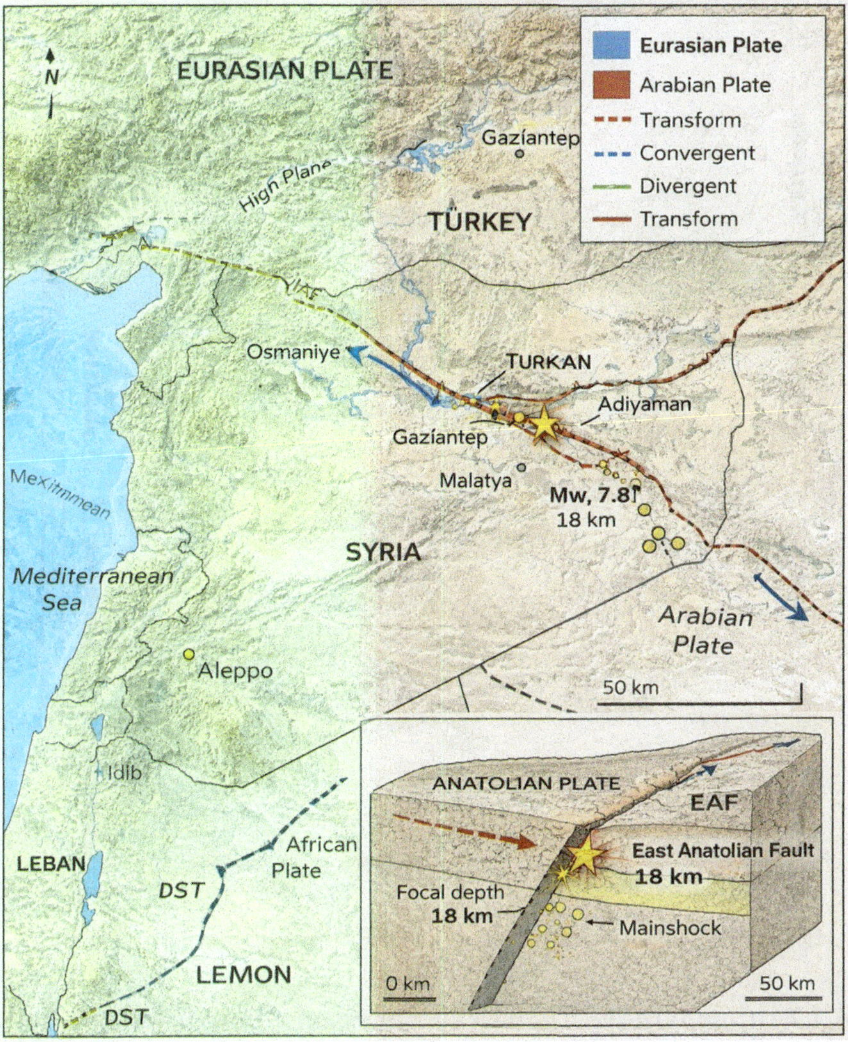

Fig. 16.1 Tectonic framework of the Türkiye–Syria region showing plate boundaries and the 2023 earthquake sequence

and obstructed transit routes severely impeded emergency medical services, while deficiencies in electricity and water infrastructure exacerbated initial humanitarian efforts. The World Bank (2023) calculated that the earthquakes resulted in damages above USD 100 billion in Türkiye and approximately USD 5.1 billion in Syria, rendering this sequence the most expensive disaster in the region's contemporary history (Table 16.2).

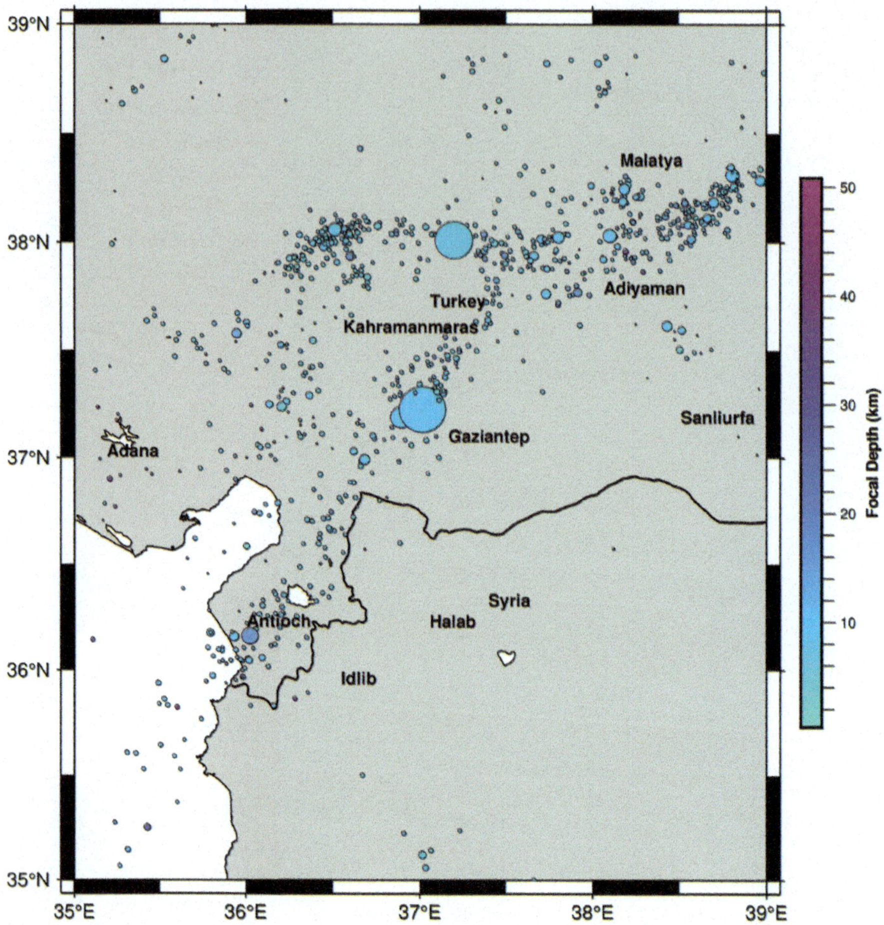

Fig. 16.2 Spatial distribution of the 6 February 2023 Türkiye–Syria earthquake sequence

16.5.2 Comparative Analysis of Emergency Response Time

The emergency response to the 2023 Türkiye–Syria earthquakes demonstrates both progress and persistent global challenges in large-scale disaster response. In Türkiye, national agencies mobilized search-and-rescue operations within hours of the mainshock, consistent with response timelines observed in the recent major earthquakes such as the 2011 Tōhoku (Japan) and 2015 Gorkha (Nepal) events. However, severe winter conditions, widespread infrastructure damage, and the geographic extent of collapse delayed access to several affected areas during the critical first 24–48 h.

In contrast, response operations in Syria were significantly delayed due to ongoing conflict, fragmented governance, and restricted cross-border access. Unlike post-earthquake responses in non-conflict settings, international assistance to northwest

Table 16.1 Catalogue of major historical and contemporary earthquakes affecting the Syria–Türkiye region (AD 115–2023), compiled from scientifically documented events with verified magnitude and impact data (*Sources* USGS, NGDC, historical earthquake catalogues)

Event #	Date/month/year	Earthquake locations			Earthquake parameters			Deaths	Injuries
		Location names	Lat	Long	Magnitude (Mw)	Intensity	Focal depth (Km)		
1	06 Feb 2023	Kahramanmaraş, Türkiye	31.17	37.03	7.8	XI	17	59000	120000
2	06 Feb 2023	Elbistan, Türkiye	38.09	37.20	7.5	X	10		44000
3	17 Aug 1999	İzmit, Türkiye	40.76	29.97	7.4	X–Xi	17	17000	
4	27 Dec1939	Erzincan, Türkiye	39.75	39.49	7.8	XI	20	32700	100000
5	13 Aug s1822	Aleppo, Syria region	36.20	37.16	7.0	X	Shallow	20000	–
6	13/12/115 AD	Antioch (Antakya)	36.20	36.16	7.3	X	Shallow	>250000	–

Table 16.2 Summary of losses and damages caused by the 6 February 2023 Türkiye–Syria earthquake sequence

Category	Türkiye	Syria	Total (Region-wide)	Sources
Human losses				
Deaths	~55,000	~4,500	~59,000	AFAD; UNDRR; CRED
Injuries	>107,000	~12,000	>120,000	AFAD; OCHA; CRED
Displaced population	~2.7 million	~0.5 million	~3.2 million	UNHCR; OCHA; CRED
People requiring humanitarian assistance	~9.1 million	~8.8 million	~18 million	UN OCHA
Housing and building damage				
Collapsed or uninhabitable buildings	>230,000	~10,600	>240,600	AFAD; OCHA
Severely damaged residential buildings	Tens of thousands	Widespread	–	AFAD; UNOSAT; OCHA
Informal/ non-engineered housing affected	Moderate–high	Very high	–	OCHA; IRC
Critical infrastructure damage				
Health facilities affected	Hospitals damaged, partial service disruption	Hospitals were severely damaged, shortages	Widespread	WHO; OCHA
Education facilities affected	Schools and universities were damaged	Schools heavily impacted	Widespread	UNICEF; OCHA
Transport infrastructure	Roads, bridges, and airports were damaged	Roads damaged, access restricted	Region-wide disruption	AFAD; OCHA
Water and sanitation systems	Localized damage	Severe system failures	Widespread	UNDRR; OCHA
Energy and communication networks	Power outages, network damage	Major service disruptions	Widespread	AFAD; UNDRR
Economic losses				
Direct economic losses	>USD 100 billion	~USD 5.1 billion	>USD 105 billion	World Bank

(continued)

Table 16.2 (continued)

Category	Türkiye	Syria	Total (Region-wide)	Sources
Housing sector losses	Major	Significant	–	World Bank
Infrastructure and services losses	Major	Significant	–	World Bank; UNDRR
Emergency response and recovery challenges				
Search-and-rescue constraints	Winter conditions, damaged transport	Conflict, restricted access	Region-wide delays	OCHA; IRC
Access to international assistance	Rapid mobilization	Delayed, limited border crossings	Uneven	UN OCHA
Recovery and reconstruction capacity	Moderate–high	Severely constrained	Uneven	World Bank; UNDP

Syria was initially limited to a single border crossing, delaying heavy search-and-rescue deployment and medical relief. Comparative analysis with earthquakes in Haiti (2010) and Nepal (2015) indicates that while international teams play an essential supporting role, effective life-saving response remains heavily dependent on local capacity an advantage severely constrained in conflict-affected regions.

16.5.3 Analysis of Response, Rescue, and Relief

The emergency response operations in Türkiye and Syria varied significantly due to differences in government competence, political environment, and international accessibility. In Türkiye, national agencies specifically AFAD and the Turkish Red Crescent, swiftly mobilized, aided by the military forces and volunteers. International collaborators reacted swiftly, with over 80 nations deploying search-and-rescue (USAR) teams, field hospitals, and logistical assistance (UNOCHA, 2023). However, extreme winter conditions, obstructed roadways, failing bridges, and the broad geographic scope of devastation caused considerable delays in initial rescue efforts. The initial 48 h crucial for rescuing trapped survivors were characterized by logistical constraints and disproportionate resource distribution between urban and rural regions. Nevertheless, many life-saving rescues occurred in the first 24–72 h by local actors, consistent with historical evidence that most live rescues are local. By contrast, comparable magnitude events (e.g. 2010 Haiti, 2015 Nepal) show similar patterns: international teams often arrive after the critical early window, supplementing but rarely substituting for local first responders. The 2023 response therefore matched the global pattern, a rapid national and local response followed by progressively larger international support (UNOCHA, 2023).

In Syria, emergency response was significantly hindered by fragmented territorial control, limited cross-border access, and persistent shortages of medical supplies and heavy rescue equipment. In the opposition-controlled northwest Syria, local civil defence organizations, like as the White Helmets, were instrumental in saving civilians despite constrained resources and external assistance (IRC, 2023). International help was initially sent through the Bab al-Hawa crossing, resulting in a delay in the delivery of critical relief supplies. Political intricacies and bureaucratic obstacles impeded collaboration between foreign organizations and local governments, exacerbating the humanitarian catastrophe and diminishing the efficacy of the overall response. International response and disaster data coordination were supported by global frameworks and databases (Centre for Research on the Epidemiology of Disasters, 2023; Applied Technology Council, 2023).

Principal search-and-rescue and relief challenges. (1) sheer scale of building collapse combined with construction non-compliance produced very high casualty loads; (2) severe winter conditions and continuing aftershocks hindered SAR operations and access; (3) logistics—damaged transport infrastructure and port/air capacity—slowed relief flows; (4) in Syria, conflict-related access, insecurity and fragmented governance impeded both SAR and relief; (5) coordination overload from many international actors stressed information management Rapid sectorization and INSARAG coordination improved interoperability among international teams; civil-society actors and well-organized national assets (AFAD, Turkish Red Crescent) provided large-scale logistics, shelter, and cash assistance; international clustering of medical, WASH and shelter resources was effective in the transition from life-saving to early recovery (INSARAG, 2023).

In spite of these obstacles, community self-organization, grassroots volunteerism, and private-sector donations were vital in both nations, underscoring the significance of local capability and social solidarity during extensive disasters. In sum, the Türkiye–Syria 2023 response exemplifies both improvements in international coordination and persistent limits: local capacity and timely national action were decisive in early rescues, while political access and structural vulnerabilities determined uneven outcomes across the affected region.

16.6 Insights Gained

The 2023 earthquake series provides significant insights for engineering, disaster management, and humanitarian efforts. The catastrophe underscored the critical necessity for stringent enforcement of seismic building regulations, enhanced construction supervision, and the prioritization of upgrading at-risk structures. Non-compliant edifices constituted an excessive proportion of structural failures, underscoring persistent challenges in urban planning and regulatory enforcement (EERI, 2023; GEER, 2023). The catastrophe underscored the necessity of including micro zonation, site-specific hazard evaluation, and risk-sensitive land-use planning, particularly in soft-soil basins like Hatay's Amik area, where amplification effects exacerbated structural damage.

The earthquakes exposed deficiencies in centralized response systems and highlighted the need for decentralized contingency planning, interoperable communication networks, and pre-positioned rescue resources in disaster governance. Enhancing early warning distribution and guaranteeing the continuity of operations for critical services, such as hospitals, emergency command centres, and transportation networks, are paramount goals.

The incident highlighted the necessity of safeguarding vulnerable populations, such as the elderly, refugees, children, and individuals displaced by conflict. It underscored the necessity for psychological first aid, ongoing mental health assistance, and community-oriented response mechanisms. The efficacy of local volunteers in Türkiye and Syria underscores the importance of utilizing local expertise, including civil society, and fostering business-community collaborations in crisis logistics.

Rapid, largely national search-and-rescue (SAR) activity led by Türkiye's AFAD and national actors combined with an unprecedented international INSARAG mobilization, demonstrated that well-prepared national systems and pre-existing international coordination frameworks materially increase the reach and speed of life-saving action (INSARAG, 2024). However, the event reaffirmed a now familiar pattern: most live rescues occur within the first 24–72 h and international teams, while essential, are rarely able to substitute for robust local capacity in that critical window (INSARAG, 2024; Manirambona et al., 2023).

Conflict-affected areas in northwest Syria suffered markedly different outcomes. Pre-existing access constraints, damaged infrastructure, and security risks delayed heavy-equipment SAR and timely relief deliveries, producing spatial inequities in coverage and outcomes. Negotiated opening of additional crossings eased but did not eliminate those challenges (UNOCHA, 2023). Structural vulnerability, widespread non-compliance with seismic codes and extensive building collapse were the proximate drivers of high casualty and displacement burdens and thus remain a primary prevention priority (Manirambona et al., 2023).

Key lessons for the way forward are threefold. First, invest systematically in local SAR capacity, rapid medical surge, and community preparedness because early response saves lives. Second, strengthen building-safety enforcement, retrofit programmes, and urban resilience to reduce exposure and collapse risk. Third, institutionalize contingency mechanisms for humanitarian access in conflict contexts prearranged cross-border modalities, negotiated humanitarian corridors, and context-specific logistics plans to reduce inequities in relief delivery. Finally, sustain interoperable information systems and cluster coordination to manage complexity during massive multi-actor responses (INSARAG, 2024; UNOCHA, 2023).

16.7 Necessity for Regional Collaboration

The 2023 earthquake sequence underscores the imperative for strong regional collaboration in mitigating seismic risks, given the common geological environment of the Eastern Mediterranean. Earthquakes that traverse national borders, as demonstrated by the Türkiye–Syria incident, necessitate cohesive systems for collaborative seismic monitoring, instantaneous data interchange, and synchronized early warning networks (Gomez et al., 2021). Implementing standardized building rules, engineering design criteria, and robust rebuilding protocols throughout Türkiye, Syria, Lebanon, Jordan, and adjacent nations would mitigate structural safety discrepancies and enhance long-term regional resilience.

Cross-border collaboration must encompass mutual aid protocols, pre-established humanitarian access corridors, and coordinated emergency response exercises to guarantee rapid mobilization in future incidents. Collaborative seismic risk mapping, joint scientific research, and the establishment of regional hazard databases can enhance preparedness, while the aggregation of financial and technical resources can facilitate reconstruction in resource-limited regions.

Notwithstanding ongoing political obstacles, the 2023 earthquake illustrated that scientific collaboration, humanitarian coordination, and data-sharing methods are both viable and indispensable. Enhancing these regional alliances is essential for mitigating potential disaster repercussions in one of the globe's most seismically active and geopolitically intricate areas.

16.8 Conclusions

The 2023 Syria–Türkiye earthquake sequence is a significant occurrence in the seismic history of the Eastern Mediterranean, exposing profound structural, social, and institutional weaknesses in one of the world's most tectonically active and politically intricate regions. The extent of devastation highlighted persistent inadequacies in seismic-resistant construction methods, risk-aware urban design, and the preparedness of disaster management systems in both Türkiye and Syria. The catastrophe illustrated the interconnected nature of hazards in conflict-affected and resource-limited settings, where deteriorated infrastructure, constricted governance, and unstable social systems exacerbate the human and economic impact of significant earthquakes.

An assessment of the event reveals that mitigating future seismic risk necessitates a comprehensive strategy encompassing scientific study, engineering resilience, government reforms, and cross-border collaboration. Enhancing enforcement of seismic building codes, broadening retrofitting initiatives for at-risk structures, and guaranteeing the operational continuity of critical infrastructure during severe earthquakes are crucial measures to reduce structural losses. Concurrent investments in early warning systems, real-time monitoring networks, and extensive micro zonation studies will improve situational awareness and facilitate proactive decision-making.

The earthquake underscored the significance of inclusive and multi-sectoral disaster governance, in addition to physical mitigation efforts. Effective preparedness must include community-based reaction training, interoperable communication networks, and ongoing psychosocial support services, especially in areas with significant displacement and trauma. The catastrophe underscored the necessity of coordinated regional frameworks, as seismic risks frequently surpass national boundaries. Improved cross-border collaboration in seismic monitoring, emergency response, data sharing, and reconstruction planning can substantially enhance collective resilience throughout the Middle East. It is also imperative to use long-term seismic and hazard data for earthquake risk reduction because they provide the empirical foundation necessary to assess hazard frequencies, magnitudes, recurrence intervals, and long-term exposure trends parameters that short-term observations alone cannot yield (Alam, 2019). Moreover, long-term data supports accurate vulnerability and exposure modelling and permits informed land-use planning, retrofitting decisions, and enforcement of building codes essential for reducing future losses.

The 2023 Syria–Türkiye earthquake sequence exposes critical shortcomings in seismic preparedness, structural resilience, and disaster governance across the Eastern Mediterranean. Strengthening enforcement of building codes, expanding retrofitting programmes, investing in early warning and monitoring systems, and institutionalizing cross-border cooperation are essential steps towards reducing future risk. The disaster further demonstrates that conflict and political instability significantly magnify earthquake impacts by weakening preparedness and delaying response. Translating the lessons of this event into sustained policy reform and regional collaboration is imperative for safeguarding vulnerable populations in one of the world's most seismically active regions.

References

Akyüz, H. S., & Meghraoui, M. (2022). Paleoseismic investigations along the East Anatolian Fault Zone. *Seismological Research Letters, 93*(3), 1550–1565. https://doi.org/10.1785/0220210291

AFAD (Disaster and Emergency Management Authority). (2023). *2023 Kahramanmaraş earthquakes situation report*.

Alam, E. (2019). Importance of long-term earthquake, tsunami and tropical cyclone data for disaster risk reduction in Bangladesh. *Progress in Disaster Science, 2*, Article 100019. https://doi.org/10.1016/j.pdisas.2019.100019

Applied Technology Council (ATC). (2023). *ATC reconnaissance report: Türkiye–Syria earthquake*. ATC.

Centre for Research on the Epidemiology of Disasters (CRED). (2023). *EM-DAT: The international disaster database*. Université catholique de Louvain. https://www.emdat.be

Cheloni, D., Atzori, S., Antonioli, A., et al. (2023). Ground deformation and fault slip distribution from the 2023 Türkiye-Syria earthquake sequence. *Geophysical Journal International, 234*(2), 1222–1239. https://doi.org/10.1093/gji/ggad236

Dong, L., & Shan, J. (2013). A comprehensive review of earthquake-induced building damage detection with remote sensing techniques. *ISPRS Journal of Photogrammetry and Remote Sensing, 84*, 85–99. https://doi.org/10.1016/j.isprsjprs.2013.06.011

Duman, T. Y., Çan, T., Emre, Ö., et al. (2023). Surface rupture characteristics of the Mw 7.5 Sürgü-Çardak earthquake (6 February 2023), Türkiye. *Tectonophysics, 845*, 229–244. https://doi.org/10.1016/j.tecto.2023.229244

Earthquake Engineering Research Institute (EERI). (2023). *Preliminary reconnaissance report: Türkiye-Syria earthquake sequence.*

Fialko, Y., Sandwell, D., Mellors, R., et al. (2023). Insights into the rupture dynamics of the 2023 Türkiye earthquake from remote sensing and field data. *Nature Communications, 14*, 4550. https://doi.org/10.1038/s41467-023-40105-6

Geotechnical Extreme Events Reconnaissance Association (GEER). (2023). *Geotechnical reconnaissance: Türkiye–Syria earthquake.* GEER Association Report.

Gomez, F., Beauchamp, W., & McClusky, S. (2021). Seismicity and tectonics of the Dead Sea Transform. *Journal of Geophysical Research: Solid Earth, 126*(4), e2020JB020769. https://doi.org/10.1029/2020JB020769

Hussain, E., Kalaycıoğlu, S., Milliner, C. W. D., & Çakir, Z. (2023). Preconditioning the 2023 Kahramanmaraş (Türkiye) earthquake disaster. *Nature Reviews Earth & Environment, 4*(5), 287–289. https://doi.org/10.1038/s43017-023-00411-2

International Rescue Committee (IRC). (2023). *Humanitarian impacts of the Türkiye–Syria earthquake.* IRC Report.

INSARAG. (2023). *INSARAG after action review: 2023 Türkiye and Syria earthquakes.* United Nations INSARAG Secretariat.

INSARAG. (2024). *INSARAG after action review: 2023 Türkiye and Syria earthquakes.* United Nations. https://insarag.org

Manirambona, E., Obnial, J. C., Musa, S. S., Miranda, A. V., Haruna, U. A., & Shomuyiwa, D. O. (2023). The need for health system strengthening in the wake of natural disasters: Lessons from the 2023 Türkiye-Syria earthquake. *Public Health Challenges, 2*(4), Article e131. https://doi.org/10.1002/puh2.131

NASA Jet Propulsion Laboratory (NASA-JPL). (2023). *InSAR damage assessment for the Türkiye–Syria earthquakes.*

Office for the Coordination of Humanitarian Affairs (OCHA). (2023). *Syria earthquake flash appeal.* United Nations.

Scholz, C. H. (1989). Mechanics of faulting. *Annual Review of Earth and Planetary Sciences, 17*, 309–334. https://doi.org/10.1146/annurev.ea.17.050189.001521

Scholz, C. H. (2019). *The mechanics of earthquakes and faulting* (3rd ed.). Cambridge University Press.

Sibson, R. H. (1989). Earthquake faulting as a structural process. *Journal of Structural Geology, 11*(1–2), 1–14. https://doi.org/10.1016/0191-8141(89)90032-1

Tan, O., & Öcal, N. (2023). Seismotectonic analysis of the Kahramanmaraş earthquake sequence. *Bulletin of the Seismological Society of America, 113*(4), 1890–1908. https://doi.org/10.1785/0120230021

Tokçer, M., Aydin, A., Çetin, K. Ö., et al. (2023). Field observations from the February 2023 earthquakes in Türkiye. *Engineering Geology, 320*, Article 107178. https://doi.org/10.1016/j.enggeo.2023.107178

UNHCR (United Nations High Commissioner for Refugees). (2023). *Türkiye–Syria earthquake emergency response update.* UNHCR.

UNICEF. (2023). *Education under emergency: Türkiye–Syria earthquake situation report.* United Nations Children's Fund.

UNDRR (United Nations Office for Disaster Risk Reduction). (2023). *Global assessment report: Türkiye–Syria earthquake special section.* Geneva.

United Nations Office for the Coordination of Humanitarian Affairs (OCHA). (2023). *Türkiye—2023 earthquakes situation reports.* United Nations. https://www.unocha.org

United States Geological Survey (USGS). (2023). *Mw 7.8 Türkiye earthquake summary*. USGS Earthquake Hazards Program.

Whittaker, J., McLennan, B., & Handmer, J. (2023). Community responses during the Türkiye earthquake disaster. *International Journal of Disaster Risk Reduction, 93*, Article 103804. https://doi.org/10.1016/j.ijdrr.2023.103804

World Bank. (2023). *Rapid damage and needs assessment: Türkiye–Syria earthquake*. World Bank Group.

World Health Organization (WHO). (2023). *Türkiye–Syria earthquake health situation report*. WHO.

Wu, Z., & Hu, M. (2024). Definitions and classification schemes for active faults and their applications. *Geosciences, 14*(3), 68. https://doi.org/10.3390/geosciences14030068

Chapter 17
A Case Study of the June 2024 Extreme Heat Disaster in Saudi Arabia: Influence of Atmospheric Shifts on Regional Heatwaves

Farahnaz Fazel-Rastgar, Sakineh Khansalari, Venkataraman Sivakumar, Masoud Rostami, and Bijan Fallah

Abstract This study investigates the atmospheric conditions that led to the extreme heatwave during the June 2024 Hajj period, focusing on the role of atmospheric shifts. The study highlighted a notable warming trend over the past three decades, with June 2024 temperatures significantly exceeding historical averages. From June 14–19, 2024, rapid changes in the low and upper atmosphere, including abnormal deeper and drier Red Sea trough (RST), anticyclonic Rossby wave breaking (AWB), and the weakening of the subtropical jet stream, contributed to intense surface heat in the region. Analysis showed a significant decrease in wind speed and a shift in the jet stream's position, leading to enhanced subsidence and a rise in surface temperatures. Synoptic analysis revealed the intensification of heat-low pressure centres and the deepening of the RST, resulting in dry, hot conditions across the region, particularly in Makkah. These changes and reduced moisture influx from the weakening Somali jet intensified the heatwave. The findings suggest that such extreme heat events are becoming more frequent and severe, stressing the need for adaptive measures,

F. Fazel-Rastgar (✉)
School of Science, University of KwaZulu-Natal, Scottsville 3209, South Africa
e-mail: farah_rast@yahoo.com

F. Fazel-Rastgar · V. Sivakumar
S.V. Raman Researchers Roadmap, Westville, Durban, South Africa

S. Khansalari
Atmospheric Science and Meteorological Research Center, Tehran, Iran

M. Rostami
Potsdam Institute for Climate Impact Research (PIK), Potsdam, Germany

Laboratoire de Météorologie Dynamique (LMD), IPSL, ENS-PSL Université, Ecole Polytechnique-Institut Polytechnique de Paris, Sorbonne Université, CNRS, Paris, France

State Key Laboratory of Deep Oiland Gas, China University of Petroleum (East China), Qingdao, China

B. Fallah
German Climate Computing Center (DKRZ), Hamburg, Germany

E. Alam (ed.), *Disaster, Crisis and Emergency Management in the Middle East & North Africa*, https://doi.org/10.1007/978-3-032-24436-9_17

including changes in pilgrimage schedules, enhanced cooling systems, and increased water availability to mitigate heat risks during the summer, especially during the Hajj pilgrimage.

Keywords Extreme heat · Dry Red Sea trough · Anticyclonic Rossby wave breaking (AWB) · Somali jet stream · Subtropical jet stream · Saudi Arabia

17.1 Introduction

Extreme weather events such as heatwaves, droughts, floods, and compound events are expected to increase due to climate change (Arias et al., 2023; Fazel-Rastgar & Sivakumar, 2024; Jaagus et al., 2024). They pose serious health risks, which can significantly raise mortality rates around the world, particularly among vulnerable populations (Caruso et al., 2024; McDermott-Levy et al., 2021; Opoku et al., 2021; Ranta et al., 2024; Yang et al., 2021). These events also lead to widespread socio-economic and ecological impacts. Recent studies show that temperatures in the Middle East are rising nearly twice as fast as the global average and other populated regions, which accelerate warming, heightening the risk of extreme weather events, and presenting significant health, socio-economic, and ecological challenges (Neira et al., 2023; Zittis et al., 2022). Extreme weather in the Middle East, Eastern Mediterranean, and North Africa is linked to regional synoptic systems and their interaction with the large-scale atmospheric flow. The summertime warming extends the thermal low at the surface from South Asia across the Middle East over North Africa, as the hot desert climate intensifies and becomes more extreme. Observations and model calculations of the recent past consistently indicate increasing heat extremes, which are projected to accelerate in future. The number of warm days and nights may increase sharply. On average in the Middle East and North Africa, the maximum temperature during the hottest days in the recent past was about 43 °C, which could increase to about 46 °C by the middle of the century and reach 56 °C by the end of the century, the latter according to the RCP8.5 (business-as-usual) scenario (Lelieveld et al., 2016; Zittis et al., 2021). This will have important consequences for human health and society. The region has faced multiple extreme heatwaves in recent years. The frequency, duration, and intensity of these events are expected to rise in the coming decades (Hochman et al., 2021; Tanarhte et al., 2015; Zittis et al., 2022).

The Red Sea trough (RST) is fundamentally an extension of the Sudan Monsoon Low a vital element of the equatorial low-pressure system (Awad & Almazroui, 2016; De Vries et al., 2013; El-Fandy, 1948; Krichak et al., 1997; Solot, 1950). The RST is most common during the transitional seasons, especially in autumn when atmospheric conditions support its formation and northward movement (Awad & Mashat, 2018). The case studies show that RSTs extend northwards when a high geopotential gradient forms over the Arabian Peninsula, and they align from west to east when the Azores High shifts eastwards and the Siberian High contracts or moves north (Awad & Mashat, 2018). In summer, the Persian Trough is more dominant, whereas in winter, reduced land heating limits its occurrence (De Vries et al., 2013; Kadhum et al., 2022; Mohammadi et al., 2021; Tsvieli & Zangvil, 2005; Ziv et al., 2022). The Persian Trough has been formed over the Middle East and the Persian Gulf which has been extended to eastern Türkiye. It is, specifically formed during the summer months, driven by the contrast in temperatures between the hot desert air and cooler air over the surrounding seas.

In spring, the orientation of the Red Sea trough is influenced by the Siberian and Azores highs, with the RST extending westwards or eastwards based on their positions, or directly northwards when both systems withdraw (Baseer et al., 2019). In spring, the RST is characterised by varying northward extensions, classified as short, moderate, or long. Its occurrence is influenced by atmospheric conditions, with stronger RSTs associated with dust transport into the Arabian Peninsula, especially during warmer springs when the trough extends further north (Mashat et al., 2021).

The weather linked to the RST is typically good-natured, but can also bring about various weather events, some of which can be severe, including intense heatwaves, rainstorms, thunderstorms, flooding, intense winds, haze, and fog. For example, the wet RST are linked to a prominent Trough at 500 hPa that extends from the Black Sea towards northern Egypt. So, in contrast, when the trough's horizontal axis tilts towards the Mediterranean in tandem with an upper-level mid-latitude Trough, extending over the Mediterranean, conditions favour the development of thunderstorms and intense rainfall as a configuration known as the "Wet" Red Sea trough (De Vries et al., 2013; Hochman et al., 2023; Krichak et al., 2012; Tsvieli & Zangvil, 2005; Ziv et al., 2022) that can result in flash floods (Abdalla et al., 2014; Baseer et al., 2019; De Vries et al., 2013; Fazel-Rastgar et al., 2025; Hochman et al., 2023). But under conditions of dominant easterly airflow over the northern side of the Red Sea, a trough typically forms on the leeward side of the surrounding mountains (Krichak et al., 1997) which causes the deepening of depressions and Troughs. This is generally linked with warm, dry conditions, resulting from near-surface easterly to southeasterly flows from the desert. This activity is often referred to as the dry RST (Tsvieli & Zangvil, 2005) and is commonly associated with events such as forest fires (Kutiel & Kutiel, 1991; Kutiel, 2012), air pollution episodes (Cusack et al., 2020; Dayan et al., 2017), and dust storms (Bodenheimer et al., 2019; Gasch et al., 2017; Mashat et al., 2021). Dry Red Sea troughs (DRST) are associated with the subtropical high that is present in the mid troposphere, and the axis of the RST lies to the west of Israel. This positioning leads to southeasterly winds at ground level, resulting in hot and dry weather conditions, which occasionally trigger heatwaves (Tsvieli & Zangvil, 2005). On an average, the CMIP5 models project increase in the meridional pressure gradient for both Wet-RST (WRST) and DRST days. This implies that both modes of the Red Sea trough may be associated with stronger horizontal winds in future decades. The WRST occurrence and persistence are projected to significantly decline, while these properties may substantially increase for DRST conditions. We specifically provide evidence that the decline in the frequency of the WRST may primarily affect the transition seasons, while the increase in the DRST frequency takes place across the wet season and an extended autumn season dominated by DRST conditions (Hochman et al., 2021). Besides, climate change can shift large-scale atmospheric circulation features, like the subtropical highs and Hadley cells. The poleward expansion of the Hadley cells affects surface-level climate, influencing precipitation patterns and regional dryness (Amaya et al., 2017; Grise et al., 2019; Schmidt & Grise, 2019; Wei et al., 2018).

Between 14 and 19 June 2024, at least 1,301 people in the Hajj pilgrimage in Makkah died due to extreme heat, with temperatures exceeding 50 °C (https://english.alarabiya.net/News/saudi-arabia/2024/06/23/saudi-health-minister-says-1-301-died-during-hajj; https://www.theguardian.com/world/article/2024/jun/18/hundreds-of-hajj-pilgrims-die-in-Makkah-from-heat-related-illness). Extreme heat caused heat strokes and dehydration, leading to the deaths. The hottest recorded temperature was reported in the Makkah's Grand Mosque and was 51.8 °C (https://www.khaleejtimes.com/world/saudi-temperatures-hit-51-8c-at-makkahs-grand-mosque-as-haj-ends). The main objective of this chapter is to examine the weather conditions associated with a record-breaking heat extreme in Makkah in June 2024.

17.2 Materials and Methods

The data used in this study were obtained from a range of high-resolution reanalysis models and observational datasets. Specifically, we utilised the Modern-Era Retrospective Analysis for Research and Applications (MERRA) dataset from NASA (Rienecker et al., 2011). We also examined ERA5 hourly data (Hersbach et al., 2020) provided by the ECMWF, which included skin temperatures during the study period. The ERA5 reanalysis dataset exhibits a spatial resolution of roughly 0.25° on a latitude–longitude grid point. In addition, we incorporated meteorological datasets from the National Centres for Environmental Prediction-National Centres for Atmospheric Research (NCEP-NCAR) reanalysis, which offers various atmospheric parameters. The data were analysed using 6-hourly mean sea level pressure records from NCEP-NCAR, available on a 2.5° × 2.5° latitude–longitude grid (Kalnay et al., 1996). Both mean and anomaly composites were created, with anomalies representing deviations from the climatological normal (1991–2020), which serves as the baseline reference period for evaluating long-term climate change.

Study Area

Figure 17.1 shows a geographical map of the Kingdom of Saudi Arabia (KSA), highlighting Makkah and its borders. Makkah's geographical location, situated in a valley surrounded by mountains can exacerbate heat. This topography can trap air, preventing cooler winds from reaching the city and intensifying the heat. Additionally, the region's proximity to the Red Sea can increase the likelihood of high humidity levels, which makes the heat feel even more extreme. This combination of high temperatures and humidity would increase the risk of heat-related health problems. Additionally, wind patterns in the region could also play a role in amplifying

Fig. 17.1 Geographical map of the Kingdom of Saudi Arabia (KSA), highlighting Makkah (21.43°N, 39,83°E) and the borders

the heat. Winds from the desert, such as the Khamsin winds, could bring in hot, dry air from the interior of the Arabian Peninsula. These winds are known for their ability to raise temperatures significantly and may affect the outdoor activities.

17.3 Results and Discussion

17.3.1 Temporal Variation of Temperature

Figure 17.2a shows the average heatwave frequency (count of days satisfying the heatwave criteria of at least three consecutive days above the 90th percentile) monthly $0.5 \times 0.625°$. Data were obtained from MERRA-2 Reanalysis M2SMNXEDI v2 and counted for June from 1980 to 2024 for Saudi Arabia. This figure clearly shows a considerable number of 15-day heat waves for June 2024 over the study area. Figure 17.2b shows an area distribution for the average heatwave frequency over the study area during June 2024. This figure shows average heatwave frequency changing from 4 to 28 days, with a maximum number mostly over the western main area. Figure 17.3 presents the time series of mean daily 2-m maximum (a) and minimum (b) air temperatures over the past three decades, from 1994 to 2024, obtained from the MERRA-2 database for Saudi Arabia. For each year, the data have been averaged for June. This figure illustrates a warming trend for maximum and minimum 2-m air temperatures, with a notably steeper positive slope for the minimum temperatures. The time-series analysis reveals recent peak values of approximately 44 °C for the maximum and ~27 °C for the minimum 2-m air temperatures in June 2024. Figure 17.4 displays the composite mean daily 2-m air temperature for the period from 14 to 19 June 2024 (a) and the long-term climatological mean for the study period (b) over Saudi Arabia. A comparison of these two figures clearly shows a warmer 2 m air temperature associated with a greater temperature gradient in the western boundaries of Saudi Arabia, including the Makkah western region. Additionally, Fig. 17.4c shows the time series of area-averaged long-term anomalies, representing departures from 1991 to 2020 for the mean daily 2-m air temperatures in June over the last three decades. This figure indicates a significant warming shift of approximately 2.3 °C in June 2024 over the study area. The ERA5 skin surface temperature data at 1200Z reveal that much of Saudi Arabia is experiencing high surface skin temperatures with values reaching about 50 °C. When these temperatures are compared to the long-term (1991–2020) climatology, the data indicate that some regions are about 2–2.5 °C warmer than the long-term average. In the central parts of Saudi Arabia, the observed temperatures during the study period are slightly lower than the long-term average. Within Makkah province, a clear spatial contrast is evident. The northern section shows skin temperatures 2 °C above the long-term average, whereas certain areas in the southern parts and northeastern corner are about 0.5–2 °C cooler.

17.3.2 Synoptic and Dynamic Analysis

In the study area, the synoptic and dynamic characteristics of the weather system associated with the extreme heat from 14 to 19 June, 2024, were investigated. Figure 17.5 presents the average daily composite mean of mean sea level (MSL) pressure (a), the climatological mean map for 1991–2020 (b), and the corresponding anomalies (c) during the study period. Figure 17.5a reveals two intensified extensions of heat-low pressure centres over most parts of the Middle East: one centres at 999 h Pa in the southeast of the Arabian Peninsula and another at 1003 h Pa in the northeast of Sudan (Sudan Low) and the middle of the Red Sea. This pattern is associated with two notable northwesterly tilted Troughs: the Persian Trough extending from the south of the Persian Gulf to the eastern Mediterranean Sea, and the narrow, dry Red Sea trough extending from northeastern Sudan to the northeastern Red Sea. A comparison between the composite mean (Fig. 17.5a) and the long-term climatological mean (Fig. 17.5b) indicates a weakening of the Persian Trough, with a maximum reduction of around 2.5 h Pa over eastern Turkey, and a deepening of the Red Sea trough, with a maximum increase of approximately 2 h Pa predominantly over the study region during the study period (see Fig. 17.5c). This suggests a more intense heat, low centred over the study area. It is noted that the intensification of the dry Red Sea trough has resulted in intense surface heating over the Arabian Peninsula and the Sudanese lowlands. This intensification has enhanced the pressure gradient between

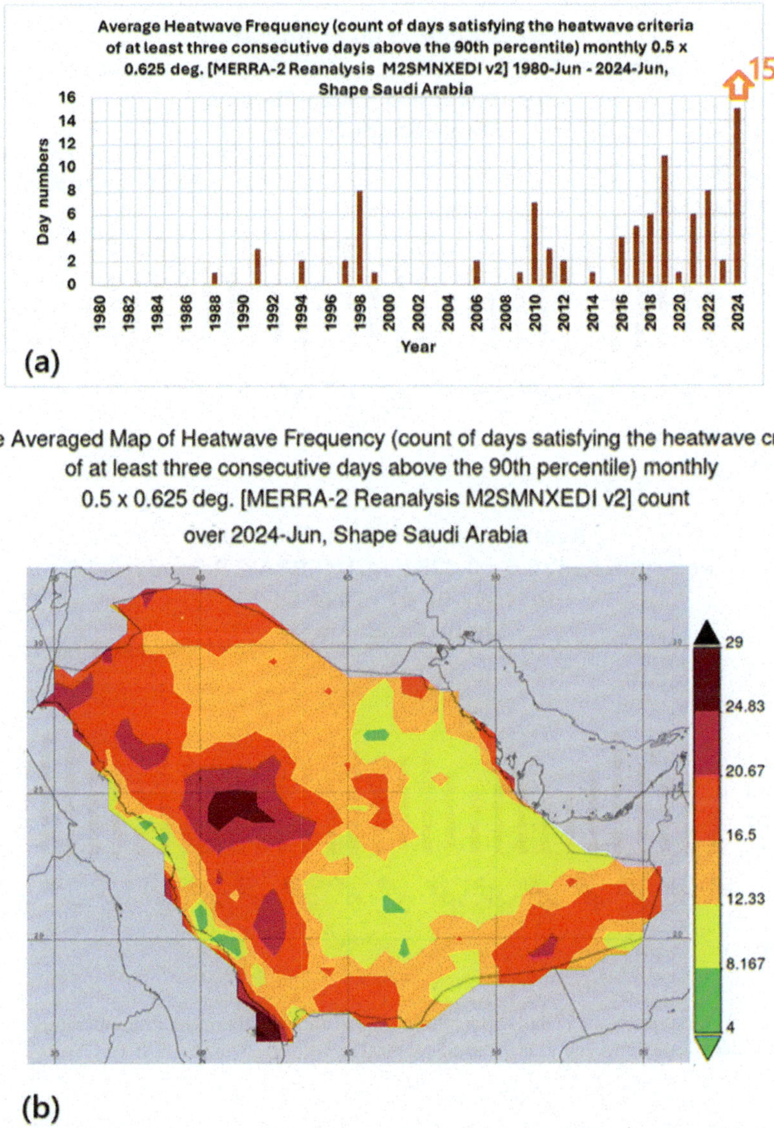

Fig. 17.2 Average heatwave frequency (count of days satisfying the heatwave criteria of at least three consecutive days above the 90th percentile) monthly 0.5 × 0.625°. [MERRA-2 Reanalysis M2SMNXEDI v2] count for Jun 1980–Jun 2024, Shape Saudi Arabia

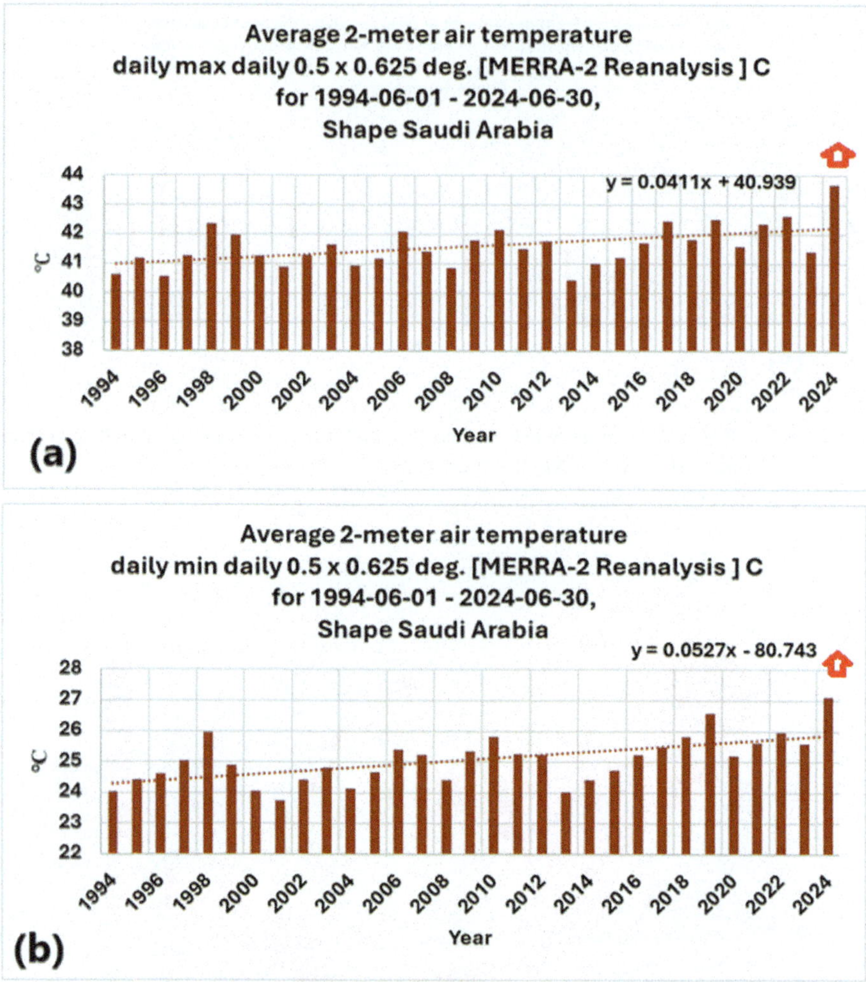

Fig. 17.3 Time series of mean daily 2-m maximum (**a**) and minimum (**b**) air temperatures over the past three decades, from 1994 to 2024 in Saudi Arabia

the hot land and the relatively cooler Red Sea (see isobaric lines in Fig. 17.5a and c) caused by a temperature gradient (see the western boundaries in Fig. 17.5a), leading to extremely drier (see Fig. 17.7b), and rather hotter conditions (see Fig. 17.3a) over the study area during the study period. Figure 17.6 illustrates the average daily composite mean (a) of 500 h Pa geopotential heights for 14–19 June, 2024, the climatological mean (b), and the anomalies (c) relative to the 1991–2020 climate normal for these specified days. Figure 17.6a depicts the formation of the subtropical ridge over the Arabian Peninsula, centred above the Red Sea during this period. A comparison between the composite means (Fig. 17.6a) and the climatological mean (Fig. 17.6b) reveals an intensification, with a maximum increase of 80 geopotential metres over

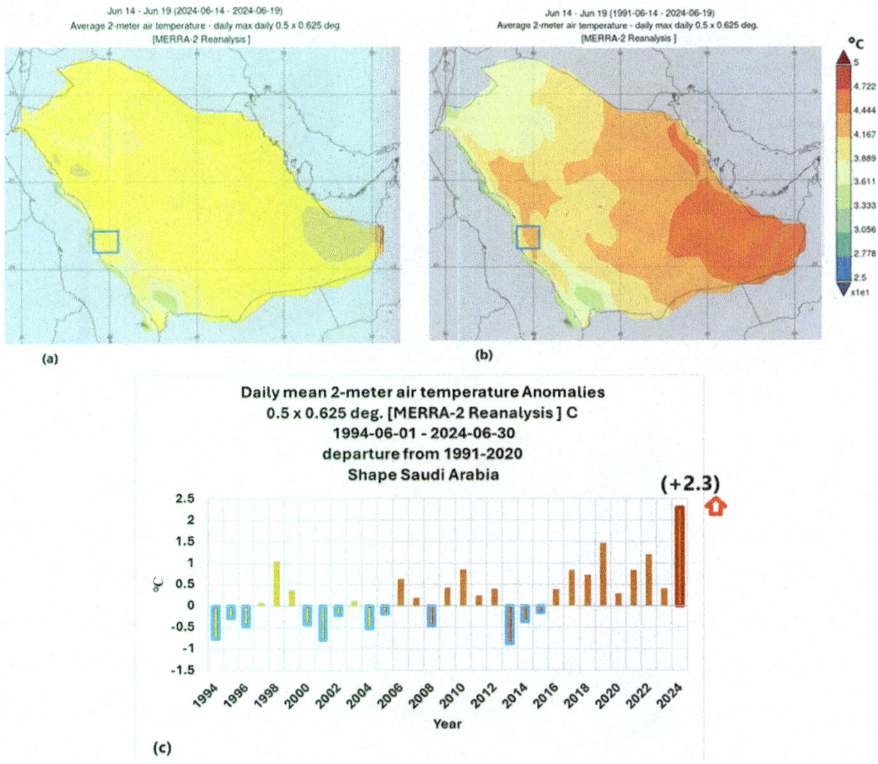

Fig. 17.4 Composite mean daily 2-m air temperature for the period from 14 to 19 June 2024 (**a**) and the long-term climatological mean for the study period (**b**). A time series of area-averaged long-term anomalies, representing departures from the 1991–2020 mean for the mean daily 2-m air temperatures in June over the last three decades over Saudi Arabia

eastern Türkiye (Fig. 17.6c), along with a northward shift of the subtropical high axis by approximately 40°. This poleward shift is a key aspect of climate change and global warming, bearing significant implications for regions in the Middle East. Figure 17.7a shows the average daily composite mean surface relative humidity for 14–19 June, 2024, while Fig. 17.7b illustrates the anomalies of this data, comparing the departure from the 1991–2020 climate normal over the specified period for the study area. The figure indicates very dry surface air, with relative humidity around 20% across most parts of Saudi Arabia. A comparison between the mean values and the long-term climate normal reveals drier conditions, particularly in the northern and western regions, where the relative humidity drops to between 2 and 12% mostly over and accompanied by the deeper RST (see Fig. 17.5a and c). Figure 17.8 shows the omega, vertical pressure velocity (a), air temperature (b), omega anomaly (c), and air temperature anomalies (d) at a level of 150 h Pa, respectively. This figure shows abnormally cold air and positive omega at an upper-tropospheric level. This figure indicates stable cold conditions above the level of a jet stream associated with strong

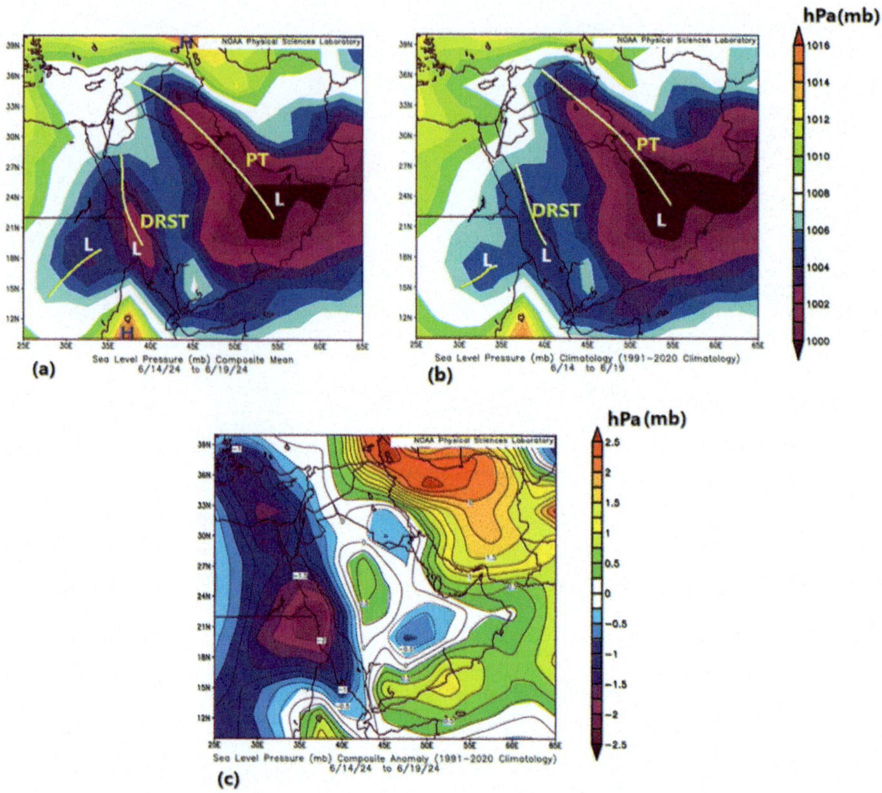

Fig. 17.5 Average daily composite mean of mean sea-level (MSL) pressure (**a**), the climatological mean map for 1991–2020 (**b**), and the corresponding anomalies (**c**) during the study period

subsidence. The strong subsidence and downward motion in the study area result in an increase in the air temperatures by descending to lower-level temperatures.

Figure 17.9a, b present the averaged upper-level Ertel potential vorticity at 300 h Pa, calculated every 3 h during 14–19 June, 2024, (the period of the extreme heat event), and 10–13 June, 2024 (prior to the event), over the study area. A comparison of these figures reveals a rapid deformation of potential vorticity contours in the upper atmosphere, transitioning from approximately 2 PVU before the event to 0.5 PVU during the event. This change is indicative of anticyclonic Rossby wave breaking (AWB), associated with the subtropical jet stream core (refer to Fig. 17.9c and d). Figure 17.9c and d display the average upper-level wind vectors at 200 h Pa, averaged over the same periods as those above. The comparison shows a significant weakening of the subtropical jet stream by approximately 8 m/s, along with a sudden northward shift of its core from 29°N to 34°N, which moves from south to north of Iraq and forms an anticyclonic curvature. This phenomenon occurs when a Rossby wave tilts in a specific manner, causing a sudden overturning of potential

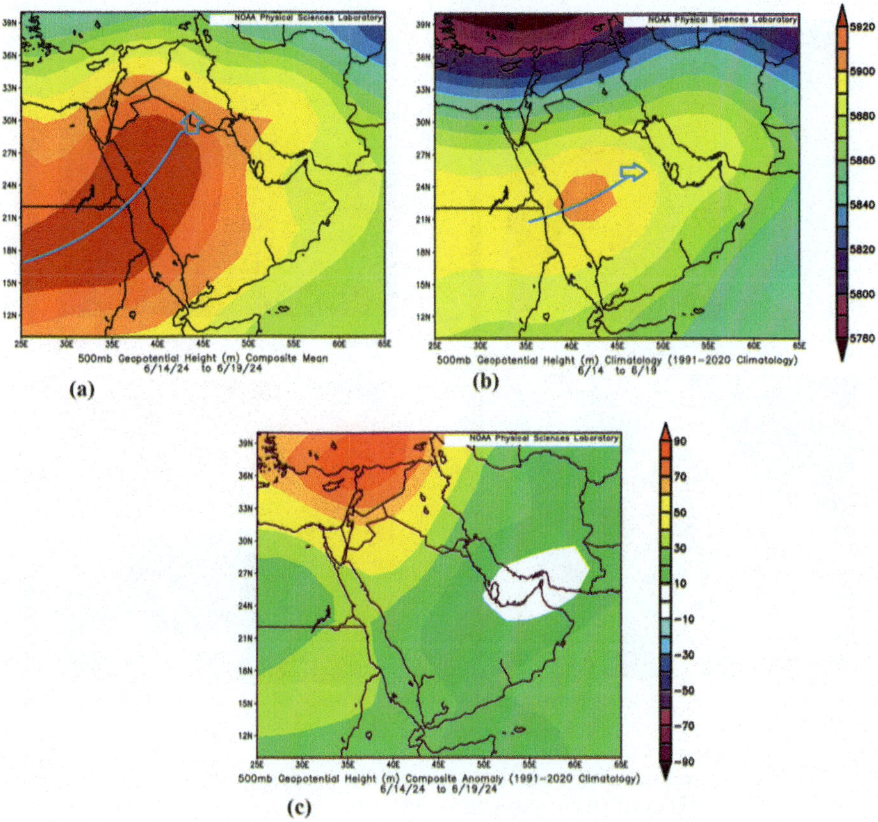

Fig. 17.6 Average daily composite mean (**a**) of 500 h Pa geopotential heights for 14–19 June, 2024, the climatological mean (**b**), and the anomalies (**c**) relative to the 1991–2020 climate normal for these specified days

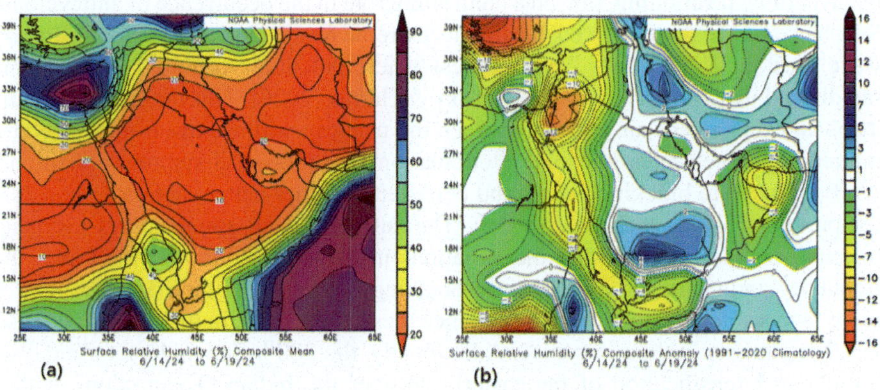

Fig. 17.7 Surface relative humidity (**a**) and anomalies (**b**) for 14–19 June, 2024, over Saudi Arabia

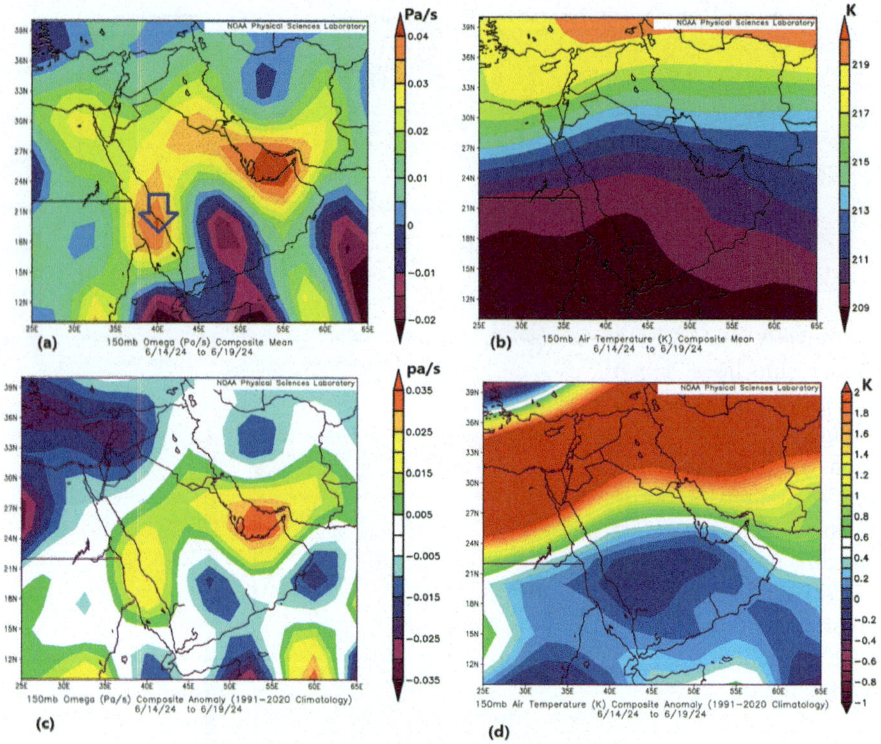

Fig. 17.8 Vertical pressure velocity (**a**), air temperature (**b**), omega anomaly (**c**), and air temperature anomalies (**d**) at a level of 150 h Pa, respectively

vorticity on the equatorial side of the jet stream, leading to the formation of an anticyclonic structure. These observations suggest that the subtropical jet stream, through its favourable positioning, played a controlling role in the occurrence of anticyclonic Rossby wave breaking (AWB) during this extreme heat event. Anticyclonic Rossby wave breaking is an atmospheric process where potential vorticity (PV) deforms rapidly and irreversibly (Tamarin-Brodsky & Harnik, 2026). This deformation facilitates an exchange of air masses between the tropics and the extra tropics, thereby influencing large-scale atmospheric circulation. In subtropical summer conditions, AWB is linked to blocking events and persistent weather patterns often responsible for prolonged heatwaves or droughts. The subtropical jet stream plays a crucial role in modulating AWB, and the variations in its intensity and position alter the frequency and characteristics of wave-breaking events, meaning that a stronger or more favourable positioned jet can enhance these events and, in turn, affect regional weather patterns in the subtropics. Additionally, the 200 h Pa wind vector analysis reveals an intensification of the easterly jet over the Indian Ocean, along with a displacement of the subtropical jet core from the Mediterranean Sea to eastern Asia. The subtropical jet has intensified in the shifted area to around 6 m/s rather than normal values during the study period (figures are not presented here). Such variations are consistent with observed changes in the upper-tropospheric wind patterns over Asia, which have been associated with both tropical and extratropical influences (Li et al., 2019). The intensification of the easterly jet over the southern Indian Ocean aligns with the findings from recent studies, which note a strengthening of the Tropical Easterly Jet post-1998 (Sharma et al., 2024). This change has been linked to alterations in the Indian Summer Monsoon Rainfall patterns. Furthermore, the displacement of the subtropical jet core from the Mediterranean Sea to eastern Asia corresponds with observed interannual variations in the Asian Subtropical Westerly Jet. Studies have documented meridional shifts and intensity changes of the ASWJ, which significantly influence East Asian climate and precipitation patterns (Du et al., 2015). These results underscore the dynamic nature of atmospheric circulation patterns and their profound impact on regional climate systems.

Figure 17.10 displays the composite mean of wind vectors (a), climatological mean (b), and anomalies (c) at 850 h Pa during 14–19 February, 2024, over the study area. A comparison of Fig. 17.10a and b clearly shows a weakening of the Somali jet stream, as depicted in Fig. 17.10c. The Somali jet forms in the western Arabian Sea, typically preceding the onset of monsoon rainfall (Halpern & Woiceshyn, 2001). It is driven by the temperature contrast between the hot landmass and the cooler Indian Ocean during the summer months. As shown in Fig. 17.10a and b, the jet flows from the southwest to the northeast, bringing moist air from the Indian Ocean into the region. This influences weather patterns, including the strength and timing

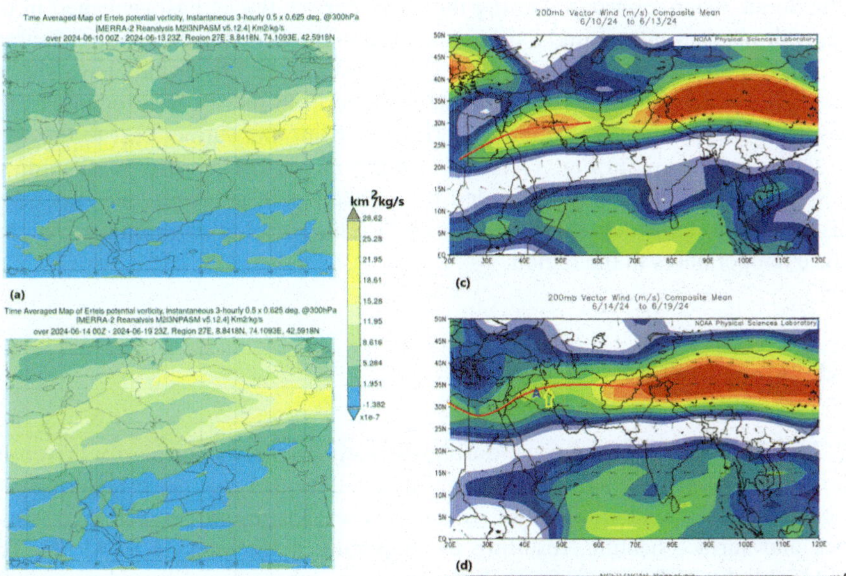

Fig. 17.9 Averaged upper-level Ertel potential vorticity at 300 h Pa, calculated every 3 h during 14–19 June, 2024 (**a**), and 10–13 June, 2024 (**a**), over the study area

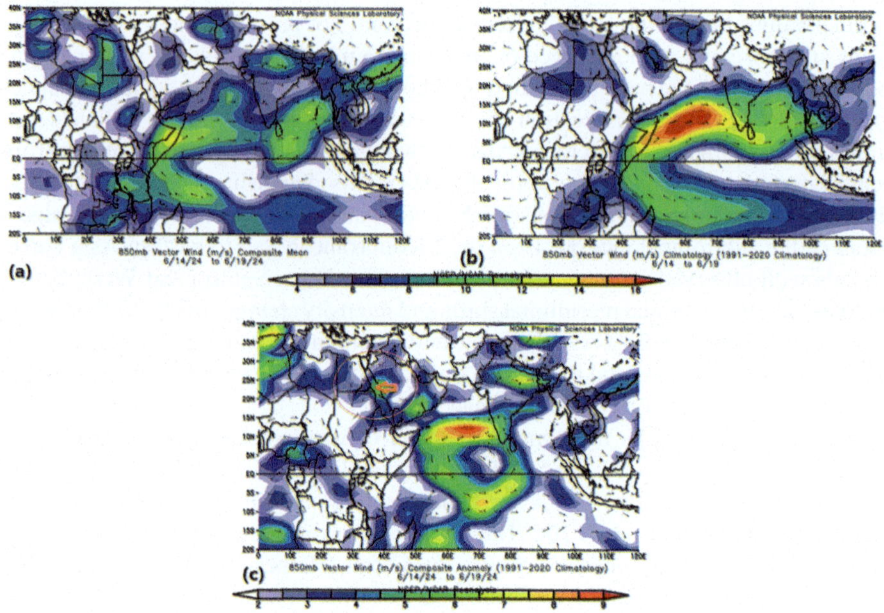

Fig. 17.10 Wind vector (**a**), climatology mean (**b**), and anomaly (**c**) composite maps at 850 h Pa over the study area during 14–19 June, 2024

of the South Asian monsoon. During the study period, the jet has weakened in the core, with wind speeds around 5 m/s, compared to the climatological normal values (Fig. 17.10a and b). Additionally, Fig. 17.10c reveals that the winds have shifted to easterlies rather than the typical westerly pattern. In the western regions of the study area, the winds now exhibit easterly flow from the warmer east to the west (see the orange arrow in Fig. 17.10c), resulting in warmer air. The weakening of the Somali jet in Saudi Arabia has led to a reduction in moisture-laden air being transported into the study area (see Fig. 17.7). This shift has contributed to higher surface temperatures in the region.

17.4 Conclusions

The June 2024 extreme heatwave during the Hajj period was primarily driven by significant atmospheric shifts, including anticyclonic Rossby wave breaking and a weakening of the subtropical jet stream. These changes led to increased subsidence and surface heating, resulting in temperatures reaching up to 44 °C. The study highlights a clear warming trend over the past three decades, with June 2024 exhibiting temperature anomalies of up to 2.3 °C above the long-term average. The study concludes that such extreme heat events will likely become more frequent and intense due to ongoing climate change. To mitigate risks during future heatwaves, especially during critical events like the Hajj pilgrimage, it is essential to implement adaptive strategies, such as modifying pilgrimage schedules, enhancing cooling infrastructure, and ensuring a reliable water supply to protect public health.

Advanced weather monitoring and forecasting are crucial for ensuring the safety of pilgrims during Hajj, especially with the increasing frequency of heat waves due to global warming. These developments aim to improve the accuracy and timeliness of weather predictions, which is vital for mitigating the impacts of extreme weather events intensified by climate change. With further research and investment in innovative technologies, authorities can accurately predict extreme temperatures and implement timely safety measures, such as adjusting the schedule of rituals to cooler times. Enhanced observational technologies provide real-time data that support strategic planning and effective crowd management. By integrating forecasting models and collaborating with international weather agencies, these advancements not only enhance the accuracy of weather predictions but also play a vital role in protecting pilgrims from heat-related risks, ensuring a safer and more comfortable Hajj experience. Educating pilgrims about heat safety is also essential, with materials provided in multiple languages and mobile apps offering real-time heat alerts and safety tips. Empowering pilgrims with knowledge and preventive strategies significantly reduces the likelihood of heat-related incidents, thereby contributing to a safer and more comfortable Hajj. Policymakers should invest in climate-resilient infrastructure, advanced weather forecasting, and heat-health monitoring systems to protect pilgrims during the Hajj. It is essential that educating pilgrims on heat safety ensure medical readiness and collaborate with global experts. Flexible ritual schedules based on weather predictions can reduce heat-related risks, providing a safer pilgrimage experience.

Acknowledgements We extend our gratitude to the European Centre for Medium-Range Weather Forecasts (ECMWF) for providing the ERA5 hourly data. We also thank the Physical Science Division in Boulder, Colorado (accessible via http://www.esrl.noaa.gov/psd/), as well as the Giovanni online data system developed and maintained by NASA's Goddard Earth Sciences Data and Information Services Center (GES DISC) and the NASA Land Processes Distributed Active Archive Center (LP DAAC) at the USGS EROS Center. We are especially grateful for the generous support from Virgin Unite USA, Inc., through the Planetary Boundary Science Laboratory project.

References

Abdalla, F., Shamy, I. E., Bamousa, A. O., Mansour, A., Mohamed, A., & Tahoon, M. (2014). Flash floods and groundwater recharge potentials in arid land alluvial basins, southern Red Sea coast, Egypt. *International Journal of Geosciences, 05*(09), 971–982. https://doi.org/10.4236/ijg.2014.59083

Amaya, D. J., Siler, N., Xie, S., & Miller, A. J. (2017). The interplay of internal and forced modes of Hadley cell expansion: Lessons from the global warming hiatus. *Climate Dynamics, 51*(1–2), 305–319. https://doi.org/10.1007/s00382-017-3921-5

Arias, P. A., Rivera, J. A., Sörensson, A. A., Zachariah, M., Barnes, C., Philip, S., et al. (2023). Interplay between climate change and climate variability: The 2022 drought in central South America. *Climatic Change, 177*(1). https://doi.org/10.1007/s10584-023-03664-4

Awad, A. M., & Almazroui, M. (2016). Climatology of the winter Red Sea trough. *Atmospheric Research, 182*, 20–29. https://doi.org/10.1016/j.atmosres.2016.07.019

Awad, A. M., & Mashat, A. S. (2018). Climatology of the Autumn Red Sea trough. *Theoretical and Applied Climatology, 135*(3–4), 1545–1558. https://doi.org/10.1007/s00704-018-2453-1

Baseer, M. N., Awad, A. M., & Almazroui, M. (2019). Climatology of the spring Red Sea trough. *International Journal of Climatology, 39*(11), 4218–4233. https://doi.org/10.1002/joc.6069

Bodenheimer, S., Lensky, I. M., & Dayan, U. (2019). Characterization of Eastern Mediterranean dust storms by area of origin; North Africa versus Arabian Peninsula. *Atmospheric Environment, 198*, 158–165. https://doi.org/10.1016/j.atmosenv.2018.10.034

Cusack, M., Arrieta, J. M., & Duarte, C. M. (2020). Source apportionment and elemental composition of atmospheric total suspended particulates (TSP) over the Red Sea coast of Saudi Arabia. *Earth Systems and Environment, 4*(4), 777–788. https://doi.org/10.1007/s41748-020-00189-z

Dayan, U., Ricaud, P., Zbinden, R., & Dulac, F. (2017). Atmospheric pollution over the Eastern Mediterranean during summer—A review. *Atmospheric Chemistry and Physics, 17*(21), 13233–13263. https://doi.org/10.5194/acp-17-13233-2017

De Vries, A. J., Tyrlis, E., Edry, D., Krichak, S. O., Steil, B., & Lelieveld, J. (2013). Extreme precipitation events in the Middle East: Dynamics of the active Red Sea trough. *Journal of Geophysical Research: Atmospheres, 118*(13), 7087–7108. https://doi.org/10.1002/jgrd.50569

Du, Y., Li, T., Xie, Z., & Zhu, Z. (2015). Interannual variability of the Asian subtropical westerly jet in boreal summer and associated with circulation and SST anomalies. *Climate Dynamics, 46*(7–8), 2673–2688. https://doi.org/10.1007/s00382-015-2723-x

El-Fandy, M. G. (1948). The effect of the Sudan monsoon low on the development of thundery conditions in Egpyt, Palestine and Syria. *Quarterly Journal of the Royal Meteorological Society, 74*(319), 31–38. https://doi.org/10.1002/qj.49707431904

Fazel-Rastgar, F., & Sivakumar, V. (2024). A case study on more recent heat wave occurred in South Africa, based on background weather synoptic and dynamic characteristics analysis. *Bulletin of Atmospheric Science and Technology, 5*(1). https://doi.org/10.1007/s42865-024-00068-9

Fazel-Rastgar, F., Rostami, M., Fallah, B., & Sivakumar, V. (2025). Torrential rainfall with severe flooding associated with a baroclinic disturbance on November 17, 2023, United Arab Emirates (UAE). *International Journal of River Basin Management, 1–10*. https://doi.org/10.1080/15715124.2025.2462575

Gasch, P., Rieger, D., Walter, C., Khain, P., Levi, Y., Knippertz, P., & Vogel, B. (2017). Revealing the meteorological drivers of the September 2015 severe dust event in the Eastern Mediterranean. *Atmospheric Chemistry and Physics, 17*(22), 13573–13604. https://doi.org/10.5194/acp-17-13573-2017

Halpern, D., & Woiceshyn, P. M. (2001). Somali jet in the Arabian Sea, El Nino, and India rainfall. *Journal of Climate, 14*(3), 434–441. https://doi.org/10.1175/1520-0442(2001)014%3c0434:sjitas%3e2.0.co;2

Hersbach, H., Bell, B., Berrisford, P., Dahlgren, P., Horányi, A., Muñoz-Sabater, J., et al. (2020). The ERA5 global reanalysis: Achieving a detailed record of the climate and weather for the past 70 years. https://doi.org/10.5194/egusphere-egu2020-10375

Hochman, A., Scher, S., Quinting, J., Pinto, J. G., & Messori, G. (2021). A new view of heat wave dynamics and predictability over the Eastern Mediterranean. *Earth System Dynamics, 12*(1), 133–149. https://doi.org/10.5194/esd-12-133-2021

Hochman, A., Plotnik, T., Marra, F., Shehter, E., Raveh-Rubin, S., & Magaritz-Ronen, L. (2023). The sources of extreme precipitation predictability; the case of the 'Wet' Red Sea trough. *Weather and Climate Extremes, 40*, Article 100564. https://doi.org/10.1016/j.wace.2023.100564

Jaagus, J., Rimkus, E., Briede, A., Sagris, V., Aasa, A., Kapilovaite, J., & Sepp, M. (2024). Long-term changes in heat wave parameters in the eastern Baltic region. *Theoretical and Applied Climatology, 155*(6), 5053–5068. https://doi.org/10.1007/s00704-024-04925-7

Kalnay, E., Kanamitsu, M., Kistler, R., Collins, W., Deaven, D., Gandin, L., … Joseph, D. (1996). The NCEP/NCAR 40-Year reanalysis project. *Bulletin of the American Meteorological Society, 77*(3), 437–471. https://doi.org/10.1175/1520-0477(1996)0770437:tnyrp2.0.co;2

Krichak, S. O., Alpert, P., & Krishnamurti, T. N. (1997). Red Sea trough/cyclone development? Numerical investigation. *Meteorology and Atmospheric Physics, 63*(3–4), 159–169. https://doi.org/10.1007/bf01027382

Kutiel, H., & Kutiel, P. (1991). The distribution of autumnal easterly wind spells favoring rapid spread of forest wildfires on Mount Carmel, Israel. *GeoJournal, 23*(2). https://doi.org/10.1007/bf00241399

Lelieveld, J., Proestos, Y., Hadjinicolaou, P., Tanarhte, M., Tyrlis, E., & Zittis, G. (2016). Strongly increasing heat extremes in the Middle East and North Africa (MENA) in the 21st century. *Climatic Change, 137*(1–2), 245–260. https://doi.org/10.1007/s10584-016-1665-6

Li, D., Chen, H., Liu, P., & Zhou, C. (2019). Zonally asymmetric mode of anomalous activity in summer Asian subtropical westerly jet and its possible sources. *Theoretical and Applied Climatology, 139*(1–2), 17–32. https://doi.org/10.1007/s00704-019-02934-5

Mashat, A. S., Awad, A. M., Assiri, M. E., & Labban, A. H. (2021). Synoptic pattern of the Red Sea trough associated with spring dust over the northern and western Arabian Peninsula. *Meteorology and Atmospheric Physics, 133*(3), 655–673. https://doi.org/10.1007/s00703-020-00771-0

McDermott-Levy, R., Scolio, M., Shakya, K. M., & Moore, C. H. (2021). Factors that influence climate change-related mortality in the United States: An integrative review. *International Journal of Environmental Research and Public Health, 18*(15), 8220. https://doi.org/10.3390/ijerph18158220

Mohammadi, Z., Lashkari, H., & Mohammadi, M. S. (2021). Synoptic analysis and core situations of arabian anticyclone in shortest period precipitation in the south and Southwest of Iran. *Arabian Journal of Geosciences, 14*(12). https://doi.org/10.1007/s12517-021-07572-8

Neira, M., Erguler, K., Ahmady-Birgani, H., AL-Hmoud, N. D., Fears, R., Gogos, C., et al. (2023). Climate change and human health in the Eastern Mediterranean and Middle East: Literature review, research priorities and policy suggestions. *Environmental Research, 216*, 114537. https://doi.org/10.1016/j.envres.2022.114537

Opoku, S. K., Filho, W. L., Hubert, F., & Adejumo, O. (2021). Climate change and health preparedness in Africa: Analysing trends in six African countries. *International Journal of Environmental Research and Public Health, 18*(9), 4672. https://doi.org/10.3390/ijerph18094672

Ranta, A., Kang, J., Saad, A., Wasay, M., Béjot, Y., Ozturk, S., et al. (2024). Climate change and stroke: A topical narrative review. *Stroke, 55*(4), 1118–1128. https://doi.org/10.1161/strokeaha.123.043826

Rienecker, M. M., Suarez, M. J., Gelaro, R., Todling, R., Bacmeister, J., Liu, E., … Woollen, J. (2011). MERRA: NASA's modern-era retrospective analysis for research and applications. *Journal of Climate, 24*(14), 3624-3648. https://doi.org/10.1175/jcli-d-11-00015.1

Schmidt, D. F., & Grise, K. M. (2019). Impacts of subtropical highs on summertime precipitation in North America. *Journal of Geophysical Research: Atmospheres, 124*(21), 11188–11204. https://doi.org/10.1029/2019jd031282

Sharma, V., Amarjeet, Sharma, S., & Chakraborty, A. (2024). Evidence of strengthening of tropical easterly jet after 1998 climate shift using ERA-5 datasets. *Theoretical and Applied Climatology, 155*(7), 5859–5874. https://doi.org/10.1007/s00704-024-04975-x

Solot, S. B. (1950). General circulation over the Anglo-Egyptian Sudan and adjacent regions. *Bulletin of the American Meteorological Society, 31*(3), 85–94. https://doi.org/10.1175/1520-0477-31.3.85

Tamarin-Brodsky, T., Harnik, N., & Falkena, S. K. (2026). On storm tracks, weather regimes, and a wave breaking recipe. *AGU Advances, 7*(1), e2025AV002049.

Tanarhte, M., Hadjinicolaou, P., & Lelieveld, J. (2015). Heat wave characteristics in the Eastern Mediterranean and Middle East using extreme value theory. *Climate Research, 63*(2), 99–113. https://doi.org/10.3354/cr01285

Tsvieli, Y., & Zangvil, A. (2005). Synoptic climatological analysis of 'wet' and 'dry' Red Sea troughs over Israel. *International Journal of Climatology, 25*(15), 1997–2015. https://doi.org/10.1002/joc.1232

Wei, W., Li, W., Deng, Y., & Yang, S. (2018). Correction to: Intraseasonal variation of the summer rainfall over the southeastern United States. *Climate Dynamics, 53*(1–2), 1185–1185. https://doi.org/10.1007/s00382-018-4415-9

Yang, J., Zhou, M., Ren, Z., Li, M., Wang, B., Liu, D. L., et al. (2021). Projecting heat-related excess mortality under climate change scenarios in China. *Nature Communications, 12*(1). https://doi.org/10.1038/s41467-021-21305-1

Zittis, G., Hadjinicolaou, P., Almazroui, M., Bucchignani, E., Driouech, F., El Rhaz, K., et al. (2021). Business-as-usual will lead to super and ultra-extreme heatwaves in the Middle East and North Africa. *NPJ Climate and Atmospheric Science, 4*(1), 20.

Zittis, G., Almazroui, M., Alpert, P., Ciais, P., Cramer, W., Dahdal, Y., et al. (2022). Climate change and weather extremes in the Eastern Mediterranean and Middle East. *Reviews of Geophysics, 60*(3). https://doi.org/10.1029/2021rg000762

Ziv, B., Shimer, R., Harpaz, T., Drori, R., Alpert, P., Raveh-Rubin, S., & Saaroni, H. (2022). Identification and classification of the wet Red Sea trough over Israel. *International Journal of Climatology, 42*(16), 10062–10082. https://doi.org/10.1002/joc.7884

Chapter 18
The 2024 Dubai Flooding: Climatic Drivers and Environmental and Socio-economic Consequences

Diana Francis and Ricardo Fonseca

Abstract The 16 April 2024 record-breaking precipitation event wreaked havoc in the United Arab Emirates, causing severe disruption to infrastructure and public life. The event was triggered by tropical-extratropical interactions, fuelled by unusually high levels of atmospheric water vapour, and potentially high dust loadings. Four people lost their lives, schools and businesses closed, more than 1,500 flights were delayed or cancelled at major airports, and a Dengue outbreak following the event placed pressure on health services. An analysis of satellite data revealed a correlation between the degree of urbanisation and population density and the effects of the flooding rains in the major urban centres. A greening of the desert was also observed, peaking 2–3 months after the precipitation event. A detection and attribution of anthropogenic climate change revealed that it is very likely enhancing the frequency of events like that of 16 April 2024, but not necessarily their intensity. The use of impact-based forecasts, public awareness campaigns, more effective early warning systems, adaptations to the existing infrastructure to better cope with high precipitation totals, and increased investment in flood defence and drainage systems are recommended to minimise the effects of future extreme rainfall episodes.

Keywords Extreme precipitation event · United Arab Emirates · Tropical-extratropical interactions · Anthropogenic climate change · Desert greening · Early warning systems

18.1 Introduction

Subtropical regions are characterised by arid and semi-arid conditions featuring meagre and irregular amounts of rainfall throughout the year. The dry conditions arise from the large-scale subsidence associated with the subtropical anticyclones in the descending branch of the Hadley Cell, and the interaction between the monsoons

D. Francis (✉) · R. Fonseca
ENGEOS Lab, Earth Science Department, Khalifa University, Abu Dhabi, UAE
e-mail: diana.francis@ku.ac.ae

E. Alam (ed.), *Disaster, Crisis and Emergency Management in the Middle East & North Africa*, https://doi.org/10.1007/978-3-032-24436-9_18

and mid-latitude westerlies (Rodwell & Hoskins, 1996). On the westward and poleward edges of subtropical arid regions, the greatest fraction of the annual rainfall falls in the colder months in association with the passage of mid-latitude weather systems. On the eastward and equatorward side, rainfall is primarily driven by tropical disturbances and is more frequent in the warm season (de Vries et al., 2024). One of the most prominent subtropical arid regions on Earth is the Arabian Peninsula, comprising the Rub' Al Khali Desert on its southeastern side that straddles over four countries: the United Arab Emirates (UAE) and Oman in the east and southeast, Yemen to the south, and Saudi Arabia to the west, southwest and northwest. In the Arabian Peninsula, most of the annual rainfall totals fall in the colder months of November to April (Hasanean & Almazroui, 2015; Nelli et al., 2022; Wehbe et al., 2017, 2020), even though in parts of southern and southeastern Oman and Yemen precipitation levels are higher in summer owing to the inland penetration of the southwest monsoon moist flow from the Arabian Sea (Kwarteng et al., 2009). Summertime convection elsewhere in the region is mostly confined to the Al Hajar mountains in eastern UAE and northern Oman (Fonseca et al., 2025) and the Hijaz mountains in Saudi Arabia (Parajuli et al., 2022), arising from the interaction of the air flows associated with topography and heat lows with the diurnal sea-land breeze circulations.

Extreme precipitation episodes in the Arabian Peninsula are typically tied to tropical-extratropical interactions, which generally feature the (i) Red Sea Trough (RST; de Vries et al., 2013; Tsvieli & Zangvil, 2007), a tongue of low pressure driven by both thermodynamic and dynamical forcing that lies over the Red Sea in spring and autumn and shifts southwards into the Ethiopian Highlands in winter; (ii) atmospheric rivers (Esfandiari & Rezaei, 2022; Francis et al., 2024), narrow and long corridors of high moisture transport that originate in the tropics/subtropics and extend into mid- and high-latitudes; and (iii) mid-latitude baroclinic systems, including fronts and cut-off lows, that are a regular occurrence in the colder months (de Vries et al., 2016; Francis et al., 2019). In addition, tropical cyclones from the Arabian Sea occasionally impact the Arabian Peninsula, bringing strong winds and torrential rainfall, as was the case for cyclone Gonu in June 2007 and Shaheen in October 2021 (Fritz et al., 2010; Francis et al., 2022; Ibrahim & Al Maghawry, 2022). On the eastern side of the Arabian Peninsula, organised convection in the form of mesoscale convective systems (MCSs), long-lived thunderstorm clusters, is more frequent in spring, mostly in March and April. It is fuelled by warmer and more moist low-level air advected from the Arabian Sea, Red Sea and/or the Arabian Gulf; the presence of a mid-level trough that, together with warmer surface temperatures, promotes a conditionally unstable stratification; an equatorward shift in the position of the subtropical jet (Nelli et al., 2021). An example of a high-impact extreme weather event in the UAE driven by MCSs occurred on 9 March 2016, when more than 240 mm of rain fell in Dubai, with winds of up to 35 m s^{-1} in Abu Dhabi (Wehbe et al., 2019). All the listed ingredients for the development of MCSs were present on this day, along with sea surface temperature (SST) anomalies of +1.5 to 2.5 K that could also have played a role. During 2001–2020, March–April MCSs in the UAE have become smaller, stronger and longer-lasting, with the latter trend statistically significant at the 95% confidence level (Nelli et al., 2021). Summertime extreme precipitation events, albeit less frequent, also occur, and can be as impactful as cold season ones (Robertson, 2024). They are linked to the interaction of sea- and land-breezes from the adjacent water bodies, the Arabian heat low (Fonseca et al., 2022, 2025), topography-driven circulations, and the southwest monsoon circulation (Fonseca et al., 2025; Francis et al., 2021a, 2021b; Parajuli et al., 2022, 2023). A recent extreme warm season rainfall event occurred in July 2022, when Fujairah in eastern UAE received ~235 mm in roughly 72 h, with a maximum rainfall intensity sustained over 15 min of ~71 mm h^{-1} (Terry et al., 2023). This event is triggered by a westward extension of the Indian Summer Monsoon into the southeastern Arabian Peninsula, enhanced inland moisture advection by cooler Sea Surface Temperatures (SSTs) just off the coast of Oman, Yemen and the western Sea of Oman, and the interaction of this moist onshore flow with topography-driven flows around the Al Hajar mountains.

The fact that springtime organised precipitation events have become more frequent and intense in the southern Arabian Peninsula in the last couple of decades (Nelli et al., 2021) raises the question of whether, and if yes, to what extent, anthropogenic climate change has played a role. The challenges involved and the importance of performing detection and attribution of climate change were first emphasised by Allen (2003). One of the earliest studies to adopt a comprehensive approach was that of Stott et al. (2004), who investigated the role of anthropogenic emissions in the 2003 summer European heatwave, and reported that human influence has, at the very least, doubled the risk of the summer temperature averaged over southern Europe exceeding the 1961–1990 long-term mean by 1.6 K at the 90% confidence level. Since then, thousands of detection and attribution studies have been published, many confirming the role of human-induced climate change (e.g., Philip et al., 2018), while some have not reached conclusive results (e.g., Vautard et al., 2019). Despite inherent challenges, the detection and attribution of anthropogenic climate change is a necessary step to better understand the internal variability of the climate system and develop effective adaptation and mitigation strategies (Bindoff et al., 2013). As climate change projections indicate an increased likelihood of extreme precipitation events in the Middle East, including in the UAE (Rao et al., 2024), it is vital to assess the potential role of anthropogenic climate change in these episodes. A quantification of the effects of human emissions will also enable more accurate and reliable climate change projections and further motivate and encourage the public and policymakers to take aggressive measures to reduce greenhouse gas emissions.

In the last four decades, the Middle East has experienced a warming trend (Zittis et al., 2022) accompanied by a general decrease in cold season and increase in warm season rainfall (Francis & Fonseca, 2024). Climate model simulations project an increase in extreme precipitation events in the region towards the middle and end of the twenty-first century (Rao et al., 2024), underscoring the need for a better understanding of the processes driving these changes. In this chapter, the focus is on the recent extreme rainfall event on 16 April 2024 in the UAE (Francis et al., 2025), which brought historical rainfall amounts to several parts of the country. The 24-h rainfall totals exceeded 250 mm in some areas and were the highest since the records began in 1949, causing widespread flooding and disruption in the region, with a total of 20 fatalities in Oman and four in the UAE (Abuhani et al., 2025; Deliso et al., 2024; Hussein et al., 2025; van Zyl, 2024; Zhuang, 2024; Zhang et al., 2025). The historical nature of the 16 April 2024 episode raises the question of whether anthropogenic forcing is at least partially responsible for its extremeness and underscores the need to understand its drivers to better predict similar occurrences in the future. The main objectives of this chapter are two-fold: (i) to shed light on the drivers and impacts of the extreme precipitation amounts that fell on this day, including a discussion of the role of anthropogenic climate change, and (ii) to put forward recommendations on how the effects of such events can be mitigated in the future. Understanding the triggering mechanisms of this historical event will enable better prediction of similar occurrences in a warming climate, while quantifying the contribution of anthropogenic emissions will enable more reliable climate change projections. The lessons learned from this highly impactful extreme weather episode are translated into a set of recommendations, focusing both on the early warning systems and the need for a more resilient infrastructure.

18.2 Datasets and Methodology

A combination of reanalysis, modelling, and observational data products listed in Table 18.1 is considered in this work. Linear regressions are computed using the Theil-Sen estimator (Sen, 1968; Theil, 1992) and statistical significance is assessed with the Mann-Kendall test (Mann, 1945). The National Oceanic and Atmospheric Administration Air Resources Laboratory Hybrid Single-Particle Lagrangian Integrated Trajectory transport and dispersion model (HYSPLIT; Stein et al., 2015) driven by ERA-5 data is used to compute back-trajectories. The trajectories are initialised at selected locations in time and space with the air parcels advected backwards in time using the reanalysis' hourly three-dimensional wind field. The effects of the flooding rains on the vegetation are assessed through the Normalised Difference Vegetation Index (NDVI), which is a reliable indicator of vegetation health and density. Following Pandya et al. (2024), four NDVI classes are considered: no vegetation ($\text{NDVI} \leq 0$); low vegetation ($0 < \text{NDVI} \leq 0.1$), moderate vegetation ($0.1 < \text{NDVI} \leq 0.2$), and dense vegetation ($\text{NDVI} > 0.2$).

The process of detection and attribution of anthropogenic climate change comprises eight steps (Philip et al., 2020; van Oldenborgh et al., 2021) that can be summarised as follows: (i) analysis trigger, in which the event is selected. This choice can be based on the event's impact and/or record or near-record breaking meteorological features; (ii) event definition, in which the extreme event is defined in a quantitative way (e.g., "*June–August maximum of 3-day averaged rainfall over Al Ain*"). Area averages are discouraged as they will make the event rare, as similar events of a stronger in magnitude but offset in space will only be partially reflected in the area average; (iii) observed probability and trends, in which observational measurements are analysed to extract the return period of the event and investigate

Table 18.1 List of observational and modelling products considered in this work

Product	Specifications
ERA-5	Hourly data at 0.25° (~27 km) spatial resolution available from 1940 to present. Performs well against ground-based observations over the UAE and Arabian Peninsula (Nelli et al., 2024; Ullah et al., 2024)
CMIP6	Climate change projections of 25 models contributing to the sixth phase of the Coupled Model Intercomparison Project (CMIP6; Eyring et al., 2016). Only those with a nominal spatial resolution of at least 100 km in latitude/longitude are selected (Francis & Fonseca, 2024)
GPM IMERG	Integrated Multi-Satellite Retrievals by the Global Precipitation Measuring mission (GPM IMERG; Hou et al., 2014; Pradhan et al., 2022) 30 min precipitation rates at a resolution of 0.1° (~11 km) from 1998 to present. Its estimates are found to be in close agreement with ground-based observations in UAE (Alsumaiti et al., 2020)
GHRSST	SSTs from a blend of satellite-derived estimates available daily on a 0.01° (~11 km) grid from 2002 to present (Martin et al., 2012). They are in close agreement with in-situ measurements in the Arabian Gulf and the Sea of Oman (Nesterov et al., 2021)
Sentinel	Sentinel-1 and Sentinel-2 satellite data are processed following Normand and Heggy (2024) to investigate the microscale effects of the flooding rains in the urban centres of Dubai and Abu Dhabi

changes in its frequency of occurrence; (iv) model evaluation, at which stage suitable models are selected for further analysis based on their performance against observational products and the successful simulation of the event in question. In general, models have to accurately represent the extreme event of interest, the statistics of the modelled extreme events should be compatible with those of observed events, and the meteorological drivers of the simulated extreme event must be realistic; (v) climate model analysis, in which the models are run for the actual (current) climate and one in which the anthropogenic emissions are removed (counterfactual climate). At this stage, metrics such as the probability ratio (PR), defined as the ratio of the probability of the event occurring in the actual to that in the counterfactual climate, are extracted; (vi) hazard synthesis, in which all the information and insight gained from the analysis of steps (iv) and (v) are combined into a single attribution statement; (vii) analysis of trends in vulnerability and exposure, which considers the hydrometeorological hazard itself, the exposure of people, property and systems, and the vulnerability of the community (defined as the characteristics or circumstances that make the community susceptible to the damaging effects of the hazard). It is important to note that an event may be impactful not because of the meteorological forcing but as a result of human factors such as population increase, lack of proper infrastructure, changes in land use land cover, poor management and warning systems, unprepared health services, etc.; (viii) communication, at which stage the findings of the study are disseminated to various groups ranging from scientists and policy-makers to media outlets, emergency management agencies, and the public in general. This rigorous eight-step procedure allows for a quantification of the effect of anthropogenic forcing on the event in question at a given confidence level.

18.3 Atmospheric Drivers

The large-scale circulation on 15–17 April 2024 is summarised in Fig. 18.1. The main drivers and effects of the flooding rains are outlined in Table 18.2.

The main ingredients of this extreme event are as follows: (i) a potential vorticity (PV) streamer, associated with anticyclonic Rossby wave breaking, which moved into the Arabian Peninsula (Fig. 18.1a–c), with a cut-off in the mid-troposphere (Fig. 18.1d–f). Colder mid-tropospheric temperatures, when combined with warmer low-level temperatures, promote a conditionally unstable stratification that favours the development of convection; (ii) a low pressure system over the southeastern Arabian Peninsula, which resulted from the three-way interaction of an active RST, an upper-level trough from the mid-latitudes, and a meridional wavetrain emanating from the tropical Indian Ocean (He & Kurcharski, 2025; Fig. 18.1g–i); (iii) advection of very moist air from the tropical Indian Ocean in response to a low–high pressure couplet, with the Arabian subtropical anticyclone located east of the referred low pressure (Fig. 18.1g–l), at a time when there was enhanced convection in the region; (iv) a southward-shifted subtropical jet (Fig. 18.1d–f). The surface low over the southeastern Arabian Peninsula is ahead of a mid-level trough (cf. Figure 18.1e with 1 h), a phase relationship indicative of baroclinic wave growth (Hoskins & James, 2014; Hoskins et al., 1985). These atmospheric processes are in line with previously studied organised convective events in the UAE (Nelli et al., 2021), and over the Middle East (de Vries et al., 2013, 2016, 2018; Taraphdar et al., 2025).

What made the event record-breaking in terms of rainfall is the extremely high amounts of atmospheric moisture, owing to the deep penetration of the PV streamer into low latitudes at a time when deep convection was occurring in the tropical Indian Ocean. The maximum hourly IVT around the UAE (22°–27°N, 51°–57°E) on 16 April 2024 is in the top 0.1% of the climatological distribution, underscoring its extreme nature on this day. HYSPLIT back-trajectories revealed that at lower levels the air masses originate predominantly from the tropical Arabian Sea between 5° and 15°N, where the satellite-derived SSTs are 1–2 K above the 2003–2023 climatological mean. The generalised warming of the Indian Ocean in the spring months following an El Nino (as is the case in 2024) is an expected response driven by both oceanic and atmospheric forcing (Schott et al., 2009; Xie et al., 2009). The resulting moist and warm air mass (Fig. 18.1j–l) fostered the occurrence of deep convection in parts of the UAE.

In addition to tapping into high levels of tropical moisture, another factor that could have contributed to the extreme event is the high levels of atmospheric dust. Dust aerosol optical depths exceeded two in parts of the UAE on 16 April 2024, comparable to those observed during severe spring dust storms in the region (e.g., Francis et al., 2023). The effects of aerosols on precipitation are highly dependent on their levels and chemical composition, and the availability of water vapour. Parajuli et al. (2022) investigated the effects of dust on precipitation events in western Saudi Arabia for the month of August. The authors found that dust generally suppresses rainfall,

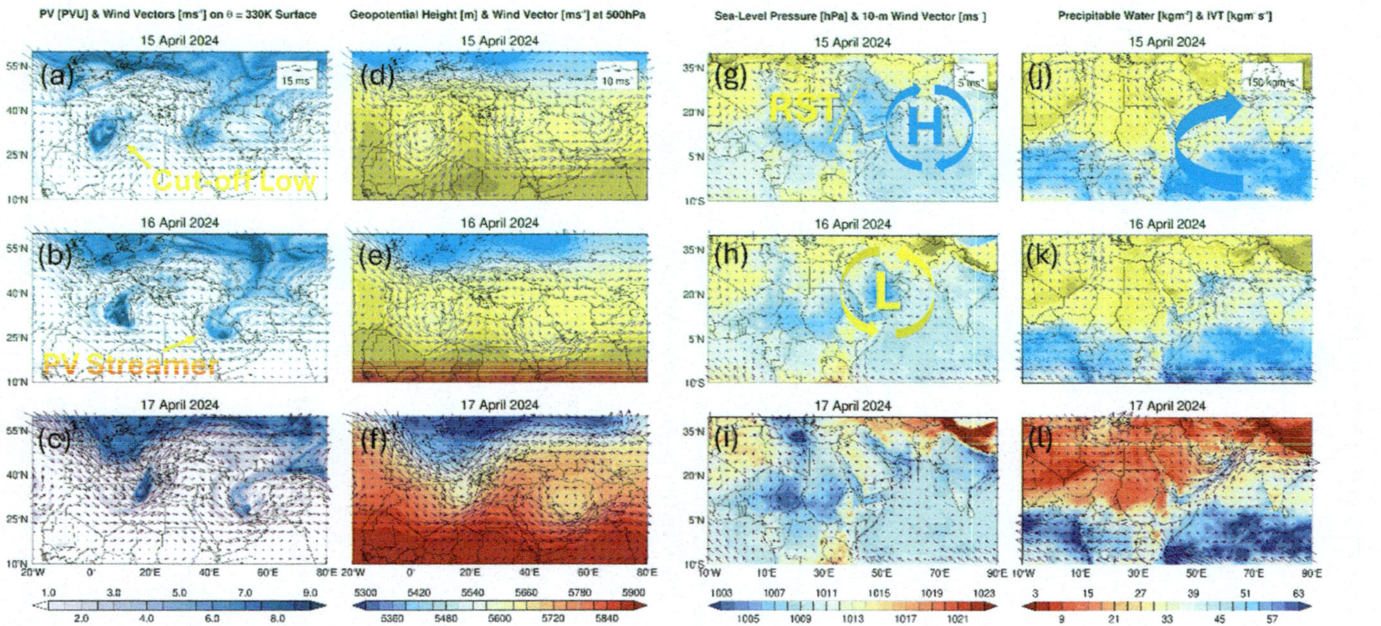

Fig. 18.1 Large-Scale Circulation during 15–17 April 2024: Potential vorticity (PV; shading; PVU; 1 PVU = 10^{-6} K m^2 s^{-1} kg^{-1}) and wind vectors (arrows; m s^{-1}) on 330 K potential temperature surface at 00 UTC on **a** 15 April, **b** 16 April and **c** 17 April 2024. **d–f** are as **a–c** but for the 500 hPa geopotential height (shading; m) and wind vectors (arrows; m s^{-1}); **g–i** are as **a–c** but for the sea-level pressure (shading; hPa) and 10 m wind vectors (arrows; m s^{-1}); **j–l** are as **a–c** but for precipitable water (shading; kg m^{-2}) and integrated vapour transport (IVT; arrows; kg m^{-1} s^{-1}). The location of the PV cut-offs and streamers is highlighted in (**a**)–(**b**), the Red Sea Trough (RST) and Arabian subtropical anticyclone are noted in (**g**), the low-pressure discussed in the text is drawn in (**h**), while in (**j**) the relevant poleward moisture transport pathway is identified

Table 18.2 Drivers and major impacts of the 16 April 2024 flooding event

<table>
<tr><th>Drivers</th><th>Impacts</th></tr>
<tr><td>Mid-tropospheric cut-off low, active Red Sea Trough, and meridional wave train from the tropical Indian Ocean promoting low pressures over the southeastern Arabian Peninsula</td><td rowspan="4">– 4 deaths in UAE and 20 deaths in Oman
– 1,500 flights delayed or cancelled at Dubai's International Airport over a 3-day period three days
– Schools closed for four days
– Severe disruption to work and services
– Insurance losses around 2.9–3.4 billion USD
– Dengue outbreak following heavy rainfall
– Greening of the desert, peaking in summer</td></tr>
<tr><td>Advection of warm and moist air at low levels from the tropical Indian Ocean by the cyclonic circulation associated with low pressure</td></tr>
<tr><td>Equatorward shift in position of subtropical jet</td></tr>
<tr><td>High dust loading over the UAE</td></tr>
</table>

as the negative direct effect, resulting in an increased static stability of the low-level atmosphere and reduced water vapour transport by the sea-breeze circulation, overwhelms the positive indirect effect, as dust aerosols support droplet growth in the presence of sufficient moisture. For extreme rainfall events, when the domain-averaged daily accumulated precipitation is in the top 10th percentile, dust acts to enhance the model-generated precipitation, with both the direct and indirect effects contributing to higher precipitation amounts. Feng et al. (2023), and for an extreme precipitation event in Beijing in July 2016, also reported that dust acted to increase the precipitation amounts by around 20–40%. Dust particles are transported upwards by the strong updrafts, where, in the presence of sufficient moisture, they promote the formation of ice nuclei, with the associated latent heat release enhancing the convective system. Both Parajuli et al. (2022) and Feng et al. (2023) stressed the importance of high-water vapour levels for intensifying precipitation events. The extremely high IVT values on 16 April 2024 around the UAE could have enabled such a mechanism to take place. This should be investigated further in coupled atmosphere-chemistry model simulations, with aerosols and pollutants playing an important role in the surface meteorology in hyper-arid regions like the UAE (Yarragunta et al., 2025).

18.4 Detection and Attribution of Anthropogenic Climate Change

The potential role of anthropogenic climate change in the intensity and frequency of occurrence of the 16 April 2024 event is assessed using the methodology described in Sect. 1.2. The event is defined as "*the maximum daily precipitation rate across the UAE (22°–27°N, 51°–57°E) during the months of March to May (MAM).*" The focus on the spring season reflects the fact that organised convection in the form of MCSs is more frequent during this time (Nelli et al., 2021). The daily precipitation on 16 April 2024 over the target domain, including at the location of eight airport stations, is given in Fig. 18.2a. The highest rainfall totals occurred in the highly populated northeastern UAE, between Abu Dhabi (station #1) and Ras Al Khaimah (station #7), with little to no rainfall in the more sparsely populated southwestern and southern UAE.

The performance of each of the 25 CMIP6 models selected is evaluated against precipitation data at the location of the eight airport stations highlighted in Fig. 18.2a for the period 1976–2014 (the historical period in CMIP6 ends in 2014), with only those that generate skilful forecasts selected for the multi-model ensemble (MME), as in Francis and Fonseca (2024). A total of 11 models are used for the CMIP6 MME, with Fig. 18.2b, illustrating the observed, reanalysis, and CMIP6 MME spring season precipitation for the common period of 1976–2014. A statistically significant increase at 95% confidence level is present in the observed spring accumulated precipitation

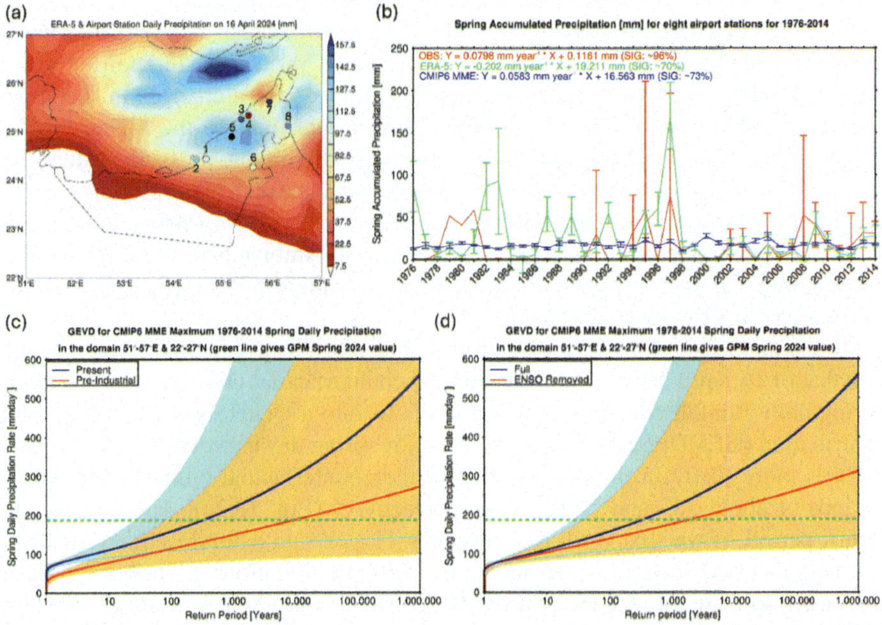

Fig. 18.2 Detection and Attribution of Anthropogenic Climate Change: **a** 16 April 2024 daily precipitation (mm) from ERA-5 over the target domain (22°–27°N, 51°–57°E). The circles give the precipitation at the location of eight airport stations: Abu Dhabi (#1), Al Bateen (#2), Dubai (#3), Sharjah (#4), Al Maktoum (#5; no data on this day), Al Ain (#6), Ras Al Khaimah (#7), Fujairah (#8). **b** Spring precipitation at the location of the eight airport stations for 1976 to 2014 from observations (red), ERA-5 (green), and the CMIP6 MME (blue). For each year, the mean and one standard deviation from the mean over the eight stations are given. The trend and intercept of the linear regression are given in the top left. **c** GEVD extracted from the daily spring rainfall over the target domain for 1976–2014 from the spring CMIP6 MME for the present (blue) and pre-industrial (red) simulations. The cyan and orange shading indicate the 95% confidence interval for the present and pre-industrial distribution, respectively. The dashed green line indicates the 16 April 2024 rainfall, as given by the Global Precipitation Measuring Mission (GPM) satellite that is of 189 mm. **d** is as (**c**) but for the present simulation (blue) and that obtained when the variability linearly related to ENSO is removed (red)

for the station data, in line with the heavier rainfall amounts associated with springtime MCSs (Nelli et al., 2021). An inspection of the time series reveals a tendency for higher precipitation values in spring seasons following El Niño winters (e.g. 1991, 1997, 2002–2003), which is consistent with a southward shifted subtropical jet and storm track during those episodes (Sandeep & Ajayamohan, 2018). The CMIP6 MME and ERA-5 capture the increasing precipitation totals across the UAE observations, even though the trend is not statistically significant. A comparison of the spatial patterns of air temperature and rainfall for the current (actual) and pre-industrial (counterfactual) climates in the CMIP6 MME over the Arabian Peninsula reveals that they are largely similar, suggesting the same processes are present in both. The higher levels of greenhouse gases in the present climate are consistent with warmer near-surface temperatures and higher rainfall amounts. For the UAE (22°–27°N, 51°–57°E), springtime air temperature is approximately 0.16 K higher, and accumulated precipitation is about 10% greater in the present climate than under pre-industrial conditions, accompanied by increased variability. In addition to the fact that warmer air has a higher moisture holding capacity, changes in the atmospheric circulation and moisture availability also contribute to the higher rainfall totals. A deepening of the Sudan Low, a thermally driven low pressure system that sits over Sudan in the colder months (Hejazizadeh et al., 2021), in the current climate, combined with lower pressures over the southeastern Arabian Peninsula, promote increased moisture advection from the Red Sea and Arabian Sea into the UAE and surrounding region. This is a well-known driver of wintertime and springtime rainfall in the area (de Vries et al., 2016, 2018).

To estimate the return periods and probability of occurrence of extreme events like that of 16 April 2024 for the current and counterfactual climates, the maximum spring daily rainfall over the target region is fit into a Generalised Extreme Value Distribution (GEVD; Easterling et al., 2016). A spring day as wet as that of 16 April 2024 is more likely in the present than in the counterfactual climate, with return periods of about 333 and 9,981 years, respectively (Fig. 18.2c). Additionally, the return period of the 16 April 2024 event is 19 years in a future climate under a business-as-usual scenario. The maximum daily precipitation on 16 April 2024 in the target domain, estimated from the GPM IMERG product, is roughly 189 mm. Taking this value as reference, the PR is 30 and the intensity change is +56 mm, with the 95% confidence interval for these parameters being 3 to 10,716 and −156 mm to +94 mm, respectively (Fig. 18.2c). Hence, a precipitation event with the magnitude of that of 16 April 2024 is more likely in the current than in the counterfactual climate, although it is not necessarily more intense. The PR and intensity changes obtained here are comparable to those reported by Zachariah et al. (2025), even though in this study, the 95% confidence interval of the former comprises values always larger than one, which allows for the attribution of the increase in frequency to anthropogenic forcing. This arises from differences in the choice of models, and in the definition of the counterfactual climate and of the event itself. Given the important role of the El Niño Southern Oscillation (ENSO; Rasmusson & Carpenter, 1982) on the UAE's wintertime and springtime rainfall (Sandeep & Ajayamohan, 2018), its role on extreme rainfall events is also explored. Figure 18.2d shows that, at the 95%

confidence level, extreme precipitation events are more likely in El Niño years, with the PR ratio varying from 1.4 to 9,339, and no robust sign for the intensity change, which is in the range −207 to +69 mm. The fact that El Niño events are linked to an increased likelihood of extreme precipitation episodes in the region during spring has been noted by several studies including Alizadeh and Mousavizadeh (2025).

18.5 Impacts

The major impacts of 16 April 2024 extreme precipitation event are listed in Table 18.2. A total of 20 fatalities in Oman and four in the UAE are recorded, causing a massive disruption to infrastructure and public life (Zhang et al., 2025). These figures, however, are lower than that in previous extreme events, which is likely a reflection of the investment made on resilient infrastructure and rapid emergency response to inclement weather (Langton, 2024). Dubai's International Airport, one of the busiest in the world, was severely impacted by the extreme weather (Fig. 18.3a), with more than 1,500 flights delayed or cancelled in a period of three days (Guy Carpenter, 2024a; The National, 2024). Several homes and offices in the city were severely flooded (Shanneik, 2024). In addition to traffic disruptions, schools were closed, and employees were requested to work from home, which highly impacted business activities (The National, 2024). Dubai was particularly affected by the torrential rains (Gambrell, 2024), even though the impacts were far from uniform throughout the city. Figure 18.3a reveals that the bulk of the damage is located in the more developed area just off and parallel to the coast, with the distance to the coast, which is linked to the degree of urbanisation and population levels in the city, an important factor in explaining the intensity and spatial extent of the damages. This is also noted by Abuhani et al. (2025), who stresses the role of urban planning and infrastructure distribution on the microscale impacts of the flooding event. Some parts of Dubai were left completely isolated with limited connectivity and reduced accessibility while others were able to maintain high reachability throughout the event, despite a comparable flooding intensity (Fig. 18.3a). The primary roads experienced mostly low to moderate damage, although some segments exhibited a high degree of structural impacts. This made it harder for vehicles to go into and out of those parts of the city. With residents trapped amidst rapidly rising flood waters and their homes and businesses experiencing major damage and disruption, substantial economic losses followed, aggravated by the fact that a large fraction of the population did not have adequate insurance coverage (Ahmed, 2024; Guy Carpenter, 2024a). Insurance losses are estimated to be in the range 2.9–3.4 billion USD (Guy Carpenter, 2024b), with an overall 15% increase in property insurance rates following the extreme event (Merani, 2025). The flooding effects in Abu Dhabi were generally less impactful than in Dubai (cf. Figure 18.3a and b), most likely because of lower rainfall totals (Fig. 18.2a). As in Dubai, however, the major impacts were concentrated in the downtown and major industrial areas where most people live and work (Elessawy, 2021).

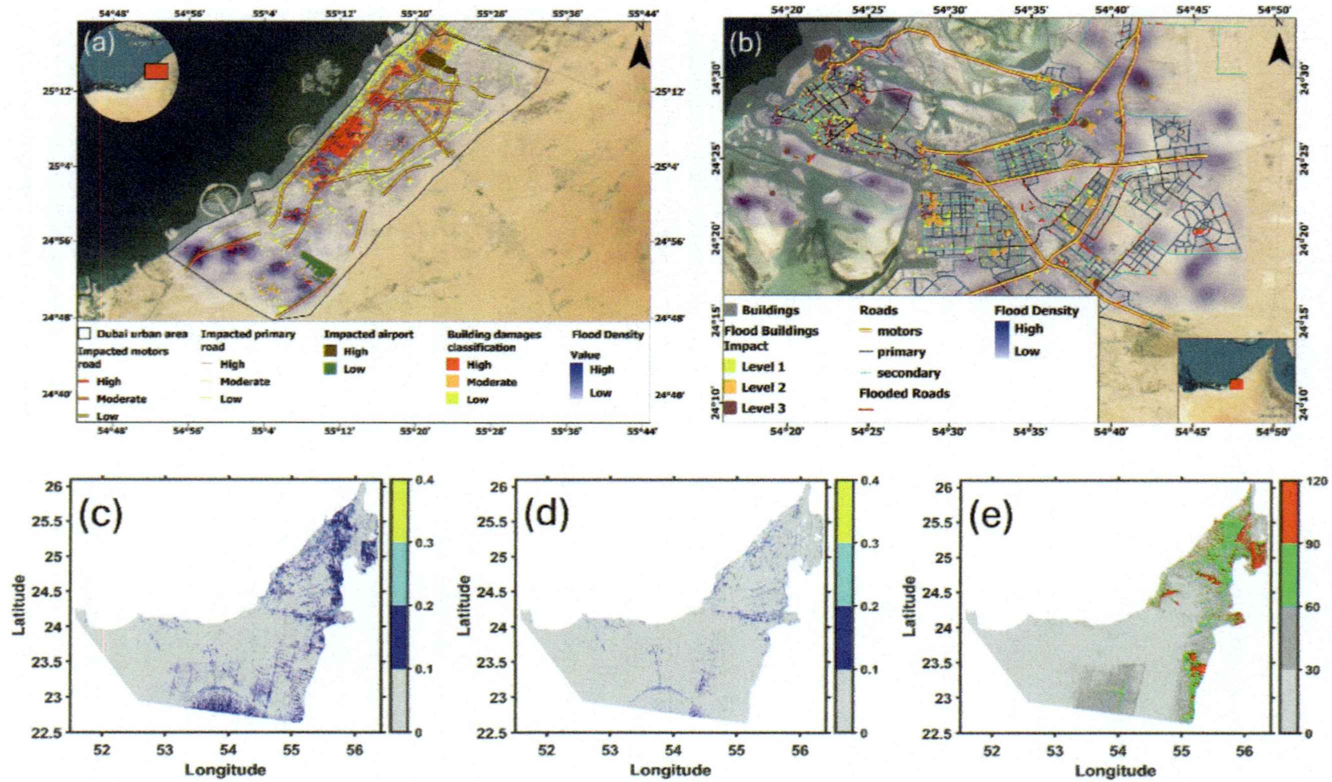

Fig. 18.3 Impacts of Extreme Event: Effects of 16 April 2024 historical rainfall event on infrastructure (roads, buildings, and airports) and flooded regions (purple shading) in the **a** Dubai and **b** Abu Dhabi urban areas, respectively. Spatial distribution of NDVI in the UAE for the month of June **c** 2024, **d** 2023, and **e** fractional change (%) between the two years

Besides direct damage to property, flooding rains can bring about mosquito-borne viruses. While such diseases are not as prevalent in the UAE as in western and southwestern Saudi Arabia where warmer winter temperatures allow the mosquitos to survive and reproduce year-round (Alyahya, 2023), they have been recorded in the recent past (Camp et al., 2019) and may become more frequent in a warming climate (Paz et al., 2021). A dengue outbreak followed the April–May 2024 flooding events in the UAE, putting pressure on health services (Webster, 2024). This stresses the need to adapt the infrastructure to make it more resilient to extreme weather conditions, ensuring a more efficient draining of the flooding waters and a rapid response of the health services to minimise the associated health risks and impacts to the population.

The heavy rain contributed to replenishing underground reservoirs, with the moister soil supporting a wide variety of shrubs as evidenced by the satellite-derived NDVI values (Fig. 18.3c–e). In the three months following the heavy rain, there is a decrease in no/low vegetation pixels by up to ~20%, and an increase in moderate and dense vegetation pixels by up to ~371% and 40%, respectively. These changes indicate a greening of the desert, with the 2–3-month lag between the precipitation event (16 April, followed by another of reduced magnitude on 02 May 2024) and the peak in vegetation response (June-July) in line with the findings of Broich et al. (2018) over the Australian Desert, and Chavez et al. (2019) over the Atacama Desert in Chile. In semi-arid and humid regions, however, the response of NDVI to rainfall is generally faster, taking up to 1.5 months (Martiny et al., 2006; Piao et al., 2003), and is more dependent on the distribution and intensity of the precipitation than on potential evapotranspiration, vegetation cover or soil type. Localised but at times heavy rainfall in eastern UAE in July and early August 2024 (Robertson, 2024) likely further increased the dense vegetated covered areas. The dry and hot conditions in the subsequent months led to the desiccation of the shrubs and vegetation, accelerated by the fact that the lower albedo of a vegetated surface promotes an increase in surface and soil temperature compared to that of a sandy (desert) surface, which encourages evapotranspiration and the drying of the soil. This is consistent with the more similar NDVI values for August-October 2024 compared to those of the previous year. Spatially, the bulk of the NDVI values between 0.1 and 0.2, indicating moderate vegetation, are found over the northeastern and eastern parts of the country (Fig. 18.3c), where most of the precipitation fell (Fig. 18.2a). Here, the fractional increase in NDVI between June 2023 and June 2024 exceeds 100% (Fig. 18.3e). The highest values are located just to the west and east of the Al Hajar mountains, suggesting higher rainfall accumulation due to orographic enhancement and subsequent downslope flow along the wadis (dry riverbeds) and the major stream networks (Al-Ruzouq et al., 2019; Alsharhan & Rizk, 2020; Elmhady et al., 2014; Sherif et al., 2010). NDVI values higher than 0.1 are also seen in the Liwa oasis in the southern part of the country, where agricultural activities have increased in the last few decades and groundwater reservoirs are present (Fragaszy & McDonnell, 2016), albeit here the differences with respect to 2023 are smaller. A comparison of Figs. 18.2a and 18.3e reveal a close link between precipitation and NDVI, stressing how rainfall episodes can have a pronounced effect on the surface properties and vegetation, besides on urban areas.

18.6 Conclusions and Recommendations

The drivers, potential role of anthropogenic climate change, and impacts of the 16 April 2024 historical flooding event in the UAE are investigated using in-situ observations, satellite-derived data, and reanalysis products. It is concluded that the destruction and loss of life did not arise only from the record-breaking precipitation amounts, made worse by anthropogenic forcing. Other factors, such as recent changes to land use land cover, infrastructure distribution, and population increases, also likely contributed to this high-impact event.

Warm and very moist air advection at low-levels, coupled with a mid-level trough, and a southward shift of the subtropical jet, are the main ingredients of the extreme event. A detection and attribution of anthropogenic climate change revealed man-made climate change has increased the likelihood of events like that of 16 April 2024: the return period decreased from 9000 years for the pre-industrial climate to 330 years for the current climate, and is projected to be around 19 for a business-as-usual climate change scenario. Recent changes to land use, in particular the increasing urbanization around the major cities of Abu Dhabi and Dubai and parts of the northwestern UAE coastline, and the rapid population growth, likely exacerbated the effects of the record-breaking rainfall amounts.

The local-scale impacts of the extreme events in the urban areas of Abu Dhabi and Dubai are further investigated using high-resolution satellite data. The degree of urbanisation, population density, and heavy precipitation are found to be linked to the effects of the flooding rains. Away from populated areas, a major consequence of the historical event is the greening of the desert, with vegetated areas increasing by up to a factor of four with respect to the previous year. Spatially, regions adjacent to mountain ranges, wadis (dry riverbeds), and stream networks exhibited the largest changes in vegetation cover, peaking 2–3 months after the rainfall event.

As the UAE, and the Arabian Peninsula in general, are particularly vulnerable to extreme precipitation events, which are projected to occur more frequently in a warming climate (Rao et al., 2024), it is important to take steps to mitigate their effects. Several recommendations are put forward following the discussion of the effects of the 16 April 2024 historical flooding episode: (i) *Investment in early warning systems and public awareness campaigns.* Despite advanced warnings by the UAE's National Centre of Meteorology (Elsa, 2024), it is possible that many people did not heed the warnings as such extreme weather episodes are uncommon in the country, which could have further aggravated their impacts. In fact, the previous widespread flooding event occurred in January 2020 (Al Serkal, 2020) and was far less impactful. As a result, it is important to raise awareness of the potential impacts of extreme weather events and the need to take preventive actions to safeguard lives and goods. More effective, direct, and targeted early warning systems must be developed and timely disseminated through all available channels to ensure everyone is aware of the impending danger; (ii) *Implementation of impact-based forecasts.* In addition to more accurate quantitative precipitation forecasts, that account for the effects of aerosols and soil dynamics, external factors such as population density and infrastructure vulnerability should be considered as well. These impact-based forecasts (Hussein et al., 2025), which can make use of artificial intelligence and machine learning algorithms, will allow for more targeted, timely and effective warnings to be issued to the public, helping to better prepare the community for the upcoming weather event; (iii) *Adaptation of Infrastructure and Investment in Flood Defences.* Buildings, roads, and houses should be modified to enhance their resilience to extreme weather episodes. Improved urban planning, which should account for the nature of the underlying terrain, building materials, and increased investments in flood defence and drainage systems, will undoubtedly assist in mitigating the effects of the extreme weather. In summary, we recommend more effort to be placed on improving the quality, accuracy, and dissemination of weather forecasts and ensuring the infrastructure can cope with high return-period precipitation events.

References

Abuhani, D. A., Aldamani, R., Zualkernan, I., Alali, R., & Grover, H. (2025). Floods' impact on cities' critical infrastructure: A case study on Dubai. *IEEE Access, 13*, 17652–17661. https://doi.org/10.1109/ACCESS.2025.3527554

Al-Ruzouq, R., Shanableh, A., Merabtene, T., Siddique, M., Ali Khalil, M., Idris, A., & Almulla, E. (2019). Potential groundwater zone mapping based on geo-hydrological considerations and multi-criteria spatial analysis: North UAE. *CATENA, 173*, 511–524. https://doi.org/10.1016/j.catena.2018.10.037

Al Serkal, M. M. (2020) UAE rainfall breaks 24-year record. Gulf news. Retrieved December 09, 2024, from https://gulfnews.com/uae/weather/uae-rainfall-breaks-24-year-record-1.1578815601808

Ahmed, A. (2024). Dubai's Flood in 2024: Historic Rainfall Brings City to Its Knees. The Gulf Entrepreneur. Retrieved December 09, 2024, from https://thegulfentrepreneur.com/dubais-flood-in-2024/

Alizadeh, O., & Mousavizadeh, M. (2025). Impact of ENSO on extreme precipitation in Southwest Asia. *Global and Planetary Change, 244*, Article 104645. https://doi.org/10.1016/j.gloplacha.2024.104645

Allen, M. (2003). Liability for climate change. *Nature, 421*, 891–892. https://doi.org/10.1038/421891a

Alsharhan, A. S., & Rizk, Z. E. (2020) Application of GIS techniques for water resources investigations in the UAE. In *Water resources and integrated management of the United Arab Emirates. World water resources*, 3. Springer. https://doi.org/10.1007/978-3-030-31684-6_23

Alsumaiti, T. S., Hussein, K., Ghebreyesus, D. T., & Sharif, H. O. (2020). Performance of the CMORPH and GPM IMERG products over the United Arab Emirates. *Remote Sensing, 12*, 1496. https://doi.org/10.3390/rs12091426

Alyahya, H. S. (2023). Prevalence of dengue fever in Saudi Arabia: Jeddah as a case study. *Entomological Research, 53*, 539–553. https://doi.org/10.1111/1748-5967.12685

Bindoff, N. L., Stott, P. A., AchutaRao, K. M., Allen, M. R., Gillett, N., Gutzler, D., Hansingo, K., Hegerl, G., Hu, Y., Jain, S., Mokhov, I. I., Overland, J., Perlwitz, J., Sebbari, R., & Zhang, X. (2013). Detection and Attribution of climate change: From Global to Regional. In T. F. Stocker, D. Qin, G.-K. Plattner, M. Tignor, S. K. Allen, J. Boschung, A. Nauels, Y. Xia, V. Bex & P. M. Midgley (Eds.), *Climate change 2013: The physical science basis. Contribution of Working Group I to the fifth assessment report of the intergovernmental panel on climate change.* Cambridge University Press.

Broich, M., Tulbure, M. G., Verbesselt, J., Xin, Q., & Wearne, J. (2018). Quantifying Australia's dryland vegetation response to flooding and drought at sub-continental scale. *Remote Sensing of Environment, 212*, 60–78. https://doi.org/10.1016/j.rse.2018.04.032

Camp, J. V., Karuvantevida, N., Chouhna, H., Safi, E., Shah, J. N., & Nowotny, N. (2019). Mosquito biodiversity and mosquito-borne viruses in the United Arab Emirates. *Parasites Vectors, 12*, 153. https://doi.org/10.1186/s13071-019-3417-8

Chavez, R. O., Moreira-Munoz, A., Galleguillos, M., Olea, M., Aguayo, J., Latin, A., Aguilera-Betti, I., Munoz, A. A., & Mariquez, H. (2019). GIMMS NDVI time series reveal the extent, duration, and intensity of "blooming desert" events in the hyper-arid Atacama Desert, Northern Chile. *International Journal of Applied Earth Observation and Geoinformation, 76*, 193–203. https://doi.org/10.1016/j.jag.2018.11.013

Deliso, M., Pack, D., Al-Tawy, A., & Khan, H. (2024) Dubai sees severe flooding after getting 2 years' worth of rain in 24 h. Retrieved December 29, 2024, from https://abcnews.go.com/International/dubai-flooding-heavy-rainfall/story?id=109321601

de Vries, A. J., Tyrlis, E., Edry, D., Krichak, S. O., Steil, B., & Lelieveld, J. (2013). Extreme precipitation events in the Middle East: Dynamics of the active Red Sea Trough. *Journal of Geophysical Research: Atmospheres, 118*, 7087–7108. https://doi.org/10.1002/jgrd.50569

de Vries, A. J., Feldstein, S. B., Riemer, M., Tyrlis, E., Sprenger, M., Baumgart, M., Fnais, M., & Lelieveld, J. (2016). Dynamics of tropical-extratropical interactions and extreme precipitation events in Saudi Arabia in autumn, winter and spring. *Quarterly Journal of the Royal Meteorological Society, 142*, 1862–1880. https://doi.org/10.1002/qj.2781

de Vries, A. J., Ouwersloot, H. G., Feldstein, S. B., Riemer, M., El Kenawy, A. M., McCabe, M. F., & Lelieveld, J. (2018). Identification of tropical-extratropical precipitation events in the Middle East based on potential vorticity and moisture transport. *Journal of Geophysical Research: Atmospheres, 123*, 861–881. https://doi.org/10.1002/2017JD027587

de Vries, A. J., Armon, M., Klingmuller, K., Portmann, R., Rothlisberger, M., & Domeisen, D. I. V. (2024). Breaking Rossby waves drive extreme precipitation in the world's arid regions. *Communications Earth & Environment, 5*, 493. https://doi.org/10.1038/s43247-024-01633-y

Easterling, D. R., Kunkel, K. E., Wehner, M. F., & Sun, L. (2016). Detection and attribution of climate extremes in the observed record. *Weather and Climate Extremes, 11*, 17–27. https://doi.org/10.1016/j.wace.2016.01.001

Elessawy, F. M. (2021). The abnormal population growth and urban sprawl of an Arabian Gulf City: The case of Abu Dhabi City. *Open Journal of Social Sciences, 9*, 245–269. https://doi.org/10.4236/jss.2021.92017

Elmhady, S. I., Mohamed, M. M., & Marghany, M. M. (2014). Mapping and classification of hydrological parameters from digital terrain data in the Musandam Peninsula, UAE and Oman. *Geocarto International, 30*, 330–345. https://doi.org/10.1080/10106049.2014.965755

Elsa, E. (2024) Heavy rain, thunder and lightning in some parts of the UAE; NCM issues weather alert. Gulf News. Retrieved December 09, 2024, from https://gulfnews.com/uae/weather/heavy-rain-thunder-and-lightning-in-some-parts-of-uae-ncm-issues-weather-alert-1.1713178972294

Eyring, V., Bony, S., Meehl, G. A., Senior, C. A., Stevens, B., Stouffer, R. J., & Taylor, K. E. (2016). Overview of the coupled model intercomparison project phase 6 (CMIP6) experimental design and organization. *Geoscientific Model Development, 9*, 1937–1958. https://doi.org/10.5194/gmd-9-1937-2016

Esfandiari, N., & Rezaei, M. (2022). Automatic detection, classification, and long-term investigation of temporal-spatial changes of atmospheric rivers in the Middle East. *International Journal of Climatology, 42*, 7730–7750. https://doi.org/10.1002/joc.7674

Feng, T., Yuang, T., Cao, J., Wang, Z., Zhi, R., Hu, Z., & Huang, J. (2023). The influence of dust on extreme precipitation at a large city in North China. *Science of the Total Environment, 901*, Article 165890. https://doi.org/10.1016/j.scitotenv.2023.165890

Fonseca, R., Francis, D., Nelli, N., & Thota, M. (2022). Climatology of the heat low and the intertropical discontinuity in the Arabian Peninsula. *International Journal of Climatology, 42*, 1092–1117. https://doi.org/10.1002/joc.7291

Fonseca, R., Francis, D., Nelli, N., Yarragunta, Y., Paparella, F., & Pauluis, O. M. (2025). Summertime secondary circulation and interaction with sea-breeze circulations. *Quarterly Journal of the Royal Meteorological Society, 1–30*. https://doi.org/10.1002/qj.4907
Fragaszy, S., & McDonnell, R. (2016). Oasis at a crossroads: Agriculture and Groundwater in Liwa, United Arab Emirates. IWMI Project Report, Groundwater governance in the Arab World. Retrieved December 10, 2024, from https://gw-mena.iwmi.org/wp-content/uploads/sites/3/2017/04/Rep.15-Groundwater-governance-in-Liwa-oasis-report_final_cover.pdf
Francis, D., Alshamsi, N., Cuesta, J., Isik, A. G., & Dundar, C. (2019). Cyclogenesis and density currents in the middle east and the associated dust activity in September 2015. *Geosciences (Switzerland), 9*(9), Article 376. https://doi.org/10.3390/geosciences9090376
Francis, D., Chaboureau, J.-P., Nelli, N., Cuesta, J., Alshamsi, N., Temimi, M., Pauluis, O., & Xue, L. (2021a). Summertime dust storms over the Arabian Peninsula and impacts on radiation, circulation, cloud development and rain. *Atmospheric Research, 250*, Article 105364. https://doi.org/10.1016/j.atmosres.2020.105364
Francis, D., Temimi, M., Fonseca, R., Nelli, N. R., Abida, R., Weston, M., & Wehbe, Y. (2021b). On the analysis of a summertime convective event in a hyperarid environment. *Quarterly Journal of the Royal Meteorological Society, 147*, 501–525. https://doi.org/10.1002/qj.3930
Francis, D., Fonseca, R., & Nelli, N. (2022) Key factors modulating the threat of the Arabian Sea's tropical cyclones to the Gulf countries. *Journal of Geophysical Research: Atmospheres,* 127, e2022JD036528. https://doi.org/10.1029/2022JD036528
Francis, D., Fonseca, R., Nelli, N., Bozkurt, D., Cuesta, J., & Bosc, E. (2023). On the Middle East's severe dust storms in spring 2022: Triggers and impacts. *Atmospheric Environment, 296*, Article 119539. https://doi.org/10.1016/j.atmosenv.2022.119539
Francis, D., & Fonseca, R. (2024). Recent and projected changes in climate patterns in the Middle East and North Africa (MENA) region. *Scientific Reports, 14*, 10279. https://doi.org/10.1038/s41598-024-60976-w
Francis, D., Fonseca, R., Bozkurt, D., Nelli, N., & Guan, B. (2024) Atmospheric river rapids and their role in the extreme rainfall event of April 2023 in the Middle East. *Geophysical Research Letters,* 51, e2024GL109446. https://doi.org/10.1029/2024GL109446
Francis, D., Fonseca, R., Nelli, N., Cherif, C., Yarragunta, Y., Zittis, G., & de Vries, A. J. (2025) From cause to consequence: examining the historic April 2024 rainstorm in the United Arab Emirates through the lens of climate change. *NPJ Climate and Atmospheric Sciences, 8*, 183. https://doi.org/10.1038/s41612-025-01073-1
Fritz, H. M., Blount, C. D., Albusaidi, F. B., & Al-Harthy, A. H. M. (2010). Cyclone Gonu storm surge in Oman. Estuarine. *Coastal and Shelf Science, 86*, 102–106. https://doi.org/10.1016/j.ecss.2009.10.019
Gambrell, J. (2024). The deser nation of UAE records its most rain ever, flooding highways and Dubai's airport. Associated Press. Retrieved December 09, 2024, from https://apnews.com/article/uae-historic-rain-storm-flooding-dubai-airport-disruption-3e838dbc169e52dcd6d11b64f79bdcb7
Guy Carpenter. (2024a). Post event report: Gulf Floods. Retrieved December 09, 2024, from https://www.guycarp.com/insights/2024/04/Gulf-Floods-April-2024.html
Guy Carpenter. (2024b). Post-event report: Gulf Floods Update—September 2024. Retrieved March 17, 2025, from https://www.guycarp.com/insights/2024/09/post-event-report-gulf-floods-update-september-2024.html
Hasanean, H., & Almazroui, M. (2015). Rainfall: Features and variations over Saudi Arabia, a review. *Climate, 3*, 578–626. https://doi.org/10.3390/cli3030578
He, C., & Kucharski, F. (2025). Tele-connected rainfall extremes over West and East Asia in April 2024 tied to Indian Ocean heating. *Climate Dynamics, 63*, 107. https://doi.org/10.1007/s00382-025-07593-z
Hejazizadeh, Z. B., Alijani, B., Saligheh, M., & Sayad, V. (2021). Analyzing the trend and extent of Sudan low pressure and its impact on precipitation in western and southwestern Iran. *Arab Journal of Geosciences, 14*, 295. https://doi.org/10.1007/s12517-021-06452-5

Hoskins, B. J., McIntyre, M. E., & Robertson, A. W. (1985). On the use and significance of isentropic potential vorticity maps. *Quarterly Journal of the Royal Meteorological Society, 111*, 877–946. https://doi.org/10.1002/qj.49711147002

Hoskins, B. J., & James, I. N. (2014). *Fluid dynamics of the midlatitude atmosphere*. Wiley.

Hou, A. Y., Ramesh, K. K., Neeck, S., Azarbarzin, A. A., Kummerow, C. D., Kojima, M., Oki, R., Nakamura, K., & Iguchi, T. (2014). The global precipitation measurement mission. *Bulletin of the American Meteorological Society, 95*, 701–722. https://doi.org/10.1175/BAMS-D-13-00164.1

Hussein, K., Alhosani, N., Al-Areeq, A. M., Al aAghbari, A. A., Elkamali, M., Alsumaiti, T., Sharif, H. O., Almurshidi, A. M. G., & Abdalati, W. (2025) Unprecedented rainfall in the United Arab Emirates: Hydrologic and flood impact analysis of the April 2024 event. *Natural Hazards*. https://doi.org/10.1007/s11069-025-07156-9

Ibrahim, O. R., & Al Maghawry, S. (2022). Review on cyclone Shaheen in the Sultanate of Oman. *Arab Journal of Geosciences, 15*, 833. https://doi.org/10.1007/s12517-022-10118-1

Kwarteng, A. Y., Dorvlo, A. S., & Vijaya Kumar, G. T. (2009). Analysis of a 27-year rainfall data (1977–2003) in the Sultanate of Oman. *International Journal of Climatology, 29*, 605–617. https://doi.org/10.1002/joc.1727

Langton, J. (2024) A look at some of the UAE's most extreme weather events after record rains and flood. *The National*. Retrieved March 17, 2025, from https://www.thenationalnews.com/news/uae/2024/04/22/extreme-weather-rain-flood-dubai/

Mann, H. B. (1945). Nonparametric tests against trend. *Econometrica, 13*, 245–259. https://doi.org/10.2307/1907187

Martin, M., Dash, P., Ignatov, A., Banzon, V., Beggs, H., Brasnett, B., Cayula, J.-F., Cummings, J., Donlon, C., Gentemann, C., Grumbine, R., Ishizaki, S., Maturi, E., Reynolds, R. W., & Roberts-Jones, J. (2012). Group for high resolution Sea surface temperature (GHRSST) analysis fields inter-comparisons. Part 1: A GHRSST multiproduct ensemble (GMPE). *Deep Sea Research Part II: Topical Studies in Oceanography, 77–80*, 21–30. https://doi.org/10.1016/j.dsr2.2012.04.013

Martiny, N., Chamberlain, P., Richard, Y., & Philippon, N. (2006). Compared Regimes of NDVI and Rainfall in Semi-arid Regions of Africa. *International Journal of Remote Sensing, 27*, 5201–5223. https://doi.org/10.1080/01431160600567787

Merani, M. (2025) Almost one year after floods, UAE insurance costs keep surging. Arabian Gulf Business Insight. Retrieved March 17, 2025, from https://www.agbi.com/banking-finance/2025/02/almost-one-year-after-floods-uae-insurance-costs-keep-surging/

Nelli, N. R., Francis, D., Fonseca, R., Abida, R., Weston, M., Wehbe, Y., & Al Hosary, T. (2021). The atmospheric controls of extreme convective events over the southern Arabian Peninsula during the spring season. *Atmospheric Research, 262*, Article 105788. https://doi.org/10.1016/j.atmosres.2021.105788

Nelli, N., Francis, D., Fonseca, R., Bosc, E., Addad, Y., Temimi, M., Abida, R., Weston, M., & Cherif, C. (2022). Characterization of the atmospheric circulation near the empty quarter desert during major weather events. *Frontiers in Environmental Science, 10*, 972380. https://doi.org/10.3389/fenvs.2022.972380

Nelli, N., Francis, D., Alkatheeri, A., & Fonseca, R. (2024). Evaluation of reanalysis and satellite products against ground-based observations in a desert environment. *Remote Sensing, 16*, 3593. https://doi.org/10.3390/rs16193593

Nesterov, O., Temimi, M., Fonseca, R., Nelli, N. R., Addad, Y., Bosc, E., & Abida, R. (2021). Validation and statistical analysis of the group for high resolution Sea surface temperature data in the Arabian Gulf. *Oceanologia, 63*, 497–515. https://doi.org/10.1016/j.oceano.2021.07.001

Normand, J. C. L., & Heggy, E. (2024). Assessing flash flood erosion following storm Daniel in Libya. *Nature Communications, 15*, 6493. https://doi.org/10.1038/s41467-024-49699-8

Pandya, U., Mudaliar, A. N., & Alvi, S. (2024). Coastal vegetation change detection using a remote sensing approach. *Environmental Sciences Proceedings, 29*, 1–17. https://doi.org/10.3390/ECRS2023-15853

Parajuli, S. P., Stenchikov, G. L., Ukhov, A., Morrison, H., Shevchenko, I., & Mostamandi, S. (2023) Simulation of dust-and-rain event across the Red Sea using WRF-Chem. *Journal of Geophysical Research: Atmospheres, 128*, e2022JD038384. https://doi.org/10.1029/2022JD038384

Parajuli, S. P., Stenchikov, G. L., Ukhov, A., Mostamandi, S., Kucera, P. A., Axisa, D., Gustafson, W. I., Jr., & Zhu, Y. (2022). Effect of dust on rainfall over the Red Sea coast based on WRF-Chem model simulations. *Atmospheric Chemistry and Physics, 22*, 8659–8682. https://doi.org/10.5194/acp-22-8659-2022

Paz, S., Majeed, A., & Christophides, G. K. (2021). Climate change impacts on infectious diseases in the Eastern Mediterranean and the Middle East (EMME)—Risks and recommendations. *Climatic Change, 169*, 40. https://doi.org/10.1007/s10584-021-03300-z

Philip, S., Kew, S. F., van Oldenborgh, G. J., Aalbers, E., Vautard, R., Otto, F., Haustein, K., Habets, F., & Singh, R. (2018). Validation of a rapid attribution of the May/June 2016 flood-inducing precipitation in France to climate change. *Journal of Hydrometeorology, 19*, 1881–1898. https://doi.org/10.1175/JHM-D-18-0074.1

Philip, S., Kew, S., van Oldenborgh, G. J., Otto, F., Vautard, R., van der Wiel, K., King, A., Lott, F., Arrighi, J., Singh, R., & van Aalst, M. (2020). A protocol for probabilistic extreme event attribution analyses. *Advances in Statistical Climatology, Meteorology and Oceanography, 6*, 177–203. https://doi.org/10.5194/ascmo-6-177-2020

Piao, S., Fang, J., Zhou, L., Guo, Q., Henderson, M., Ji, W., Li, Y., & Tao, S. (2003). Interannual variations of monthly and seasonal normalized difference vegetation index (NDVI) in China from 1982 to 1999. *Journal of Geophysical Research, 108*, 4401. https://doi.org/10.1029/2002JD002848

Pradhan, R. K., Markonis, Y., Godoy, M. R. V., Villalba-Pradas, A., Andreadis, K. M., Nikolopoulos, E. I., Papalexiou, S. M., Rahim, A., Tapiador, F. J., & Hanel, M. (2022). Review of GPM IMERG performance: A global perspective. *Remote Sensing of the Environment, 268*, Article 112754. https://doi.org/10.1016/j.rse.2021.112754

Rao, K. K., Al Mandous, A., Al Ebri, M., Al Hameli, N., Rakib, M., & Al Kaabi, S. (2024). Future changes in the precipitation regime over the Arabian Peninsula with special emphasis on UAE: Insights from NEX-GDDP CMIP6 model simulations. *Scientific Reports, 14*, 151. https://doi.org/10.1038/s41598-023-49910-8

Rasmusson, E. M., & Carpenter, T. H. (1982). Variations in tropical sea surface temperature and surface wind fields associated with the Southern Oscillation/El Nino. *Monthly Weather Review, 110*, 354–384. https://doi.org/10.1175/1520-0493(1982)110%3c0354:VITSST%3e2.0.CO;2

Robertson, A. (2024). UAE weather: Heavy rain and hail recorded amid unsettled conditions. Retrieved December 10, 2024, from https://www.thenationalnews.com/news/uae/2024/08/05/uae-weather-cloudy-week-ahead-with-cooler-temperatures-and-chance-of-rain/

Rodwell, M. J., & Hoskins, B. J. (1996). Monsoons and the dynamics of deserts. *Quarterly Journal of the Royal Meteorological Society, 122*, 1385–1404. https://doi.org/10.1002/qj.49712253408

Sandeep, S., & Ajayamohan, R. S. (2018). Modulation of winter precipitation dynamics over the Arabian Gulf by ENSO. *Journal of Geophysical Research: Atmospheres, 123*, 198–210. https://doi.org/10.1002/2017JD027263

Schott, F. A., Xie, S.-P., & McCreary Jr., J. P. (2009). Indian Ocean circulation and climate variability. *Reviews of Geophysics, 47*, RG1002. https://doi.org/10.1029/2007RG000245

Sen, P. K. (1968). Estimates of the regression coefficient based on Kendall's Tau. *Journal of the American Statistical Association, 63*, 1379–1389. https://doi.org/10.1080/01621459.1968.10480934

Sherif, M. M., Mohamed, M. M., ASCE, A. M., Shetty, A., & Almulla, M. (2010). Rainfall-runoff modeling of three Wadis in the Northern area of UAE. *Journal of Hydrological Engineering, 16*. https://doi.org/10.1061/(ASCE)HE.1943-5584.0000280

Stein, A. F., Draxler, R. R., Rolph, G. D., Stunder, B. J. B., Cohen, M. D., & Ngan, F. (2015). NOAA's HYSPLIT atmospheric transport and dispersion modeling system. *Bulletin of the American Meteorological Society, 96*, 2059–2077. https://doi.org/10.1175/BAMS-D-14-00110.1

Stott, P., Stone, D., & Allen, M. (2004). Human contribution to the European heatwave of 2003. *Nature, 432*, 610–614. https://doi.org/10.1038/nature03089

Taraphdar, S., Gopalakrishnan, D., Liu, C., Pauluis, O. M., Xue, L., Ajayamohan, R. S., Ruby Leung, L., Hagos, S., Grabowski, W. W., Chen, S., Rasmussen, R. M., & Tessendorf, S. A. (2025). Subtropical jet regulates Arabian winter precipitation: A viable mechanism. *Journal of the Atmospheric Sciences*. https://doi.org/10.1175/JAS-D-23-0213.1

Terry, J. P., Al Ruheili, A., Almarzooqi, M. A., Almheiri, R. Y., & Alshehhi, A. K. (2023). The rain deluge and flash floods of summer 2022 in the United Arab Emirates: Causes, analysis and perspectives on flood-risk reduction. *Journal of Arid Environments, 215*, Article 105013. https://doi.org/10.1016/j.jaridenv.2023.105013

The National. (2024). Dubai flights: all arrivals diverted away from airport amid floods and rain in UAE. Retrieved March 17, 2025, from https://www.thenationalnews.com/news/uae/2024/04/16/abu-dhabi-and-dubai-lashed-by-heavy-rain-thunder-and-lightning/

Theil, H. (1992). A rank-invariant method of linear and polynomial regression analysis. In B. Raj & J. Koerts (Eds.), *Henri Theil's contributions to economics and econometrics. Advanced studies in theoretical and applied econometrics*, 23. Springer, Dordrecht. https://doi.org/10.1007/978-94-011-2546-8_20

Tsvieli, Y., & Zangvil, A. (2007). Synoptic climatological analysis of Red Sea Trough and non-Red Sea Trough rain situations over Israel. *Advances in Geosciences, 12*, 137–143. https://doi.org/10.5194/adgeo-12-137-2007

Ullah, W., Alabduoli, K., Ullah, S., Al-Ghamdi, S. G., Alhebsi, K., Almazroui, M., Assiri, M. E., Azeem, W., Abuelgasim, A., & Hagan, D. F. T. (2024). Comparison of 2-m surface temperature data between reanalysis and observations over the Arabian Peninsula. *Atmospheric Research, 311*, Article 107725. https://doi.org/10.1016/j.atmosres.2024.107725

van Oldenborgh, G. J., van der Wiel, K., Kew, S., Philip, S., Otto, F., Vautard, R., King, A., Lott, F., Arrighi, J., Singh, R., & van Aalst, M. (2021). Pathways and pitfalls in extreme event attribution. *Weather Attribution, 166*, 13. https://doi.org/10.1007/s10584-021-03071-7

van Zyle, G. (2024) Highest rain levels recorded in UAE since records began in 1949. Gulf Business. Retrieved November 25, 2024, from https://gulfbusiness.com/uae-record-breaking-rain-1949/

Vautard, R., van Oldbenborgh, G. J., Otto, F. E. L., You, P., de Vries, H., van Meijgaard, E., Stepek, A., Soubeyroux, J.-M., Philip, S., Kew, S. F., Costella, C., Singh, R., & Telbadi, C. (2019). Human influence on European winter wind storms such as those of January 2018. *Earth System Dynamics, 10*, 271–286. https://doi.org/10.5194/esd-10-271-2019

Webster, N. (2024) UAE steps up anti-mosquito measures after heavy rain. Retrieved December 09, 2024, from https://www.thenationalnews.com/news/uae/2024/05/04/uae-steps-up-anti-mosquito-measures-to-limit-spread-of-disease/

Wehbe, Y., Ghebreyesus, D., Temimi, M., Milewski, A., & Al Mandous, A. (2017). Assessment of the consistency among global precipitation products over the United Arab Emirates. *Journal of Hydrology: Regional Studies, 12*, 122–135. https://doi.org/10.1016/j.ejrh.2017.05.002

Wehbe, Y., Temimi, M., Weston, M., Chaouch, N., Branch, O., Schwitalla, T., Wulmeyer, V., Zhan, X., Liu, J., & Al Mandous, A. (2019). Analysis of an extreme weather event in a hyper-arid region using WRF-hydro coupling, station, and satellite data. *Natural Hazards and Earth System Sciences, 19*, 1129–1149. https://doi.org/10.5194/nhess-19-1129-2019

Wehbe, Y., Temimi, M., & Adler, R. F. (2020). Enhancing precipitation estimates through the fusion of weather radar, satellite retrievals, and surface parameters. *Remote Sensing, 12*, 1342. https://doi.org/10.3390/rs12081342

Xie, S.-P., Hu, K., Hafner, J., Tokinaga, H., Du, Y., Huang, G., & Sampe, T. (2009). Indian Ocean capacitor effect on Indo-Western Pacific climate during the summer following El Nino. *Journal of Climate, 22*, 730–747. https://doi.org/10.1175/2008JCLI2544.1

Yarragunta, Y., Francis, D., Fonseca, R., & Nelli, N. (2025). Evaluation of the WRF-Chem performance for gaseous pollutants over the United Arab Emirates. *Atmospheric Chemistry and Physics, 25*, 1685–1709. https://doi.org/10.5194/egusphere-2024-959

Zachariah, M., Kimutai, J., Barnes, C., Gryspeerdt, E., Seneviratne, S. I., Almazroui, M., Vautard, R., Zhang, X., Pinto, I., Vahlberg, M., Sengupta, S., Saeed, F., & Otto, F. E. L. (2025). Heavy precipitation hitting vulnerable communities in the UAE and Oman becoming an increasing threat as the climate warms. *World Weather Attribution*. Retrieved March 13, 2025, from https://www.worldweatherattribution.org/heavy-precipitation-hitting-vulnerable-communities-in-the-uae-and-oman-becoming-an-increasing-threat-as-the-climate-warms/

Zhang, W., Zhou, T., Ye, W., Zhang, T., Zhang, L., Wolski, P., Risbey, J., Wang, Z., Min, S.-K., Ramsay, H., Brody, M., Grimm, A., Clark, R., Ren, K., Jiang, J., Chen, X., Fu, S., Li, L., Tang, S., & Hu, S. (2025). A year marked by the extreme precipitation and floods: Weather and climate extremes in 2024. *Advances in Atmospheric Sciences*. https://doi.org/10.1007/s00376-025-4540-4

Zittis, G., Almazroui, M., Alpert, P., Ciais, P., Cramer, W., Dahdal, Y., Fnais, M., Francis, D., Hadjinicolaou, P., Howari, F., Jrrar, A., Kaskaoutis, D., Kulmala, M., Lazoglou, G., Lin, X., Mihalopoulos, N., Rudich, Y., Sciare, J., Stenchikov, G., Xoplaki, E., & Lelieveld, J. (2022) Climate change and weather extremes in the Eastern Mediterranean and Middle East. *Reviews of Geophysics, 60*, e2021RG000762. https://doi.org/10.1029/2021RG000762

Zhuang, Y. (2024). Heavy Rain and Floods Kill 19 in Oman and Disrupt Dubai Airport. Retrieved December 29, 2024, from https://www.nytimes.com/2024/04/17/us/dubai-flood-rain-oman-deaths.html

Chapter 19
The 2020 Beirut Port Explosion: A Wake-up Call for Disasters in Lebanon

Maroun El Rayess

Abstract This chapter provides an overview of both natural hazards and human-caused disasters to which Lebanon is vulnerable and highlights how chronic political instability, mismanagement, and weak institutions have deepened the country's exposure to risk. Attention is paid to recent events such as the 2020 Beirut port explosion, as well as other events to show how poor governance, unregulated urban growth and accountability issues have undermined Lebanon's ability to successfully identify, prevent, prepare for, and respond to different types of threats. The Beirut port explosion serves as an example of institutional failure, and both scientific research and media reports point out regulatory gaps, poor risk management strategies, and lack of accountability. The chapter will also highlight efforts and progress made over time by the Lebanese government to advance disaster risk reduction and management, including the creation of the Disaster Risk Management Unit and attempts to align with international standards. It further highlights how challenges such as weak enforcement and heavy reliance on foreign aid continue to hamper real progress. Lebanon continues to focus on responding to disasters after they happen rather than working to identify and prevent them, which is evident in the analysis, implying that this approach is still mostly reactive. The Lebanese case offers valuable takeaways for the MENA region as it capitalises the importance of building stronger, more coordinated, and accountable systems to identify, avoid, prepare for, manage risks, and improve resilience.

Keywords Emergency preparedness · Crisis management · Ammonium nitrate explosion · Beirut blast · Lebanon

M. El Rayess (✉)
Rabdan Academy, Abu Dhabi, United Arab Emirates
e-mail: melrayess@ra.ac.ae

E. Alam (ed.), *Disaster, Crisis and Emergency Management in the Middle East & North Africa*, https://doi.org/10.1007/978-3-032-24436-9_19

19.1 Introduction

Lebanon occupies a relatively small geographical location on the east coast of the Mediterranean, bordered by Syria and Israel, with a population exceeding five million. The Lebanese economy encountered serious contractions with many attributing it to years of mismanagement, unforeseen crises, political instabilities, and large number of refugees. In the last 25 years, a series of events have contributed to institutional decay and hampered efforts to implement effective and efficient disaster risk reduction and emergency preparedness strategies. These events were driven either by social, economic, political, military, or sectarian changes. For instance, ongoing conflicts in South Lebanon, economically driven protests, the assassination of Prime Minister Rafic Hariri, the Cedar Revolution, all played a role. These were followed by the war with Israel in 2006 and a wave of political assassinations targeting Lebanese leaders, members of the Parliament, journalists, and senior intelligence officials. The outbreak of the Syrian crisis in 2011 resulted in mass waves of displaced populations entering Lebanon. These immigrants mainly ended up settling in the country's most vulnerable regions, eventually making Lebanon host to the largest number of refugees per capita and per square kilometre in the world (United Nations High Commissioner for Refugees, 2024). In 2014, violent clashes erupted in Arsal, a Lebanese town on the Syrian border, between the Lebanese Security Forces (LSF) and al-Nusra Front fighters, resulting in the arrest of the al-Nusra Front commander by LSF. At least sixteen policemen were captured in retaliation by al-Nusra Front fighters, engaging in deadly confrontations with the Lebanese army.

In 2017, fighting broke out in Nahr El-Barid, a Palestinian camp near Tripoli in north Lebanon, between the Lebanese Armed Forces and Fatah al-Islam, prompting a prolonged counterterrorism operation in the Anti-Lebanon mountain range. The year 2019 witnessed another series of events starting with terrorist attacks, rising tensions on the southern border, and over one hundred large wildfires. The same year also witnessed the eruption of the October 17 revolution following a cabinet resolution to impose taxes on WhatsApp. The revolution resulted in a large-scale protest that lasted for months with clashes between protesters and security forces. As a result, the country was shut down for a prolonged period, leading to the resignation of the entire government. Lebanon's economy slid into a rapid and severe downfall. With the local currency losing over 98 per cent of its value and foreign currency inflows declining significantly, particularly with the outflow of the U.S. dollars, Lebanon's situation worsened further in 2020. The COVID-19 pandemic, the Beirut port explosion, and the ongoing political instability made matters even worse, posing an unprecedented challenge for the Lebanese government.

The 2020 Beirut port explosion remains a harrowing symbol of institutional decay and the catastrophic consequences of systemic negligence killing hundreds, injuring thousands, displacing hundreds of thousands, and causing billions of dollars in damages. Occurring amid the country's worst economic collapse, political paralysis, and the pandemic, this catastrophic event revealed the lethal interplay of bureaucracy, neglect, lack of accountability, and the absence of a proactive emergency preparedness and crisis management framework. This chapter synthesises a selection of local newspaper narratives, scientific analyses, and Lebanon's broader disaster preparedness, reduction, and management initiatives. It also examines the explosion's causes, impacts, and enduring lessons.

19.2 Methods

To examine the 2020 Beirut Port explosion as a case of systemic failure and government negligence, a qualitative, document-based methodology was employed. This chapter draws primarily on two sources of information: academic literature published in the aftermath of the explosion and newspaper coverage. A review of academic literature published after 2020 was conducted to provide an overview of disasters in Lebanon and to examine important aspects related to the Beirut Port explosion, such as the storage and handling of ammonium nitrate, the physical and environmental aftermath, public health implications, and the administrative failures that resulted in the disaster.

To complement the literature review, headlines from nine local newspapers in Arabic, English, and French were analysed to reflect the diverse political and ideological perspectives that characterise Lebanon's fragmented media landscape. The review of headlines was equally important as newspapers acted as first responders, documenting the impact of the tragedy, public reactions, and early officials' statements. Newspapers were selected from the recognised publications listed on the Lebanese Ministry of Information website. The combined use of scholarly sources and local media provides contextual understanding and a comprehensive view of both the immediate reporting and the subsequent analysis related to the Beirut Port explosion.

19.3 An Overview of Natural Hazards and Human-Caused Disasters in Lebanon

Lebanon's exposure to both natural hazards and human-caused disasters is the result of a complex intersection between environmental vulnerability and structural weaknesses. Lebanon's geography mostly consists of high elevations with urban concentrations, and it is situated along major underground fault lines. This makes it susceptible to various natural hazards and disasters, which are further intensified by human-driven pressures such as the absence of effective land-use planning. Concurrently, other factors including chronic political instability and institutional fragmentation have weakened the country's ability to manage and prepare for potential risks effectively. While disasters can occur unexpectedly and, in any country, planning and preparedness contribute to minimising the impact of disasters, this is not the case in Lebanon, given the recurrence of disasters that reflect a broader pattern of governance lapse and infrastructural neglect. This section provides an overview of the major natural and human-induced disasters that have shaped Lebanon's recent history, emphasising the interdependencies between ecological and political stressors and institutional shortcomings.

19.4 Natural Hazards

Wildfires: Lebanon has suffered numerous wildfires over the years. According to "Forest fires in Lebanon: a recurring disaster" (2019), a study by the Issam Fares Institute for Public Policy and International Affairs at the American University of Beirut, in 2007 a total of 275 wildfires were recorded across Lebanon, resulting in the destruction of approximately 4031 hectares of land. Just a few years later, Lebanon experienced an even more severe wildfire season in 2010, with 320 fires impacting around 4661 hectares, one of the most damaging years in terms of both frequency and scale of fire damage. The fire incidents declined in 2014 with 156 wildfires reported; however, these still led to the burning of 1852 hectares of land. Two years later, the number rose again with 260 wildfires igniting across the country, causing damage to nearly 1871 hectares. Lebanon experienced significant wildfires in 2017 and 2018, according to a 2020 report published by the Department of Ecosystems at the Ministry of Environment, in collaboration with the Land and Natural Resources Programme at the University of Balamand's Institute of the Environment. The report also notes that 194 wildfires were reported in 2019, destroying approximately 3155 hectares of land, with mountainous regions suffering the greatest damage. The size and damage caused by the fires escalated by the growing impacts of climate change and the country's limited firefighting resources. The scale of the disaster revealed weaknesses in Lebanon's local emergency response capabilities, prompting the need to solicit help from neighbouring countries to assist in battling the flames, including Cyprus, Jordan, Turkey, and Greece. A continual trend can be observed, despite year-to-year fluctuations in the number of fires and areas affected.

Earthquakes: Despite experiencing low to moderate seismic activity, Lebanon lies along important fault lines that have historically produced strong and devastating earthquakes. The country is considered among the most vulnerable in the Mediterranean due to its susceptibility to seismic hazards. Significant vulnerabilities are presented by the high concentration of urban development in coastal areas, as demonstrated by Beirut, the capital, which is home to more than 40% of the country's population and acts as the centre of governmental, economic, and administrative activities. Additionally, the country lacks comprehensive and up-to-date building codes that fully incorporate current earthquake-resistant construction standards (Libris, 2019). The Dead Sea Fault, a transform plate boundary poses a significant seismic risk stretching approximately 1200 kilometres from the Gulf of Aqaba to Turkey passing through Lebanese territory. The Dead Sea Fault consists of three primary branches, each historically responsible for powerful earthquakes exceeding magnitude 7 on the Richter scale (Libris, 2019). A powerful example is the earthquake and tsunami of the year AD 551 off the coast of Mount Lebanon (in the sea), and the destructive earthquake along the Yammouneh Fault in AD 1202. According to speleologists examining the Jeita and Kanaan caves north of Beirut, the area is susceptible to strong earthquakes with peak ground accelerations between 0.2 and 0.6 g (gravitational acceleration). Paleoseismic data shows that the Yammouneh Fault and the slopes of Mount Lebanon are still active enough to rupture in the future, despite the moderate seismic activity observed in recent years. Lebanon's earthquake preparedness measures continue to be inadequate. The Public Safety Law was fully enacted in 2018, after its initial publication in 2005 and after undergoing several revisions, a delay that is indicative of a pattern of a slow institutional response to a recurring natural threat.

Floods: Floods normally occur between December and March, particularly after heavy storms or at the beginning of snowmelt. Most of these floods are related to anthropogenic factors, especially uncontrolled urban sprawl, causing damage estimated to amount to 15 million US$ per year (Abdallah, 2019). The author notes that urban development remains uncontrolled, with little to no enforcement of regulations. For example, roads and bridges are often built across waterways, and buildings are frequently constructed along their edges. These areas are also commonly used as dumping grounds for rocks and waste—practices that significantly heighten their vulnerability to environmental hazards. The North of Lebanon and urban areas in the Beqaa region are at high risk of flooding. A recent example is the Northern Lebanon's 2024 floods which submerged 8000 hectares of farmland, displaced 10,000 people, and unearthed land mines from Syria's civil war and then scattered them around (UNDRR, 2024).

Landslides and Soil Erosion: Lebanon's rugged topography makes it vulnerable to landslides, which can be triggered by different factors including earthquakes, rainfall, and human activities and can contribute to human and economic losses. The West Beqaa is known to be a mountainous area and thus is prone to landslides. Similarly, the northern coastal area of Lebanon is also at risk. In January 2019, a major road that connects Beirut to Tripoli was partially blocked due to heavy rains that caused a substantial landslide, injuring at least three civilians (Antar, 2020). Recent studies confirm a lack of government planning for soil erosion, landslides, and hazard maps in Lebanon (Nuwayhid et al., 2022).

Tsunamis: Lebanon is categorised as a moderate tsunami hazard zone. There is a 10% probability, or more on a rolling basis, of a potentially destructive tsunami occurring within the next 50 years, as outlined by a 2020 ThinkHazard report (ThinkHazard, 2020). The National Center for Geophysics, under the National Council for Scientific Research in Lebanon (CNRS-L), serves as the country's official tsunami warning agency. History shows that different tsunami events occurred on the Lebanese coast. Probably the one with the most significant impact was recorded during the AD 551earthquake, which John of Ephesus (c. 507–586/588), a Miaphysite bishop and prominent historian of early Syriac Christianity describes it as follows, according to excerpts from the English translation (Guidoboni et al., 1994, p. 334):

"In the year 870 [of the Greeks, i.e. 558-559 AD] there was a severe earthquake, and Beirut collapsed, as did many coastal cities and villages in Galilee, Arabia, Palestine and Samaria. Along the whole Phoenician coast, too, the sea withdrew and retreated nearly two miles. As for the terrible disaster and the great and remarkable portent which happened in the city of Beirut in Phoenicia, when the earthquake took place and the cities collapsed, we had decided to make it a warning sign for the knowing of posterity. For when the earthquake came from heaven, the sea withdrew and retreated from Beirut and the other coastal cities of Phoenicia for a distance of nearly two miles; the dreadful depths of the sea became visible and various and amazing sights were revealed: sunken ships full of different cargoes and other ones which suddenly, when the sea withdrew from the land, were moored in the harbours, settled on the ground and they were broken to pieces when the sea left them and withdrew on God's command […] Then, by a secret command, a tremendous surge of the sea rushed up to return to its original depth, overwhelmed and consumed all these wretched people in the depths of its swirling waters".

Hazard preparedness efforts are still insufficient in Lebanon, despite this historical record and the country's known tsunami risk. The planning and development of coastal infrastructure does not consistently incorporate tsunami risk consideration, except for occasional projects and activities. One notable exception occurred in 2014, when the city of Byblos, which is located on the coast of the Mediterranean Sea about 30 km north of Beirut, organised a tsunami simulation exercise that envisioned a hypothetical scenario involving an earthquake-induced tsunami. The initiative was in collaboration with the United Nations Development Programme (UNDP), the Lebanese Army, the National Committee for Disaster Management and the Response Committee of the Caza of Byblos. This programme was carried out regionally in the Caza of Byblos and within the parameters of the National Response Plan.

19.5 Human-Caused Disasters

The impact of human-caused disasters can be catastrophic, and Lebanon's recent history is a testament to that impact. The consequences of the different conflicts, political dysfunction, and systemic mismanagement have impacted nearly every aspect of life in Lebanon. The country's vulnerability is not only due to being at the crossroads of never-ending regional tensions, but also the product of decades of internal unrest, entrenched transparency and accountability issues, and institutional paralysis. From the devastation wrought by internal and external wars and armed conflicts to the persistent threats posed by political violence and economic collapse, Lebanese society has been repeatedly tested by crises of human origin. These crises have claimed thousands of lives, and the country is left with worn-out infrastructure and a decreased trust in government institutions. The short and long-term consequences have included increasing poverty levels and straining the resilience of Lebanese people already burdened by the influx of refugees and the decay of essential services.

Wars and Armed Conflicts

The Lebanese Civil War (1975–1990) is one of the bloodiest and most devastating chapters in the history of this conflict-ridden country which resulted in many victims (Paul et al., 2013). This prolonged period of conflict also contributed to the disruption of services and the destruction of the country's infrastructure. Shortly after the start of the Lebanese civil war, Syria began its occupation of Lebanon. This was also a start of a period of interference in Lebanon's internal affairs, troubling the relationship between the two countries. As a result of a troubling period, political divisions, and different interferences, the Lebanese parliament was unable to elect a new president in 1988. Consequently, the Commander of the Lebanese Armed Forces assumed the role of acting prime minister. The latter launched a war against the Lebanese Forces, a previously armed political party that later transitioned into a political party with no military wing, after playing a significant role during the civil war. The acting prime minister further declared the 'war of liberation' to expel the Syrian army, end their occupation and weaken their influence. The results of the two wars were catastrophic, leading to substantial number of casualties, destruction, and strengthening rather than weakening the Syrian influence. Additionally, Lebanon has endured countless military operations along its southern border and in Beirut's southern suburbs between Israel and Hezbollah, a Lebanese Shia Islamist political party with a military wing. These include the occupation of south Lebanon and the wars in 1982, 2000, and 2006, and most recently in 2023 till 2025. These wars have had the most significant impact over many decades and have weakened the entire country as well as hampered efforts towards reform and planning.

Bombings, Assassinations, and Political Violence

Lebanon has also been the site of numerous intense bombings and assassinations of high-profile figures. While the list is long, president-elect Bachir Gemayel is only one of these examples in Lebanon's recent history. He was assassinated by a massive bomb blast in 1982 just 21 days after being elected. This was followed by the bombings of the embassy of the United States in Lebanon in 1983 and 1984. Another more recent example is the assassination of Prime Minister Rafik Hariri in 2005, which triggered widespread unrest that included a series of bombings targeting mainly political figures and journalists. The bombings also targeted civilians, such as the attack on a bus carrying 26 people in one of the mountainous villages in 2007.

Systemic Failures

Governance failures have paralysed effective planning and decision-making across public institutions in Lebanon, with no efficient preventive strategies pertaining to threats and hazards. This systemic neglect floated further to the surface following the 2020 Beirut port explosion, where officials disregarded sixteen formal warnings pertaining to the unsafe storage of ammonium nitrate (AN). This purposeful negligence ultimately led to one of the most devastating non-nuclear explosions in history. Pervasive mismanagement, potential misuse of power, and entrenched issues with transparency and accountability have also played a central role in the country's economic collapse, as evidenced by hyperinflation reaching 1000 percent in 2023 and the rapid devaluation of the local currency. Together, these factors resulted in a substantial percentage of the population struggling to afford necessities to their families. Compounding all the above was the decayed infrastructure, including ageing electrical grids, debilitated public services, unregulated construction, and the absence of proper waste management systems. The significant influx of Syrian refugees has further worsened the deficiencies, particularly in the last 15 year. The 2015 garbage crisis, which saw 15,000 tonnes of waste accumulate on the streets of Beirut, exemplified how infrastructural neglect amplifies public health and environmental risks, further destabilising Lebanese society.

19.5.1 The Case Study of the 2020 Beirut Port Explosion

As stated earlier in this chapter, the Beirut port explosion is perceived as one of the largest non-nuclear disasters in history according to national and international sources. On August 4, 2020, a stockpile of approximately 2750 tonnes of ammonium nitrate ignited, unleashing a blast equivalent to 1.1 kilotonnes of trinitrotoluene, more commonly known as TNT. The explosion killed 220 people, injured over 6500, displaced over 300,000 residents, and created a massive blast that produced a 140-metre-wide crater and an earthquake of a 3.3 magnitude on Richter scale (Al-Hajj et al., 2021). The blast also caused extensive damage to historic buildings and urban areas, posing significant challenges to recovery and rehabilitation (Kallas & Napolitano, 2023). Following the explosion, local community organisations stepped in to help with disaster management. However, the lack of communication, collaboration, and government support made non-profit organisations' interventions ineffective (Haddad & Sakr, 2023). The following subsections will examine key dimensions of the explosion as addressed in academic literature, particularly the storage conditions of the ammonium nitrate, and regulatory failures that preceded the blast. The explosion itself will also be examined, including its causes, immediate consequences, and broader public health implications. Lastly, the failures that contributed to the tragedy will be briefly discussed, including the role of government negligence, systemic lapses, and institutional accountability.

19.5.2 Storage Conditions

In 2013, the ammonium nitrate (AN) was confiscated from a Russian-owned ship called MV Rhosus, which was originally destined for Mozambique. However, due to financial difficulties that also created unrest among the ship's Russian and Ukrainian crew, the ship stopped in Beirut and the AN was stored in warehouse/hangar12, which was near residential areas. The ammonium nitrate was abandoned for six years and stored alongside other combustible materials, including fireworks (Dysz et al., 2023). Prugh (2020) noted that, AN is stable under normal or standard conditions which refers to storage and temperature. However, the author notes that AN can become highly explosive, particularly when mixed with impurities or exposed to fire. It appears that the warehouse lacked adequate air circulation, security, and fire suppression systems, leading to the degradation of the AN overtime into explosive consolidated masses due to humidity and poor storage conditions. Its prolonged storage near residential zones, inadequate fire safety protocols, and lack of regulatory enforcement reveal systemic failures, and the blast exposed gaps in Lebanon's adherence to international standards like the Seveso Directive (Sivaraman & Varadharajan, 2021).

19.5.3 The Explosion

The explosion was triggered by a fire that ignited the stored ammonium nitrate. The uncontrolled fire and the improper storage conditions caused the ammonium nitrate to explode. This explosion contributed to the formation of a 140-metre crater and an earthquake with a magnitude of 3.3 on the Richter scale, and the blast was classified as the third most destructive urban explosion of all time (Dysz et al., 2023). Apart from the short-term impact that included lost lives, injuries, and psychologically impacted individuals, the explosion released a huge amount of chemicals, including ammonium nitrate, raising concerns about the long-term effects on air quality and public health (Zhao et al., 2023). The blast occurred during Lebanon's economic collapse and the COVID-19 crisis, revealing various gaps, including in chemical safety laws and emergency preparedness. While investigations are not concluded, various media reports pointed out the non-existence of buffer zones and the absence of a comprehensive plan that addressed the regulation of such dangerous materials.

19.5.4 Government Negligence

There was significant government negligence in managing and regulating the storage of hazardous materials confiscated from the Russian-owned ship, and despite being aware of the dangerous arrangement, no effective measures were taken to prevent the disaster (Skinner, 2022). Lebanese customs, port officials, security agencies, and other officials repeatedly ignored warnings about the risks, despite various memos and documents pleading for removal of the AN to avert 'serious danger'. One memo issued in 2014 explicitly warned officials that the nitrate could 'destroy Beirut'. Yet no regulatory action took place to prevent this disaster.

Many senior officials, including those in the Ministry of Public Works and Transportation and General Directorate of State Security, were informed of the risks of storing ammonium nitrate in Hangar 12. However, no adequate precautionary measures were taken to prevent the disaster. It is as if all those who knew about the danger chose to bury their heads in the sand, hoping that the danger will go away by itself. The explosion exacerbated an existing severe economic crisis and political instability, leading to widespread protests and calls for government accountability (Ziade, 2023). The Lebanese authorities rushed to act after the blast, when it was too late.

19.5.5 Lebanese Newspapers' Response to the Beirut Port Explosion

Shortly after the catastrophic explosion, the shock, grief, and political frustration that engulfed the country were reflected in various newspapers headlines. As the scale of the devastation unfolded with more than 200 lives lost, thousands injured, and entire neighbourhoods reduced to rubble. Lebanese press played a critical role in both documenting and reporting the devastation and expressing the people's anger and pain. Front pages of almost every newspaper progressively transformed into space not only to report and document facts, but to reflect the collective sorrow, demand accountability, and to confront a rooted crisis of governance.

Newspapers also tracked and reported the investigation process, and it is worth noting that at of the time of this writing, the investigation into the Beirut port explosion remains inconclusive. The initial investigative judge charged two former ministers with negligence, prompting swift political backlash. On the grounds of 'perceived bias', the judge was removed from one of the most important cases in the country's history with many claiming that the fact that his residence was damaged in the explosion compromised his impartiality. His successor was not in a better position, and he also faced persistent political and legal obstruction. Multiple lawsuits filed by political figures that he sought to question have resulted in repeated suspensions of the inquiry. The investigation resumed in early 2025 following a prolonged hiatus. This section examines headlines published on August 5 (Ministry of Information, 2020a) and 6 (Ministry of Information, 2020b) drawn from a selection of Arabic, English, and French local newspapers as critical texts that capture the emotional intensity, historical resonance, and political weight of one of the darkest episodes in Lebanon's recent history.

19.5.6 Headlines from August 5, 2020

The press mirrored the national trauma, institutional collapse, and collective grief immediately after the explosion. Arabic-language newspapers headlines articulated physical and emotional damage and the wider political consequences of the blast. 'The Great Collapse', as headlined by Al-Akhbar, went beyond the material damage to evoke a deeper systematic disintegration of the government. Al-Binaa newspaper compared the magnitude of the damage to that incurred during the 33-day war with Israel in 2006 but compressed into a matter of minutes, emphasising the unprecedented scale of the tragedy. The newspaper also cited Prime Minister Hassan Diab's characterisation of the event as a national catastrophe that would not pass without accountability. Ad-Diyar newspaper reported on the government's declaration of Beirut as a 'disaster-stricken city' and the imposition of a state of emergency. The newspaper described the explosion of ammonium nitrate stored in Warehouse 12 and reported its immediate human toll, over 70 fatalities and thousands of injuries. Nidaa Al-Watan responded with a touching, almost mournful headline—'Oh Beirut'—that captured the collective anguish of the nation. Al-Liwaa described the tragedy as one that 'shakes the world'. Criticising the Lebanese government, An-Nahar newspaper suggested that the political establishment had effectively turned its destructive capacities inward. In harsh yet clear terms, Al-Joumhouria newspaper proclaimed, 'Beirut... blood and destruction', summarising the violence and enduring scars of the explosion. Complementing these were Lebanon's English and French-language dailies, which conveyed the events to an international audience. The Daily Star headlined its coverage with 'Huge explosion rocks Beirut, at least 73 killed', reflecting the blunt immediacy of the tragedy, while L'Orient-Le Jour described the city as a 'scene of apocalypse', invoking imagery of absolute destruction. These publications did not merely report on a disaster, they also documented a moment of rupture in which a nation's longstanding grievances, vulnerabilities, and hopes were laid bare.

19.5.7 Headlines from August 6, 2020

By August 6, two days following the explosion, Lebanese newspapers began to expand their coverage beyond the immediate destruction, delving more into the root cause of this tragedy and its different implications. Arabic-language newspapers adopted a more investigative and interpretive approach. For instance, Al-Akhbar newspaper referred to the catastrophe as 'after the earthquake', alluding to both its physical and figurative aftershocks. The newspaper uncovered a decision that was taken in 2015 by senior officials to re-export the ammonium nitrate, raising critical questions of administrative negligence. It also reported the depletion of medical supplies, even as international aid began to arrive, possibly signalling the need of partially lifting the diplomatic isolation which Lebanon was experiencing due to issues related to ongoing internal political issues and the influence of Hezbollah.

Ad-Diyar emphasised the emergence of global solidarity, noting that the 'earthquake in Beirut' had galvanised international humanitarian responses, including the anticipated arrival of French President Emmanuel Macron, two days after the explosion. The newspaper questioned whether this might be a 'deliberate attack' suggesting that the investigation into the causes of the blast remained unclear. Nidaa Al-Watan headlines the 'Beirut Harbor' debacle and accused the Lebanese President and Prime Minister of disregarding explicit warnings from senior officials regarding the AN. The governmental failure appeared in Al-Liwaa newspaper, which reported on the arrival of international humanitarian aid while showing disappointment of the Lebanese authorities' focus on limited administrative investigations. The newspaper also mentioned that former prime ministers called for an independent inquiry under international or Arab auspices. An-Nahar emphasised Macron's visit as a moment of symbolic impeachment, depicting it as an encounter with Lebanese political irresponsibility. Al-Binaa considered that the disaster had opened a window for increased international oversight, as calls for humanitarian aid and financial assistance became intertwined with demands for external involvement in the investigation. Al-Joumhouria newspaper was of the view of collective blame, using the headline 'They are all criminals', which points the finger to the entire political class. Meanwhile, Lebanon's international-language newspapers provided continued coverage to the global audience. The Daily Star reported the government's declaration of a state of emergency in Beirut, while L'Orient-Le Jour emphasised Macron's message of solidarity and fraternity towards the Lebanese people.

19.6 Lebanon's Approach to Disaster Preparedness, Response, and Recovery

Substantial efforts have been done by the Lebanese government to improve disaster preparedness, response and recovery capabilities. Yet, apart from being a lengthy and costly process that requires thorough communication between all stakeholders, the government's efforts have been shaped and hindered by different factors. These factors, which also contributed to the intensification of a variety of threats, include political instability, large numbers of refugees, unplanned urban growth, inadequate infrastructure, and insufficient funding. Other factors include sectarian divisions among the Lebanese communities and the domination of religio-political organisations (Baytiyeh, 2017). This section will briefly highlight some of the government's efforts towards disaster management.

The Lebanese government's interest in disaster management goes back to 1976, following the civil war, when the High Relief Commission (HRC) was established. However, the HRC's role focused mainly on receiving and distributing aid to affected populations, rather than comprehensive disaster preparedness and risk reduction. In 2010, the Disaster Risk Management Unit (DRMU) was established under the Prime Minister's office following a joint project proposal in 2009 between the United

Nations Development Programme (UNDP) and the Lebanese Government titled 'Strengthening Disaster Risk Management Capacities in Lebanon'. The unit was tasked with coordinating all stages that pertain to DRR (UNDP, 2014). However, the unit's role was not fully institutionalised and remained heavily reliant on external funding and support. Two years following the establishment of the DRMU, a Disaster Risk Reduction (DRR) strategy was developed and supported by UNDP. Between the establishment of the DRMU in 2010 and 2013, two National Committees were also established, with the first being technical (National Committee for Emergency Planning) and the latter being advisory (National Coordinating Committee). A list of key actors in Lebanon's disaster management system is presented in Table 19.1. The list is not meant to be exhaustive and does not capture all entities contributing to disaster preparedness, response, and recovery.

Lebanon worked on aligning itself with international frameworks such as its endorsement and adoption of the Hyogo Framework for Action in 2005 and its successor, the Sendai Framework. The aim was to build resilience, develop national

Table 19.1 Key actors in Lebanon's disaster management system

#	Key institution/authority
1	Council of Ministers/Presidency of the Council of Ministers
2	The General Directorate of General Security
3	Disaster Risk Management Unit (DRMU)
4	General Directorate of Civil Defense (formerly Lebanese Civil Defense; Civil Defense Higher Committee)
5	High Relief Commission (HRC)
6	Internal Security Forces (ISF)
7	Joint Rescue Coordination Centre (JRCC)
8	Lebanese Armed Forces (LAF)
9	Lebanese Red Cross (LRC)
10	Local authorities, and regional governors.
11	Ministry of Defense (MoD)
12	Ministry of Education and Higher Education (MEHE)
13	Ministry of Environment (MoE)
14	Ministry of Information
15	Ministry of Interior and Municipalities (MOIM)
16	Ministry of Public Health (MOPH)
17	Ministry of Public Works and Transport (MPWT)
18	National Council for Scientific Research—Lebanon (CNRS-L)
19	National Emergency Response Committee (NERC)
20	National Geophysical Center (CNG)
21	United Nations Development Programme (UNDP) Lebanon

strategies, and participate in capacity-building initiatives (United Nations International Strategy for Disaster Reduction 2012). Being aligned with such frameworks means that the country and its officials recognise the seriousness of the impact of disasters, as indicated in the Sendai Framework. The framework states that 'It is urgent and critical to anticipate, plan for and reduce disaster risk in order to more effectively protect persons, communities and countries, their livelihoods, health, cultural heritage, socioeconomic assets and ecosystems, and thus strengthen their resilience' (UNDRR, 2015, p. 10). Furthermore, Lebanon's National Strategy for Forest Fire Management was established in 2009 (subsequently amended), and the Ministry of Environment has been running campaigns to increase preparedness against fire. In 2022, the ministry launched the National Emergency Plan for fire prevention, awareness and readiness. Most recently in 2025, the Ministry of Environment and its stakeholders published the Standard Operating Procedures for Post Disaster Rubble Management. The Standard Operating Procedures serve as a comprehensive and actionable framework for all stages of rubble management resulting from natural hazards, disasters and conflicts in Lebanon.

19.6.1 Early Warning System

Lebanon's National Early Warning System Platform (NEWSP) and Sustainable Environmental Early Warning Platform (SuNaR) were established in 2015 as part of the National Council for Scientific Research in Lebanon (CNRS-L) through funding from the World Bank. The NEWSP covers natural hazards (floods, landslides, drought, wildfires, storms) with interactive dashboards and hazard maps (Figs. 19.1 and 19.2). It is connected to the Disaster Risk Management Unit (DRMU) at the Prime Minister's Office and to governorate-level operations rooms. The NEWSP uses satellite imagery, modelling, geo-information systems, critical infrastructure mapping, among others to monitor resources and predict upcoming natural hazards. It includes an extensive database that allows decision-makers to easily look up vital infrastructure and sensitive sites, such as schools, universities, and hospitals that are currently at risk or could be in the future (Fig. 19.3).

Fig. 19.1 NEWSP fires dashboard. *Note* Screenshot of national early warning system platform (NEWSP) fires dashboard on ArcGIS online by Esri. *Source* National early warning system platform

Fig. 19.2 SuNaR-CNRS-L hazard maps *Note* Screenshot of SuNaR-CNRS-L showing different hazard maps. *Source* SUNAR-CNRS-L

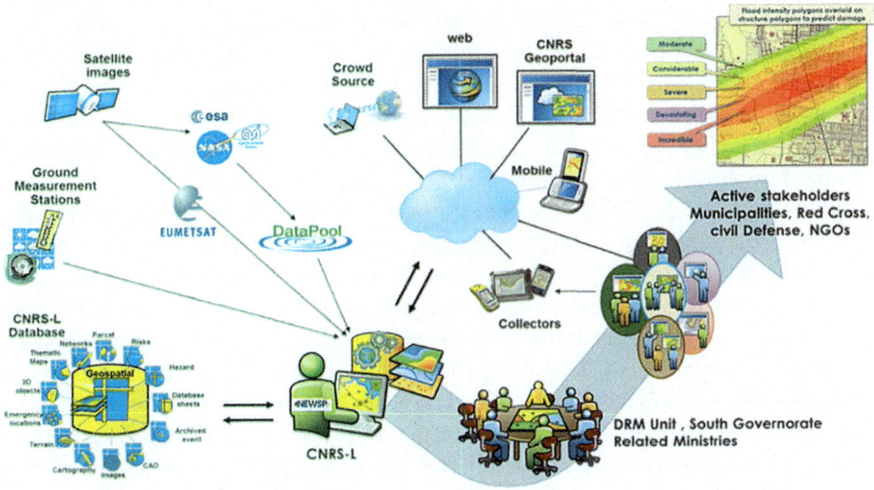

Fig. 19.3 NEWSP information flow and stakeholder coordination. *Note* Adapted from aim and scope, national early warning system platform

19.7 Conclusion

Lebanon has extensive experience, though not always efficient, in managing small-scale and recurring emergencies as they occur. However, recent disasters, particularly the Beirut port explosion, have revealed significant shortcomings in the country's ability to address large-scale or high-impact crises. These gaps are most evident in areas related to disaster preparedness and risk reduction. Shaw et al. (2022, p. 304) note that 'disaster preparedness takes place at different levels of society from the national, through the community, to the individual level' and plays a critical role in mitigating the adverse effect of a potential crisis. The weaknesses unveiled by the port explosion emphasise the need for more comprehensive measures consistent with international frameworks and guidelines.

The Sendai Framework for Disaster Risk Reduction 2015–2030 outlines seven clear targets and four priorities for action to prevent new and reduce existing disaster risks. One of these targets is 'strengthening disaster risk governance to manage disaster risk' (UNDRR, 2015). The framework highlights the importance of governance for an effective and efficient management of disaster risk. This includes a clear national vision, plans, competence, guidance, and coordination. In this regard, Lebanon could continue enhancing disaster governance by addressing fragmented institutional coordination. This could be through having a centralised body to oversee, communicate, and coordinate all the efforts of government ministries, local authorities, civil society, and scientific organisations. Additionally, Lebanon has been mainly focusing on response to crises as they happen. Historically, most resources were directed towards relief, rehabilitation, and reconstruction but insufficient efforts to address risk identification, prevention, and mitigation. One of the most recent examples was when the Speaker of the Parliament called in 2023 for discussion of a national emergency plan to enhance preparedness against Israeli aggressions 21 days after the outbreak of hostilities with Hezbollah. Shortly after, a national emergency plan was developed based on several assumptions derived from a benchmarking comparison with the July 2006 Israeli aggression.

Over the years, Lebanon has demonstrated political interest and commitment to improve disaster preparedness, response, and recovery capabilities particularly through the Prime Minister's Office and the Presidency of the Council of Ministers. While it is known that disasters are sometimes unavoidable, had there been more effective implementation of threat mitigation measures, the Beirut Port tragic event might have been prevented. The country's authorities have overlooked identification, reduction, and preparedness during times of peace, acting only in direct response to events, the most recent of which have been the Beirut port explosion and the military tensions on the Lebanese border with Israel. As Lebanon continues to face a range of natural and human-caused crises, the urgency for concrete and immediate action plans has never been greater. It is time to move beyond justification for inaction and burying our heads in the sand and instead prioritise meaningful, efficient, and preventive efforts.

References

Abdallah, C. (2019). Risks (II): Floods and human activities. In E. Verdeil, G. Faour, & M. Hamzé (Eds.), *Atlas of Lebanon*. Presses de l'Ifpo. https://doi.org/10.4000/books.ifpo.13272

Al-Hajj, S., Dhaini, R. H., Mondello, S., et al. (2021). Beirut ammonium nitrate blast: Analysis, review, and recommendations. *Frontiers in Public Health, 9*, 1–11. https://doi.org/10.3389/fpubh.2021.657996

Antar, O. (2020). Why the highway connecting Tripoli to Beirut still hasn't been repaired nearly two years later. L'Orient Today. https://betaot.lorientlejour.com/article/1237356/why-the-highway-connecting-tripoli-to-beirut-still-hasnt-been-repaired-nearly-two-years-later.html. Last accessed 6 07 2025.

Baytiyeh, H. (2017). Socio-cultural characteristics: The missing factor in disaster risk reduction strategy in sectarian divided societies. *International Journal of Disaster Risk Reduction, 21*, 63–69. https://doi.org/10.1016/j.ijdrr.2016.11.012

Department of Ecosystems at the Ministry of Environment and the Land and Natural Resources Program at the Institute of the Environment, University of Balamand. (2020). State of Lebanon's wildfires in 2019. http://ioe-firelab.balamand.edu.lb/ForestFires/ForestFires2019.pdf. Last accessed 6 06 2025.

Dysz, K., Poszwald, B., Kwak, A., et al. (2023). Thermal stability of ammonium nitrate in two-component mixtures with powdered and fine-grained materials. *Materials Research Proceedings, 34*, 127–138. https://doi.org/10.21741/9781644902691-16

Guidoboni, E., Comastri, A., & Traina, G. (1994). *Catalogue of ancient earthquakes in the Mediterranean area up to the 10th century*. ING, Roma-SGA,. https://deadseaquake.info/pdfs/Catg1.pdf. Last accessed 6 06 2025.

Haddad, T., & Sakr, T. (2023). Interorganisational relation in disaster response in developing context: Assessing response to Beirut explosion. *Public Organiz Rev, 23*, 1167–1193. https://doi.org/10.1007/s11115-022-00650-9

Issam Fares Institute for Public Policy and International Affairs—AUB. (2019). Forest fires in Lebanon: A recurring disaster. https://www.aub.edu.lb/ifi/Documents/publications/infographics/2019-2020/20191220_october_forest_fires.pdf. Last accessed 6 07 2025.

Kallas, J., & Napolitano, R. (2023). Image-based 3D modeling as a damage prioritisation tool for historic buildings in post-disaster areas: The case of the 2020 Beirut blast. *Journal of Cultural Heritage, 62*, 314–321. https://doi.org/10.1016/j.culher.2023.06.007

Libris, A. (2019). A country facing a high seismic risk. In Verdeil, E., Faour, G., Hamzé, M (Eds.), *Atlas of Lebanon*, Presses de l'Ifpo. https://doi.org/10.4000/books.ifpo.13274

Ministry of Information. (2020a). Lebanese newspaper headlines for Wednesday, August 5, 2020. https://www.ministryinfo.gov.lb/53971. Last accessed 6 07 2025.

Ministry of Information. (2020b). Lebanese newspaper headlines for Thursday, August 6, 2020. https://www.ministryinfo.gov.lb/54029. Last accessed 6 07 2025.

Nuwayhid, I., Samia, G., & Ali, S. (2022). Disaster management in Lebanon: Time to break the deadlock. In *A dynamic and responsive needs assessment (DARINA)—A Khaddit Beirut sustainable model for rebuilding Lebanon*. Khaddit Beirut & TBHF, Beirut, Lebanon. https://bpb-eu-w2.wpmucdn.com/sites.aub.edu.lb/dist/9/287/files/2024/02/Environment-Report-bd79c10ac24e50d0.pdf. Last accessed 6 07 2025.

Paul, C., Clarke, C. P., Grill, B., et al. (2013). Lebanese civil war, 1975–1990: Case outcome: COIN loss (mixed, favoring Iisurgents). In *Paths to victory: Detailed insurgency case studies* (pp. 383–393). RAND Corporation. http://www.jstor.org/stable/10.7249/j.ctt5hhsjk.45

Prugh, R. W. (2020). Historical record of ammonium nitrate disasters. *Process Safety Progress, 39*(4), 1–5. https://doi.org/10.1002/prs.12210

Shaw, R., Bismark, A., & Vibhas, S. (2022). Disaster preparedness and risk reduction: An Asian perspective. In *McGee T, penning-Rowsell E (Eds) Routledge handbook of environmental hazards and society*. https://doi.org/10.4324/9780367854584-23

Sivaraman, S., & Varadharajan, S. (2021). Investigative consequence analysis: A case study research of Beirut explosion accident. *Journal of Loss Prevention in the Process Industries, 69*, 1–10. https://doi.org/10.1016/j.jlp.2020.104387

Skinner, C. (2022). The spatial void could be a chance to recreate: Queer visual activism in the fallout of the Beirut blast. *Journal of Cultural Analysis and Social Change, 7*(2), 12. https://doi.org/10.20897/jcasc/12753

ThinkHazard. (2020). Lebanon. GFDRR. https://thinkhazard.org/en/report/141-lebanon/LS. Last accessed 6 07 2025.

United Nations Office for Disaster Risk Reduction. (2015). The Sendai Framework for Disaster Risk Reduction 2015–2030, UN World Conference in Sendai, Japan. https://www.undrr.org/publication/sendai-framework-disaster-risk-reduction-2015-2030. Last Accessed 6 07 2025.

United Nations Development Programme. (2014). Strengthening disaster risk management capacities in Lebanon. https://info.undp.org/docs/pdc/Documents/LBN/SDC%202014%20Report%20-%20revised%20after%20SDC%20recommendations.pdf. Last accessed 6 07 2025.

United Nations High Commissioner for Refugees. (2024). Lebanon: Needs at a glance—2024. https://www.unhcr.org/lb/wp-content/uploads/sites/16/2024/03/At-a-glance-2024.pdf. Last accessed 6 07 2025.

United Nations International Strategy for Disaster Reduction. (2012). Making Lebanon resilient: Achieving disaster risk reduction in the Arab States—Good practice country brief. https://www.undrr.org/publication/making-lebanon-resilient-achieving-disaster-risk-reduction-arab-states-good-practice. Accessed 6 07 2025.

United Nations Office for Disaster Risk Reduction. (2024). Lebanon floods, 2024 – forensic analysis. https://www.undrr.org/resource/lebanon-floods-2024-forensic-analysis. Last accessed 6 07 2025.

Zhao, Y., Hayeck, N., Saliba, N. A., et al. (2023). Any long-term effect of the Beirut port explosion on the airborne particulate matter? *Aerosol and Air Quality Research, 23*(5), 1–15. https://doi.org/10.4209/aaqr.220395

Ziade, L. (2023). *My port of Beirut*. Pluto Press. https://research.ebsco.com/linkprocessor/plink?id=a33e7d16-9120-3f72-9ac4-39736d304ea4. Last accessed 6 07 2025.

Chapter 20
Desert Earthquakes and Biblical Narratives in the Southern Levant

Amos Frumkin

Abstract This study explores the potential geological foundations of environmental hazard events described in the later part of the biblical Exodus desert narrative. Using an interdisciplinary approach combining textual analysis, paleoseismology, hydrogeology, and historical accounts, we investigate whether these narratives could reflect actual seismic events in the southern Araba valley during the second millennium BCE. Our analysis reveals striking parallels between the biblical descriptions and phenomena associated with major earthquakes, including surface ruptures and changes in groundwater systems. Paleoseismic data from the region indicate significant tectonic activity during the period traditionally associated with the Exodus. We propose that the accounts of the earth swallowing Korah's company and water gushing from a rock at Kadesh may reverberate ancient memories of earthquake-induced ground ruptures and spring activation, respectively. The study also considers other seismic-like events described in the biblical narrative, such as the stopping of the Jordan River's flow and the fall of Jericho's walls, as possible aftershocks along the same fault system. This research suggests that some elements of the narrative may be grounded in actual environmental hazard events, offering a new perspective on the interplay between ancient texts and natural hazard phenomena.

Keywords Historical seismicity · Paleoseismology · Exodus · Araba desert · Geomythology · Dead Sea Rift

---

A. Frumkin (✉)
Institute of Earth Sciences, The Hebrew University of Jerusalem, Jerusalem, Israel
e-mail: amos.frumkin@mail.huji.ac.il

E. Alam (ed.), *Disaster, Crisis and Emergency Management in the Middle East & North Africa*, https://doi.org/10.1007/978-3-032-24436-9_20

20.1 Introduction

20.1.1 Ancient Disaster Narratives and Collective Memory

Throughout human history, natural hazards have profoundly shaped societies, cultures, and collective memory. The documentation and interpretation of these catastrophic events in ancient texts provide valuable insights into how past civilizations understood, processed, and transmitted knowledge about environmental hazards. Archaeological and historical evidence demonstrates that ancient peoples were keen observers of natural phenomena, often developing sophisticated systems for recording, interpreting, and responding to disasters (Nur & Cline, 2000).

The relationship between natural hazards and cultural memory in ancient societies reveals a complex interplay between empirical observation and cosmological interpretation. While modern scientific understanding distinguishes between natural processes and supernatural intervention, ancient worldviews typically integrated both perspectives, viewing disasters as manifestations of divine will while simultaneously developing practical responses based on empirical knowledge (Chester, 2005; Torrence and Grattan, 2002). This dual framework enabled ancient societies to preserve accurate observational details within theological or mythological narratives, creating a rich repository of environmental information embedded in religious and literary texts. The documentation of disaster-related knowledge through oral traditions and written records served multiple functions, such as practical guidance for future generations, religious instruction, and cultural identity formation. Archaeological evidence indicates that ancient civilizations developed sophisticated strategies for disaster preparedness, response, and recovery.

Earthquakes have a tremendous impact on human society, attracting various responses through the ages (Alam & Dominey-Howes, 2016; Bluestone, 2010; Khan, 2018). In seismically active regions such as the ancient Near East, Bronze Age, and Iron Age, settlements demonstrate adaptations to earthquake hazards, including architectural modifications, site selection criteria, and emergency response protocols (Ambraseys, 2009; Nur & Burgess, 2008). Yet, even today, scientific achievements for mitigating earthquake hazards and risks often fail in protecting and safe-guarding human life and property. Historic data from the last 2000 years are commonly used for reconstructing earthquake magnitude and return period (e.g., Salamon, 2010). The credibility of such records increases with time (e.g.,Ambraseys, 1971; Ben-Menahem, 1991; Karcz, 2004; Ken-Tor et al., 2001). These historic documentations are often compared with destruction layers in archaeological sites (e.g., Ambraseys, 2006; Drews, 1993; Nur & Burgess, 2008).

The ancient Near East, characterized by seismic activity along the Dead Sea Transform and other fault systems, provides particularly rich examples of disaster documentation in historical texts. Cuneiform records from Mesopotamia, Egyptian hieroglyphic accounts, and biblical narratives all contain descriptions of catastrophic events that can be correlated with geological and archaeological evidence (Guidoboni & Comastri, 2005). These sources reveal that ancient scribes and chroniclers possessed detailed knowledge of environmental processes, even when describing them within supernatural frameworks.

The preservation of earthquake memories in ancient Near Eastern literature follows recognizable patterns observed in other seismically active regions worldwide. Comparative studies of disaster narratives from Japan, Greece, and the Americas demonstrate that societies in earthquake-prone areas developed consistent methods for encoding seismic information in cultural traditions (McGuire et al., 2000). These include specific linguistic formulations for describing ground rupture, characteristic narrative structures for recounting disaster sequences, and frameworks for integrating multiple related events into coherent accounts.

Research into paleoseismology and archaeoseismology has validated many ancient disaster accounts previously dismissed as mythological, revealing their value as historical sources for understanding long-term seismic patterns (Hinzen et al., 2011; Stiros and Jones, 1996). The temporal gaps between major earthquake cycles, often spanning centuries or millennia, make ancient written records essential for comprehensive hazard assessment, as they extend the historical record far beyond modern instrumental measurements.

The integration of ancient textual sources with modern earth science data requires careful methodological consideration of both disciplines' strengths and limitations. Textual analysis must account for literary conventions, editorial processes, and transmission histories that may have modified original accounts (Finkelstein & Silberman, 2001). Simultaneously, geological interpretations must consider the inherent uncertainties in dating ancient seismic events and the complex relationships between fault systems and surface manifestations (Klinger et al., 2015; Wechsler et al., 2009).

Recent advances in paleoseismology, using techniques such as trenching across fault scarps, analysis of disturbed sedimentary sequences, and correlation of archaeological destruction layers, have provided increasingly precise chronological frameworks for ancient earthquake activity (Ken-Tor et al., 2001; Marco et al., 1996). These developments enable more rigorous testing of hypotheses linking geological events with historical narratives, moving beyond speculative correlation toward evidence-based reconstruction of ancient disaster sequences.

20.1.2 Earthquakes in the Bible and Ancient Levantine Sources

The application of disaster history methodologies to biblical narratives represents a significant development in interdisciplinary scholarship, offering new perspectives on long-standing questions regarding the historical foundations of scriptural accounts. Rather than approaching these texts solely through literary or theological lenses, disaster history research examines them as potential repositories of environmental information, preserved through oral and written transmission processes that prioritized accurate documentation of significant events (Dever, 2017; Finkelstein, 2007).

This approach does not require acceptance of biblical narratives as historically accurate in all details but rather recognizes their potential value as sources of information about ancient environmental conditions and catastrophic events. By comparing textual descriptions with independent geological and archaeological evidence, researchers can identify areas of convergence that may reflect underlying historical realities, while simultaneously acknowledging the complex processes through which such information was preserved and transmitted across generations.

Here we check the possibility of extending the historic earthquake records back in time using biblical narrative. The earthquakes were documented reliably in later parts of the biblical literature, termed "ra'ash" (Amos 1,1; Zechariah 14,5) (e.g., Austin, 2000; Roberts, 2012). However, older narratives, such as the Exodus stories, are usually regarded as myths or legends. This is in common with other religions: the further back in time the inquiry goes, the more descriptions of catastrophes have been incorporated into myths (Dilek & Kahya, 2023). The stories' fantastical elements, as well as their often orally transmitted nature have discouraged most scientists from using them in modern research. Some investigators include the stories in their scientific works, but searching for documentable, quantifiable, and verifiable data is difficult or impossible, precluding further research (Bluestone, 2010). Here the Exodus narrative is used as a test case, checking it through the lens of modern geoscience evidence.

20.1.3 The Exodus Narrative

The narrative of the Israelite Exodus through the desert is one of the most enduring stories. It is a foundational event for the western world religions and cultures, with its timeless themes of redemption and deliverance. Despite its unique legacy, the research attitude toward the historicity of the biblical Exodus narrative is complex and varied among scholars (Albertz, 2011; Finkelstein, 2015; Germany, 2016; Hughes, 1990; Janzen, 2021; Levy et al., 2015; Merrill, 2008; Na'aman, 2011; Schmid, 2015; Stripling et al., 2021).

The Exodus as described in the Bible versus current archaeological and historic evidence is still debated. Many scholars point to the difficulty of finding material evidence in Egypt and the Sinai Peninsula for this event. Some researchers focus on potential historical kernels within the narrative, while others examine its literary and cultural significance regardless of historicity; both attitudes usually discuss the beginning of the Exodus in Egypt (Davies, 2004; Grabbe, 2014; Hendel, 2001; Hoffmeier, 1996; Malamat, 2001; Sparks, 2015).

The various approaches to the Exodus narrative may be divided into the following groups:

1. Traditional view: Accepts the Exodus account as historically accurate.
2. Maximalist approach: Argues for a historical core to the narrative, with some exaggeration.

3. Minimalist approach: Views the Exodus story as largely or entirely non-historical.
4. Middle ground: Suggests a smaller-scale migration or series of events that were later mythologized.
5. Symbolic interpretation: Sees the narrative as primarily theological, not historical.

It is hard to ignore, however, that the narrative contains realistic details in the Egyptian and the desert parts of the Exodus story (e.g., Kitchen, 1998; Römer, 2015; Shupak, 2001). The narrative probably preserved ancient oral traditions for a long time (e.g., Avner, 2021; Bietak, 2015; Hoffmeier, 1996; Rendsburg, 2015; Schneider, 2015). The various studies that attribute some historicity to the Exodus narrative commonly refer to a calendar date within the 17–twelfth centuries BCE (e.g., Jack, 2018; Johnstone, 2005; Laato, 2015; Stripling et al., 2021). The narrated "miracles" of the Exodus have been discussed in various ways, also emphasizing the Egyptian, early part of the narrative (e.g., Byko, 2004; Dürr, 2017; Gertz, 2014; Humphreys, 2004).

20.1.4 The Enigmatic Route of the Exodus Through the Desert

Studies that attempt to reconstruct the geographic route of the Exodus narrative are rich in reference to desert sites that have long intrigued scholars. While some locations mentioned in the latter part of the journey, particularly from Moab onwards, have been identified based on textual and archaeological evidence, the desert portion of the travels remain largely enigmatic (e.g., Tebes, 2017). This has given rise to numerous theories attempting to reconstruct the hypothetical path taken by the Israelites, particularly through the wilderness portions of their journey.

Previous studies have taken various approaches to unraveling this geographical puzzle. Some have relied primarily on close readings and analysis of the biblical text itself, cross-referencing with extra-biblical sources when possible (e.g., Anati, 2015). Others have endeavored to correlate the narrative with the physical geography and topography of the Sinai Peninsula and surrounding regions (e.g., Har-el, 1968; Kerkeslager, 1998; Shanks, 2005). Archaeological excavations and surveys in these areas have also contributed some data (e.g., Ben-Yosef et al., 2014; Bruins & van der Plicht, 2007). However, despite the valuable insights offered by such investigations, many unresolved questions remain.

Here the later part of the desert route is compared with Earth science and geographic evidence, attempting to evaluate the historicity of catastrophic events and their representation in the biblical text.

20.2 Objectives, Approach, and Methodology

Exploring the historicity and potential realistic foundations associated with earthquakes within the later part of the Exodus narrative, this study may offer insights into the interpretation of ancient earthquakes, as well as reconstructing the possible Israelites' route and distinguishing between plausible and implausible elements in this ancient text.

The study focuses on two key events:

1. The rebellion of Korah and his followers, who were allegedly engulfed by the earth opening beneath them.
2. The "Waters of Meribah" incident, where Moses reportedly struck a rock, causing water to gush forth abundantly.

Rather than relying solely on textual analysis or archaeology, this study adopts an interdisciplinary approach informed by modern research into earthquake phenomena and their effects, as well as the region's realities. By comparing these data with the biblical narrative, the research attempts to propose naturalistic explanations for these events.

The study area encompasses mainly the Araba valley region, situated at the border of ancient Edom (presently Jordan/Israel border), which is believed to be part of the Israelites' route during their exodus from Egypt, according to the biblical account (Fig. 20.1).

The methodology integrates textual analysis of relevant biblical passages with data and interpretations through the lens of their environment, geology, geography, seismology, and hydrology, supplemented by historical and archaeological information.

Particularly, the research methods include:

1. Textual analysis: Examination of key passages from the Book of Numbers, focusing on geographic sites and descriptions that may correspond to geological phenomena.
2. Paleoseismic investigations: Review of trench excavations across fault lines in southern Araba, as well as seismic history of the Dead Sea basin using drilled cores and outcrops.
3. Groundwater observations: Analysis of documented cases linking a major earthquake to groundwater level changes and spring flow variations in the Araba area.
4. Historical accounts: Examination of ancient sources and archaeological evidence for potential correlations and site identification.

The analysis synthesizes various geological and hydrogeological datasets to evaluate the possible occurrence of major earthquakes capable of causing surface ruptures and groundwater outbursts along the Araba valley during the second millennium BCE. Groundwater responses are examined by comparing documented modern cases to the biblical accounts.

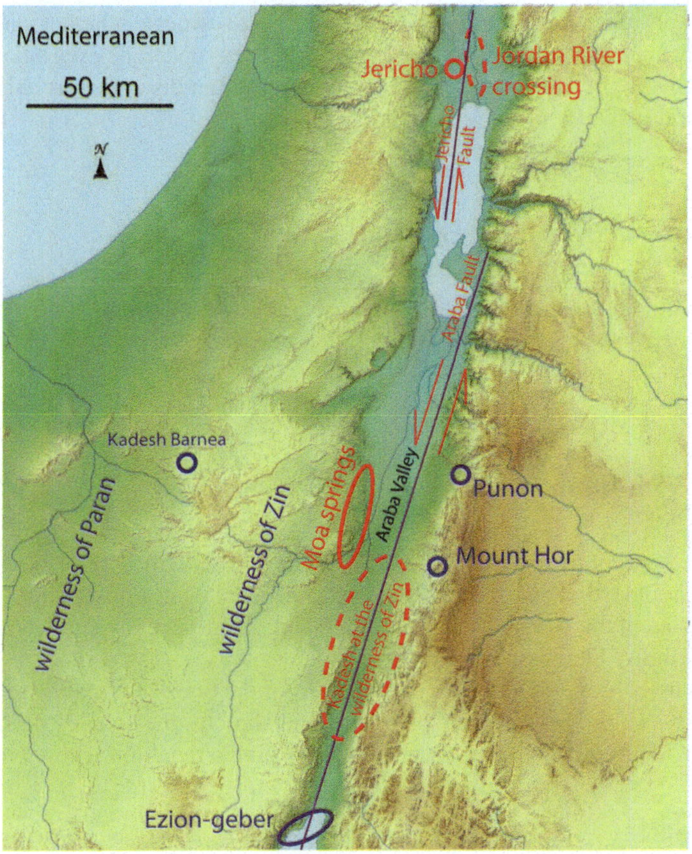

Fig. 20.1 Topographic-tectonic map showing the main sites mentioned in the text. Sites associated with Earthquakes are indicated in red

By analyzing the timing and geographic locations of identified seismic events in the context of the biblical narrative's sequence and itinerary through the Araba region, the study aims to determine if these events could provide a plausible, realistic background for the described episodes.

The data and interpretations from these investigations provide critical context regarding the timing and magnitude of powerful earthquakes affecting the region during the second millennium BCE, the approximate period commonly suggested for the Exodus.

By synthesizing the textual evidence with geologic and geographic data, locations and timelines are proposed that could plausibly reconcile the biblical narratives with the region's seismic and hydrogeological history during the relevant era. Focus is placed on the area of the southern Araba valley where the main two biblical accounts (in the book of Numbers) intersect geographically. Additional evidence is gathered further north along the same tectonic line, in the Dead Sea and Jericho areas.

Ultimately, this interdisciplinary approach aims to evaluate whether a naturalistic geological explanation could offer a compelling and coherent exegesis for earthquake events in the later part of the Exodus story, often dismissed as ahistorical or supernatural in nature.

20.3 Interdisciplinary Evaluation

Here we attempt to analyze and compare various interdisciplinary fields associated with the later part of the Exodus route, checking for possible indications of earthquakes.

20.3.1 The Biblical Geographic Framework

The studied part of the exodus route is schematically described in Deuteronomy 2:8 "…we passed beyond our brethren, the descendants of Esau who dwell in Seir, away from the road of the Araba, away from Elath and Ezion-geber, we turned and passed by way of the Wilderness of Moab."

The route is further detailed in two sections of the book of Numbers (Fig. 20.1).

The first section, Numbers chapters 16–21, elaborates the detailed narrative with various incidents along the way, starting with Korah's rebellion whose culmination is:

> Numbers 16: 31 "…the ground split apart under them… 32 and the earth opened its mouth and swallowed them…".

The next narrated episode is the "Waters of Meribah":

> Numbers 20: 1 "The whole Israelite community arrived at the wilderness of Zin in the first month, and they stayed at Kadesh… 11 Then Moses raised his arm and struck the rock twice with his staff. Water gushed out, and the community and their livestock drank… 22 They set out from Kadesh and the Israelites, the whole community, came to Mount Hor…".

The death of Aaron on Mount Hor is followed by the copper snake incident:

> Numbers 21: 4 "They set out from Mount Hor by way of the Red Sea… 8 The Lord said to Moses, "Make a seraph figure and mount it on a standard. And if anyone is bitten looks at it, he shall recover". 9 And Moses made a copper snake… 10 The Israelites marched on and camped at Oboth. 11 They marched from Oboth and camped at Iye-Habarim, in the wilderness at the outskirts of Moab…".

The second section, Numbers 33, portrays the travel stops with hardly any detail of particular events:

> Numbers 33: 36 "They set out from Ezion-geber and camped in the wilderness of Zin (that is, Kadesh). 37 They set out from Kadesh and camped at Mount Hor, on the edge of the land

of Edom... 41. They set out from Mount Hor and camped at Zalmonah. 42 They set out from Zalmonah and camped at Punon. 43 They set out from Punon and camped at Oboth. 44 They set out from Oboth and camped at Iye-Habarim at the border of Moab."

20.3.2 Geographic Anchors

The two geographic narratives of the route in Numbers, presented above, are compatible with each other albeit slight differences, in the following geographic and textual order. The narrative suggests several geographic anchors (Fig. 20.1):

1. Ezion-geber (Numbers 33: 36) was a well-known port at the northern part of the Gulf of Aqaba of the Red Sea during the Iron Age, used by King Solomon (1 Kings 9:26, 2 Chronicles 8:17) and King Jehoshaphat (1 Chronicles 22:48–49; 2 Chronicles 20:36) as a gateway to the south (e.g., Finkelstein, 2015).
2. Kadesh at the wilderness of Zin is where Moses struck the rock, and water gushed out (Numbers 20: 7–20). It became known as the "Waters of Meribah" (Numbers 20:13, 27:14; Deuteronomy 32:51). The wilderness of Zin is located at the southern border of the land of Canaan (Numbers 13:21, 34:2–4, Joshua 15:1–3). The close association with Ezion-geber and Mount Hor indicates that Kadesh at the wilderness of Zin is not identical with Kadesh-barnea at the wilderness of Paran, whose geographic attributes (Barnea and the wilderness of Paran) point to the conventional site in the present Ein Qedeis/Qudeirat area (Numbers 13:21–26, 34:4, Joshua 15:3) (Fig. 20.1). In contrast, Kadesh at the wilderness of Zin according to the order of travels is located further to the south-east (Numbers 33: 36–37), to the north of Ezion-geber. Only in a later period, the prophet Ezekiel (47:19; 48:28) extended the future borders of the land to "Waters of Meribah."
3. Mount Hor, the site of Aaron's death (Numbers 20: 22–25, 33: 37–39; Deuteronomy 32: 50). The site of Mount Hor has been identified for (at least) 2000 years near Petra, as noted by Flavius in Antiquities 4 [82] [7] (trans. A. Shure): "...the general led his army through the desert, till he came to a place called Areke, a city so-called now by the Arabs, though it was formerly called Petra..." [83] "This place is surrounded with high mountains, and with a passage admitting only one at a time. On this mountain Aaron ascended, when Moses had foretold that he was to die...".

   The same site is known during the last millennium until today at Jabal Harun, A prominent sandstone mountain, 3 km SW of Petra (Fig. 20.2) (Kouki & Tenhunen, 2013). This long-term tradition is compatible with both Mount Hor's and Jabal Harun's location between Ezion-geber and Punon, so there is no reason to question Mount Hor's traditional identification.
4. Punon (Πηινovin the Septuagint; Phinon in the Vulgate), is a station between Mount Hor - Zalmonah and Oboth (Numbers 33: 42–43). It is commonly identified at Khirbat Faynan, the largest copper mining center in the southern Levant during the Iron Age (Avner, 2021; Ben-Yosef et al., 2014; Finkelstein, 2015; Levy et al., 2015). The narrative of Moses making a copper snake (Numbers 21:

Fig. 20.2 The Araba valley and Jabal Harun, traditionally identified as Mount Hor for over 2000 years. (a) View from the Araba valley toward the traditional Aaron's grave at the top of Jabal Harun (photo: Amos Frumkin). (b) nineteenth century drawing of Mount Hor (Kitto, 1841)

9) between Mount Hor and Oboth is probably associated with the copper ores at Punon, similarly located between Mount Hor and Oboth (Numbers 33: 42–43).

Out of the four sites mentioned above, three can be geographically identified today with a highly probable accuracy: Ezion-geber at the northern edge of the Gulf of Aqaba; Mount Hor at Jabal Harun; and Punon at Khirbat Faynan. These three anchor points, threaded along the topographically convenient route of the

Araba valley, clearly define this segment of the Exodus journey. This ancient, topographically easy route was first described by Burckhardt (1822), and recently by Avner (2021).

Based on these anchor points, the Waters of Meribah at Kadesh in the Wilderness of Zin, according to the order of travels, should be located between Ezion-geber and Mount Hor (Fig. 20.1).

20.3.3 The Waters of Meribah: Bible Versus Hydrogeology

The Waters of Meribah (Numbers 20: 11–13, 24; 27: 14; Deuteronomy 32: 51; 33: 8; Psalms 81: 8; 95: 8; 106: 32) is an important episode in the biblical narrative, whose traumatic memory remained significant for generations to come, traditionally demonstrating several points:

1. God's power, providing water miraculously from a rock, and demonstrating His ability to sustain the Israelites in the desert.
2. Test of faith: It represents a crucial moment where the Israelites' faith in God and Moses' leadership was tested during their journey through the wilderness.
3. Moses' disobedience and punishment: At Meribah, Moses struck the rock instead of speaking to it as God commanded, which led to his punishment of not entering the Promised Land.
4. Symbol of rebellion: The place became synonymous with the Israelites' complaining and lack of trust in God, despite His previous miracles and provisions.
5. The consequences of disobedience and the importance of trusting God.

The main description of the incident is: "Then Moses raised his arm and struck the rock twice with his staff. Water gushed out, and the community and their livestock drank." (Numbers 20:11)

Can this incident be associated with natural processes? While porous rocks in general may store water and this may generally explain the story (e.g., Humphreys, 2004), an explicit modern analog in the right geographic context has never been suggested. It is thus interesting to check the behavior of many small springs along the Araba valley (Fig. 20.3) that had negligible or no discharge under the arid climate of recent millennia. The springs of the Moa area began flowing abundantly following a 7.1 magnitude earthquake in 1995 (epicenter at Nuweiba, Sinai). The rise in groundwater level was explained hydrogeologically by the opening of new fissures due to strong ground shaking, which allowed the ascent of artesian water from the Mishash Formation, underlying the Ghareb-Taqiye impervious caprock (Enmar, 1999; Enmar et al., 1998; Yechieli & Bein, 2002) (Fig. 20.3c). The groundwater is confined in the aquifer and only exposed at structural domes. The artesian springs that discharge around these structures create seepage areas around the domes (Fig. 20.3c), which are usually dry. Rising groundwater levels, as occurred here following the 1995

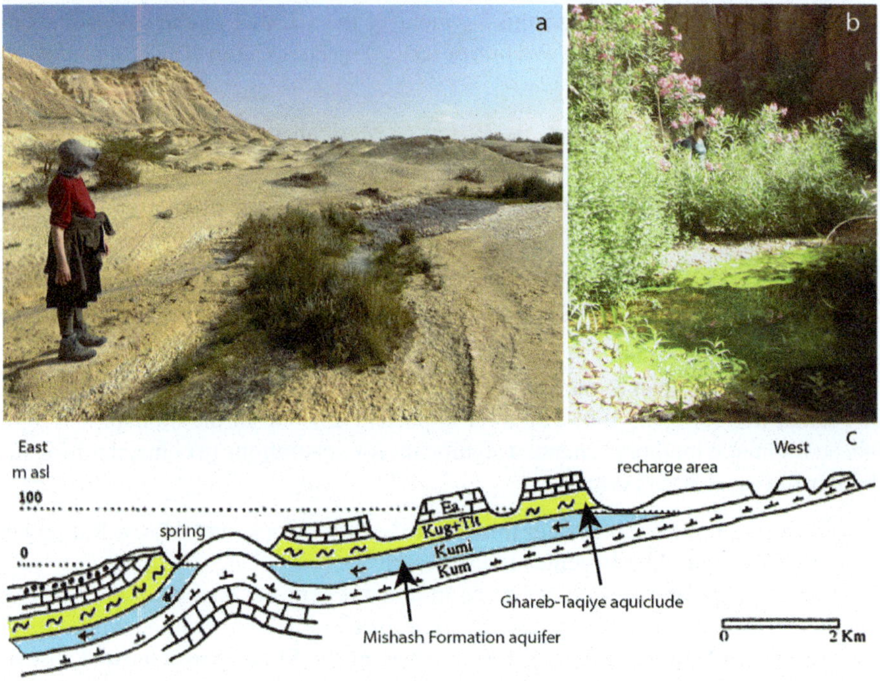

Fig. 20.3 Springs across the central Araba. (a) Artesian spring near Moa (Fig. 20.1); the surrounding marls belong to the confining Ghareb-Taqiye aquiclude. (b) Wadi Musa Spring (nicknamed "Moses Spring"), at the close vicinity of Jabal Harun—Mount Hor. (c) Hydrogeological section through Moa springs, modified after Enmar (1999) (with permission). Water flows from the recharge area (west) eastward, confined by the Ghareb-Taqiye aquiclude. Normally, the water hardly seeps at the spring. The Mag. 7.1 1995 Nuweiba earthquake caused the water to gush forth abundantly

earthquake, were also documented in other locations following 6.8–7.7 magnitude modern earthquakes (e.g., Kim et al., 2019; Manga & Rowland, 2009).

20.3.4 The Episode of Korah and His Company: Bible Versus Geology

In the Korah episode, the culminating phrase "the earth opened its mouth and swallowed them up" (Numbers 16: 32) employs root words and imagery explicitly connoting the gaping and consuming actions of the ground surface. This phrasing corresponds remarkably well to observational descriptions of earthquake-induced ground ruptures, where tensile cracking causes fissures to open spans of earth (Figs. 20.4 and 20.5). The anthropomorphic metaphor of the earth "opening its

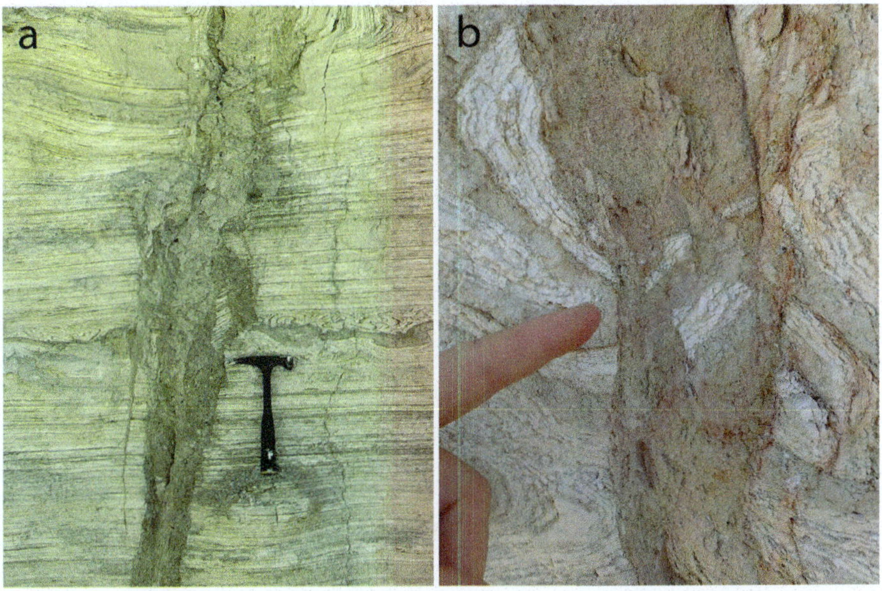

Fig. 20.4 Lisan sediments above the Dead Sea fault, ruptured and deformed by Holocene earthquakes, and filled by the fractured layers. (a) At the central Dead Sea; (b) At the southern Dead Sea

mouth" evokes the violent, destructive consumption of the rebellious company in highly vivid terms.

The preceding phrase, "the ground that was under them split open" (16: 31) further emphasizes the physical fracturing and parting of solid terrain in a fashion characteristic of high-magnitude seismic events capable of generating surface ruptures many kilometers in length.

The passage uses the word "adama" (16: 31) meaning "ground" or "soil" to describe what opened up, reinforcing the sense of a physical, surficial rupturing as opposed to a more metaphysical or symbolic parting (as with waters or space).

The interpretation/tradition of Josephus adds: "When Moses had said this, in a modest tone of voice, a prodigious noise of thunder was suddenly heard, the earth shook violently, and there was a terrible hurricane and violent storms of wind…" (Antiquities 4 [51] [3], trans. A. Shure).

Both descriptions above fit a strong earthquake. At magnitude 6, surface rupture above the epicenter is expected in 50% of earthquakes (Fig. 20.4). At magnitude 8 and above, rupture is expected in 100% cases (Biasi & Weldon, 2006).

The episode of Korah is not ascribed to a particular year within the Exodus biblical narrative, but in the geographic context, the implicated location of the Korah incident along the line of events is prior to the Waters of Meribah, i.e., between Ezion-geber and Kadesh at the wilderness of Zin. The modern earthquake evidence discussed above seems to associate the two events with major earthquakes. If we assume

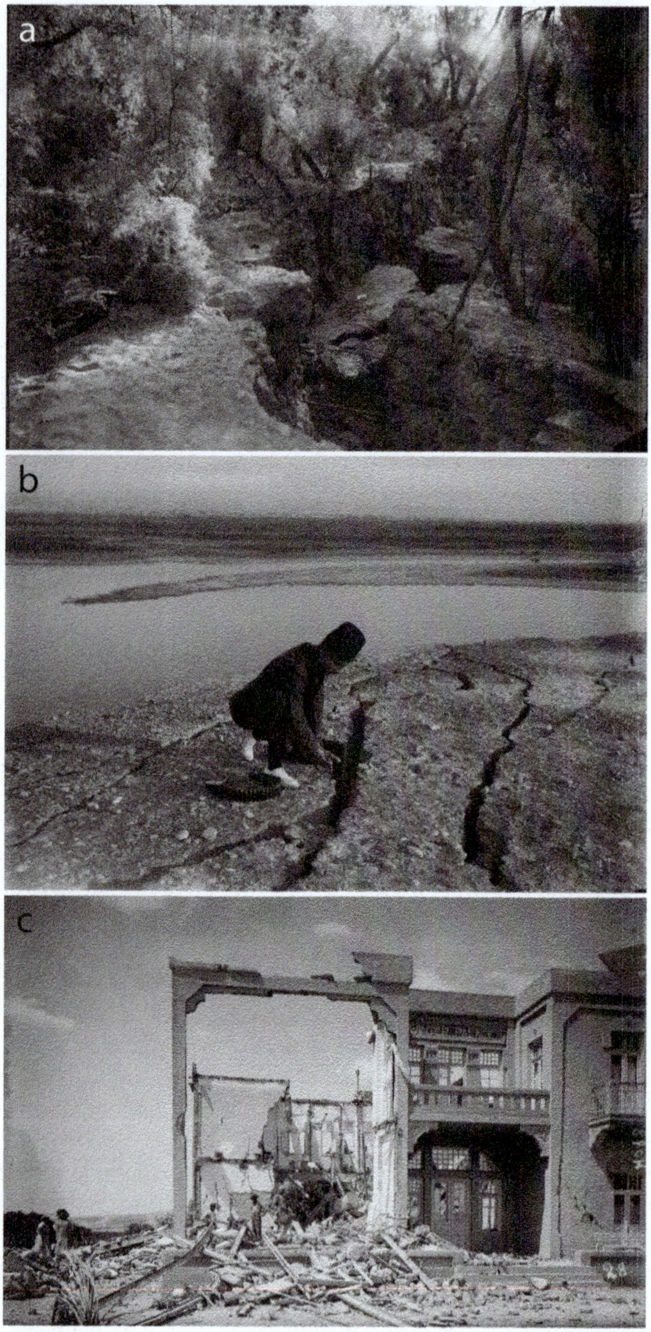

Fig. 20.5 Effects of 1927 Mag. 6.25 Jericho earthquake in the Dead Sea-Jericho area (Library of Congress, Matson collection). **a** Surface rupture at the lower Jordan River area. **b** Surface rupture at the Dead Sea area. **c** Collapsed Winter Palace Hotel in Jericho

historicity, then it is highly probable that these events were two manifestations of a large earthquake and its aftershocks, or an earthquake cluster, a common feature along the Dead Sea transform (Lefevre et al., 2018; Marco et al., 1996; Nur & Cline, 2000).

We now turn to look at the southern Araba region, for indications of surface ruptures by earthquakes, implied by the Korah incident.

20.3.5 Surface Ruptures by Earthquakes at the Araba: The Scientific Record

The major Araba fault is an active plate boundary between the Arabian Plate and the Sinai sub-plate (e.g., Ben-Avraham et al., 2008; Garfunkel, 2014; Ginat et al., 1998). These plates move laterally to the left (sinistral transform fault) by ~ 5 mm/year (Klinger et al., 2000; Masson et al., 2015). The southern Araba fault segment is locked ~12–20 km below surface, thus producing intermittent large earthquakes with quiet interseismic periods (Hamiel et al., 2018; Le Beon et al., 2008). Large earthquakes ruptured the surface during recent millennia (Kanari et al., 2020; Klinger et al., 2015). Paleoseismic data from the Dead Sea rift indicates the region was very active tectonically during the second millennia BCE—the general era associated with the biblical Exodus tradition.

Migowski et al. (2004) demonstrated the occurrence of numerous large earthquakes greater than magnitude 6 during this period, based on identification and dating of seismic-triggered sedimentary disturbances and deformation of Dead Sea layers. Major earthquakes are attributed to ca. 1050–1100, 1365, and 1560 BCE, with inferred magnitudes of 7.0, 7.8, and 6.8, respectively (Migowski et al., 2004). Kagen et al. (2011) inferred from disturbed Dead Sea sediments a major earthquake between 2020 and 1635 BCE. The major Holocene earthquakes ruptured also the late Pleistocene Lisan Formation, filling the ruptured sediments with deformed layers (Fig. 20.5).

Further to the south, the Gulf of Aqaba is an extremely active part of the Dead Sea transform (Engelberge et al., 2024). At its northern end, the southern Araba valley between Ezion-geber and Kadesh/Mount Hor inferred locations, surface rupturing earthquakes were investigated by digging trenches across faults, within which the infill of surface ruptures along the fault was dated (Amit et al., 1999, 2002). A major fault rupture of the surface occurred ca. 3700 ± 300 yr ago, displacing the top of the fault by 50–60 cm. Five similar major tectonic events occurred between 14,200 ± 300 and 3700 ± 300 yr ago.

Lacking other explanations, the effects of such ancient major earthquakes were apparently recorded in early biblical narratives as the wrath of God (Dilek & Kahya, 2023; Nur & Burgess, 2008). Another example is the Sodom and Gomorrah upheaval and Lot's Wife pillar formation in the book of Genesis (Freund & Nur, 2001; Frumkin,

2009; Harris & Beardow, 1995; Neev & Emery, 1995). We confine the present discussion to the possible repercussions of additional earthquakes at the end of the Exodus narrative.

20.3.6 Possible Records of Additional Exodus Earthquakes: Bible Versus Geology

The events described in the biblical book of Joshua, specifically the crossing of the Jordan River and the fall of Jericho's walls, have been subjects of both religious interpretation and scientific inquiry. Some researchers have proposed that these events could be explained by seismic activity in the region (e.g., Bentor, 1989; Neev & Emery, 1995; Nur & Burgess, 2008), and several archaeological destruction layers were attributed to major earthquakes, including during the second millennium BCE (Ferry et al., 2011).

20.3.6.1 Crossing of the Jordan River

The sudden halt of the Jordan River's flow (Joshua 3:15–16) could potentially be explained by an earthquake-induced collapse of the high Lisan Formation cliffs along the river. The Jordan Rift Valley is seismically active, and a tremor could have caused a section of the riverbank to collapse, temporarily damming the river.

This phenomenon has been observed in modern times. The most relevant modern analog is the July 11, 1927 Jericho earthquake (magnitude 6.25), centered in the northern Dead Sea, that affected Jericho and many other sites (Fig. 20.5), and also caused a cliff-collapse that blocked the Jordan for about 21 h (Ambraseys & Melville, 1988; Avni, 1999; Zohar et al., 2014).

20.3.6.2 Fall of Jericho's Walls

The collapse of Jericho's walls (Joshua 6:20) could potentially be attributed to seismic activity (e.g., Neev & Emery, 1995; Nur & Burgess, 2008). Jericho is located near the seismically active Jericho fault (Fig. 20.1). The city walls, made of mud-brick, would have been extremely vulnerable to ground shaking. Some researchers suggest that the Israelites' marching and shouting, as described in the text, could have been a narrative device to explain the coincidental timing of an earthquake that undermined the city walls.

Both the sudden stopping of the Jordan River's flow and the collapse of Jericho's walls occurred along the major strike-slip Jericho Fault (Fig. 20.1), a continuation of the Araba, Dead Sea Fault system, and their chronology is close enough to the Korah and Waters of Meribah incidents, to be considered belonging to a single cluster of

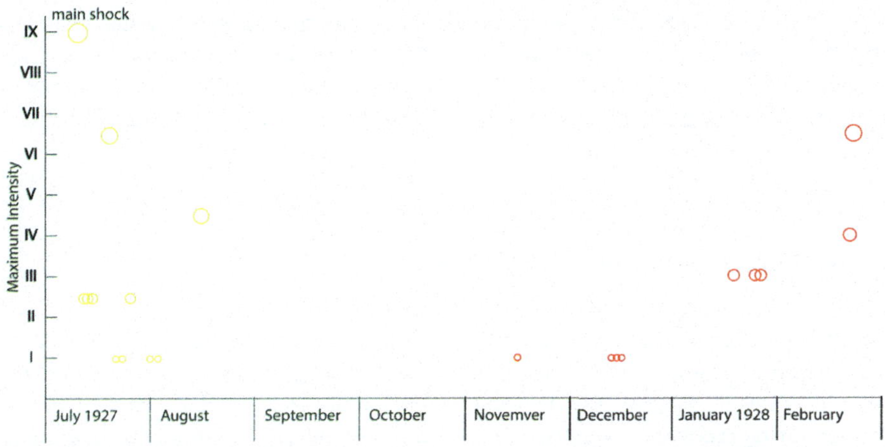

Fig. 20.6 The 1927 Mag. 6.25 Jericho earthquake and its aftershocks for 8 months. Circles' area corresponds to earthquake intensity, based on macroseismic reports (modified after Avni, 1999)

seismic events, or to aftershocks of the same event (Nur & Cline, 2000). The modern analog is again the 1927 earthquake whose aftershocks continued for eight months (Avni, 1999) (Fig. 20.6).

20.3.6.3 Poetic Earthquake Descriptions?

Psalm 114 is a poetic description of the Exodus, using vivid imagery that some researchers have interpreted as resonating an ancient memory of seismic activity (e.g., Johnstone, 2005; Nur et al., 1996). They suggest that the poetic language in Psalms 114 might be describing real geological events that occurred during the Exodus and had passed through the generations as an ancient liturgical song. The events occurring during the Exodus ("When Israel went out of Egypt," verse 1) are portrayed as a major earthquake or series of earthquake tremors ("The mountains leaped like rams," verse 4; "Tremble, O earth, at the presence of the Lord," 7). The earthquakes allegedly caused various phenomena such as the provision of water from rocks ("Who turned the rock into a pool of water, the hard rock into a spring of water," 8), and drying of the river Jordan ("The Jordan turned backwards," 5). It is interesting to note that this poetic description of the Exodus includes earth trembling in addition to other earthquake-associated events discussed in the present paper. The motif of trembling earth appears also in Psalms 29, where God is acclaimed to make the wilderness of Kadesh tremble with fear (29:8). The association between the trembling earth and the wilderness of Kadesh renders this text particularly relevant to our discussion. Does it resonate a distant memory of a dramatic event such as the "Waters of Meribah" incident-earthquake?

Another biblical poetic description connects between Edom, Se'ir, and earthquakes, using the known term for earthquake—the land "ra'asha" (Judges 5,4).

While these texts attract attention, it is important to note that the Psalm is primarily a poetic and theological text, using powerful natural imagery to convey the idea of God's might and involvement in Israel's history. While it's intriguing to consider potential geological explanations, the text itself is not intended as a scientific description of events.

20.4 Discussion and Conclusions

The Bible can be divided into two periods by the cultural recognition of earthquakes: a later period during which a seismic event was recognized as a natural phenomenon termed "ra'ash" (Judges 5,4; Amos 1,1; Zechariah 14,5), and an ancient period whose earthquakes were reported as divine interventions, or miracles. The interdisciplinary analysis presented in this study offers a possible naturalistic explanation for such ancient key events described in the later part of the biblical Exodus narrative. By synthesizing the geological and hydrological realities of the region as revealed through modern research with a close reading of the scriptural accounts, we propose that these incidents can be understood against the backdrop of powerful seismic activity.

Close examination of the Hebrew texts describing the incidents of Korah's rebellion (Numbers 16:31–35) and the "Waters of Meribah" at Kadesh (Numbers 20:1–13) reveal striking linguistic parallels to phenomena associated with major earthquake activity. These textual elements were also compared against other biblical passages describing geological and hydrological upheavals, to assess similarities in terminology and metaphor usage.

In particular, the biblical narrative of the culmination of the Korah incident is: "the earth opened its mouth and swallowed them" (Numbers 16: 31–32) resembles the rupturing of the earth above faults during strong earthquakes, as is indeed observed in the Araba valley.

The subsequent episode of Moses striking the rock at Kadesh to yield abundant waters exhibits a subtler but no less suggestive seismic connection through analogous spring discharge phenomena. The propagation of seismic waves from strong earthquakes can dramatically uplift groundwater levels by creating new fissures and pathways for flow. This process of increasing subsurface permeability can reinvigorate previously low-flowing springs or induce new ones to appear where they did not exist before. In this light, the biblical description of copious waters bursting forth from the struck rock takes on a distinctly naturalistic dimension. Rather than embodying a purely supernatural hydrological event, the narrative can be interpreted as reflecting an acute observer's experiential perspective of aquifer discharge activated by the effects of powerful seismic shaking on regional groundwater systems. This image is compatible with tectonic hydrological models validated by modern hydrologic effects at the Araba valley in the wake of a large earthquake.

As established by several studies, the associated section of the Araba valley has been repeatedly struck by major earthquakes of magnitude 6.5 and higher over historical and prehistorical time frames—well exceeding shaking intensities that could prompt enhanced spring flow by raising pore water pressures and remobilizing aquifer networks.

The sedimentary evidence documented by researchers indicates that the Dead Sea rift zone experienced numerous large earthquakes of magnitude 6 and greater during the second millennium BCE, the period associated with the Exodus tradition. As demonstrated by more recent quakes like the 1927 Jericho and the 1995 Nuweiba events, earthquakes of this scale have the capacity to rupture the ground surface and disrupt groundwater flow systems, leading to new or reinvigorated spring discharge.

When viewed through this geological viewpoint, the biblical portrayal of ground splitting and swallowing up of the rebellious company led by Korah takes on a realistic aura. The subsequent account of Moses striking the rock at Kadesh and unleashing a torrent of water appears plausibly rooted in historical fact—the eruption of groundwater to the surface caused by shaking and shifting subsurface pressures and pathways during a major earthquake.

Following these events, the entry to the promised land was accompanied by the stopping of the Jordan River's flow and Jericho walls' collapse, events that may be associated with aftershocks along the continuation of the Dead Sea tectonic line. The spatial and chronological occurrence of all seismic-like events along the same faulted valley and within the time frame of ~one year supports their possible association with a cluster of major earthquakes. Such earthquake cluster in the region around 1200 BCE is also indicated by archaeological evidence (Nur & Cline, 2000). Liturgical songs in the books of Psalms and Judges may resonate the ancient memory of these dramatic events near the end of the Late Bronze Age.

If our hypothesis is correct, it would bolster the argument that the biblical writers were narrating a core memory of real events, albeit cloaked in theological perspectives and literary motifs. The spatial and chronological proximity of these incidents hint at their being aspects of the same underlying seismic paroxysms.

Moreover, the proposed geological scenario strengthens the prospect that the biblical writers possessed at least some knowledge of the broader geographic realities of the region. This may be an important consideration for theories attempting to reconstruct the travel route of the Israelites based on correspondence between description, geology, and topography. Our analysis suggests that the biblical writers were attuned to their surroundings and incorporated realistic environmental depictions, at least from the point of approaching the southern Levant.

Of course, our study represents just one facet of an overarching investigation into the credibility and historicity of the Exodus account that spans multiple academic disciplines. While the geological data appears to offer a plausible naturalistic explanation for scriptural details often dismissed as literary fabrication, other components of the narrative still strain credulity. The precise locations of places like Kadesh in the wilderness of Zin remain tantalizingly unclear. The overall geographical coherence of the complete journey from Egypt to Canaan as described will continue to fuel debates.

Nonetheless, by providing a rational exegesis grounded in sound science for at least few episodes, this study demonstrates the value of an interdisciplinary approach that considers both textual and physical evidence. As additional archaeological and scientific data accumulates, a more comprehensive understanding of ancient attitudes to earthquakes and the Exodus narrative's historicity and geography may yet emerge. What remains clear is that the biblical writers possessed an intimate familiarity with the natural world they inhabited—a world that has now begun revealing some of its seismic secrets in a way that may shed new light on ancient realities.

It is important to note that while these geological explanations offer potential natural mechanisms for the events described in the Bible, they remain hypothetical. It is suggested here that the interpretation of biblical texts and their relation to historical events is a complex and often contentious field of study, involving considerations of various scientific disciplines such as geology, hydrology, and archaeology.

Acknowledgements The manuscript has benefited from discussions with Prof. Yoel Elitzur and Dr. Yoel Bin Nun.

References

Alam, E., & Dominey-Howes, D. (2016). Earthquakes in the north-eastern coastline of Indian Ocean with a particular focus on the Bay of Bengal—A synthesis and review. *Natural Hazards, 81*, 2031–2102.

Albertz, R. (2011). The late exilic book of Exodus (Exodus 1–34*): a contribution to the Pentateuchal discussion. In: T.B. Dozeman, K. Schmid, & B. J. Schwartz (Eds.), *The Pentateuch. International perspectives on current research.* Forschungen zum Alten Testament 78 (pp. 243–256). Mohr Siebeck.

Ambraseys, N. N. (1971). Value of historical records of earthquakes. *Nature, 232*(5310), 375–379.

Ambraseys, N. N. (2006). Earthquakes and archaeology. *Journal of Archaeological Science, 33*(7), 1008–1016.

Ambraseys, N. (2009). *Earthquakes in the Mediterranean and Middle East: A multidisciplinary study of seismicity up to 1900.* Cambridge University Press.

Ambraseys, N. N., & Melville, C. P. (1988). An analysis of the Eastern Mediterranean earthquake of 11 July 1927. *Natural Hazards, 1*(4), 351–379.

Amit, R., Zilberman, E., Porat, N., & Enzel, Y. (1999). Relief inversion in the Avrona Playa as evidence of large-magnitude historical earthquakes, Southern Arava Valley, Dead Sea Rift. *Quaternary Research, 52*(1), 76–91.

Amit, R., Zilberman, E., Enzel, Y., & Porat, N. (2002). Paleoseismic evidence for time dependency of seismic response on a fault system in the southern Arava Valley, Dead Sea rift, Israel. *Geological Society of America Bulletin, 114*(2), 192–206.

Anati, E. (2015). Har Karkom: Archaeological discoveries in a holy mountain in the desert of Exodus. In *Israel's Exodus in transdisciplinary perspective: Text, archaeology, culture, and geoscience* (pp. 449–456). Springer International Publishing.

Austin, S. A., Franz, G. W., & Frost, E. G. (2000). Amos's earthquake: An extraordinary Middle East seismic event of 750 BC. *International Geology Review, 42*(7), 657–671.

Avner, U. (2021). The desert's role in the formation of early Israel and the origin of Yhwh. *Entangled Religions, 12*(2).

Avni, R. (1999). *The 1927 Jericho earthquake.* Ph.D. dissertation, Ben Gurion University, Beer Sheva.

Ben-Menahem, A. (1991). Four thousand years of seismicity along the Dead Sea Rift. *Journal of Geophysical Research, 96*(B12), 20195–20216.

Ben-Avraham, Z., Garfunkel, Z., & Lazar, M. (2008). Geology and evolution of the southern Dead Sea fault with emphasis on subsurface structure. *Annual Review of Earth and Planetary Sciences, 36*(1), 357–387.

Bentor, Y. K. (1989). Geological events in the Bible. *Terra Nova, 1*(4), 326–338.

Ben-Yosef, E., Levy, T. E., Higham, T., Knudeson, M., & Najjar, M. (2014). *The beginning of Iron Age copper production in the southern Arabah Valley: The copper mines of Khirbat Al-Nahas*. Antiquity Publications.

Biasi, G. P., & Weldon, R. J. (2006). Estimating surface rupture length and magnitude of paleoearthquakes from point measurements of rupture displacement. *Bulletin of the Seismological Society of America, 96*(5), 1612–1623.

Bietak, M. (2015). On the historicity of the exodus: What Egyptology today can contribute to assessing the biblical account of the sojourn in Egypt. *Israel's Exodus in Transdisciplinary Perspective: Text, Archaeology, Culture, and Geoscience*, 17–37.

Bluestone, J. R. (2010). *Why the earth shakes: Pre-modern understandings and modern earthquake science*. Ph.D. dissertation, University of Minnesota.

Bruins, H. J., & van der Plicht, J. (2007). Radiocarbon dating the "Wilderness of Zin." *Radiocarbon, 49*(2), 481–497.

Burckhardt, J. L. (1822). *Travels in Syria and the Holy Land*. John Murray.

Byko, M. (2004). The miracles of exodus: A materials science approach to biblical study. *JOM Journal of the Minerals Metals and Materials Society, 56*(1), 64.

Chester, D. K. (2005). Theology and disaster studies: The need for dialogue. *Journal of Volcanology and Geothermal Research, 146*(4), 319–328.

Davies, G. (2004). Was there an Exodus? *Journal for the Study of the Old Testament Supplement Series*, 23–40.

Dever, W. G. (2017). *Beyond the texts: An archaeological portrait of ancient Israel and Judah*. SBL Press.

Dilek, Y., & Kahya, Ö. (2023). Flood and earthquake as punishment of gods in antiquity. *Afet Ve Risk Dergisi, 6*(3), 819–828.

Drews, R. (1993). *The end of the Bronze Age: Changes in warfare and the catastrophe ca. 1200 BC*. Princeton University Press.

Dürr, R. (2017). Mapping the miracle: Empirical approaches in the exodus debate of the eighteenth century. *Past and Present, 237*(1), 93–133.

Engelberg, S., Sagy, A., Shaar, R., Matmon, A., Hidy, A. J., Ginat, H., & Marco, S. (2024). Northward propagation of the Gulf of Elat-Aqaba constrained by cosmogenic burial ages and magnetostratigraphy of onshore sediments. *Tectonophysics, 871*, Article 230178.

Enmar, L. (1999). *The travertines in the northern and central Arava: Stratigraphy, petrology and geochemistry*. Geological Survey of Israel Report, GSI/1/99, Jerusalem.

Enmar, L., Heimann, A., Starinsky, A., & Katz, A. (1998). The Arava travertines: Lithology and stratigraphy. *Geol. Surv. Isr. Current Research, 11*, 17–22.

Ferry, M., Meghraoui, M., Abou Karaki, N., Al-Taj, M., & Khalil, L. (2011). Episodic behavior of the Jordan Valley section of the Dead Sea fault inferred from a 14-ka-long integrated catalog of large earthquakes. *Bulletin of the Seismological Society of America, 101*(1), 39–67.

Finkelstein, I. (2007). *David and Solomon: In search of the Bible's Sacred kings and the roots of the western tradition*. Simon and Schuster.

Finkelstein, I. (2015). The wilderness narrative and itineraries and the evolution of the Exodus tradition. In *Israel's Exodus in transdisciplinary perspective: Text, archaeology, culture, and geoscience* (pp. 39–53).

Finkelstein, I., & Silberman, N. A. (2001). *The Bible Unearthed: Archaeology's New Vision of Ancient Israel and the Birth of Sacred Texts*. Simon and Schuster.

Freund, R. A., & Nur, A. (2001). The Jordan Valley and the Dead Sea: The destruction of Sodom and Gomorrah. In E. Loon (Ed.), *Natural disasters and cultural change* (pp. 115–123). Routledge.

Frumkin, A. (2009). Formation and dating of a salt pillar in Mount Sedom diapir, Israel. *Geological Society of America Bulletin, 121*(1–2), 286–293.

Garfunkel, Z. (2014). Lateral motion and deformation along the Dead Sea transform. *Dead Sea transform fault system: Reviews*, 109–150.

Germany, S. M. (2016). *Parallel narratives in Exodus, numbers, and deuteronomy and the identification of pre-priestly material in the Pentateuch*. Ph.D. dissertation, Emory University.

Gertz, J. C. (2014). The miracle at the sea: Remarks on the recent discussion about origin and composition of the Exodus narrative. In *The book of Exodus* (pp. 91–120). Brill.

Ginat, H., Enzel, Y., & Avni, Y. (1998). Translocated Plio-Pleistocene drainage systems along the Arava fault of the Dead Sea transform. *Tectonophysics, 284*(1–2), 151–160.

Grabbe, L. L. (2014). Exodus and history. In *The book of Exodus* (pp. 61–87). Brill.

Guidoboni, E., & Comastri, A. (2005). *Catalogue of earthquakes and tsunamis in the Mediterranean area from the 11th to the 15th century*. Istituto Nazionale di Geofisica e Vulcanologia, Romehinzen.

Hamiel, Y., Masson, F., Piatibratova, O., & Mizrahi, Y. (2018). GPS measurements of crustal deformation across the southern Arava Valley section of the Dead Sea Fault and implications to regional seismic hazard assessment. *Tectonophysics, 724*, 171–178.

Har-El, M. (1968). *The Sinai journeys*. Am Oved, Tel-Aviv (Hebrew).

Harris, G. M., & Beardow, A. P. (1995). The destruction of Sodom and Gomorrah: A geotechnical perspective. *Quarterly Journal of Engineering Geology and Hydrogeology, 28*(4), 349–362.

Hendel, R. (2001). The Exodus in biblical memory. *Journal of biblical Literature, 120*(4), 601–622.

Hinzen, K. G., Fleischer, C., Reamer, S. K., Schreiber, S., Schütte, S., & Yerli, B. (2011). Quantitative methods in archaeoseismology. *Quaternary International, 242*(1), 31–41.

Hoffmeier, J. K. (1996). *Israel in Egypt: The evidence for the authenticity of the Exodus tradition*. Oxford University Press.

Hughes, J. (1990). *Secrets of the times: Myth and history in biblical chronology*. Sheffield Academic Press.

Humphreys, C. (2004). *Miracles of Exodus: A scientist's discovery of the extraordinary natural causes of the biblical stories*. Bloomsbury Publishing.

Jack, J. W. (2018). *The date of the Exodus: In the light of external evidence*. Wipf and Stock Publishers.

Johnstone, W. (2005). The miracles of Exodus: A scientist's discovery of the extraordinary natural causes of the biblical stories. *Journal of Semitic Studies, 50*(2), 373–379.

Kagan, E., Stein, M., Agnon, A., & Neumann, F. (2011). Intrabasin paleoearthquake and quiescence correlation of the late Holocene Dead Sea. *Journal of Geophysical Research: Solid Earth, 116*(B4).

Kanari, M., Niemi, T. M., Ben-Avraham, Z., Frieslander, U., Tibor, G., Goodman-Tchernov, B. N., Wechsler, N., Abueladas, A., Al-Zoubi, A., Basson, U., & Marco, S. (2020). Seismic potential of the dead sea fault in the northern gulf of aqaba-elat: New evidence from liquefaction, seismic reflection, and paleoseismic data. *Tectonophysics, 793*, Article 228596.

Karcz, I. (2004). Implications of some early Jewish sources for estimates of earthquake hazard in the Holy Land. *Annals of Geophysics*.

Ken-Tor, R., Agnon, A., Enzel, Y., Stein, M., Marco, S., & Negendank, J. F. (2001). High-resolution geological record of historic earthquakes in the Dead Sea basin. *Journal of Geophysical Research: Solid Earth, 106*(B2), 2221–2234.

Kerkeslager, A. (1998). Jewish Pilgrimage and Jewish Identity in Hellenistic and Early Roman Egypt. In A. Kasher, U. Rappaport, & G. Fuks (Eds.), *Greece and Rome in Eretz Israel*. Yad Izhak Ben Zvi, Jerusalem.

Khan, A. A. (2018). The geo-genetic status of earthquake-related hazards and the role of human and policy dimensions in impact mitigation. *Earthquake Disasters, 17*, 8.

Kim, J., Lee, J., Petitta, M., Kim, H., Kaown, D., Park, I. W., Lee, S., & Lee, K. K. (2019). Groundwater system responses to the 2016 ML 5.8 Gyeongju earthquake, South Korea. *Journal of Hydrology, 576*, 150–163.

Kitchen, K. A. (1998). Egyptians and Hebrews, from Ra'amses to Jericho. In *The origin of early Israel-Current debate* (pp. 65–132). Routledge.

Kitto, J. (1841). *Palestine: The physical geography and natural history of the Holy Land*. C. Knight and Company.

Klinger, Y., Le Béon, M., & Al-Qaryouti, M. (2015). 5000 yr of paleoseismicity along the southern Dead Sea fault. *Geophysical Journal International, 202*(1), 313–327.

Klinger, Y., Avouac, J. P., Abou Karaki, N., Dorbath, L., Bourles, D., & Reyss, J. L. (2000). Slip rate on the Dead Sea transform fault in northern Araba valley (Jordan). *Geophysical Journal International, 142*(3), 755–768.

Kouki, P., & Tenhunen, T. (2013). The environmental setting of the Jabal Hārūn area. In *Petra-The Mountain of Aaron III: The Archaeological Survey*. Societas Scientiarum Fennica.

Laato, A. (2015). *Guide to biblical Chronology*. Sheffield Phoenix Press.

Le Beon, M., Klinger, Y., Amrat, A. Q., Agnon, A., Dorbath, L., Baer, G., Ruegg, J. C., Charade, O., & Mayyas, O. (2008). Slip rate and locking depth from GPS profiles across the southern Dead Sea Transform. *Journal of Geophysical Research: Solid Earth, 113*(B11).

Lefevre, M., Klinger, Y., Al-Qaryouti, M., Le Béon, M., & Moumani, K. (2018). Slip deficit and temporal clustering along the Dead Sea fault from paleoseismological investigations. *Scientific Reports, 8*(1), 4511.

Levy, T. E., Schneider, T., & Propp, W. H. (2015). Israel's Exodus in transdisciplinary perspective. In *Text, archaeology, culture, and geoscience*. Springer.

Malamat, A. (2001). The Exodus: Egyptian analogies. In *History of biblical Israel* (pp. 57–67). Brill.

Manga, M., & Rowland, J. C. (2009). Response of Alum Rock springs to the October 30, 2007 Alum Rock earthquake and implications for the origin of increased discharge after earthquakes. *Geofluids, 9*(3), 237–250.

Marco, S., Stein, M., Agnon, A., & Ron, H. (1996). Long-term earthquake clustering: A 50,000-year paleoseismic record in the Dead Sea Graben. *Journal of Geophysical Research: Solid Earth, 101*(B3), 6179–6191.

Masson, F., Hamiel, Y., Agnon, A., Klinger, Y., & Deprez, A. (2015). Variable behavior of the Dead Sea Fault along the southern Arava segment from GPS measurements. *Comptes Rendus Geoscience, 347*(4), 161–169.

McGuire, B., Griffiths, D. R., Hancock, P. L., & Stewart, I. S. (2000). *The archaeology of geological catastrophes* (p. 171). Geological Society Special Publication.

Merrill, E. H. (2008). *Kingdom of priests: A history of Old Testament Israel*. Baker Academic.

Migowski, C., Agnon, A., Bookman, R., Negendank, J. F., & Stein, M. (2004). Recurrence pattern of Holocene earthquakes along the Dead Sea transform revealed by varve-counting and radiocarbon dating of lacustrine sediments. *Earth and Planetary Science Letters, 222*(1), 301–314.

Na'aman, N. (2011). The Exodus story: Between historical memory and historiographical composition. *Journal of Ancient near Eastern Religions, 11*(1), 39–69.

Neev, D., & Emery, K. O. (1995). *The destruction of Sodom, Gomorrah, and Jericho: Geological, climatological, and archaeological background*. Oxford University Press.

Nur, A., & Burgess, D. (2008). *Apocalypse: Earthquakes, archaeology, and the wrath of God*. Princeton University Press.

Nur, A., & Cline, E. H. (2000). Poseidon's horses: Plate tectonics and earthquake storms in the Late Bronze Age Aegean and Eastern Mediterranean. *Journal of Archaeological Science, 27*(1), 43–63.

Nur, A., Ron, H., Stiros, S., & Jones, R. (1996). *And the walls came tumbling down: Earthquake history in the Holy Land* (Vol. 7, pp. 75–85). British School at Athens, Athens.

Rendsburg, G. A. (2015). Moses the Magician. In *Israel's Exodus in transdisciplinary perspective: Text, archaeology, culture, and geoscience* (pp. 243–258).

Roberts, R. N. (2012). *Terra Terror: An interdisciplinary study of earthquakes in ancient near eastern texts and the Hebrew Bible*. University of California.

Römer, T. (2015). *The invention of God*. Harvard University Press.

Salamon, A. (2010). Patterns of seismic sequences in the Levant—interpretation of historical seismicity. *Journal of Seismology, 14*, 339–367.

Schmid, K. (2015). Distinguishing the world of the Exodus narrative from the world of its narrators: The question of the priestly Exodus account in its historical setting. In *Israel's Exodus in transdisciplinary perspective: Text, archaeology, culture, and geoscience* (pp. 331–344).

Schneider, T. (2015). Modern scholarship versus the Demon of Passover: An outlook on Exodus research and Egyptology through the lens of Exodus 12. In *Israel's Exodus in transdisciplinary perspective: Text, archaeology, culture, and geoscience* (pp. 537–553).

Shanks, H. (2005). The mystery of the Exodus from Egypt. *Biblical Archaeology Review, 31*(3), 30–41.

Shupak, N. (2001). The god from Teman and the Egyptian sun god: A reconsideration of Habakkuk 3: 3–7. *Journal of the Ancient Near Eastern Society, 28*(1).

Sparks, B.C. (2015). Egyptian texts relating to the Exodus: Discussions of Exodus parallels in the egyptology literature. In T. Levy, T. Schneider, & W. Propp (Eds.), *Israel's Exodus in transdisciplinary perspective. Quantitative methods in the humanities and social sciences.* Springer.

Stiros, S. C., & Jones, R. E. (Eds.). (1996). *Archaeoseismology*. Fitch Laboratory Occasional Paper 7, British School at Athens, Oxbow Books, Oxford.

Stripling, S., Hoffmeier, J. K., Rendsburg, G. A., Feinman, P., & Hendel, R. (2021). *Five views on the Exodus: Historicity, chronology, and theological implications.* Zondervan Academic.

Tebes, J. (2017). Desert place-names in Numbers 33: 34, Assurbanipal's Arabian wars and the historical geography of the biblical wilderness toponymy. *Journal of Northwest Semitic Languages, 43*(2), 65–96.

Torrence, R., & Grattan, J. (Eds.). (2002). *Natural disasters and cultural change.* Routledge.

Wechsler, N., Katz, T., Dray, T., Levy, Y., & Marco, S. (2009). Estimating location and size of historical earthquake by combining archaeology and geology in Umm-el-Qanatir, Dead Sea Transform. *Natural Hazards, 50*(2), 27–43.

Yechieli, Y., & Bein, A. (2002). Response of groundwater systems in the Dead Sea Rift Valley to the Nuweiba earthquake: Changes in head, water chemistry, and near-surface effects. *Journal of Geophysical Research: Solid Earth, 107*(B12), ETG-4.

Zohar, M., Rubin, R., & Salamon, A. (2014). Earthquake damage and repair: New evidence from Jerusalem on the 1927 Jericho earthquake. *Seismological Research Letters, 85*(4), 912–922.

Chapter 21
Drought and Water Management in Morocco: Statistical Analysis in the Sebou Basin and Adaptation Strategies Through Inter-Basin Transfer

Zohair Qadem, Lamia Elbezzari, and Abdelghani Qadem

Abstract Since the 1980s, Morocco has faced recurrent and intense droughts, significantly affecting water resources and socio-economic sectors. This study examines drought variability in the Sebou basin based on a methodology that combines the standardized precipitation index (SPI) at multiple scales (3, 6, 12, and 24 months) and long-term trends with the Mann–Kendall test. Results show alternating wet and dry periods, with notable episodes in the 1980s, 1990s, and early 2000s impacting agricultural productivity and groundwater recharge. The Mann–Kendall test indicates significant negative trends in the Atlas and Middle Atlas on the SPI12 and SPI24, reflecting progressive drying, while the Saïs and Gharb plains show no significant trends, highlighting spatial contrasts. Exceptional wet periods occurred in 1996–1997 and 2009–2012, confirming strong regional climate variability. In this context, Morocco has implemented inter-basin transfers, such as the "water highway" linking the Sebou to the Bouregreg, and maintains a modern institutional framework, including Law 36–15, Water Basin Agencies, and local commissions, to support integrated and participatory water governance. However, challenges remain regarding sustainability, equity, and coordination, emphasizing the need for integrated, multi-scalar, and participatory approaches to secure water resources under climate change.

Keywords Morocco · Drought · Standardized Precipitation Index (SPI) · Climate trend · Mann–Kendall test · Inter-basin transfer

Z. Qadem (✉) · L. Elbezzari
Department of Geography, Ibn Tofail University, Kenitra, Morocco
e-mail: Zohair.qadem@uit.ac.ma

A. Qadem
Departement of Geography, Faculty of Arts and Humanities—Sais, Sidi Mohamed Ben Abdellah University, Fez, Morocco

E. Alam (ed.), *Disaster, Crisis and Emergency Management in the Middle East & North Africa*, https://doi.org/10.1007/978-3-032-24436-9_21

21.1 Introduction

Today, climate change is a major challenge that increases the frequency and intensity of droughts, linked to land degradation and desertification (Sheffield et al., 2012). On the other hand, the increase in greenhouse gases disrupts the hydrological cycle and exacerbates water scarcity (Vengosh & Weinthal, 2022; Mélia, 2022; Mwanake et al., 2023). Drought, defined as a prolonged water deficit (Tallaksen et al., 2023; Wilhite, 2000), is exacerbated by rising temperatures and rainfall variability, leading to a reduction in water resources, with impacts on agriculture, food security, and ecosystems (Cook et al., 2015).

In Morocco, drought is a structural component of the climate, which has been increasing since the 1960s (Driouech, 2010; Hakam et al., 2022a; Oubaha et al., 2024; Goumi et al., 2024; Qadem & Ettaqy, 2024; Bijaber et al., 2024; Qadem et al., 2025b). In this context, this article focuses on assessing drought in one of Morocco's largest river basins, the Sebou basin. The Standardized Precipitation Index (SPI) is the essential tool for this assessment and remains the most widely used at different time scales (Abeysingha & Rajapaksha, 2020; Guttman, 1998; Mckee et al., 1993; Mehr et al., 2020; Moccia et al., 2022; Qadem & Tayfur, 2024; Qadem et al., 2025b; Türkeş & Tatli, 2009; Venturi et al., 2024).

Secondly, trend analysis was performed using the nonparametric Mann–Kendall test and its variants (Kendall, 1975; Mann, 1945). This test is widely used in hydrological and climate studies to detect the existence of monotonic increasing or decreasing trends in time series (Hamed & Ramachandra Rao, 1998; Güçlü, 2018; Ashraf et al., 2021; Zeybekoglu, 2023), without assuming any particular distribution of data. Its robustness in the face of outliers and missing data makes it a valuable tool for assessing changes in drought indices and characterizing climate dynamics at the regional level.

The Sebou basin occupies a strategic position in national water policy, partly because of its relatively abundant water potential (Qadem, 2020) and, secondly, due to its agricultural importance. These strategies contribute to Rabat's drinking water supply via transfer to the Bouregreg basin, as well as supporting agriculture in the Saïs plain by replenishing overexploited groundwater resources. However, this national role increases local pressure, where agricultural and domestic needs are already significant.

Despite the growing number of studies on drought in Morocco, most existing work has focused on national or regional assessments without incorporating multi-scale temporal analyses or linking drought severity to hydroclimatic trends at the watershed level. Furthermore, few studies have specifically examined the Sebou basin (Hakam et al., 2022b), despite its strategic importance for agriculture, water transfers, and national water security. This study fills these gaps by combining the Standardized Precipitation Index (SPI) at multiple temporal scales with the nonparametric Mann–Kendall trend test to provide a detailed spatiotemporal characterization of drought dynamics. By focusing on the Sebou basin, this research provides new insights into the temporal evolution of drought, identifies spatial contrasts between stations, and highlights implications for sustainable water management and climate adaptation policies in Morocco.

In this context of increasing climatic and anthropogenic pressures, it is essential to quantify drought using robust indicators such as the Standardized Precipitation Index (SPI), which is widely used to characterize the severity and duration of dry spells. Furthermore, assessing long-term climate trends using the nonparametric Mann–Kendall test makes it possible to identify the direction and statistical significance of hydroclimatic changes. Integrating these two approaches into the study of the Sebou basin provides a relevant framework for understanding drought dynamics and supporting the implementation of adaptation and sustainable water resource management strategies. The objective of this work is therefore to analyze the temporal variability of drought across different SPI time steps, identify hydroclimatic trends at the scale of stations in the basin, and discuss the implications of these results for water management and adaptation strategies, particularly through inter-basin transfers.

21.2 Data and Methodology

21.2.1 Study Area

The study area corresponds to the Sebou river basin (Fig. 21.1), which covers an area of approximately 40,000 km^2 and accounts for nearly one-third of Morocco's surface runoff. This basin is home to 5.9 million inhabitants and is the richest in surface and groundwater resources in Morocco (Qadem, 2020). The basin's average annual inflow is 5560 million cubic meters per year, or an average of 887 cubic meters per capita. Its water resources represent approximately 25% of Morocco's usable potential (Sebou River Basin Agency).

Like surface water resources, groundwater resources in the Sebou basin are very significant. It contains several aquifers, the most important of which are: The Fez-Taza corridor aquifer (phreatic and deep); The aquifers of the Fez-Meknes basin (phreatic and deep); The Mamora aquifer; The Gharb aquifer; The Bou Agba aquifer; The aquifers of the Middle Atlas Causse; The aquifer of the Middle Atlas fold.

21.2.2 Data Used

The data used in this paper come from 12 weather stations located in different parts of the Sebou basin (Fig. 21.1 and Table 21.1), covering a wide range of altitudes from 5 m in Kenitra to 1660 m in the Middle Atlas Mountains upstream (Table 21.1).

According to Table 21.1, the series include average annual precipitation for the period from 1970–2017, as well as their dispersion and extremes. Average precipitation varies, e.g., from 397 mm/year in Souk El Had to 945 mm/year in Ifrane in

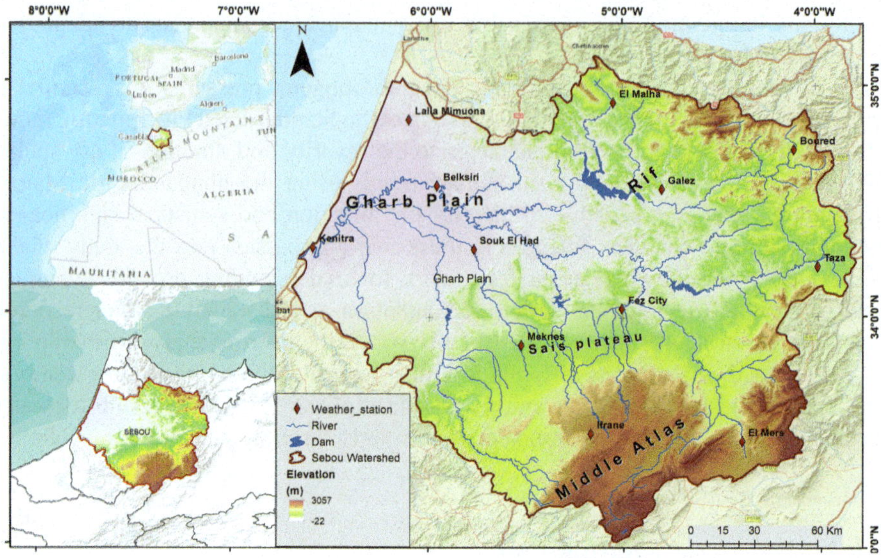

Fig. 21.1 Study area

the Middle Atlas Mountains, illustrating the strong influence of orography. Stations located in the Rif, such as El Malha (801 mm/year) and Galez (684 mm/year), also show higher variability (standard deviation of 273 mm and 252 mm, respectively), while stations in the Gharb plain, such as Kenitra (548 mm/year) and Belksiri (502 mm/year), have more moderate and regular precipitation (Table 21.1).

21.2.3 Data Analysis

The methodology adopted in this study aims to provide a comprehensive and in-depth analysis of precipitation variability and its impacts in the Sebou basin. It is based on two main steps. First, drought was assessed using the SPI index, calculated at different time scales: short term (3 and 6 months) to understand the effects on soil moisture and crops, and long term (12 and 24 months) to analyze the influence on water resources and groundwater recharge.

Secondly, trends were examined using conventional statistical methods, notably the Mann–Kendall (MK) test, in order to reinforce the robustness of the results. In a second step, based on this assessment, the analysis also incorporates the adaptation strategies implemented in the Sebou basin, in particular inter-basin transfer projects designed to mitigate the effects of drought, both for drinking water supply and agricultural needs.

Table 21.1 Statistical characteristics of the stations operated

Station	Period	Longitude	Latitude	Elevation (m)	Mean (mm)	Std (mm)	Q25 (mm)	Q50 (mm)	Q75 (mm)	Min (mm)	Max (mm)	CV (%)	IQR (mm)
Kenitra	1970–2017	389,553	411,513	5	548	175	447	527	643	256	1132	32	196
Belksiri	1970–2017	448,350	440,850	15	502	174	354	487	602	221	985	35	248
Lalla Mimuona	1970–2017	435,000	472,600	15	576	184	453	548	695	222	1096	32	242
Souk El Had	1970–2017	466,100	410,483	34	397	116	324	392	467	193	753	29	143
Galez	1970–2017	555,154	439,338	214	684	252	505	638	814	343	1445	37	309
El Malha	1970–2017	532,210	480,860	340	801	273	598	757	946	307	1699	34	348
Fez City	1970–2017	536,459	381,970	415	428	147	308	411	532	170	835	34	223
Taza	1970–2017	628,978	402,502	500	573	195	446	556	697	181	1195	34	251
Meknes	1970–2017	488,645	364,558	560	487	157	387	478	540	150	946	32	153
Boured	1970–2017	617,600	458,700	815	557	206	422	526	634	282	1363	37	212
El Mers	1970–2017	593,100	318,620	1210	442	136	366	431	497	205	833	31	131
Ifran	1970–2017	521,675	322,197	1660	945	298	745	895	1065	555	2011	32	321

21.2.4 Standardized Precipitation Index (SPI)

By enabling the identification of drought's onset, duration, and severity, the SPI is a useful instrument for monitoring it (Guttman, 1998; Liu et al., 2021; Qadem et al., 2025a; Tayfur, 2021; Yu et al., 2023). It is a good indicator for agricultural, hydrological, and meteorological droughts because of its adaptability and simplicity, especially the ability to be used to various time periods.

The SPI index is calculated by adjusting precipitation to a probability distribution (often gamma), then normalizing it to obtain a zero mean over the reference period (Hu et al., 2015). It can be calculated at different scales (1, 3, 6, 12, 24, and 48 months) to assess the intensity and duration of droughts (McKee et al., 1993; Guttman, 1999; Smakhtin & Hughes, 2004).

The SPI is calculated by adjusting cumulative precipitation (1, 3, 6, 12, 24 months) to a probability distribution, usually the gamma distribution, due to its ability to represent the asymmetry of rainfall data (Guttman, 1999). This distribution is then transformed into a standardized normal distribution to obtain the IPS values. The mathematical formulation of this normalization is given as follows:

$$SPI_{kl} = \frac{a_{k,l} - \mu_l}{\sigma_l} \tag{21.1}$$

Let $a_{k,l}$ represents the transformed precipitation for the l th month (1 = 1, 3, …, 24) of the k th year ($k = 1, 2, \ldots, n$). The mean value of the transformed precipitation for the l th month is denoted as μ_l, while σ_l represents the standard deviation for the same month. The SPI values are computed for each time scale $c = 1, 3 \ldots 24, 48$ months individually, resulting in a distinct SPI_c time series for each c (McKee et al., 1993; Aksoy & Cavus, 2022).

The SPI is interpreted using a seven-category scale ranging from "Extremely wet" to "Extremely dry" (Table 21.2). This classification, widely used in water management, agriculture, and climate studies, is a valuable tool for decision-making in regions vulnerable to precipitation variations (McKee et al., 1993).

Table 21.2 Drought classes according to SPI values (McKee et al., 1993)

SPI value	Classification	% of probability
>2.0	Extremely wet	2.3
1.5–1.99	Very wet	4.4
1.0–1.49	Moderately wet	9.2
(–0.99)–0.99	Near normal	68.2
(–1.0)–(–1.49)	Moderately dry	9.2
(–1.5)–(–1.99)	Very dry	4.4
<–2	Extremely dry	2.3

21.2.5 *Test Mann–Kendall*

A statistical technique for identifying upward or downward trends in a time series is Mann–Kendall (MK) trend analysis. (Alashan, 2023). Introduced by Mann (1945), and further developed by Kendall (1975), it is widely used in climatology and hydrology to analyze hydrometeorological time series (Acar Deniz & Gönençgil, 2015; Güçlü, 2018a; Qadem & Tayfur, 2024; Qadem et al., 2025a; Yacoub & Tayfur, 2020; Zeybekoglu, 2023;). Based on a rank correlation test, this test considers the absence of a trend or serial correlation between observations as the null hypothesis (Hamed, 2008). The MK test calculation is presented as follows (Kendall, 1975; Mann, 1945).

$$\operatorname{sgn}(x_i - x_j) = \begin{cases} 1; & \text{IF } x_j > x_i \\ 0; & \text{IF } x_j = xi \\ -1; & \text{IF } x_j < xi \end{cases}$$

$$S = \sum_{l=1}^{n-1} \sum_{j=i+1}^{n} \operatorname{sgn}(x_i - x_j) \tag{21.2}$$

The data values x_i and x_j correspond, respectively, to time points i and j, with n representing the total number of observations. A positive result for S indicates an upward trend, while a negative value indicates a downward trend.

21.3 Results and Discussion

21.3.1 *The History of Drought in Morocco*

Drought is a significant and ongoing environmental issue in Morocco that impacts several areas, including agriculture, water resources, and the economy. The semi-arid climate of Morocco makes drought episodes more frequent and severe, which has serious social repercussions. Historical data reveal a marked upward trend in temperatures (Qadem & Tayfur, 2024) and a decrease in precipitation, which has exacerbated the severity and frequency of droughts in recent decades.

According to several studies (El Qorchi et al., 2023; Esper et al., 2007), periods of drought in Morocco occur cyclically, with an average frequency of 5–6 years. Morocco has experienced several major drought episodes since the 1940s (Table 21.3), with significant impacts on water resources, agriculture, and the economy. The recurrent droughts of 1975–1985 and 1994–1995 left their mark on the country due to their intensity and lasting effects (El-Khantoury et al., 2022; Hafid, 2017). Those of 1999–2001 and 2005–2008 confirmed a growing trend toward aridity and rainfall irregularity (Hadri et al., 2024). More recently, the periods 2015–2017

Table 21.3 History of drought episodes in Morocco and their impacts

Period	Climate characteristics and main impacts	References
1940–1950	Morocco has experienced several episodes of severe drought, notably in 1944–1945, which have had a significant impact on water resources, agricultural production, and the national economy	Hafid (2017)
1975–1985	Between 1980 and 1985, the country experienced persistent drought, which continued until the mid-1990s, interrupted by exceptionally wet years (notably in 1995 and 2002). This period was characterized by high interannual variability in precipitation, leading to significant hydrological and agricultural instability	Hafid (2017), Ouharba et al. (2024), Qadem and Ettaqy (2024)
1994–1995	This episode was one of the most severe droughts ever experienced in northern Morocco, with an estimated reduction in water resources of between 62 and 82%. It caused severe water shortages and weakened local water infrastructure	Qadem (2019), El-Khantoury et al. (2022), Qadem and Ettaqy (2024)
1999–2001	The drought of 1998–1999 was classified as severe in central Morocco. This period is part of a general trend of worsening drought conditions observed between 1980 and 2019, marked by a continuous reduction in precipitation and a decline in water flows	Qadem (2019)
2005–2008	The year 2005 was identified as one of the driest of the century, marked by a major rainfall deficit and prolonged dry spells. Studies show that 75% of droughts between 2005 and 2008 lasted between 6 and 20 months, illustrating the increasing irregularity of rainfall patterns	Hadri et al. (2024)
2015–2017	The Ouergha basin, as a tributary of the greater Sebou basin, experienced severe drought during this period, with a statistically significant increase in severity detected by drought indices	Mounir et al. (2023)
2020	In 2020, Morocco experienced a major drought across the country, leading to a reduction in groundwater and surface water resources and increased migration of the rural population	Naim and Bonaccorso (2023)

and 2020 illustrated the persistence of severe droughts, particularly in the Sebou basin and at the national level (Khrabcha et al., 2025; Laachrate & Fadil, 2022; Mounir et al., 2023; Naim & Bonaccorso, 2023).

21.3.2 *Standardized Precipitation Index*

The assessment of drought in the Sebou basin using SPI indices (3, 6, 12, and 24 months) reveals significant spatial and temporal variability in hydroclimatic conditions between the Rif, El Gharb, the Sais Plateau, and the Middle Atlas in the upper Sebou basin for the period 1970 to 2017 (Fig. 21.2).

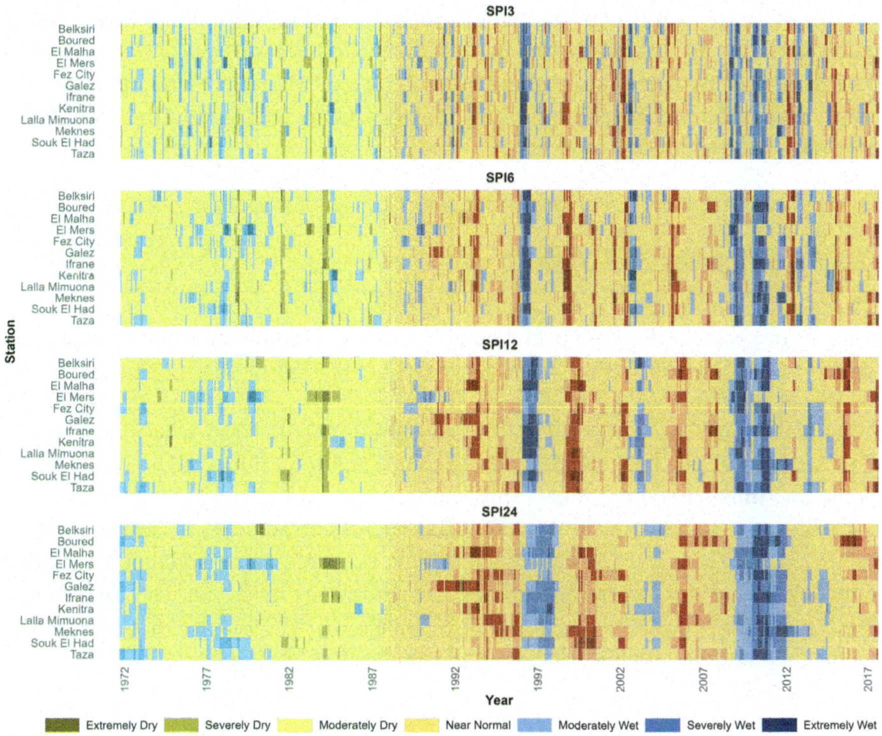

Fig. 21.2 Temporal drought characteristics for SPI 3, 6, 12, and 24 months

In the short term (SPI3 and SPI6), fluctuations between dry and wet periods appear frequent and irregular from one year to the next. Moderate to severe droughts were particularly pronounced in the early 1980s and mid-1990s (Fig. 21.2), affecting all stations and even the entire Sebou basin. Similarly, the period 2000–2007 was marked by a succession of rainfall deficits, reflecting a high recurrence of short-term droughts, often associated with immediate impacts on rain-fed agriculture.

In the longer term (SPI12 and SPI24), clearer trends emerge, influencing groundwater recharge and river flow at the basin level. SPI12 shows a clear alternation between dry spells (early 1980s, late 1990s, early 2000s) and wet periods (particularly 2008–2013). SPI24 highlights prolonged sequences of drought, with notable intensity during the 1980s and early 2000s, contrasting with an exceptionally wet phase observed between 2008 and 2013 (Fig. 21.2).

In spatial terms, all stations are affected by hydroclimatic variability, but with differences in intensity: for example, stations in the Atlas Mountains (Ifran, Taza) appear to be more exposed to extreme events (both dry and wet), while stations in Saïs and Gharb (Meknes, Fez City, Lalla Mimouna) show slightly more moderate patterns.

Overall, these results confirm that the Sebou region is subjected to marked alternation between droughts and wet periods, with drought severity worsening since the 1980s. This dynamic, highlighted by multi-scale SPI, reflects the importance of considering different time scales to fully understand the impacts of drought on water resources and agricultural systems.

21.3.3 Trend by MK Test

Analysis of precipitation trends in the Sebou basin, based on the Mann–Kendall test applied to the SPI index at different time scales (SPI3, SPI6, SPI12, SPI24), reveals significant regional variability (Fig. 21.3). Some stations on the Saïs Plateau, such as Fez City (SPI12 Tau = −0.12, P 0.001) and Boured (SPI24 Tau = −0.19, P 0.001), show significant negative trends in the medium and long term, reflecting increased aridity and intensified droughts in this region.

Conversely, stations in the Middle Atlas such as Ifrane (SPI12 Tau = 0.004, P = 0.898) and Galez (SPI24 Tau = −0.037, P = 0.196) show no significant trend, reflecting stable rainfall probably linked to altitude (Fig. 21.3). The Taza station stands out with widespread drying at all SPI scales (SPI12 Tau = −0.144, P 0.001), while other stations, such as Belksiri and Kenitra, remain stable.

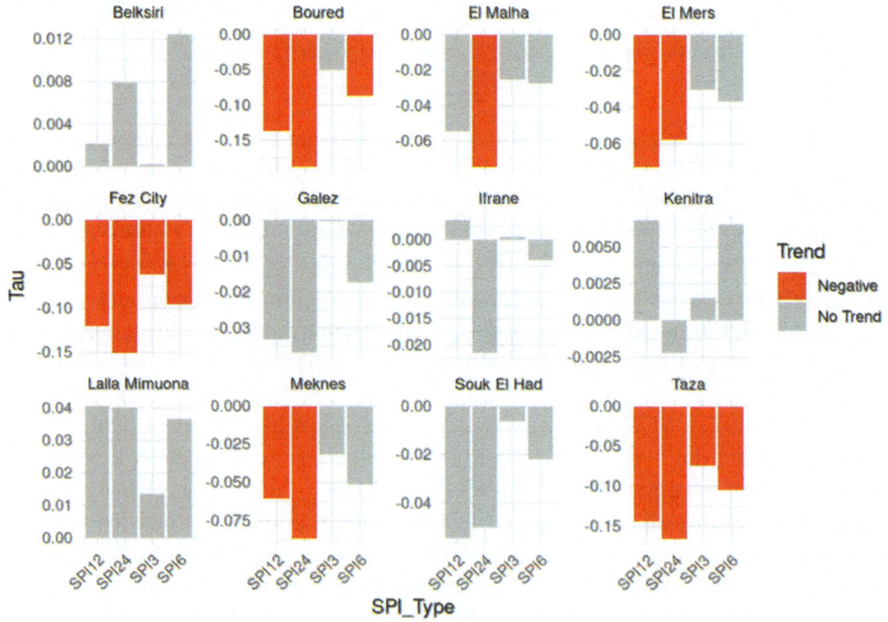

Fig. 21.3 Drought trends using the Mann–Kendall test

21.3.4 Water Transfer Between Basins in Morocco

Faced with the growing impacts of climate change, and in particular the intensification of droughts, Morocco is continuing to roll out structural projects aimed at strengthening its water security. These initiatives include the development of inter-basin transfer infrastructure, particularly between the Sebou and Bouregreg basins. This project, considered one of the most important ever undertaken in the country, is part of the national program for drinking water supply and irrigation. It is a response to the growing pressure on water resources, particularly in regions most vulnerable to water scarcity.

The inter-basin transfer project is a major initiative in the national water security strategy. It provides for an annual transfer volume of 1.2 billion cubic meters of water, contributing significantly to the balance between surplus and deficit basins. Beyond the figures, the impacts will be major. This system will strengthen the supply of drinking water in densely populated urban areas, particularly Rabat and its surrounding region.

Regarding the Sais plateau, it should be noted that periods of drought in the study area have increased demand for irrigation water since the 1970s.

Given the high demand for irrigation water in this plateau, the Saïs region benefited from a major irrigation water transfer project (22,000 ha via a gravity-fed system and 23,000 ha via pumping) (Fig. 21.4). The direct water intake was built in the M'Dez dam, and the transfer was carried out by a 90 km pipeline, including 12 km of underground tunnels (Fig. 21.4). This major project transfers water from the M'Dez-Ain Timedrine hydraulic complex, whose theoretical storage capacity is estimated at 760 million m^3.

Secondly, a project known as the "water highway" linking the Sebou basin to the Bouregreg basin is one of Morocco's most strategic water projects (Photo 1). It aims to mobilize part of the surplus resources available in Sebou and transfer them to areas under high water pressure in Bouregreg and, beyond that, to Oum Er-Rbia (Moroccan Minister of Equipment).

The main objective is to secure the supply of drinking water for large urban centers on the Atlantic coast. Due to its technical scope and transfer volume, this project represents an adaptation response to the combined effects of climate change, declining rainfall, and overexploitation of groundwater in deficit regions, particularly the Sais plateau and the El Gharb plain. It illustrates the desire to integrate inter-basin management into a broader approach to water resilience and sustainable water resource planning (Photo 21.5).

Overall, water resource management in Morocco faces increasing pressures linked to the effects of climate change and growing socio-economic demand. Several studies highlight the importance of integrated water resource management (IWRM) and participatory governance, although challenges remain due to institutional fragmentation and social inequalities (Ait Kadi & Ziyad, 2018; Rbaibi & SahibEddine, 2024; Silva-Novoa Sánchez et al., 2022).

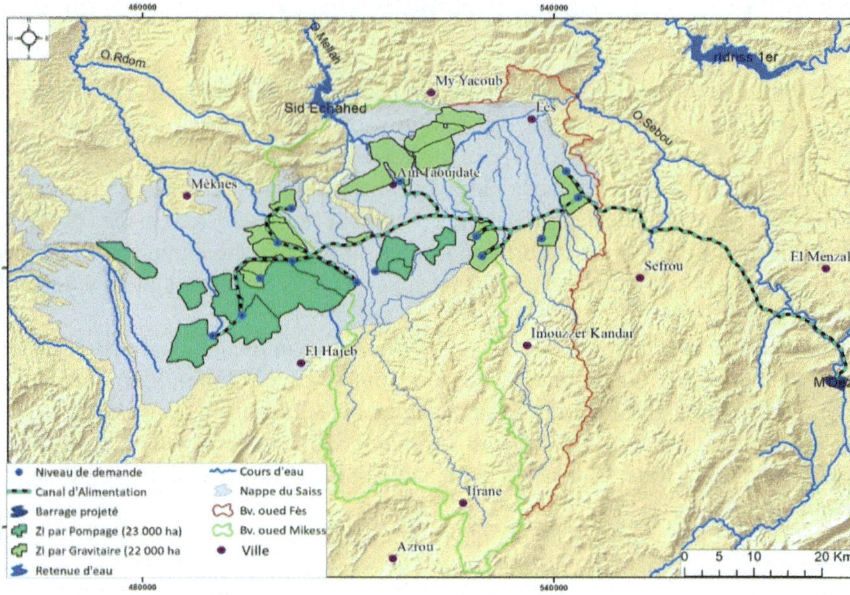

Fig. 21.4 Project to transfer water to the Saïs (Qadem, 2020)

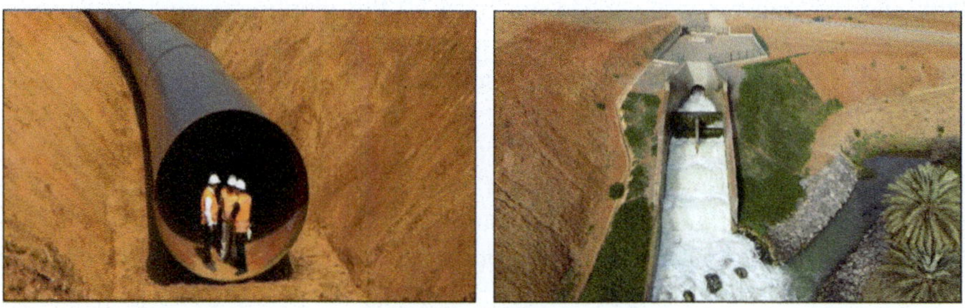

Photo 21.5 Water highway between the Sebou basin and the Bouregreg basin

In Morocco, the agricultural sector, the main consumer of water, remains at the heart of the debate, with a gradual transition toward more efficient irrigation techniques, such as drip irrigation, to reconcile productivity and sustainability (Guemouria et al., 2023). The overexploitation of groundwater remains a major issue, particularly in the Sais plateau and the El Ghrab plain (Qadem, 2020; Qadem et al., 2023; Zohair et al., 2020), giving rise to innovative approaches such as water table contracts and participatory management (Houdret & Heinz, 2022).

At the same time, climate variability and increased frequency of droughts are exacerbating tensions over resources, reinforcing the need for hydrological models, decision support tools, and monitoring technologies (Kalpakian & Jenks, 2025). The recent studies confirm the seriousness of the situation, revealing a significant decline in water resources in several basins in the north and center of the country, exacerbated by increased evapotranspiration and recurring severe droughts (Ait Dhmane et al., 2024). These dynamics directly affect ecosystem sustainability and food security, particularly in rural areas that are heavily dependent on rain-fed agriculture.

In this context, inter-basin transfers appear to be a key strategy for reducing regional water imbalances. The so-called water highway project, linking the Sebou basin to the Bouregreg basin and, ultimately, to the Oum Er-Rbia, illustrates this approach. This project aims to mobilize an annual volume of 1.2 billion cubic meters of water, with the second section alone enabling a transfer of 800 million cubic meters per year, or a flow rate of 45 m^3/s (Ministry of Equipment and Water, 2024). Beyond its technical importance, this project is part of the National Drinking Water Supply and Irrigation Program and responds to the need to ensure equitable redistribution of resources between surplus and deficit basins.

The limitations and controversies surrounding this type of project, inter-basin transfers, while helping to secure supply in deficit areas, can put new pressure on donor basins, alter ecological balances, and raise questions of equity. Furthermore, these projects require significant financial investment and inclusive governance to avoid reproducing existing socio-economic imbalances.

At the same time, technological advances offer promising prospects for strengthening proactive drought management. The combined use of climatic and biophysical indices, such as the Standardized Precipitation Index (SPI) and trend tests such as Mann–Kendall, allows drought to be assessed on different time scales.

Similarly, remote sensing tools, machine learning, and cloud computing are opening new avenues for monitoring and forecasting extreme events (Goumi et al., 2024). These approaches, integrated into participatory management models and multi-scale governance, can support more resilient and sustainable resource planning.

Finally, the literature highlights socio-political dimensions and traditional knowledge, emphasizing that policy effectiveness depends on technological innovation, social equity, and the promotion of local practices (Azemzi & Erraoui, 2021; Erraoui & Azemzi, 2021; Habib et al., 2022; Simon, 2021; Smida et al., 2022). Combining heavy infrastructure such as inter-basin transfers with innovative and inclusive management tools appears to be the best way to address the challenges of drought and water scarcity in Morocco.

21.4 Institutional Framework and Water Governance in Morocco

Since the 1980s, Morocco has been facing recurrent and intense droughts that have profoundly affected water resources and socio-economic sectors, particularly agriculture and groundwater recharge. In this context of increased climate variability, the establishment of a solid institutional and legal framework has become a priority at the national level to ensure sustainable, equitable, and integrated water management.

The institutional framework and governance of water in Morocco are based primarily on Law No. 36–15 on water, which replaced Law No. 10–95, and on coordination between several state and local agencies. This system aims to promote integrated, decentralized, and participatory management of water resources in response to growing pressures related to climate change and water demand.

21.5 Legal and Institutional Framework

Moroccan water governance is based on clear legislative and constitutional foundations. Article 31 of the 2011 Constitution enshrines the right of all citizens to drinking water and a healthy environment, while committing the State and local authorities to guaranteeing this right. This principle is reinforced by Law No. 36–15 on water, which is the cornerstone of the current legal framework. This law modernizes water sector management by introducing mechanisms for integrated planning, conservation, and protection of resources, while strengthening institutional capacities to deal with prolonged droughts and the effects of climate change. It also promotes governance at the watershed level, which is essential in contrasting hydrological contexts such as that of the Sebou basin, marked by significant spatial disparities between the wetlands of the Gharb and the dry regions of the Middle Atlas.

21.6 Key Players and Organizations

The implementation of water policy in Morocco is based on a multi-level institutional structure. The Ministry of Equipment and Water is the central body responsible for defining and implementing national water policy, particularly through the planning, mobilization, and management of water resources, such as the construction of dams and the establishment of desalination plants.

The nine Water Basin Agencies (ABH) are the cornerstone of decentralized management. They are responsible for planning, regulating, and protecting water resources at the watershed level, promoting a territorial approach tailored to regional climatic conditions. In the case of the Sebou basin, the Sebou Water Basin Agency, headquartered in the city of Fez, plays a central role in coordinating actions to combat drought, regulate water withdrawals, and preserve the quality of surface and groundwater.

À un niveau stratégique, le Conseil Supérieur de l'Eau et du Climat (CSEC), présidé par le Roi, oriente la politique nationale de l'eau et veille à la cohérence intersectorielle des stratégies, notamment dans le cadre des programmes d'adaptation au changement climatique. La Commission Interministérielle de l'Eau (CIE), quant à elle, assure la convergence des politiques sectorielles relatives à la ressource hydrique.

At the local level, local authorities and management operators are actively involved in drinking water distribution and sanitation. In large cities such as Casablanca, Rabat, and Tangier, these services are often delegated to private operators, while the National Office for Electricity and Drinking Water (ONEE) plays a key role in production and distribution at the national level, particularly in rural areas vulnerable to drought.

21.7 Governance Principles and Challenges Related to Climate Variability

Water governance in Morocco is based on several fundamental principles: integrated and participatory management, decentralization, long-term planning, and adaptation to climate change. Integrated management relies on consultation between various public and private stakeholders, as well as the involvement of water users through local committees. Decentralization, led by water basin agencies, allows for better consideration of regional hydrological realities, particularly in areas exposed to significant rainfall variability, such as the Sebou basin.

Long-term planning is mainly carried out through the National Water Plan (PNE) and the Integrated Water Resources Management Master Plans (PDAIRE), which aim to reconcile water security, environmental sustainability, and socio-economic development. These plans include measures to adapt to the recurring droughts observed since the 1980s, confirmed by analyses of the Standardized Precipitation Index (SPI) and the Mann–Kendall test, which show a significant negative trend in the mountainous regions of the Middle Atlas.

In response to these findings, Morocco has launched several structural programs, including seawater desalination, treated wastewater reuse, and inter-basin transfers, such as the "water highway" connecting Sebou to Bouregreg. These initiatives illustrate the national commitment to strengthening water resilience and ensuring more equitable and sustainable resource governance in the face of current and future climate challenges.

- Considering the recurrence and intensification of droughts in the context of climate change, several recommendations are needed to strengthen Morocco's water resilience.
- It is essential to consolidate the integrated approach to water resource management by improving coordination between the various institutional and territorial actors, particularly between the Water Basin Agencies and local authorities.
- The development of adaptation strategies must be based on better anticipation of droughts through early warning systems and enhanced hydrometeorological monitoring.
- The diversification of water sources, particularly through desalination, the reuse of treated wastewater, and artificial groundwater recharge, must be accelerated.
- Water use efficiency, especially in agriculture, must be improved through water-saving techniques and the promotion of crops that consume less water.

– More inclusive governance, involving local stakeholders, researchers, and civil society, is needed to promote a culture of sustainable water management in the face of the growing challenges of climate change.

21.8 Conclusion

This research highlights the recurrence and increasing severity of droughts in the Sebou basin, confirming their structuring role in Morocco's hydroclimatic dynamics. Analysis of SPI indices, coupled with the Mann–Kendall test, reveals both interannual variability in precipitation and contrasting regional trends between the upper and lower Sebou basin, with significant drying in certain areas of the Saïs Plateau, while other areas, particularly at higher altitudes, remain relatively stable, notably in the Rif and Middle Atlas Mountains. These results highlight the need to consider different temporal and spatial scales to fully understand the impacts of droughts on water resources and agricultural systems.

The adaptation strategies adopted, such as intra- and inter-basin water transfers, offer concrete solutions for securing drinking water supplies and supporting irrigated agriculture. The "water highway" project, which transfers 1.2 billion cubic meters annually, illustrates the national ambition to rebalance resources between surplus and deficit basins. However, the implementation of these infrastructures must be accompanied by inclusive governance, sustainable planning, and consideration of environmental and social impacts to avoid further pressure on ecosystems and territories.

Ultimately, combining statistical drought assessment with analysis of water transfer strategies highlights the importance of integrated, multi-level water resource management. Such an approach, combining technological innovation, social equity, and ecological resilience, appears essential to strengthening Morocco's water security in the face of the challenges posed by climate change.

Finally, Morocco has made significant progress in strengthening its water governance framework, yet recurrent droughts and increasing climate pressures require deeper institutional coordination, more efficient water use, and accelerated diversification of water resources. Ensuring resilient and sustainable water management will depend on better anticipation of droughts and stronger involvement of local stakeholders.

Acknowledgements Our special thanks go to the Sebou River Basin Authority for providing the database used in this study.

Contributions Conceptualization ZA, Investigation, Methodology, Data analysis, and Visualization are carried out by Z.Q., L.E. Supervision and reviewing are done by Q.A.

Code Availability The models or codes used to develop this study are available from the corresponding author upon reasonable request.

Availability of Data and Materials The datasets used and/or analyzed during the current study are available from the first author (Z. Qadem) on reasonable request.

Ethical Approval Not applicable.

Declaration of Competing Interest The authors declare that they have no known competing financial interests or personal relationships that could have appeared to influence the work reported in this paper.

References

Abeysingha, N. S., Rajapaksha, U. R. L. N. (2020). SPI-based spatiotemporal drought over Sri Lanka. *Advances in Meteorology*. https://doi.org/10.1155/2020/9753279

Acar Deniz, Z., & Gönençgil, B. (2015). Trends of summer daily maximum temperature extremes in Turkey. *Physical Geography, 36*, 268–281. https://doi.org/10.1080/02723646.2015.1045285

Ait Dhmane, L., Saidi, M. E., Moustadraf, J., et al. (2024). Spatiotemporal characterization and hydrological impact of drought patterns in northwestern Morocco. *Frontiers in Water, 6*, 1463748. https://doi.org/10.3389/FRWA.2024.1463748/FULL

Ait Kadi, M., & Ziyad, A. (2018). Integrated water resources management in Morocco. In *Water resources development and management*

Aksoy, H., & Cavus, Y. (2022). Discussion of "Drought assessment in a south Mediterranean trans-boundary catchment." *Hydrological Sciences Journal, 67*, 150–156. https://doi.org/10.1080/02626667.2021.2009838

Alashan, S. (2023). Non-monotonic trend analysis using Mann–Kendall with self-quantiles. *Theoretical and Applied Climatology*, 1–10. https://doi.org/10.1007/S00704-023-04666-Z/METRICS

Ashraf, M. S., Ahmad, I., Khan, N. M., et al. (2021). Streamflow variations in monthly, seasonal, annual and extreme values using Mann-Kendall, Spearmen's Rho and innovative trend analysis. *Water Resources Management, 35*. https://doi.org/10.1007/s11269-020-02723-0

Azemzi, H., & Erraoui, E. H. (2021) Irrigation water management and collective action: Understanding the shift from community management to participatory management in Souss-Massa (Morocco). *Euro-Mediterranean Journal for Environmental Integration, 6*. https://doi.org/10.1007/s41207-020-00207-6

Bijaber, N., Rochdi, A., Yessef, M., & El Yacoubi, H. (2024). Mapping the structural vulnerability to drought in Morocco. *International Journal of Engineering and Geosciences, 9*, 264–280. https://doi.org/10.26833/IJEG.1404507

Cook, B. I., Ault, T. R., & Smerdon, J. E. (2015). Unprecedented 21st century drought risk in the American Southwest and Central Plains. *Science Advances, 1*

Driouech, F. (2010). Distribution des précipitations hivernales sur le Maroc dans le cadre d'un changement climatique: descente d'échelle et incertitudes.

El Goumi, S., Namous, M., Elaloui, A., et al. (2024). Exploring drought monitoring in Morocco: A review of remote sensing and machine learning techniques. In *EGU24*. https://doi.org/10.5194/EGUSPHERE-EGU24-731

El-Khantoury, I., Obda, K., & Achiban, H. (2022). Effects of meteorological droughts on water resources in Northern Morocco: Case study of the Martil watershed. *IOP Conference Series: Earth and Environmental Science, 1090*, Article 012038. https://doi.org/10.1088/1755-1315/1090/1/012038

El Qorchi, F., Yacoubi Khebiza, M., Omondi, O. A., et al. (2023). Analyzing temporal patterns of temperature, precipitation, and drought incidents: A comprehensive study of environmental trends in the Upper Draa Basin, Morocco. *Water, 15*, 3906. https://doi.org/10.3390/W15223906

Erraoui, E. H., Azemzi, H. (2021). Participatory management of irrigation water in Morocco: Lessons from a case study in Souss-Massa Region. In *Environmental Science and Engineering*.

Esper, J., Frank, D., Büntgen, U., et al. (2007). Long-term drought severity variations in Morocco. *Geophysical Research Letters, 34*. https://doi.org/10.1029/2007GL030844

Güçlü, Y. S. (2018). Multiple Şen-innovative trend analyses and partial Mann-Kendall test. *Journal of Hydrology (Amst), 566*, 685–704. https://doi.org/10.1016/j.jhydrol.2018.09.034

Guemouria, A., El Harraki, W., Elhassnaoui, I., et al. (2023). Opportunities and challenges of irrigation in Morocco, Spain, and India: A critical analysis. *World Water Policy, 9*, 682–701. https://doi.org/10.1002/WWP2.12148

Guttman, N. B. (1999). Accepting the standardized precipitation index: A calculation algorithm. *JAWRA Journal of the American Water Resources Association, 35*, 311–322. https://doi.org/10.1111/J.1752-1688.1999.TB03592.X

Guttman, N. B. (1998). Comparing the palmer drought index and the standardized precipitation index. *Journal of the American Water Resources Association, 34*. https://doi.org/10.1111/j.1752-1688.1998.tb05964.x

Habib, A., Bouchaib, B., & Housni, S. (2022). A critical analysis of water economics in public water and sanitation services in Morocco. *Arab Gulf Journal of Scientific Research, 40*, 280–301. https://doi.org/10.1108/AGJSR-04-2022-0034/FULL/PDF

Hadri, A., Ndiaye, A. S., Khadir, L., et al. (2024). Spatio-temporal analysis of meteorological drought return periods in a Mediterranean arid region, the center of Morocco. *Journal of Water and Climate Change, 15*, 4573–4595. https://doi.org/10.2166/WCC.2024.192

Hafid, E. D. (2017). Etude des sequences de secheresse dans le bassin d'isly (maroc oriental) study of drought sequences in the ilsly basin (east morocco). *Larhyss Journal, 31*, 83–94.

Hakam, O., Baali, A., Azennoud, K., et al. (2022a). Spatiotemporal evolution of droughts and their teleconnections with large-scale climatic indices in the Lower Sebou Basin in northwestern Morocco. *Acta Geographica Slovenica, 62*. https://doi.org/10.3986/AGS.10508

Hakam, O., Baali, A., El Kamel, T., et al. (2022b). Comparative evaluation of precipitation-temperature based drought indices (DIs): A case study of Moroccan Lower Sebou basin. *Kuwait Journal of Science, 49*. https://doi.org/10.48129/KJS.13911

Hamed, K. H. (2008). Trend detection in hydrologic data: The Mann-Kendall trend test under the scaling hypothesis. *Journal of Hydrology (Amst), 349*, 350–363. https://doi.org/10.1016/j.jhydrol.2007.11.009

Hamed, K. H., & Ramachandra Rao, A. (1998). A modified Mann-Kendall trend test for autocorrelated data. *Journal of Hydrology (Amst), 204*, 182–196. https://doi.org/10.1016/S0022-1694(97)00125-X

Houdret, A., & Heinz, R. (2022). Groundwater governance through institutional bricolage? Participation in Morocco's Chtouka aquifer contract. *Water International, 47*, 565–582. https://doi.org/10.1080/02508060.2022.2049545

Hu, Y. M., Liang, Z. M., Liu, Y. W., et al. (2015). Uncertainty analysis of SPI calculation and drought assessment based on the application of Bootstrap. *International Journal of Climatology, 35*, 1847–1857. https://doi.org/10.1002/JOC.4091

Kalpakian, J., & Jenks, H. (2025). Applying the eightfold path to Morocco's Cabinet Order 2.23.105. *World Water Policy, 11*, 583–590. https://doi.org/10.1002/WWP2.12252

Kendall, M. (1975). *Rank Correlation Method*. Charless Griffin.

Khrabcha, A., Oumou, A., Essahlaoui, A., et al. (2025). Meteorological drought assessment using SPI and RDI indices in the Ourika Basin, Morocco. In *E3S Web of Conferences* (Vol. 607, p. 04017). https://doi.org/10.1051/E3SCONF/202560704017

Laachrate, H., & Fadil, A. (2022). Assessment of drought using earth observation data and cloud computing in Morocco for 2010–2020. *International Journal of Applied Geospatial Research, 13*, 1–20. https://doi.org/10.4018/IJAGR.298260

Liu, C., Yang, C., Yang, Q., & Wang, J. (2021). Spatiotemporal drought analysis by the standardized precipitation index (SPI) and standardized precipitation evapotranspiration index (SPEI) in Sichuan Province, China. *Scientific Reports, 11*. https://doi.org/10.1038/s41598-020-80527-3

Mann, H. B. (1945). Nonparametric tests against trend. *Econometrica, 13*, 245. https://doi.org/10.2307/1907187

McKee, T. B., Nolan, J., & Kleist, J. (1993). The relationship of drought frequency and duration to time scales. Preprints, *Eighth Conference on Applied Climatology*. American Meteorological Society.

Mehr, A. D., Sorman, A. U., Kahya, E., et al. (2020). Climate change impacts on meteorological drought using SPI and SPEI: Case study of Ankara, Turkey. *Hydrological Sciences Journal, 65*, 254–268. https://doi.org/10.1080/02626667.2019.1691218

Mélia, D. S. Y. (2022). Les principaux enseignements du 6e rapport du groupe I du GIEC. *Annales des Mines - Responsabilite et environnement, 106*, 11–16. https://doi.org/10.3917/RE1.106.0011

Moccia, B., Mineo, C., Ridolfi, E., et al. (2022). SPI-based drought classification in Italy: Influence of different probability distribution functions. *Water (Switzerland), 14*. https://doi.org/10.3390/w14223668

Mounir, K., Jeunesse, I. L., Sellami, H., Elkhanchoufi, A. (2023). Spatiotemporal analysis of drought occurrence in the Ouergha catchment, Morocco. *AIMS Environmental Science, 10*. https://doi.org/10.3934/environsci.2023023

Mwanake, R. M., Gettel, G. M., Wangari, E. G., et al. (2023). Anthropogenic activities significantly increase annual greenhouse gas (GHG) fluxes from temperate headwater streams in Germany. *Biogeosciences, 20*, 3395–3422. https://doi.org/10.5194/BG-20-3395-2023

Naim, M., & Bonaccorso, B. (2023). Evaluation of satellite products for drought characterization and impact assessment on water resources in the Tensift Basin (Morocco). In *EGU23*. https://doi.org/10.5194/EGUSPHERE-EGU23-8975

Oubaha, A., Ongoma, V., Hssaine, B. A., et al. (2024). Evaluation of the performance of meteorological drought indices in Morocco: A case study of different climatic zones. *International Journal of Climatology*. https://doi.org/10.1002/JOC.8565

Ouharba, E. L. H., Mabrouki, J., & Triqui, Z. E. L. A. (2024). Assessment and future climate dynamics in the Bouregreg Basin, Morocco—Impacts and adaptation alternatives. *Ecological Engineering & Environmental Technology, 25*, 51–63. https://doi.org/10.12912/27197050/177823

Qadem, Z. (2019). Variabilité Climatique et son Impact sur l'Organisation Spatiotemporelle des Précipitations. Cas des Bassins Versants des Oueds Mikkes et Fès. Maroc. *Environmental and Water Sciences, Public Health & Territorial Intelligence, 4*, 253–260.

Qadem, Z., & Tayfur, G. (2024). In-depth exploration of temperature trends in Morocco: Combining traditional methods of Mann Kendall with innovative ITA and IPTA approaches. *Pure and Applied Geophysics, 2024*, 1–23. https://doi.org/10.1007/S00024-024-03535-8

Qadem, Z., Hadria, R., Ettaqy, A., et al. (2025b). Analysis of drought in upper Moulouya Basin (Morocco): A combination of drought indices, classic and innovative trend tests. *Theoretical and Applied Climatology, 156*, 1–18. https://doi.org/10.1007/S00704-025-05699-2/METRICS

Qadem, Z., & Ettaqy, A. (2024). Exploring climatic variability and drought characteristics in Morocco's Sais Plateau and Middle Atlas: A statistical and spectral analysis. In *BIO Web of Conferences*. EDP Sciences

Qadem, A., Lebaut, S., & Qadem, Z. (2023). *Exploration of Karst groundwater and surface water in Central Middle Atlas, Morocco.*

Qadem, Z., Tayfur, G., & Kankal, M. (2025a). In-depth analysis of drought trend in semiarid Saïs Plateau and Middle Atlas region in Morocco. *Acta Geophysica, 2025*, 1–17. https://doi.org/10.1007/S11600-025-01630-5

Qadem, Z. (2020). Hydrologie des oueds Fès et Mikkès drainant le plateau de Saïss et ses bordures moyennes atlasiques.

Rbaibi, O., & SahibEddine, A. (2024). Sustainability in Integrated water resources management: Systematic literature review. *World Sustainability Series Part, F2854*, 195–202. https://doi.org/10.1007/978-3-031-56292-1_15

Sheffield, J., Wood, E. F., & Roderick, M. L. (2012). Little change in global drought over the past 60 years. *Nature, 491*. https://doi.org/10.1038/nature11575

Silva-Novoa Sánchez, L. M., Bossenbroek, L., Schilling, J., & Berger, E. (2022). Governance and Sustainability Challenges in the Water Policy of Morocco 1995–2020: Insights from the Middle Draa Valley. *Water (Switzerland), 14*. https://doi.org/10.3390/w14182932

Simon, S. (2021). Reviving indigenous water management practices in Morocco: Alternative pathways to sustainable development. In *Reviving indigenous water management practices in Morocco: Alternative pathways to sustainable development* (pp.1–133). https://doi.org/10.4324/9781003104179/REVIVING-INDIGENOUS-WATER-MANAGEMENT-PRACTICES-MOROCCO-SANDRINE-SIMON/RIGHTS-AND-PERMISSIONS

Smakhtin, V. U., & Hughes, D. A. (2004). Review, automated estimation and analyses of drought indices in South Asia.

Smida, H., Dassi, L., Boukhachem, K., & Masrouhi, A. (2022). Satellite remote sensing and GIS-based multi-criteria analysis for the assessment of groundwater potentiality in fractured limestone aquifer: Case study of Maknassy Basin, central Tunisia. *Journal of African Earth Sciences, 195*, Article 104643. https://doi.org/10.1016/J.JAFREARSCI.2022.104643

Tallaksen, L. M., Van Lanen, H. A. J., Hannaford, J., et al. (2023). Hydrological drought: Processes and estimation methods for streamflow and groundwater. In *EGU23*. https://doi.org/10.5194/EGUSPHERE-EGU23-13352

Tayfur, G. (2021). Discrepancy precipitation index for monitoring meteorological drought. *Journal of Hydrology (Amst), 597*, Article 126174. https://doi.org/10.1016/J.JHYDROL.2021.126174

Türkeş, M., & Tatli, H. (2009). Use of the standardized precipitation index (SPI) and a modified SPI for shaping the drought probabilities over Turkey. *International Journal of Climatology, 29*, 2270–2282. https://doi.org/10.1002/JOC.1862

Vengosh, A., & Weinthal, E. (2022). Integration. In *Water quality impacts of the energy-water nexus* (pp. 215–261). https://doi.org/10.1017/9781107448063.006

Venturi, S., Dunea, D., Mateescu, E., et al. (2024). SPEI and SPI correlation in the study of drought phenomena in Umbria region (central Italy). *Environmental Science and Pollution Research, 32*, 168–188. https://doi.org/10.1007/S11356-024-35740-2/FIGURES/13

Wilhite, D. (2000). *Chapter 1 Drought as a natural hazard: Concepts and definitions*. Faculty Publications.

Yacoub, E., & Tayfur, G. (2020). Spatial and temporal of variation of meteorological drought and precipitation trend analysis over whole Mauritania. *Journal of African Earth Sciences, 163*, Article 103761. https://doi.org/10.1016/J.JAFREARSCI.2020.103761

Yu, H., Wang, L., Zhang, J., & Chen, Y. (2023). A global drought-aridity index: The spatiotemporal standardized precipitation evapotranspiration index. *Ecological Indicators, 153*, Article 110484. https://doi.org/10.1016/J.ECOLIND.2023.110484

Zeybekoglu, U. (2023). Temperature series analysis of the Hirfanli Dam Basin with the Mann-Kendall and sequential Mann-Kendall tests. *Turkish Journal of Engineering, 7*, 306–313. https://doi.org/10.31127/tuje.1145716

Zohair, Q., Khalid, O., Abdelghani, Q., & Mohamed, L. (2020). L'impact de la sécheresse sur les réservoirs souterrains dans le plateau de saïs, Maroc. *Geomaghreb, 16*, 22–33.

Chapter 22
Seismogeodynamics and Seismic Risk of the Middle East and North Africa (MENA) Region

V. Svalova

Abstract The Middle East and North Africa (MENA) region is highly seismically active because it lies at the convergence of the Arabian, Eurasian, and African plates. Despite advances in geoscience, reliable forecasting and early warning for earthquakes remain unresolved challenges. Each major event, however, contributes new insight into regional seismic behavior and helps refine future hazard assessments. On April 23, 2025, an earthquake of magnitude 6.2 occurred in the Sea of Marmara, underscoring the ongoing seismic threat in this zone. The aim of this study is to enhance earthquake prognosis for the MENA region. The research applies systematic analyses of earthquakes and regional seismicity, supported by the development of geodynamic models for areas exposed to natural hazards and disasters. The methods include examining the relationship between the 2025 Marmara earthquake, earlier large earthquakes along the North Anatolian Fault between 1939 and 1999, and historical earthquakes in the Istanbul region in 1509 and 1766. Additional approaches include seismic risk assessment, construction of geodynamic and geothermal models of the deep crust and upper mantle, and comparative analyses of geological and geophysical datasets from active continental margins and seismogenic fault systems. Earthquake generation at depth is linked to two principal mechanisms: activation of seismogenic faults driven by horizontal lithospheric movements, and the rise of mantle material in diapiric form due to gravitational instability. Strong earthquakes propagating westward along the North Anatolian Fault have already passed the vicinity of Istanbul, and the 2025 event west of the city suggests a potential reduction in immediate seismic risk related to horizontal fault motion. Nevertheless, the probability of a major earthquake resembling the historical events of 1509 and 1766 remains significant. Based on the recurrence interval of approximately 250 years, a strong earthquake near Istanbul remains plausible within the coming years.

V. Svalova (✉)
Sergeev Institute of Environmental Geoscience, Russian Academy of Sciences, Moscow, Russian Federation
e-mail: v-svalova@mail.ru

Geophysical Institute, Vladikavkaz Scientific Center, Russian Academy of Sciences, Vladikavkaz, Russian Federation

E. Alam (ed.), *Disaster, Crisis and Emergency Management in the Middle East & North Africa*, https://doi.org/10.1007/978-3-032-24436-9_22

Keywords MENA region · Turkey · Earthquakes · Geodynamics · Seismicity · Geothermy · Seismic tomography · Forecast

22.1 Introduction

The Middle East and North Africa (MENA) region encompasses a geologically diverse landscape extending from the Atlantic coast of Morocco to the mountains of Iran. It lies across several major tectonic boundaries, including the convergence zones of the African, Arabian, and Eurasian plates. This tectonic complexity renders many countries within the region—such as Turkey, Iran, Morocco, Algeria, and parts of the Levant—highly susceptible to seismic hazards. Earthquakes in these areas have historically caused significant human and economic losses, underscoring the region's acute vulnerability and the pressing need for enhanced seismic risk assessment, preparedness, and resilient urban planning. Together, the countries boost a strong economy with more than 380 million residents. The area also holds 60% of the total global oil reserve, making it quite significant for the rest of the world (Fig. 22.1).

22.2 Methods of Research: Analyses of Seismicity in the MENA Region

The Middle East and North Africa (MENA) region is characterized by high seismicity due to its location at the intersection of several major tectonic plates. This activity is primarily driven by the collision of the Arabian, Eurasian, and African plates, along with the Anatolian block. Subduction zones and active fault lines,

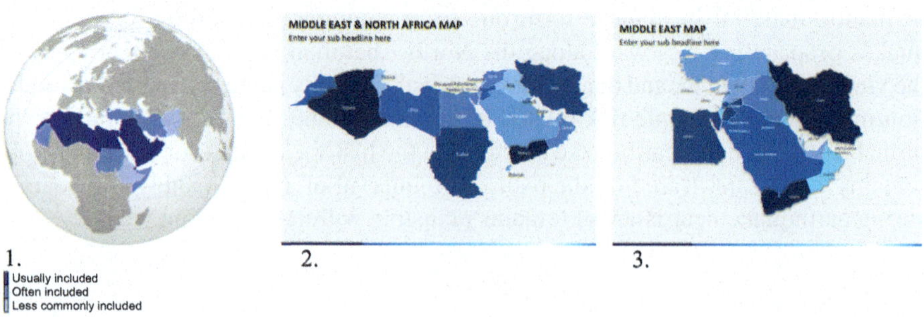

Fig. 22.1 MENA countries. 1—https://upload.wikimedia.org/wikipedia/commons/thumb/b/b0/Middle_East_and_North_Africa_classifications.png/250px-Middle_East_and_North_Africa_classifications.png. 2—https://cdn.sketchbubble.com/pub/media/catalog/product/cache/1/image/720x540/c96a280f94e22e3ee3823dd0a1a87606/m/e/mena-map-slide1.png. 3—https://cdn.sketchbubble.com/pub/media/catalog/product/cache/1/image/720x540/c96a280f94e22e3ee3823dd0a1a87606/m/i/middle-east-map-slide1.png

like the Dead Sea Fault and the North and East Anatolian Faults, also contribute to the region's seismic hazard. The Middle East region is tectonically and seismically active part of the Alpine-Himalayan orogenic belt. Many major earthquakes have occurred in this region over the years causing casualties in the millions (Figs. 22.2, 22.3, 22.4 and 22.5).

Key Factors Contributing to Seismicity in MENA:

- **Tectonic Plate Interactions**

The collision and movement of the Arabian, Eurasian, and African plates, along with the Anatolian block, create significant stress and strain within the Earth's crust, leading to earthquakes.

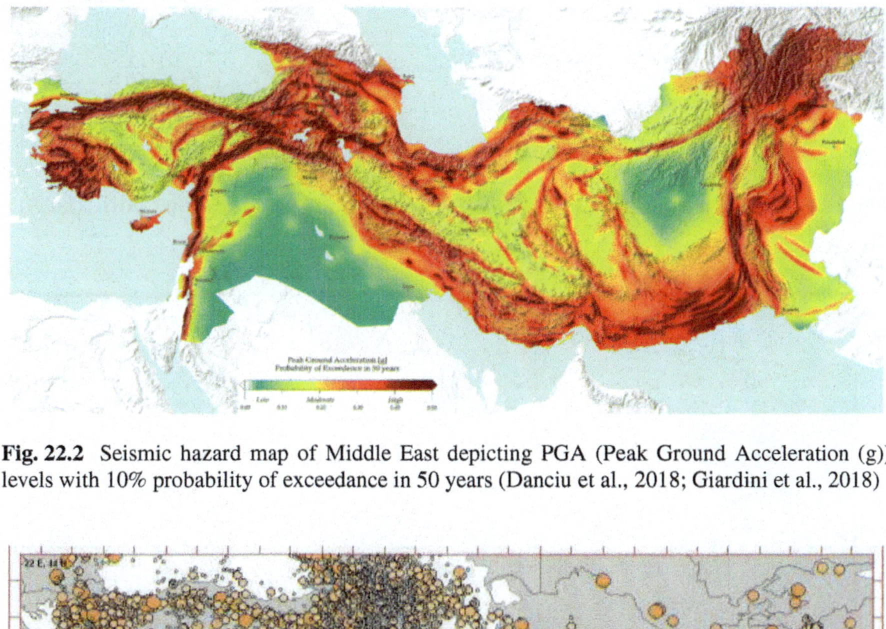

Fig. 22.2 Seismic hazard map of Middle East depicting PGA (Peak Ground Acceleration (g)) levels with 10% probability of exceedance in 50 years (Danciu et al., 2018; Giardini et al., 2018)

Fig. 22.3 Seismicity of Middle East (Erdik et al., 2012; Zare et al., 2014)

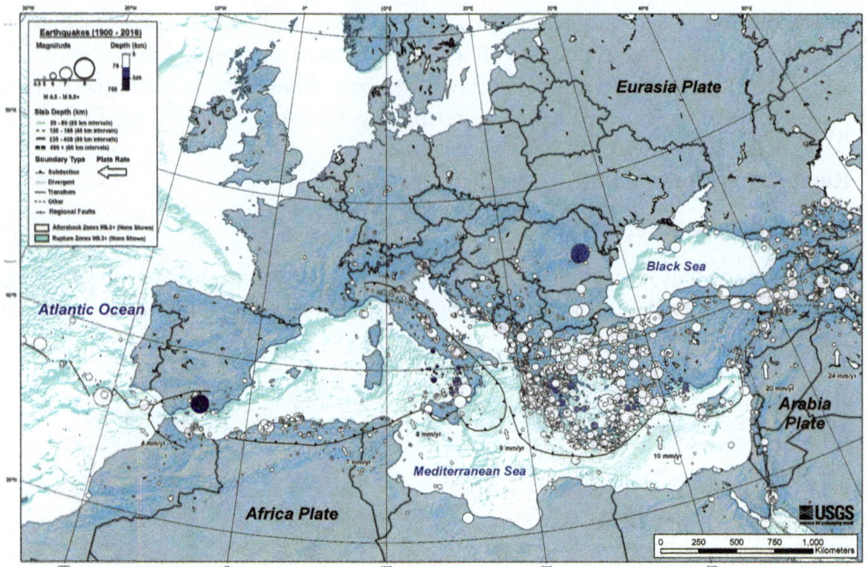

Fig. 22.4 Map of earthquakes in Europe, Mediterranean, and North Africa 1900–2016 (USGS)

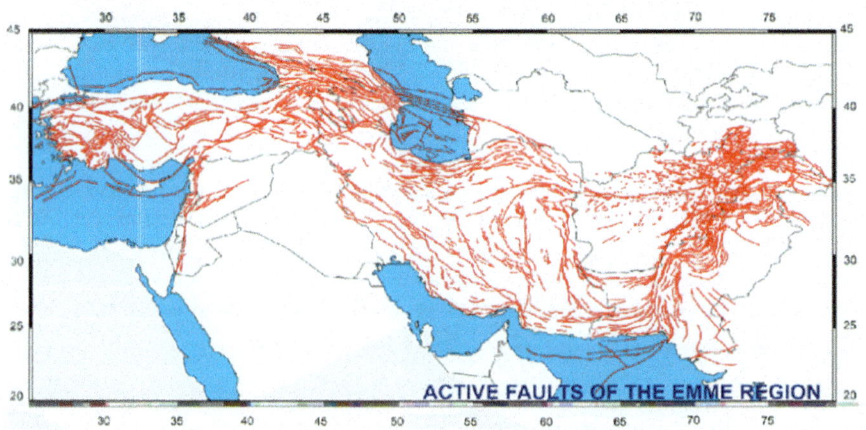

Fig. 22.5 Active fault map of the EMME (Earthquake Model of Middle East) Project region (Erdik et al., 2012)

- **Subduction Zones**

The Hellenic and Cyprian subduction zones, where the African plate subducts beneath the Anatolian block, and the Makran subduction zone, where the Arabian plate subducts beneath the Eurasian plate, are areas of high seismic activity.

- **Fault Lines**

Numerous active fault lines, such as the North Anatolian Fault, East Anatolian Fault, and Dead Sea Fault, accommodate the plate movements and contribute to the region's seismicity.

- **Regional Variations**

Seismicity varies across the MENA region, with some areas experiencing more frequent or larger earthquakes than others. For example, the Zagros Mountains in Iran, the Caucasus region, and the Anatolian region of Turkey are known for high seismic activity.

The MENA region's complex tectonic setting, characterized by plate boundaries, subduction zones, and active fault lines, results in significant seismic activity. Together with the "Ring of Fire," the Alpine-Himalayan belt is one of the most active and a high-risk area for earthquakes. Understanding and mitigating the risks associated with earthquakes is crucial for this region (Table 22.1, (USGS)).

The MENA region has a long history of destructive earthquakes, and seismic risk can severely affect modern society, as shown by the tragic events of Turkey and Syria (2023), Izmit (Turkey, 1999), Bam (Iran, 2003), and Van (Turkey, 2011). Minimization of the loss of life, property damage, and social and economic disruption due to earthquakes depends on reliable estimates of earthquake-related hazards (e.g., ground shaking, tsunami, earthquake-triggered landslides, and soil liquefaction). Earthquake catalogs, seismogenic sources, and ground shaking estimates serve as input to develop strategies for seismic risk governance and earthquake-resistant design for different applications—ranging from land use planning, improved building design and construction, adoption of modern building construction codes, emergency response preparedness plans, economic forecasts, housing and employment decisions, and other measures for risk mitigation.

22.3 Seismic Hazard and Risk Assessment and Seismic "Hot Spots" Identification

Due to the combination of seismic activity and high population density, earthquakes in MENA region pose a significant hazard and risk. The MENA's seismicity is a result of its complex tectonic setting, with major fault lines and subduction zones being the primary sources of seismic activity.

One of the fruitful approaches to the problem of earthquake hazard reduction, along with the development of earthquake-resistant construction, is the use of risk assessment and management methods. It is also important to single out the most dangerous regions according to the degree of risk in order to organize systems for

Table 22.1 List of major earthquakes in the MENA region

Event number	Date (Year-Month-Day	Earthquake locations	Epicenter latitude-longitude	Magnitude/intensity	Deaths	Injury	Damage	Sources
1	February 6, 2023	Turkey-Syria	37.166°N 37.032°E 38.089°N 37.239°E 17.9 km (11.1 mi) 10 km (6.2 mi)	7.8, 7.7/MMI XII (*Extreme*)	53,537 inTurkey 5,951–8,476 in Syria	107,213 inTurkey 14,500 in Syria	>US $157.8 billion (estimated)	USGS earthquake catalog
2	A: October 23, 2011 B: November 9, 2011	2011 Van earthquakes, Turkey	38.628°N 43.486°E	A: 7.1 M_{ww} B: 5.6 M_{ww} A: MMI VIII (*Severe*)[5] B: MMI VII (*Very strong*)	A: 604	A: 4,152		USGS earthquake catalog
3	December 26, 2003	2003 Bam earthquake, Iran	28.995°N 58.311°E 15 km (9.3 mi)	M_w 6.6 M_s 6.8–7.0 MMI X (*Extreme*)	34,000	200,000		USGS earthquake catalog
4	August 17, 1999	Izmit, Turkey	40.748°N 29.864°E 15.0 km (9.3 mi)	7.6 M_w 7.8 M_s 15.0 km (9.3 mi)	17,127–18,373	43,953–48,901 injured 5,840 missing	US$12–20 billion ($23–38.4 billion in 2025, adjusted for inflation	USGS earthquake catalog
5	November 22, 1995	1995 Gulf of Aqaba earthquake Egypt	28.826°N 34.799°E 18 km (11 mi)	7.3 M_w MMI VIII (*Severe*)	9–12	30–69		USGS earthquake catalog
6	May 20, 1990 to July 9, 1990	1990 South Sudan earthquakes	5.121°N 32.145°E	7.2 M_w, 6.5 M_w 7.1 M_w 6.6 M_w	31 dead			USGS
7	October 10, 1980	1980 El Asnam earthquake Algeria	36.2°N 1.37°E[1] 10 km (6.2 mi)	7.1 M_w	2,633–5,000	8,369–9,000		USGS
8	December 27, 1939	1939 Erzincan earthquake Turkey	39.77°N 39.58°E 20 km (12 mi)	7.8 M_w MMI XII (*Extreme*)	32,700–32,968	100,000	116,720 buildings were seriously damaged	USGS
9	May 22, 1766	1766 Istanbul earthquake Turkey	40.8°N 29°E[1]	7.1 M_s	4,000			USGS

(continued)

Table 22.1 (continued)

Event number	Date (Year-Month-Day	Earthquake locations	Epicenter latitude-longitude	Magnitude/intensity	Deaths	Injury	Damage	Sources
10	August 5, 1766	1766 Marmara earthquake Dardanelles Turkey	40.6°N 27°E	7.4–7.6 M_w	>5,000 fatalities			USGS
11	October 30, 1759 November 25, 1759	1759 events in the Levant, **1759 Near East earthquakes** Yammouneh fault, Ottoman Syria	October 30, 33.1°N 35.6°E November 25, 33.7°N 35.9°E	October 30, 6.6 M_s November 25, 7.4 M_s	2,000[] – 20,000			USGS
12	August 17, 1668	1668 North Anatolia earthquake Ottoman Empire	40.9°N 36.0°E[]	7.8–8.0 M_s MMI IX (*Violent*)	8,000 dead			USGS
13	September 10, 1509	1509 Constantinople earthquake	40.75°N 29.00°E	M_s7.2 ± 0.3	1,000 to 13,000	10,000 + injured	1070 houses destroyed	USGS
14	May 20, 1202	**1202 Syria earthquake** Ayyubid Sultanate Kingdom of Jerusalem	33.5°N 36.0°E	7.6 M_s	30,000			USGS
15	August 12, 1157	**1157 Hama earthquake, Syria**	35.1°N 36.5°E	7.4 M_s	8,000			USGS
16	October 11,1138	1138 Aleppo earthquake, Syria	36°13′N 37°10′E	M_w7.1	230,000			USGS
17	December 5,1033	1033 CE Jordan Rift Valley earthquake modern-day Israel and State of Palestine	32.5°N 35.5°E	~7.3 M_w	70,000			USGS
18	December 22, 856	**856 Damghan earthquake** Iran	36.2°N 54.3°E	7.9 Ms MMI X (*Extreme*)	200,000			USGS
19	January 18, 749	**749 CE earthquake** Bilad al-Sham province, Umayyad Caliphate (modern-day Israel, Syria, Palestinian Territories (West Bank), Jordan and Lebanon)	32°N 35.5°E	7.0	unknown, may have exceeded 100,000			USGS
20	July 9, 551	551 Beirut earthquake Lebanon	33.9°N 35.5°E	7.5 M_w	Above 30,000			USGS
21	May 18 and 19, 363	Galilee earthquake Syria-Palaestina province of Byzantine Empire	32.999°N 35.366°E					USGS

monitoring and warning the population, for which the concept of risk "hot spots" is introduced (Svalova, 2022).

"Hot spots" according to the degree of geo-ecological risk are a superposition of areas of maximum probability of a dangerous event and areas of maximum potential damage in case of an event. Geo-ecological risk is considered as a probabilistic measure of danger. One of the most common definitions of geological risk is that risk is the mathematical expectation of damage. Or the risk is equal to the product of the probability of a possible dangerous event and the damage produced:

$$R = P \times D$$

where R is the risk; P is the probability; D—damage.

Damage is meant both purely economic, to buildings and structures, and the damage expressed in money from the loss of human lives. By analyzing all the geological and geophysical data and parameters of potential damage to the MENA region, it is possible to identify and define the "hot zone" and "hot spots" of the MENA.

The main parameter of damage can be the population density of the territories, as it determines all possible parameters of damage as infrastructure density, industrial objects, price of land, price of houses, historical buildings, and others (Svalova, 2022).

For the MENA region, one of the main "hot spots" of risk is Istanbul, as well as other large cities along the North Anatolian fault, East Anatolian fault, and other seismogenic faults (Figs. 22.2 and 22.5).

A population density map of the Middle East and North Africa (MENA) region would visually represent the distribution of people across the region's countries, with areas of higher population density appearing darker or more intensely colored compared to areas with lower density. Generally, population density is highest along the coasts and in river valleys, such as the Nile River Valley in Egypt, and lowest in desert and mountainous areas (Fig. 22.6).

Here is a more detailed breakdown:

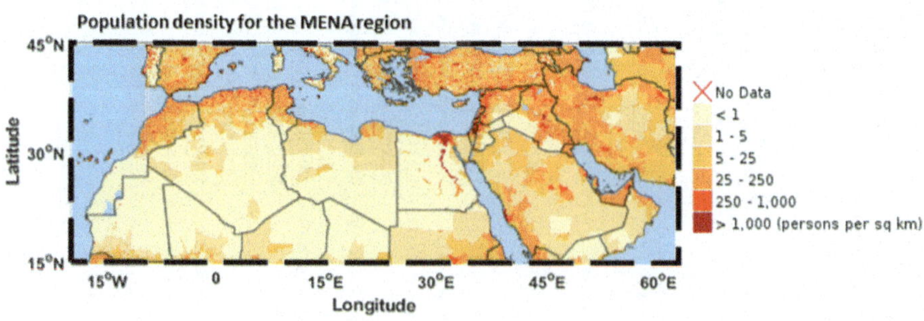

Fig. 22.6 Population density map of the MENA (Middle East and North Africa) region (Mousavi et al., 2023)

- **High Density Areas**

Coastal regions, fertile river valleys, and areas with access to water resources tend to have higher population densities. For example, Egypt, with the Nile River Valley, has a very high population density concentrated along the river.

- **Low Density Areas**

Deserts, mountainous regions, and areas with limited water resources typically have low population densities. Examples include the Sahara Desert in North Africa and the Empty Quarter in the Arabian Peninsula.

- **Urban Areas**

Major cities and urban centers will appear as clusters of high density on the map, reflecting the concentration of people in these areas.

- **Regional Variations**

The population density varies significantly across the MENA region, with some countries like Egypt and Lebanon having high densities, while others like Saudi Arabia and Algeria have much lower densities.

- **Data Sources**

Population density maps are typically created using data from national censuses, satellite imagery, and other geospatial data sources. Important source of population estimation can be night photos of the Earth.

- **Practical Applications**

Population density maps are useful for various purposes, including urban planning, resource management, disaster preparedness, and understanding demographic trends. Also these maps are very important for ecological risk assessment.

As a result, it is possible to stress that it is not necessary to estimate seismic risk in detail for the earthquakes prognosis, forecast, and prediction. It is enough to define the seismic "hot spots" for the region. One of the main "hot spots" in the MENA region is Istanbul.

22.4 The Earthquakes Prognosis and Prediction

The earthquakes prognosis and prediction is one of the most important task and problem of the modern geoscience and national economy in many countries, which is not decided yet. Seismic disasters cause enormous economic damage and lead to loss of life. Seismic disaster prediction involves a system of measures to determine the probability of major earthquakes, their magnitude, and the possible consequences in their impact zones for a specific region.

Earthquake prediction involves solving three main problems:

1. Predicting the probable location or region where an earthquake is expected.
2. Predicting the probable magnitude (or strength) of the expected event.
3. Determining the probable time of the expected seismic disaster.

The most complex and important task is predicting the exact time of an expected seismic disaster, and there are still no generally accepted methods for solving this problem, as the sources of large earthquakes are located at great depths and occupy enormous volumes that cannot be modeled using conventional rock fracture methods.

Earthquake forecasting is divided into long-term, medium-term, and short-term, as well as immediate and very-early forecasts. Long-term forecasts determine the likelihood of earthquakes over decades and centuries, based on historical data and seismic activity. Medium-term forecasts cover periods of several months to several years, using data on precursors such as ground changes and seismic activity. Short-term forecasts predict earthquakes in a period of several hours to several days, often based on studies of precursors such as foreshocks, abnormal animal behavior, or changes in groundwater levels. An immediate forecast (hours to days) helps in taking measures to protect the population and minimize consequences after precursors are detected. An immediate forecast (minutes to hours) is issued immediately before or during an earthquake and is used to warn the population of the need to evacuate. Earthquake precursors include various changes in the environment that may indicate the likelihood of an earthquake. These include:

- Foreshocks

  Small earthquakes that precede a stronger tremor.

- Changes in groundwater levels

  Fluctuations in water levels in wells and boreholes.

- Deformations of the earth's surface

  Changes in relief that can be measured using specialized instruments.

- Abnormal animal behavior

  Observations of unusual animal behavior before an earthquake.

- Changes in atmospheric pressure

  Anomalous atmospheric phenomena associated with increased stress in the earth's crust.

All types of forecasts for a given region are based on an analysis of changes in the seismic regime over time, quantitative electromagnetic field data, changes in the chemical composition and flow rate of groundwater, as well as subsurface gas and animal behavior. Hydrochemical, hydrodynamic, and geomagnetic data are used. The location of expected events is identified by seismic "gaps" (where they have not previously occurred) and by the density of seismogenic ruptures. These studies identify anomalies or earthquake precursors, the total number of which exceeds 300. Not all precursors are reliable indicators. Modern technologies allow for improved forecasting, but complete earthquake prediction is still impossible.

This makes every new step in earthquake forecasting, prediction, and prevention, based on new scientific approaches and research, all the more important. Thus, an important new approach to earthquake forecasting could be the construction of a geodynamic model of the region where a hazardous natural process or disaster may occur, based on the collection and analysis of all available geological and geophysical data, as well as the development and construction of seismodynamic and mechanical-mathematical models of the region's geological medium and earthquake focus.

The next steps and scheme is suggested for approach to the prognosis problem decision (Svalova, 2023):

Concept and Algorithm of the Earthquakes Forecast, Prognosis, Prediction, and Risk Reduction

1. To combine all possible geological-geophysical, seismotomographic, and precursors data in the region of possible natural hazards.
2. To construct geodynamic model of the region of possible geological catastrophe.
3. To determine the places of highest stresses and velocities in the lithosphere of potential natural hazards.
4. Detailed analysis and monitoring of stresses and velocities in concrete places of possible earthquakes.
5. Mechanical-mathematical modeling of geological structures evolution in the regions of events in connection with deep geodynamics of the lithosphere and asthenosphere.
6. Monitoring systems construction.
7. Seismic and earthquake risk assessment, management, and reduction.
8. Development and use of seismic-resistant constructions.

22.5 Earthquake in the Sea of Marmara (Turkiye) on April 23, 2025, and Seismicity of Turkey

On April 23, 2025, at 12:49:10 local time (UTC + 3:00), an earthquake with a magnitude of M 6.2 occurred in the Sea of Marmara. Its epicenter was located 21 km southeast of Marmara Ereglisi district, Tekirdag province, Turkey, 73 km southwest of Istanbul, with coordinates 40°49′58″ N 28°13′36″ E (Fig. 22.7).

According to a report from Istanbul Technical University (ITU), the earthquake lasted 13 s and triggered 291 aftershocks. Plate displacement was only 30 cm—significantly less than the expected 3.7 m.

The aftershocks were distributed along a 40 × 12 km band, reaching depths of up to 30 km.

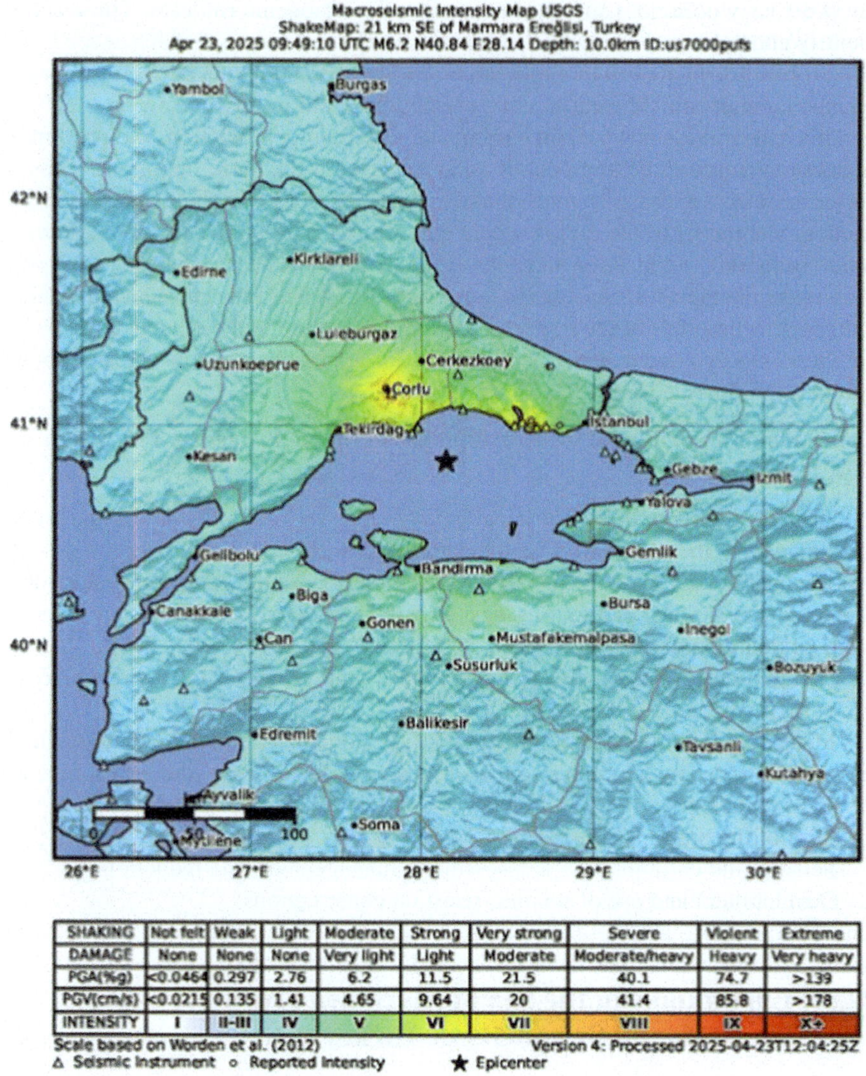

Fig. 22.7 Epicenter of the earthquake in the Marmara Sea on April 23, 2025 (USGS)

The maximum intensity rating on the modified Mercalli scale reached VII (very strong). According to various sources, the focal depth ranged from 10 to 16.9 km. This aftershock was preceded by a magnitude 3.9 earthquake, which was recorded at 12:13 PM in the Sea of Marmara, near the town of Silivri.

It was followed by three more aftershocks, including a main shock with a magnitude of 6.2. Two more earthquakes struck the Büyükçekmece region (41°01′12″N

28°34′39″E) within 11 min of each other. Their magnitudes were 4.4 and 4.9, respectively.

According to the US Geological Survey, approximately 188,000 people were affected by intensity VII tremors. Intensity VI (strong) tremors were felt by approximately 2.28 million people, primarily in the western parts of Istanbul Province. Intensity IV (moderate) tremors were recorded in the central parts of Istanbul and Bursa, as well as in Evros (Greece), and in the cities of Burgas and Yambol (Bulgaria). Tremors were reported from as far away as Gaziantep, Bucharest (Romania), and Sofia (Bulgaria). At least 359 people were injured in the mass panic caused by jumping from heights, including 236 in Istanbul. An abandoned three-story building collapsed in the Fatih district. Damage to several buildings and internet outages were reported in parts of Istanbul. At least 3,597 rescue workers, including 1,443 in Istanbul, and 250 pieces of equipment were deployed in the Marmara region.

The city of Istanbul, with a population of approximately 16 million and once the capital of the Byzantine and Ottoman empires, is located dangerously close to the North Anatolian Fault, which runs 20 km (12.4 miles) south of Istanbul across the Sea of Marmara. Because Turkey lies almost entirely on seismic fault lines, thousands of earthquakes of varying magnitudes occur there each year. In 2024, the Turkish Disaster and Emergency Management Authority recorded more than 30,400 tremors. The year with the highest number of recorded earthquakes was 2023, when more than 74,000 tremors were recorded.

A magnitude 7.8 earthquake struck southeastern Turkey on February 6, 2023, at 4:17 a.m. local time (Svalova, 2023; Svalova & Zaalishvili, 2024a). The epicenter was located 37 km west-northwest of Gaziantep. According to the U.S. Geological Survey, the hypocenter was at a depth of 10.0 km. This earthquake was followed by a magnitude 7.7 earthquake at 1:24 p.m. This earthquake occurred 95 km north-northeast of the first.

The magnitude 7.8 earthquake is the largest to hit Turkey, along with the 1939 Erzincan earthquake of the same magnitude. More than 52,800 deaths have been confirmed: more than 46,100 in Turkey and more than 6,700 in Syria. It is the deadliest disaster in Turkey's modern history. The earthquakes caused over $100 billion in damage.

As East Anatolian fault produced large earthquakes in 1789 (M 7.2), 1795 (M 7.0), 1872 (M 7.2), 1874 (M 7.1), 1875 (M 6.7), 1893 (M 7.1), 2020 (M 6.8), and 2023 (M 7.8, M 7.7), it means the frequency of groups (1790–1880–2020) near 100 years. But inside the groups, there are 2–4 close in time large earthquakes. As there were many aftershocks after 2023 earthquakes and they continue, the next large earthquake there will be in 100 years only. It is possible to forecast that if large earthquake in East Anatolian fault will not occur in 1–2 years, it will happen in 100 years only (Svalova, 2023).

Over the past century, the country has experienced thousands of tremors, 17 of which had a magnitude greater than 7. The 1939 Erzincan earthquake, known as the Great Erzincan Earthquake, was one of the most devastating in Turkish history. It struck the eastern province of Erzincan on December 27 at 1:57 a.m. local time, with a magnitude of 7.8 and a duration of approximately 52 s. It killed approximately

33,000 people and injured 100,000. The earthquake, which destroyed approximately 116,000 buildings, is considered one of the largest in the world.

In 1999, earthquakes occurred in Izmit (Gölçuk) and Düzce on August 17 and November 12, respectively. The Kozcaeli/Gölçuk earthquake occurred at 3:02 AM on August 17, 1999, with a magnitude of M7.4–7.6. It is the strongest earthquake in recent years, causing enormous loss of life and property. According to official reports, approximately 18,000 people were killed and approximately 25,000 were injured. Unofficial estimates put the death toll at approximately 50,000 and injury at approximately 100,000. This earthquake is one of the largest earthquakes of the last century in terms of strength and damage. The August 17 earthquake was felt throughout the Marmara region, spanning a vast area from Ankara to Izmir.

Between 1939 and 1999, a series of destructive M7+ strike-slip earthquakes propagated westward along the North Anatolian Fault (NAF) zone, beginning with the M7.8 Erzincan earthquake in 1939 at the eastern end of the NAF system and ending with the M7.4–7.6 İzmit earthquake in 1999 near the Sea of Marmara margins (Allen, 1969; Ambraseys, 1970; Ambraseys & Finkel, 1987; Ambraseys & Jackson, 1998, 2000; Barka, 1996; Stein, 1997).

The NAF, the largest strike-slip fault in Turkey, has produced 11 major earthquakes in the 20th century. Along the North Anatolian Fault, strong earthquakes occur every 1–10 years (1939, 1942, 1943, 1944, 1953, 1957, 1966, 1967, 1971, 1992, and 1999) with the distance between hypocenters being about 100 km in the east–west direction (Fig. 22.8).

Turkey is located in a seismically active zone, crossed by the active North Anatolian Fault Zone (NAFZ), which separates the Eurasian and Anatolian tectonic plates, sliding relative to each other at a rate of 2 cm per year. Earthquakes occur regularly along this zone, and activity has gradually shifted further west, into the Sea of Marmara, toward Istanbul. The segment of this zone adjacent to the city has not been active for 250 years.

The historic Istanbul (Constantinople) earthquake occurred in the Bosphorus and Sea of Marmara region on September 10, 1509. The epicenter was located at 40°54′ N, 28°42′ E. The magnitude of the earthquake was 7.2 ± 0.3. Due to its severe consequences, it was nicknamed the Little Doomsday Earthquake (or the Doomsday

Fig. 22.8 North Anatolian Fault and earthquakes of 1939–1999

Earthquake). Sultan Bayezid II was forced to move the capital to Edirne for several months. Estimates of the death toll range from 4,000 to 10,000.

The earthquake caused extensive damage to Istanbul (the Fatih Mosque, Topkapi Palace, and the Galata quarter) and the fortresses along the Bosphorus. The Princes' Islands were also damaged by the tremors and the resulting tsunami, which reached up to 6 m in height.

One of the most devastating earthquakes in the history of the Ottoman Empire occurred on May 22, 1766. Two strong aftershocks struck the Istanbul area. The first, with an estimated magnitude of 7.1 and an epicenter at 40°48′N 29°00′E/40.8°N 29°E, occurred in the eastern part of the Sea of Marmara, in the Çınarcık Depression near the Princes' Islands. Between 4,000 and 5,000 people died. The two-minute earthquake destroyed numerous buildings. Aftershocks that followed the earthquake lasted for three months. The Fatih Mosque, built in 1463 by Sultan Mehmed II, known as Fatih (the Conqueror), was completely destroyed.

The strong historical earthquakes of 1509 and 1766 near Istanbul have led to the assumption that the recurrence period for strong earthquakes here is 250 years.

22.6 Analyses of Seismicity of the Sea of Marmara

Each new strong earthquake provides additional information about seismicity in the region and allows for more accurate forecasting and prevention of future destructive events. The regular distribution of strong earthquakes along the North Anatolian Fault from east to west makes the Marmara earthquake a very important link in this chain and provides new data for the analysis of seismicity in the Istanbul region (Stein & Sevilgen, 2025).

It is important to understand and examine how the 2025 Sea of Marmara earthquake compares to the large earthquakes along the North Anatolian Fault from 1939–1999, as well as to earlier historical earthquakes that caused significant damage to Istanbul and heavy casualties.

This requires a detailed analysis of the seismicity of the Sea of Marmara, as well as the deep geodynamics and geothermal environment of the entire Anatolian-Caucasian and surrounding regions. The major earthquakes in the Sea of Marmara region are presented in Table 22.2 and Fig. 22.9.

The North Anatolian Fault is a right-lateral strike-slip fault approximately 1,200 km long. It extends from the Gulf of Saros in the Aegean Sea in the west to the Karlovo Triple Fault at its junction with the East Anatolian Fault in the east. The fault began to form approximately 13–11 million years ago in eastern Anatolia and eventually spread westward. The fault in the Sea of Marmara region was fully formed approximately 200,000 years ago, although strike-slip movement in this zone began as early as the late Miocene.

In the eastern part of the North Anatolian Fault up to Izmit, the fault zone is a clearly defined morphological structure that follows the boundary between the

Table 22.2 Earthquakes in the Sea of Marmara region from 1500–1999 with magnitudes M ≥ 6.0. The list is complete for earthquakes with magnitudes M ≥ 6.8, but not for weaker events (Ambraseys & Jackson, 2000)

Year	Date	Time	Lat	Long	M
1509	9–10	2200	40.9	28.7	7.2
1556	5–10	2400	40.3	27.8	7.2
1719	5–25	1200	40.7	29.8	7.4
1737	3–6	0730	40.1	27.3	7.0
1754	9–2	0330	40.8	29.2	6.8
1766	5–22	0500	40.8	29.0	7.1
1766	8–5	0530	40.5	26.6	7.4
1809	2–7	0000	40.0	27.0	6.1
1826	2–8	2030	39.8	26.4	6.2
1841	10–6	0230	40.85	29.05	6.1
1850	4–19	2330	40.1	28.3	6.1
1855	2–28	0230	40.1	28.6	7.1
1855	4–11	1940	40.2	28.9	6.3
1859	8–21	1130	40.3	26.1	6.8
1860	8–22	1009	40.5	26.0	6.1
1893	2–9	1716	40.5	26.2	6.9
1894	7–10	1224	40.7	29.6	7.3
1912	8–9	0128	40.7	27.2	7.3
1912	8–10	0923	40.8	27.5	6.2
1912	9–13	2331	40.7	27.0	6.8
1935	1–4	1441	40.50	27.60	6.4
1935	1–4	1620	40.55	27.75	6.3
1943	6–20	1532	40.68	30.48	6.4
1944	10–6	0234	39.7	26.5	6.8
1953	3–18	1906	40.0	27.4	7.1
1956	2–20	2031	39.84	30.41	6.2
1957	5–26	0633	40.6	31.0	7.2
1963	9–18	1658	40.70	28.95	6.4
1964	10–6	1431	40.1	28.2	6.8
1967	7–22	1657	40.7	30.7	7.2
1975	3–27	0515	40.45	26.12	6.5
1983	7–5	1201	40.28	27.76	6.1
1999	8–17	0001	40.7	30.0	7.4

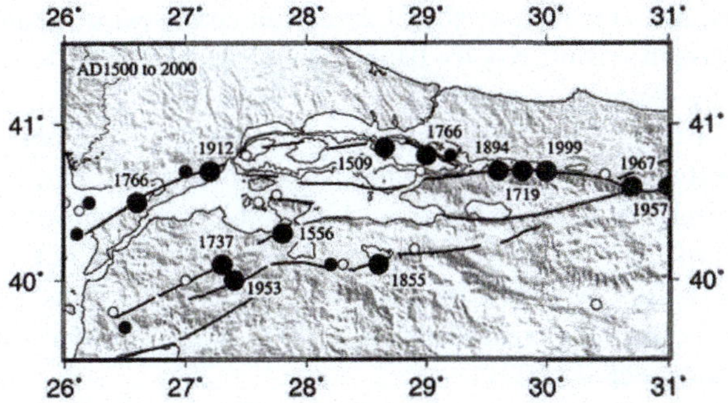

Fig. 22.9 Earthquakes in the Sea of Marmara region during the period 1500–1999 with magnitude M ≥ 6.0. The main earthquakes in the Sea of Marmara region are from Table 22.1. Large black circles are events with M ≥ 7.0, small black circles are those with 7.0 > M ≥ 6.6, and white circles are those with 6.6 > M ≥ 6.0. Only those with M ≥ 7.0 are labeled with dates (Ambraseys & Jackson, 2000)

Anatolian and Eurasian plates. The zone is fairly narrow and localized, effectively separating the relatively rigid regions of the Black Sea and Central Anatolia.

Right-lateral strike-slip faulting continues west of Izmit but becomes widespread in several subparallel belts in the Sea of Marmara, northwestern Turkey, and the northern Aegean (Barka & Kadinsky-Kade, 1988; Taymaz, 1991). Seismic studies in the Sea of Marmara reveal numerous faults with large normal components (Okay, 1999), and earthquakes with normal fault mechanisms are observed around its margins. The Sea of Marmara was presumably formed by pull-apart crustal extension. GPS studies indicate that east–west slip across the entire Fig. 22.7 region is about 23 ± 3 mm/yr (McClusky, 2000; Straub, 1997).

An important point is the transition of events to the Sea of Marmara region, which alters the rheological environment due to a decrease in crustal thickness and the dynamics of sedimentary basin formation and evolution. The observed seismicity patterns in the Sea of Marmara contribute to several earthquake prediction and warning issues. For example, it has been noted that the length of active elements of the North Anatolian Fault in the sea is shorter than onshore in the eastern part of the fault. Earthquakes in the sea are weaker than onshore. There are also seismic windows (gaps) where earthquakes have not occurred for many years. For example, the section of the North Anatolian Fault adjacent to Istanbul has not been active since 1766. A key point of the analysis is that the earthquake chain along the North Anatolian Fault, moving westward, skipped the city of Istanbul, which fits well with the concept of progradation, as the previous strong earthquake was the 1999 earthquake in the eastern Sea of Marmara. The 2025 earthquake has a lower magnitude of 6.2 compared to the magnitude of 7.4–7.6 in 1999, reflecting the transition to thinner marine crust with different rheology (Garzón et al., 2020; Akin, 2026; Artemieva & Meissner, 2012).

If we assume that strong earthquakes propagating along the North Anatolian Fault from east to west have passed the Istanbul area and are moving further west toward the Aegean Sea, as evidenced by the 2025 earthquake west of Istanbul, this could indicate a decrease in seismic risk for Istanbul in the near future in terms of horizontal movements along the fault.

The historical 1509 m earthquakes of 1766 near Istanbul are not associated with a consistent westward movement of earthquakes along the fault, but reflect other important patterns. For example, it should be noted that seismicity in many sedimentary basins of the Alpine-Himalayan and Pacific belts is concentrated on the periphery of the basins, while the deepest parts of the depressions are characterized by reduced seismicity and increased heat flow (Svalova, 2019; Svalova & Zaalishvili, 2024a; Zabarinskaya et al., 2021).

These patterns are well explained by the upwelling of mantle diapirs during the formation of sedimentary basins. In the initial stages of basin formation, a domed uplift structure forms on the Earth's surface, followed by a deep depression, reflecting the geodynamics of diapir upwelling and cooling and the spreading of material above it due to gravitational instability (Svalova, 2019; Svalova & Zaalishvili, 2024b). A similar mechanism is evident in the Sea of Marmara. The upwelling of mantle masses thins the crust and increases heat flow beneath the basin. It is important to note that the entire Marmara Sea basin consists of individual depressions, which may reflect both the manifestation of gravitational instability in the form of the rise of individual diapirs, and the formation of the basin by the pull-apart mechanism of fault extension, when the rise of mantle matter in the center of the forming basins plays a passive role, filling the vacated space (Seeber, 2010; Mchugh et al., 2014; Schmittbuhl, 2015).

Heat flow in the center of depressions is increased, and the depression boundaries are in a more brittle state, susceptible to earthquakes. The hotter environment in the center of depressions promotes lower viscosity and is less conducive to the accumulation of elastic stress energy for a subsequent earthquake, when the rock's ultimate strength is reached and stress release occurs (Bulletin of Mineral Research & Exploration, 2014; Chamorro et al., 2014; Göktürkler et al., 2003).

It is important to mark the position of earthquakes in Marmara Sea in April of 2025 (Fig. 22.10).

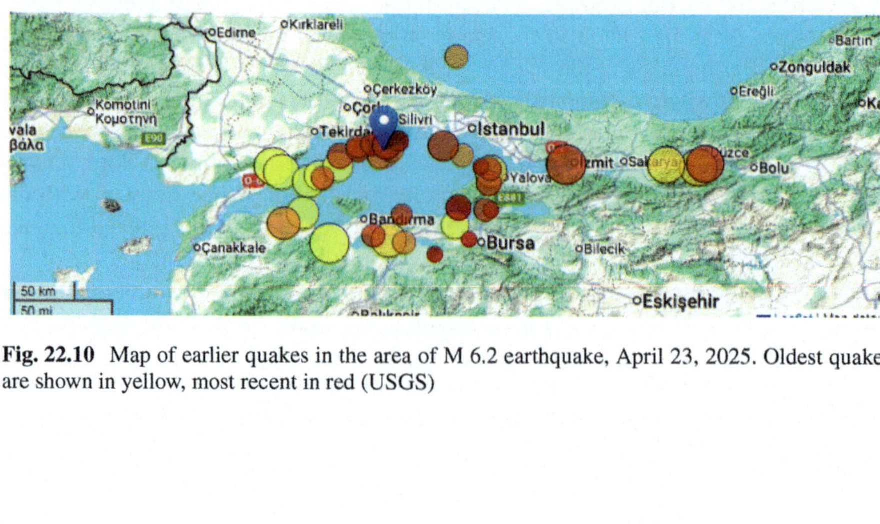

Fig. 22.10 Map of earlier quakes in the area of M 6.2 earthquake, April 23, 2025. Oldest quakes are shown in yellow, most recent in red (USGS)

22.7 Construction of Geodynamic and Geothermal Model of the Regions of Natural Hazards and Catastrophes of the MENA Region

The movement of earthquakes along the East and North Anatolian Faults may be caused by deep mantle flow rising from the core beneath the Great African Rift (Baranov et al., 2023; Becker & Boschi, 2002; Bijwaard et al., 1998; Ershov & Nikishin, 2004; Grand et al., 1997; Zhao, 2001; Zhou, 1996).

Geological-geophysical and seismic tomography research give possibility to connect complicated geodynamics and stress–strain state of the region with upwelling of mantle flow from the core boundary (Fig. 22.11).

This upwelling of huge mantle flow is confirmed by magmatism and basalt volcanism (Ershov & Nikishin, 2004; Sharkov, 2012; Sharkov & Lebedev, 2021).

Deep mantle flows extend north to the Caucasus region, where they divide into three branches: west to the Anatolian Plate, east to the Arabian Plate, and north to the Eurasian Plate and the Caucasus structure (Allen et al., 2004; Li et al., 2008; Trifonov & Sokolov, 2018). The Zagros zone characterizes the collision of the Arabian, Anatolian, and Eurasian plates (Fig. 22.12).

Two mechanisms can be considered for reaching the ultimate strength of rocks at depth, which is necessary for the formation of an earthquake source. These mechanisms are related to the geodynamics of material movement in the lithosphere. These mechanisms include the activation of seismogenic faults due to the horizontal movement of lithospheric plates and the vertical upwelling of material in the form

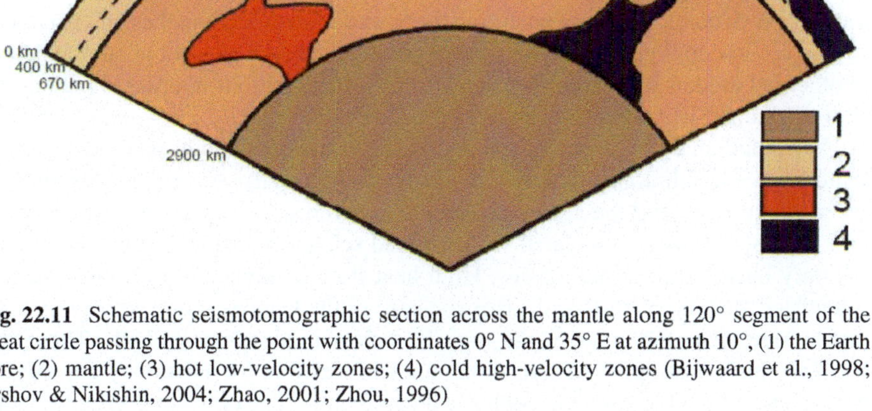

Fig. 22.11 Schematic seismotomographic section across the mantle along 120° segment of the great circle passing through the point with coordinates 0° N and 35° E at azimuth 10°, (1) the Earth core; (2) mantle; (3) hot low-velocity zones; (4) cold high-velocity zones (Bijwaard et al., 1998; Ershov & Nikishin, 2004; Zhao, 2001; Zhou, 1996)

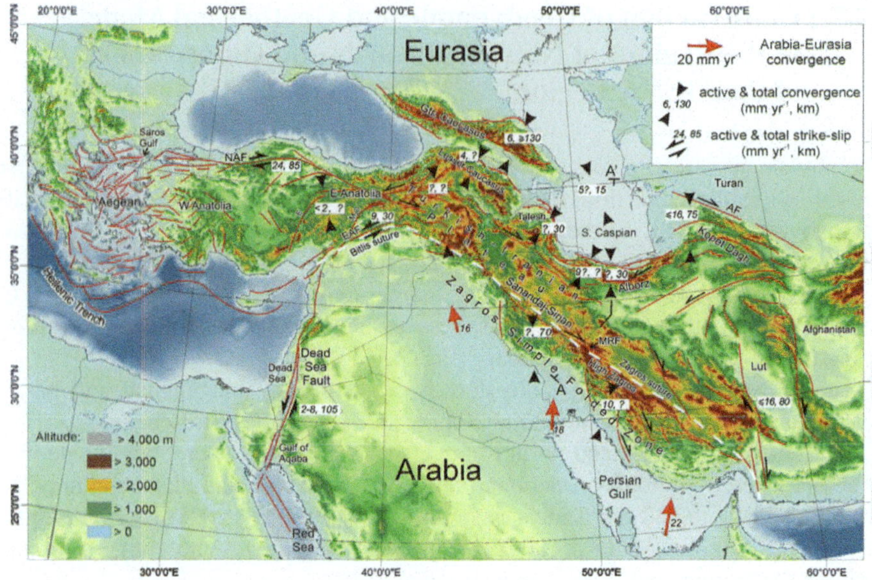

Fig. 22.12 Topography, structure, current deformation rates, and finite strain of the Arabia-Eurasia collision. Abbreviations are as follows: AF, Ashgabat Fault; E, Ecemiş Fault; EAF, East Anatolian Fault; M–O, Malatya-Ovacik Fault; MRF, Main Recent Fault; NAF, North Anatolian Fault (Allen et al., 2004)

of mantle diapirs due to gravitational instability. These mechanisms combine plate tectonics and plume tectonics. Both of these mechanisms are realized in the Sea of Marmara near Istanbul: activation of the North Anatolian Fault during the east–west movement of earthquake sources and geothermal contrast in the lithosphere at the boundary of a rising mantle diapir off the coast of the Bosphorus, which can also be accompanied by the formation of faults along the basin boundaries (2007–2011 term report of Turkish national geodesy commission, 2012; Chamorro et al., 2014).

Knowing the general patterns of mechanical behavior of lithospheric matter in various geodynamic settings and during the preparation of earthquakes, it is necessary to analyze in detail geological and geophysical data and construct geodynamic and geothermal models of seismic and aseismic zones and regions in order to predict and prevent dangerous natural processes (Table 22.3).

A detailed analysis of geological and geophysical data allows us to identify key patterns in the relationships between geothermal activity, seismicity, lithospheric layer thicknesses, and the age of geological structures. A combination of thin crust and high heat flow is associated with reduced seismicity, which is typical of sedimentary basins and marine basins. High heat flow combined with a thick crust is associated with high seismicity, as is the case in the Caucasus and Turkey. A thick crust strengthens the crust, while high heat flow reduces its strength. These two mechanisms act in opposite directions, but the outcome for seismicity depends on the balance of geological and geophysical parameters. The crust must be sufficiently

Table 22.3 Geological-geophysical data for Alpine and Pacific belts

Structures	Thickness of the sedimentary cover (km)	Thickness of crystalline crust (km) at center (numerator) surrounded (denominator)	Heat flux (mW/m ^{2})	Seismicity + high—low Center (numerator) Surrounded (denominator)
1. Tyrrhenian sea	6	16/45	30–160	–/+
2. Aegean sea	3	15/23	100	–/+
3. black sea	16	15/45	30–35	–/+
4. Pre-Caspian depression	24	12/40	40	–/–
5. South-Caspian depression	20	12/45	40–50	–/+
6. Ionian sea (South)	12	10/35	30–40	–/+
7. Ionian sea (North)	8	10/30	50–70	–/+
8. Balearic sea	8	10/40	50	–/+
9. Levantine basin	6	8/30	30	–/+
10. Pannonian basin	9	18/27	90	–/–
11. Aleutian basin	5	10/26	60–80	–/+
12. Sea of Okhotsk	4	8/20	100	–/+
13. Sea of Japan	2	12/29	100–120	–/+
14. Philippine sea	2	6/13	60	–/+
15. Sea of Marmara	3	15/35	150	–/+
16. Turkey	0–2	40/40	70–120	+/+
17. Caucasus	0–2	60/40	70	+/+

thick, and high heat flow reflects high geodynamic activity in the region, such as plate collisions in the Caucasus or the upwelling of deep mantle flows, which is typical of Turkey. The relationship between heat flow and the age of tectonic–magmatic activation is well known. The relationship between the density of the "surface" heat flow and the age of tectono-magmatic activity in the continental crust has been studied in detail (Polyak & Smirnov, 1966, 1968; Khutorskoy, 2023; Khutorskoy et al., 2022). The higher the heat flow, the younger the crust. Also, thin crust is associated with high heat flow, as seen at the Teisser-Tornquist boundary, which separates thin (young) and thick (ancient) crust in Europe. The Teisser-Tornquist zone (also called the Trans-European Suture Zone or TESZ in English terminology) is a demarcation line between the Precambrian East European Craton and the Phanerozoic orogens of southwestern Europe. The zone extends from the North Sea to the Black Sea. Seismic and gravity data confirm a strong contrast in crustal thickness, from 28–35 km west of the suture zone to 42–47 km east.

This also explains the reduced heat flow in the Black Sea, with lower temperatures recorded at greater depths east of the Teisser-Törnquist line (Artemieva & Meissner, 2012; Gee & Zeyen, 1996; Starostenko et al., 2014).

22.8 Conclusions

Two possible mechanisms for reaching the ultimate strength of rocks at depth, leading to the formation of an earthquake foci, are possible. These are the activation of seismogenic faults due to horizontal movement of lithospheric plates and the vertical upwelling of material in the form of mantle diapirs due to gravitational instability. These mechanisms combine plate tectonics and plume tectonics. Both of these mechanisms are realized in the Sea of Marmara near Istanbul: activation of the North Anatolian Fault during earthquake foci advance from the east to the west and geothermal contrast in the lithosphere at the boundary of the rising mantle diapir off the coast in the Bosphorus region.

The propagation of earthquakes along the North Anatolian Fault may be caused by deep mantle flow rising from the core beneath the Great African Rift, known on the Earth's surface as the Afar Plume. Deep mantle flows extend northward toward the Caucasus region, where they divide into three branches: westward toward the Anatolian Plate, eastward toward the Arabian Plate, and northward toward the Eurasian Plate and the Caucasus structure.

It can be assumed that strong earthquakes propagating along the North Anatolian Fault from east to west have passed the vicinity of Istanbul and are moving further west toward the Aegean Sea, as evidenced by the 2025 earthquake west of Istanbul. This suggests a decrease in the seismic risk for Istanbul in the near future in terms of horizontal movement along the fault.

The risk of a strong earthquake propagating along the North Anatolian Fault for Istanbul is reduced by the 2025 Marmara earthquake. At the same time, the risk of a possible earthquake similar to historical geothermal earthquakes remains high due to gravitational instability and the formation of an earthquake source near the depression boundaries. Such an earthquake, with a 250-year period (1509, 1766), could occur in the vicinity of Istanbul within the next few years.

The population in major cities like Istanbul resides in structures that are a mix of vulnerable and earthquake resistant construction. Earthquake prediction and prevention require processing large amounts of data (Big DATA) and organizing a comprehensive system of geological, geophysical, and geo-ecological monitoring. Progress can be achieved through the use of artificial intelligence. The challenges of earthquake-resistant construction, seismic zoning, risk assessment, the creation of a mechanical-mathematical model of earthquake sources, and geodynamic models of regions prone to hazardous natural processes and catastrophes remain pressing.

Acknowledgements The chapter was prepared in response to a government assignment №122022400105-9 on the topic "Forecast, modeling and monitoring of endogenous and exogenous geological processes to reduce their negative consequences".

Conflict of Interest The author declares no conflict of interest.

References

2007–2011 term report of Turkish national geodesy commission. (2012). TR-06100—Cebeci, Ankara—TURKEY, pp. 1–20. http://www.hgk.msb.gov.tr

Allen, M., Jackson, J., & Walker, R. (2004). Late Cenozoic reorganization of the Arabia-Eurasia collision and the comparison of short-therm and long-therm deformation rates. *Tectonics, 23*, TC2008. https://doi.org/10.1029/2003TC001530

Allen, C. R. (1969). Active faulting in northern Turkey. California Institute of Technology, California, pp. 32–34.

Ambraseys, N. N. (1970). Some characteristic features of the North Anatolian fault zone. *Tectonophysics, 9*, 143–165.

Ambraseys, N. N., & Finkel, C. (1987). The Saros-Marmara earthquake of 9 August 1912. *Journal of Earthquake Engineering Structure and Dynamics, 15*, 189–211.

Ambraseys, N. N., & Jackson, J. A. (1998). Faulting associated with historical and recent earthquakes in the eastern Mediterranean region. *Geophysical Journal International, 133*, 390–406.

Ambraseys, N. N., & Jackson, J. A. (2000). Seismicity of the Sea of Marmara (Turkey) since 1500. *Geophysical Journal International, 141*(3), F1–F6. https://doi.org/10.1046/j.1365-246x.2000.00137.x

Artemieva, I., & Meissner, R. (2012). Crustal thickness controlled by plate tectonics: A review of crust–mantle interaction processes illustrated by European examples. *Tectonophysics, 530–531*(B7), 18–49 https://doi.org/10.1016/j.tecto.2011.12.037

Baranov, A. A., Lobkovskii, L. I., & Bobrov, A. M. (2023). Global geodynamic model of the modern earth and its application to the Antarctic region. *Doklady RAS Earth Sciences, 512*(1), 854–858. https://doi.org/10.1134/S1028334X23601086

Barka, A. (1996). Slip distribution along the North Anatolian Fault associated with the large earthquakes of the period 1939 to 1967. *Bulletin of the Seismological Society of America, 86*, 1238–1254.

Barka, A., & Kadinsky-Cade, K. (1988). Strike-slip fault geometry in Turkey and its impact on earthquake activity. *Tectonics, 7*, 663–684.

Becker, T. W., & Boschi, L. (2002). A comparison of tomographic and geodynamic mantle models. *Geochemistry, Geophysics and Geosystems, 3*. https://doi.org/10.1029/2001GC000168

Bijwaard, H., Spakman, W., & Engdahl, E. R. (1998). Closing the gap between regional and global travel time tomography. *Journal of Geophysical Research: Solid Earth, 103*, 30055–30078.

Bohnhoff, M., Bulut, F., Dresen, G., Malin, P. E., Eken, T., & Aktar. M. (2013). An earthquake gap south of Istanbul. *Nature Communications, 4*(1), 1999. https://doi.org/10.1038/ncomms2999. www.nature.com/naturecommunications

Bulletin of Mineral Research and Exploration. (2014). General Directorate of Mineral Research and Exploration, Turkey.

Chamorro, C. R., García-Cuesta, J. L., Mondéjar, M. E., & Pérez-Madrazo, A. (2014). Enhanced geothermal systems in Europe: An estimation and comparison of the technical and sustainable potentials. *Energy, 65*, 250–263.

Danciu, L., Kale, Ö., & Akkar, S. (2018). The 2014 earthquake model of the Middle East: Ground motion model and uncertainties. *Bulletin of Earthquake Engineering, 16*, 3497–3533.

Erdik, M., Sesetyan, K., Demircioğlu, M. B., Tuzun, C., Giardini, D., Gulen, L., Akkar, D. S., & Zare, M. (2012). *Assessment of Seismic Hazard in the Middle East and Caucasus: EMME (Earthquake Model of Middle East) Project. Conference 15 WCEE*, Lisbon, Portuga.

Ershov, A. V., & Nikishin, A. M. (2004). Recent geodynamics of the Caucasus-Arabia-East Africa region. *Geotectonics, 2*, 55–72.

Garzón, P., Bohnhoff, M., & Nurlu, M. (2020). Crustal thickness variation across the Sea of Marmara region, NW Turkey: A reflection of modern and ancient tectonic processes. *Tectonics, 39*, e2019TC005986.

Gee, D. G., & Zeyen, H. J. (Eds.). (1996). EUROPROBE 1996—Lithosphere Dynamics: Origin and Evolution of Continents. Uppsala University, p. 138.

Giardini, D., Danciu, L., Erdik, M., Sesetyan, K., Demircioglu, M., Akkar, S., Gülen, L., & Zare, M. (2018). Seismic Hazard Map of the Middle East. *Bulletin of Earthquake Engineering, 16*(8), 1–4. https://doi.org/10.1007/s10518-018-0347-3

Gokturkler, G., Salk, M., & Sari, C. (2003). Numerical modeling of the conductive heat transfer in western Anatolia. *Journal of the Balkan Geophysical Society, 6*(1), 1–15.

Grand, S. P., van der Hilst, R. D., & Widiyantoro, S. (1997). Global seismic tomography: A snapshot of convection in the Earth. *GSA Today, 7*(4), 1–7.

Khutorskoi, M. D., Antonovskaja, G. N., Basakina, I. M., & Teveleva, E. A. (2022). Seismicity and heat flow in the frame of Eastern European platform. *Volcanology and Seismology, 2*, 74–92. https://doi.org/10.31857/S0203030622020043

Khutorskoy, M. D. (2023). Regional thermal field and geothermal resources of the Caucasus region. *Geology and Geophysics of the South of Russia, 13*(3), 73–88. https://doi.org/10.46698/VNC.2023.51.69.006

Li, C., van der Hilst, R. D., Engdahl, E. R., & Burdick, S. (2008). A new global model for P wave speed variations in Earth's mantle. *Geochemistry, Geophysics, Geosystems, 9*(5), Q05018. https://doi.org/10.1029/2007GC001806

McClusky, S., Balassanian, S., Barka, A., Demir, C., Ergintav, S., Georgiev, I., Gurkan, O., Hamburger, M., Hurst, K., Kahle, H., Kastens, K., Kekelidze, G., King, R., Kotzev, V., Lenk, O., Mahmoud, S., Mishin, A., Nadariya, M., Ouzounis, A., … Veis, G. (2000). Global positioning system constraints on plate kinematics and dynamics in the eastern Mediterranean and Caucasus. *Journal of Geophysical Research, 105*(B3), 5695–5719.

Mchugh, C. M. G., Seeber, L., Cormier, M.-H., & Hornbach, M. (2014). Submarine paleoseismology along populated transform boundaries: The Enriquillo-Plantain-Garden Fault, Canal du Sud, Haiti, and the North Anatolian Fault, Marmara Sea, Turkey. *Oceanography, 27*(2), 118–131. https://doi.org/10.5670/oceanog.2014.47

Mousavi, S. V., Karami, K., Tilmes, S., Muri, H., Xia, L., & Rezaei, A. (2023). Future dust concentration over the Middle East and North Africa region under global warming and stratospheric aerosol intervention scenarios. *Atmospheric Chemistry and Physics, 23*, 10677–10695. https://doi.org/10.5194/acp-23-10677-2023

Okay, A. I., Demirbag, E., Kurt, H., Okay, N., & Kuscu, I. (1999). An active, deep marine strike-slip basin along the North Anatolian Fault, Turkey. *Tectonics, 18*, 129–147.

Polyak, B. G., & Smirnov, Y. B. (1966). Heat flow on the continents. *Dokl. Akad. Nauk SSSR, 168*(1), 170–172.

Polyak, B. G., & Smirnov, Y. B. (1968). Relationship of deep heat flow with the tectonic structure of continents. *Geotectonics, 4*, 3–19.

Schmittbuhl, J., Karabulut, H., Lengliné, O., & Bouchon, M. (2015). Seismicity distribution and locking depth along the Main Marmara Fault, Turkey. *Geochemistry, Geophysics, Geosystems, 17*(3). https://doi.org/10.1002/2015GC006120

Seeber, L., Sorlien, C. C., & Steckler, M. S. (2010) Marie-Helene Cormier continental transform basins: Why are they asymmetric? *Eos Transactions American Geophysical Union, 91*(4), 21–40. https://doi.org/10.1029/2010EO040001

Sharkov, E. V., & Lebedev, V. A. (2021). Caucasian-Arabian segment of alpine-himalayan convergence: an example of continental collision above mantle plume. In V. Svalova, (Ed.), *Heat-Mass Transfer and Geodynamics of the Lithosphere*. Springer book: 89082526, pp. 381–390.

Sharkov, E. V., Lebedev V. A., Rodnikov A. G., Chugaev A. V., Sergeeva N. A., & Zabarinskaya, L. P. (2012). Features of Caucasian segment of the Alpine-Himalayan convergence zone: geological, volcanological, neotectonical, and geophysical data. In *Tectonics—Recent Advances. Editor Sharkov E. Ser. Earth and Planetary Sciences Rijeka, Croatia*, pp. 37–50.

Starostenko, V. I., Dolmaz, M. N., Kutas, R. I., Rusakov, O. M., Oksum, E., Hisarli, Z. M., Okyar, M., Kalyoncuoglu, U. Y., Tutunsatar, H. E., & Legostaeva, O. V. (2014). Thermal structure of the crust in the Black Sea: Comparative analysis of magnetic and heat flow data. *Marine Geophysical Research, 35*, 345–359. https://doi.org/10.1007/s11001-014-9224-x

Stein, R., Barka, A., & Dietrich, J. (1997). Progressive earthquake failure on the North Anatolian Fault since 1939 by stress triggering. *Geophysical Journal International, 128*, 594–604.

Stein, R.S., & Sevilgen, V. (2025). A magnitude 6.2 quake strikes the Marmara Fault at the site of large historic earthquakes near Istanbul, Temblor. https://doi.org/10.32858/temblor.361

Straub, C., Kahle, H.-G., & Schindler, C. (1997). GPS and geologic estimates of the tectonic activity in the Marmara Sea region, NW Anatolia. *Journal of Geophysical Research, 102*, 27587–27601.

Svalova, V.B., & Zaalishvili, V.B. (2024). Geodynamics and Geoecology. Moscow—Vladikavkaz: GPI VNC RAS, p. 346. ISBN 978-5-904868-29-1.

Svalova, V. B., & Zaalishvili, V. B. (2024a). Deep geodynamics and seismicity of the Caucasus-Anatolian region. *Geology and Geophysics of the South of Russia, 14*(2), 77–89. https://doi.org/10.46698/VNC.2024.97.46.006

Svalova, V. B. (2019). Geothermy and seismicity of the Caucasus region and the inverse problem of geodynamics. *Geology and Geophysics of the South of Russia, 9*(3), 77–93. https://doi.org/10.23671/VNC.2019.3.36485

Svalova, V. B. (2022). Management of geoenvironmental risk and problems of sustainable development of mountain territories. *Geology and geophysics of the South of Russia, 12*(1), 129–147. https://doi.org/10.46698/VNC.2022.48.61.010

Svalova, V. (2023). Geodynamics of the Caucasus—Anatolian-Arabian region and Turkey-Syria Earthquakes 2023. *Journal of Basic & Applied Sciences, 19*, 40–59.

Taymaz, T., Jackson, J., & McKenzie, D. (1991). Active tectonics of the north and central Aegean Sea. *Geophysical Journal International, 106*, 433–490.

Trifonov, V. G., & Sokolov, S. Y. (2018). Structure of the mantle and tectonic zoning of the central Alpine-Himalayan belt. *Geodynamics & Tectonophysics, 9*(4), 1127–1145. https://doi.org/10.5800/GT-2018-9-4-0386

Ugur AKIN. (2016). Investigation of the seismic velocity distribution and crustal structure of Turkey by means of gravity data. *Bulletin of the Mineral Research and Exploration, 153*, 185–202. https://doi.org/10.19111/bmre.01024. http://bulletin.mta.gov.tr

Zabarinskaya, L., Sergeyeva, N., Rashidov, V., Nisilevich, M., & Krylova, T. (2021). Structure and dynamics of the lithosphere for the Eurasia-Pacific transition zone. In V. B. Svalova, (Ed.), *The Collection: Heat-mass transfer and geodynamics of the lithosphere. Ser. Innovation and Discovery in Russian Science and Engineering* (pp. 411–428). Springer, Switzerland.

Zare, M., Amini, H., Yazdi, P., Sesetyan, K., Demircioglu, M. B., Kalafat, D., Erdik, M., Giardini, D., Asif Khan M., & Tsereteli, N. (2014). Recent developments of the Middle East catalog. *Journal of Seismol, 18*, 749–772. https://doi.org/10.1007/s10950-014-9444-1

Zhao, D. (2001). Seismic structure and origin of hotspots and mantle plumes. *Earth Planet Science Letter, 192*, 423–436.

Zhou, H. (1996). A high-resolution P wave model for the top 1200 km of the mantle. *Journal of Geophysics Research, 101*, 27791–27810.

Chapter 23
Reflecting from Crisis to Resilience Pathways

Edris Alam

23.1 Introduction

This volume has examined disaster risk, climate extremes, and resilience pathways across the Middle East and North Africa (MENA) through diverse empirical, institutional, and disciplinary lenses. Taken together, the chapters reveal a region confronting intensifying and interlinked hazards while undergoing uneven transitions in governance, planning, and risk management. The evidence assembled across environmental, urban, institutional, and conflict-affected contexts allows several overarching conclusions to be drawn regarding dominant themes, shared findings, structural challenges, research gaps, and future directions.

23.2 Synthesis of Findings

23.2.1 Climate Change, Extreme Events, and Food Security

Hazard intensity and frequency are increasing, particularly heatwaves, droughts, sand-dust storms, floods, and compound climate extremes, all of which are associated with measurable drivers of agricultural decline and food insecurity across the MENA region and beyond. Empirical analyses of soil degradation, agricultural decline, extreme rainfall, heat stress during the Hajj, and seismic recurrence confirm long-term upward trends in risk. Specifically, rising temperatures and the increasing occurrence of droughts and extreme rainfall have accelerated soil degradation, further threatening food systems. Empirical analyses using panel data for MENA countries

E. Alam (✉)
Disaster Management Program, Faculty of Resilience, Rabdan Academy, Abu Dhabi, United Arab Emirates
e-mail: ealam@ra.ac.ae

E. Alam (ed.), *Disaster, Crisis and Emergency Management in the Middle East & North Africa*, https://doi.org/10.1007/978-3-032-24436-9_23

indicate that a 1 °C rise in average temperature reduces food availability by about 8.2 kcal per person per day, reflecting direct impacts of warming on crop productivity and caloric supply (panel regression, 2000–2022) (Elkhalfi et al., 2025). The region's predicted temperature increase of up to 4 °C by 2050, alongside declining rainfall, will accelerate desertification, shrink arable lands, and further constrain food production capacity, particularly in already water-stressed North Africa and the Nile Basin.

Staple crops such as wheat may experience yield declines of up to 35% by 2050 under high-emission scenarios, compounding a heavy reliance on imports and elevating vulnerability to supply shocks (Mohammed Qasem & Scholz, 2025). Furthermore, rising sea levels and saline groundwater have created additional threats to coastal areas and have compromised soil integrity, which in turn has decreased agricultural production by 37% since 1961.

Concurrently, climate pressures interact with structural constraints, such as limited natural resources, high import dependence, and demographic growth, to amplify food insecurity. Estimates show over 51 million people in the region are already food insecure, a figure expected to rise as climate extremes intensify (Mohammed Qasem & Scholz, 2025). It has been noted that the regions of Morocco, Algeria, Tunisia, and Egypt show substantial evidence of degradation, such as aridity. This lost agricultural productivity (approximately a 34% reduction since 1961) indicates a need for integrated adaptation strategies. These include soil conservation and agroecological practices combined with improved water management, enhanced governance, and climate finance to best support the development of tailored resilience strategies for the MENA region. This underscores the need for context-specific adaptation strategies, including resilient crop varieties, improved irrigation, and integrated risk management to sustain agricultural production and food security in a warming MENA.

23.2.2 Water Scarcity, Converging, and Compound Risks

A defining feature of climate risk in the MENA is the convergence of multiple environmental stressors, where climate change amplifies existing vulnerabilities such as water scarcity, land degradation, urban expansion, and food insecurity. Climate models project a 5–20% reduction in annual precipitation across much of the MENA under high-emission scenarios, exacerbating baseline aridity and water stress in a region that receives only about 2% of the world's renewable water despite housing 6% of the global population (Mohammed Qasem & Scholz, 2025).

This intensifies drought frequency and severity, contributing to groundwater depletion and heightened competition for scarce water between the urban, agricultural, and energy sectors. Structural scarcity, rooted in extremely low per-capita renewable freshwater endowments, is increasingly intensified by climatic trends.

These include extended dry spells and higher temperatures that accelerate moisture loss from surface and subsurface water stores. Concurrently, rising temperatures—which have increased at approximately twice the global average in the recent decades—are linked to more frequent heatwaves, floods, and sea-level rise, leaving coastal lowlands and port cities highly exposed to inundation. Several chapters demonstrate how climatic extremes interact with socio-economic pressures, such as rapid urbanization, food import dependence, and fragile governance frameworks. This produces compound and cascading risks that threaten agricultural productivity, public health, infrastructure, and livelihoods across the region.

23.2.3 Persistence of Reactive Disaster Management

Despite formal commitments to frameworks such as the Sendai Framework for Disaster Risk Reduction (2015–2030), disaster governance in much of the MENA remains predominantly reactive, with limited shifts toward prevention and preparedness. Countries frequently invest more in emergency response than in structural risk reduction, contributing to repeated avoidable losses and high disaster costs.

Regional analyses show that over 120 disasters in the past five years have caused nearly US$1 billion in annual damages, reflecting response-oriented systems rather than risk mitigation. Fragile states with lower governance capacity exhibit correspondingly lower resilience indices; for example, Iraq, Syria, and Yemen maintain more reactive approaches compared to their Gulf counterparts. While early warning systems and national DRR strategies exist in several Arab states, practical prevention and anticipatory action measures remain underdeveloped. This underscores ongoing institutional and implementation gaps in risk governance across the MENA and highlights the need to transition from crisis response toward integrated risk management and preparedness.

23.2.4 Governance, Institutions, and Coordination

Institutional quality and governance capacity substantially shape disaster resilience outcomes in any jurisdiction. Empirical disaster resilience studies show that Gulf Cooperation Council (GCC) countries—such as the UAE and Bahrain—demonstrate markedly higher institutional resilience, infrastructure strength, and emergency workforce capacity. Centralized yet adaptive governance models can enhance resilience when supported by legal clarity, professional capacity, and technology. The experiences of the UAE and Bahrain illustrate how coordinated institutions, training systems, and digital platforms improve preparedness and response.

In contrast, conflict-affected states like Syria and Iraq exhibit the lowest governance scores and elevated vulnerability due to fragile institutions and limited coordination capacity. National assessments reveal that many MENA states maintain emergency management structures focused on response rather than integrated disaster risk reduction, reflecting underdeveloped DRR institutionalization and limited mainstreaming into development planning.

Jordan: Jordan has instituted formal governance arrangements, but their implementation is limited by a reliance on external financing and insufficient capacity at the municipal level.

Morocco: Morocco is making progress in aligning fiscal mechanisms with territorial governance, yet more substantive decentralization is necessary.

Lebanon: Lebanon largely depends on externally funded pilot initiatives, reflecting fragmented institutional responsibilities and underdeveloped climate-related budgetary processes.

Regional efforts, such as the Arab Coordination Mechanism for DRR and political commitments at the 6th Arab Regional Platform for Disaster Risk Reduction, underscore ongoing efforts to strengthen multi-stakeholder coordination. However, persistent fragmentation and enforcement deficits continue to constrain risk-informed governance.

23.2.5 Urbanization and Spatial Planning

Rapid and often poorly regulated urbanization in the MENA region heightens disaster vulnerability, as dense cities and informal settlements expand into hazard-prone zones. Urban growth—often rapid and poorly regulated—is a recurrent driver of exposure. Evidence from GCC cities, flood-prone urban centers, and seismic zones highlights how land-use decisions, zoning, infrastructure design, and building code compliance determine disaster impacts.

Urban planning is repeatedly identified as a critical but underutilized instrument for climate adaptation and risk reduction. For example:

Egypt: The gap between flood risk assessment and spatial planning contributes to recurrent urban flood risk, as flood hazard analysis is not yet mandatory in local planning.

Al-Ain City (UAE): Spatial modeling shows that dense urbanization and inadequate drainage infrastructure significantly magnify flood vulnerability (Alhosani et al., 2024).

Evidence from Egypt, Türkiye, Syria, and Lebanon shows that non-compliance with building codes, informal development, and inadequate infrastructure magnifies the impacts of floods, earthquakes, and heat extremes. These patterns affirm that spatial planning remains an underutilized instrument for disaster risk reduction and climate adaptation in MENA urban environments.

23.2.6 Community Agency and Localization

In conflict-affected MENA contexts, community agency and localization are empirically linked to improved disaster outcomes, where formal systems are weak or absent. In Yemen and Syria, community networks have organized local early warning dissemination and informal search and rescue, reducing mortality when state capacity collapsed. Case studies in Iraq show that neighborhood committees coordinated shelter, food distribution, and risk mapping, complementing limited official responses. Research on Somalia highlights that community disaster committees significantly increase preparedness for drought and flood hazards, a key factor in local resilience. These findings align with broader literature on localization in fragile states, which notes that investing in community structures enhances risk reduction and adaptive capacity where institutional fragmentation persists.

23.2.7 Conflict and Political Instability Deepen Vulnerability

Protracted conflict and political instability significantly heighten disaster vulnerability in MENA by disrupting essential services, displacing populations, and deteriorating infrastructure. In Syria, civil war has displaced over 13 million people, undermined water systems, and intensified famine and environmental stress that amplify hazard impacts. In Iraq, decades of conflict and weak governance have left obsolete irrigation and water treatment infrastructure, worsening drought and flood exposure. In Lebanon, ongoing political turmoil and economic collapse compound vulnerability by weakening health systems and reducing capacity to respond to hazards, particularly for displaced and informal populations. These dynamics constrain recovery, limit humanitarian access, and intensify exposure to climate extremes and disasters.

23.2.8 Natural Hazards, Conflicts, and War Shaping Evolution of Disaster Management

The evolution of disaster and crisis management authorities in the MENA region has been shaped by unique historical, political, and environmental contexts. In the early twentieth century, disaster management was largely reactive, with limited institutional capacity and ad-hoc responses led by local communities or international aid organizations.

The mid-twentieth century marked a turning point, with several countries establishing dedicated disaster management authorities. For instance, in the aftermath of the 1960 Agadir earthquake in Morocco, which killed over 12,000 people, the Moroccan government-initiated efforts to develop a national disaster management framework. Egypt's disaster management has also evolved significantly since the

1992 Cairo earthquake, where in the aftermath, they expanded early warning systems and established the National Center for Crisis and Disaster Risk Reduction.

These early initiatives primarily focused on post-disaster response and recovery, with limited emphasis on risk reduction. In Iraq, the disaster management architecture evolved directly from war-time institutions, such as the National Operations Center and Crisis Action Cell. In Kuwait, the legacy of the 1990–91 Gulf War's environmental and humanitarian emergencies prompted institutional attention to large-scale environmental risk mitigation. Together, these experiences show how conflict shaped early response capacities, prioritizing security-driven reactive coordination over structured prevention.

The late twentieth and early twenty-first centuries saw a shift toward more comprehensive disaster management systems, influenced by global frameworks like the Hyogo Framework for Action (2005–2015) and the Sendai Framework (2015–2030). These frameworks emphasized integrating risk reduction into development planning and building community resilience. In response, many countries modernized their authorities; for example, the UAE launched the National Emergency Crisis and Disaster Management Authority (NCEMA) in 2007.

23.2.9 Current Institutional Framework for Disaster and Crisis Management

The current institutional framework varies significantly across the region, reflecting differences in governance, resources, and risk profiles. Common elements include national disaster management authorities, inter-agency coordination mechanisms, and international partnerships.

In many countries, disaster management is overseen by a central authority responsible for coordinating preparedness, response, and recovery efforts. For example, in Egypt, the Egyptian Cabinet's Information and Decision Support Center (IDSC) plays a key role in disaster management, while in Saudi Arabia, the General Directorate of Civil Defense is the primary agency responsible for emergency response. These authorities often work in collaboration with other government agencies, as well as local governments and non-governmental organizations (NGOs). International partnerships also play a significant role in the region's disaster management framework. Organizations such as the United Nations Office for the Coordination of Humanitarian Affairs (OCHA), the World Health Organization (WHO), and the Red Cross and Red Crescent Movement provide technical and financial support to national authorities, particularly in low-resource settings. Regional organizations, such as the Arab League and the Gulf Cooperation Council (GCC), have also established platforms for sharing best practices and coordinating disaster response efforts.

23.2.9.1 Key Agencies and Stakeholders in Disaster Mitigation and Response

Responsibility for disaster response is shared among government authorities, military and civil defense units, NGOs, and international organizations. At the national level, agencies like NCEMA in the UAE and the National Committee for Civil Emergency Management in Bahrain develop policies and conduct risk assessments. These agencies often work closely with local governments, which play a critical role in implementing disaster management plans at the community level.

Military and civil defense units are also key players in disaster response, particularly in countries with limited institutional capacity. For example, in Lebanon, the Lebanese Armed Forces (LAF) have been instrumental in responding to disasters such as the 2020 Beirut port explosion, providing search and rescue services, medical assistance, and logistical support. In Egypt, where the Egyptian Armed Forces play a central role in disaster response and emergency relief. The military is routinely deployed during floods, severe storms, earthquakes, and public health emergencies to support evacuation, operate field hospitals, restore critical infrastructure, and deliver aid to remote and affected communities, particularly where civilian response capacity is limited.

NGOs and civil society organizations (CSOs) are increasingly recognized as important stakeholders in disaster management, particularly in conflict-affected countries where government capacity is limited. Organizations such as the Syrian Arab Red Crescent (SARC) and the Iraqi Red Crescent Society (IRCS) have played a critical role in emergency response, humanitarian relief, and displacement support during floods, heatwaves, and protracted conflict, frequently operating in insecure environments and compensating for limited state capacity in affected and hard-to-reach areas.

23.2.9.2 Legal and Policy Frameworks Governing Disaster Management

The legal and policy frameworks governing disaster management in the MENA region are diverse and reflect the varying levels of institutional development and risk profiles across countries. Many countries have enacted national disaster management laws and policies to provide a legal basis for disaster preparedness, response, and recovery. For example, in the United Arab Emirates, disaster management is governed by Federal Law No. 2 of 2011 on Civil Defense and subsequent executive regulations, under which the National Emergency, Crisis and Disaster Management Authority (NCEMA) coordinates preparedness, response, and recovery across federal and emirate levels. Similarly, in Algeria, the National Plan for Disaster Risk Reduction (2015–2029) provides a comprehensive framework for addressing natural and human-induced hazards. While many countries have adopted national DRR strategies, climate adaptation plans, and legal frameworks aligned with the Sendai

Framework and SDGs, operational delivery remains uneven. Fragmented institutional arrangements, unclear authority, and limited sub-national capacity undermine execution.

International frameworks and agreements also influence the legal and policy landscape in the MENA region. For instance, many countries have adopted the principles of the Sendai Framework for Disaster Risk Reduction, which emphasizes the importance of integrating disaster risk reduction into national development plans. Regional initiatives, such as the Arab Strategy for Disaster Risk Reduction (ASDRR), have also been developed to promote cooperation and knowledge sharing among MENA countries. Despite these advancements, challenges remain in the implementation of legal and policy frameworks, particularly in conflict-affected countries where governance structures are weak or fragmented. In such contexts, international humanitarian law (IHL) and human rights law play a critical role in guiding disaster response efforts and protecting the rights of affected populations.

23.2.9.3 Challenges and Weaknesses in Disaster Management

While significant progress has been made in developing disaster management systems, the region continues to grapple with persistent challenges and weaknesses. Disaster management authorities in the MENA region face a multitude of challenges in dealing with past disasters, many of which stem from the region's unique socio-political and environmental context. A primary issue is high exposure to overlapping hazards, such as earthquakes in Syria and Türkiye or recurrent droughts in North Africa. These are exacerbated by climate change, which strains existing management capacities. Rapid urbanization and population growth in coastal cities and informal settlements also pose significant problems. Urban areas are often ill-equipped to handle disasters due to inadequate infrastructure and poor land-use planning. The 2020 Beirut port explosion highlighted these vulnerabilities, showing how weak governance and a lack of safety enforcement can lead to catastrophe.

The following sections summarize the major challenges faced by disaster management authorities, identify political and governmental failures, explore systemic weaknesses, evaluate the effectiveness of past efforts, and highlight gaps in coordination among stakeholders.

Lack of National Plans: Countries like Iraq still lack a comprehensive national disaster management plan.

Top-Down Approaches: A need for proactive community engagement and public awareness can lead to a disconnect between authorities and the population.

Institutional Fragmentation: Fragmented frameworks, with different agencies responsible for different aspects of disaster management, can lead to poor coordination and duplication of effort.

Data Gaps: Many countries lack coherent datasets on hazards, exposure, losses, and vulnerabilities. Sparse observations, weak data-sharing arrangements, and limited analytical capacity reduce the effectiveness of early warning systems and risk-informed planning. Early warning systems, seismic monitoring, climate diagnostics, geospatial platforms, and impact-based forecasting are presented as essential enablers of preparedness. However, uneven access to data, limited integration, and capacity disparities constrain their effectiveness across the region.

Technological Disparities: Technological and capacity disparities across the region create unequal preparedness. Advanced monitoring and geospatial platforms coexist with basic early warning gaps, particularly in fragile and low-income contexts. Gulf States combine robust infrastructure with high digital skills, whereas lower-income and conflict-affected states lag in both access and governance of digital systems.

Financial Constraints: Prevention and risk reduction remain underfunded relatives to response and recovery, while additionally, climate finance, insurance mechanisms, and risk-sensitive budgeting are unevenly developed. National-level climate financing is also absent in several countries including Iraq, Syria and Lebanon and resource constraints pose a significant challenge, particularly in low-income and conflict-affected countries. Limited financial resources, outdated equipment, and insufficient training for disaster response personnel often hinder effective disaster management. In countries like Lebanon and Syria, ongoing conflicts have further depleted resources and disrupted institutional capacities, making it difficult to respond to disasters effectively.

Social and political fragility in the MENA region, manifested through conflict, corruption, and low public trust, has measurably undermined institutional performance and resilience. Countries such as Iraq and Syria are classified as "fragile" due to weak governance, repeated conflict, and institutional breakdowns. Pervasive corruption among these countries erodes trust in public institutions, contributing to civic disengagement and reinforcing a reliance on external humanitarian aid. Continued conflict correlates with lower financial development and institutional capacity, further weakening long-term resilience and prompting dependence on international support.

Political and governmental failures have compromised the capacity of disaster management institutions in several cases such as Syria, Lebanon and Libya in the MENA region. In some cases, weak governance, corruption, and a lack of accountability have undermined disaster preparedness and response efforts. One instance of this can be seen through the 2023 earthquakes in Syria, which revealed significant systemic weaknesses in disaster preparedness and response. Prior to the earthquakes, decades of conflict had left critical infrastructure such as hospitals, roads, and emergency communication networks severely damaged; this combined with fragmented coordination between local authorities, national agencies, and humanitarian organizations led to the prominent human and material toll of the earthquakes. Similarly, the 2020 Beirut port explosion revealed deep-rooted issues of corruption and negligence, as hazardous materials were stored improperly for years despite warnings from experts.

In conflict-affected countries, political instability and fragmented governance structures have further weakened disaster management capacities. In Syria for instance, the ongoing civil war has disrupted government functions, leaving disaster response efforts largely dependent on international aid organizations and local NGOs. It can be observed that the lack of a unified government within the country can complicate disaster response, particularly in areas controlled by different factions.

The MENA region faces significant challenges and weaknesses in disaster management, stemming from a combination of environmental, socio-economic, and political factors, leaving scope for improvement. While progress has been made in developing institutional frameworks and response mechanisms, persistent issues such as resource constraints, weak governance, and lack of coordination continue to hinder effective disaster preparedness, response, recovery, and mitigation. Another issue is the lack of information sharing and communication among agencies. In the absence of centralized platforms for data collection and dissemination, critical information about hazards, vulnerabilities, and resources may not reach all the relevant stakeholders in a timely manner. This lack of coordination can be particularly problematic in complex emergencies, where multiple agencies and organizations are meant to be involved in the response.

International organizations and NGOs often play a crucial role in disaster response, particularly in conflict-affected countries. However, their efforts are sometimes hampered by a lack of coordination with national authorities. For example, in the aftermath of the 2020 Beirut explosion, there were reports of duplication of recovery efforts and a lack of alignment between international aid organizations and local authorities, which slowed down the delivery of aid and services. Addressing these challenges will require a multi-faceted approach that includes strengthening governance, investing in risk reduction, enhancing community engagement, and improving coordination among stakeholders. By learning from past failures and building on existing strengths, the region can move toward a more resilient and proactive approach to disaster management, better equipped to handle the complex and evolving risks of the twenty-first century.

The findings from Stern (Chap. 12) suggest that overall, Egypt demonstrates clear strengths in national strategy development, climate-finance innovation, and cross-sectoral policy alignment. The remaining gaps relate primarily to inconsistent local delivery, fragmented coordination between national and governorate institutions, and the need for more integrated monitoring systems to translate policy intent into measurable outcomes.

23.3 Research Gaps in Disaster Management

The following sections will explore and discuss the key areas where research is lacking, the specific challenges and emerging threats that have not been adequately studied, as well as how research can contribute to improving disaster resilience in the region.

23.3.1 Integrated Risk Assessment and Multi-hazard Approaches

One of the most significant gaps in disaster management research in the MENA region is the lack of integrated risk assessments that consider multiple hazards and their interactions. Most studies focus on single hazards, such as earthquakes or floods, without accounting for the compounding effects of overlapping risks. For example, a region prone to both droughts and floods requires a nuanced understanding of how these hazards may interact and exacerbate vulnerabilities. Research that adopts a multi-hazard approach can provide a more comprehensive understanding of the risks and inform more effective mitigation and preparedness strategies.

23.3.2 Climate Change and Disaster Risk

While climate change is widely recognized as a major driver of disaster risk, its specific impacts on the MENA region have not been sufficiently studied. Research is needed to understand how rising temperatures, changing precipitation patterns, and sea levels rising will affect the frequency and intensity of natural hazards such as droughts, floods, and sand-dust storms. Additionally, there is a lack of investigation on the socio-economic impacts of climate-induced disasters, particularly on vulnerable populations such as rural communities, women, and children. It can be inferred that an increase in research on the impacts of climate change can inform policies on adaptation and mitigation, while studies on urban vulnerabilities can guide urban planning and infrastructure development.

23.3.3 Conflict and Disaster Risk

The MENA region is home to several conflict-affected countries, where disaster risk is compounded by political instability and violence. While interactions between climate hazards, conflict, health crises, and economic shocks are acknowledged, integrated modeling and longitudinal studies are limited. Compound and cascading risks also remain insufficiently studied.

However, there is limited research on the intersection of conflict and disaster risk. Studies are needed to understand how conflicts exacerbate vulnerabilities and hinder disaster response efforts. Additionally, research on the role of humanitarian aid and international cooperation in conflict settings can provide insights into how to address the unique challenges of disaster management in these contexts.

23.3.4 Technological Hazards and Industrial Risks

As the MENA region undergoes rapid industrialization, the risk of technological hazards such as chemical spills, industrial accidents, and nuclear incidents is increasing. Since these risks have not been adequately studied as of yet, more research is needed to assess the potential impacts of technological hazards and develop strategies for risk reduction and emergency response. Additionally, studies on the regulatory frameworks governing industrial safety will also be able to provide effective insights as to how to prevent and mitigate technological disasters.

23.3.5 Urbanization and Disaster Risk

Rapid urbanization in the MENA region has led to the growth of informal settlements and inadequate infrastructure, increasing exposure to disasters. However, there is currently limited research on the specific vulnerabilities of urban areas and how they can be addressed. Studies are needed to explore the role of urban planning, building codes and infrastructure development in reducing disaster risk. Additionally, research on the social dimensions of urban vulnerability, such as poverty and inequality, can provide insights into how to build more resilient cities. Seismic risk governance and compliance represent a persistent gap; despite advances in geoscience and modeling, the enforcement of seismic standards and translation of scientific knowledge into planning also remain inconsistent.

23.3.6 Community-Based Disaster Risk Reduction

Disaster risk management approaches designed for orderly institutional contexts are currently poorly suited to conflict-affected environments; where authority is dispersed, access for humanitarian actors is constrained and confidence in public institutions is weak or absent. Community engagement is a critical component of effective disaster management, yet there is a significant lack of research on community-based approaches in the MENA region. In such contexts, members of communities themselves emerge as the true first responders, leveraging local knowledge, networks, and resources to meet urgent needs. Chapter 6 examines empirical case studies drawn from Syria, Yemen, Palestine (Gaza), Lebanon, Sudan, Iraq, and Morocco to demonstrate the added value of locally driven initiatives in producing responsive outcomes to social, political and cultural contexts. The chapter synthesizes cross-cutting lessons from these cases, critically assesses persistent constraints, and advances actionable strategies to institutionalize and enhance community engagement within disaster preparedness and response systems across the MENA region. In fragile and conflict-affected settings, meaningful involvement of local communities is fundamental to shaping emergency responses that are sustainable, contextually appropriate and effective in practice. By understanding the specific needs and capacities of communities, research can inform the development of targeted interventions that enhance local resilience. Studies are needed to understand how local knowledge and practices can be integrated into disaster risk reduction strategies. Research on the role of social networks, traditional coping mechanisms, and community-led initiatives are also useful in providing valuable insights into how to enhance local resilience.

23.3.7 Addressing Emerging Threats

Increased research on emerging threats such as technological hazards and health emergencies can provide the knowledge needed to develop proactive strategies for risk reduction and preparedness. By understanding the specific risks and vulnerabilities associated with these threats, research can inform the development of regulatory frameworks, emergency response plans, and public awareness campaigns. Additionally, research on the intersection of conflict and disaster risk can provide insights into how to address the unique challenges of disaster management in conflict-affected countries.

23.3.8 Disaster Governance and Institutional Frameworks

The effectiveness of disaster management in the MENA region is heavily influenced by governance structures and institutional frameworks. However, there is limited research on how these factors shape disaster preparedness, response, and recovery. More studies are necessary in order to explore the role of political will, accountability, and transparency in disaster governance. Additionally, research on the coordination mechanisms among different levels of government, as well as between governmental and non-governmental actors, can provide insights into how to improve institutional effectiveness. By identifying the best practices and lessons learned, research can inform the development of policies and mechanisms that enhance coordination, accountability, and transparency. Furthermore, research on the role of international cooperation and humanitarian aid can also provide insights into how to strengthen partnerships and improve disaster response in conflict-affected settings.

23.4 Recommendations and Future Strategies

Reducing disaster-related deaths in the MENA region requires a multi-faceted approach that addresses both immediate responses and long-term risk reduction. Key measures include:

1. Adopting a Proactive Approach to Risk Reduction: Shift from reactive response to proactive risk management, prioritizing prevention, preparedness, and anticipatory action. Many countries in the region focus on post-disaster response rather than preventative risk reduction. Shifting to a risk-based approach involves integrating disaster risk assessments into development planning, investing in mitigation measures and prioritizing preparedness. National and local governments should develop comprehensive DRM plans that address all phases of the disaster cycle.
2. Building Local Capacities: Communities are often the first responders during disasters. Training programs that equip residents with skills in first aid, search and rescue, and emergency communication can enhance their ability to respond effectively. Establishing community disaster response teams can also improve local resilience.
3. Embed conflict sensitivity within disaster governance–Revise national disaster risk management policies to reflect the realities of fragile settings, institutionalize community participation in conflict-affected areas, establish dedicated coordination mechanisms for affected regions, and advocate for the explicit inclusion of conflict dynamics within global disaster risk reduction frameworks.
4. Strengthening Early Warning Systems: Early warning systems are critical for saving lives during disasters, as many countries in the region lack robust systems to detect and communicate hazards such as earthquakes, floods, and sand-dust storms. Investments in advanced technologies, such as seismic monitoring networks and satellite-based weather forecasting, can improve the accuracy and timeliness of warnings. Implement robust *Early Warning for All* (EW4All) framework that brings the MENA region together to develop a Joint Arab Roadmap anchored in strengthening risk knowledge and EWS capacities, fostering inter-agency coordination and alignment of national and regional priorities, enhancing inclusive early warning systems for complex and cascading risks, and promoting preparedness and resilience to protect lives, livelihoods, and critical infrastructure. Additionally, public awareness campaigns are needed to ensure that communities understand how to respond to these alerts.
5. Improving Emergency Response Capacities: A rapid and effective emergency response is essential for minimizing casualties. This requires well-trained and equipped search and rescue teams, medical personnel and emergency services. Therefore, countries should invest in training programs, modern equipment, and logistical support to enhance their response capabilities. Regional cooperation, such as joint training exercises and resource sharing, can also strengthen emergency responses.

6. Enhancing Infrastructure Resilience: Poor infrastructure is a major contributor to disaster-related deaths. Building codes and standards must be updated and enforced to ensure that critical infrastructure such as hospitals, schools, and transportation networks, can withstand disasters. Retrofitting existing structures and constructing disaster-resilient buildings can significantly reduce casualties during events like earthquakes and floods.
7. Targeting Vulnerable Populations: Vulnerable groups such as women, children, the elderly and people with disabilities, are disproportionately affected by disasters. Tailored interventions, such as accessible evacuation plans and targeted relief efforts, are needed to protect these populations, as community-based approaches can help identify and address the specific needs of vulnerable groups.
8. Mainstream risk into development and urban planning, integrating climate and hazard data into land-use decisions, infrastructure design and public investment: DRM policies must incorporate climate change adaptation strategies such as water resource management, drought-resistant agriculture and coastal protection. Aligning DRM with national climate action plans can enhance long-term resilience.
9. Invest in data systems and early warning, emphasizing interoperability, impact-based forecasting, and last-mile communication: Big data and AI can improve risk assessment, early warning, and decision-making. For instance, AI algorithms can analyze satellite imagery to predict flood risks or assess damage after an earthquake. Data-sharing platforms that integrate information from multiple sources can enhance situational awareness and coordination.
10. Scaling up ecosystem-based and nature-based approaches: Ecosystem-based adaptation (EbA) and nature-based solutions (NbS) should be systematically mainstreamed into national planning frameworks and public financing mechanisms, rather than confined to isolated pilot initiatives. Institutionalizing these approaches can improve cost efficiency, deliver simultaneous benefits for biodiversity conservation and livelihood security, and contribute to more durable and adaptive resilience outcomes over the long term. Several GCC countries including Saudi Arabia and the UAE has adopted EbA and NbS which could be scaled in the countries.
11. Institutionalize community-based approaches, embedding local actors within national DRM systems, particularly in fragile contexts.
12. Build professional and technical capacity through education, training and partnerships with academic and research institutions.
13. Mobilize sustainable finance for resilience, including climate-responsive budgeting, layered risk finance, and targeted prevention investment.
14. Strengthen regional cooperation: Climate extremes and disasters often transcend national borders, requiring regional collaboration. The MENA has experienced major sand-dust storms, floods and earthquakes which require regional level interventions. The 2023 earthquake along the Syria–Turkey border demonstrated how large seismic events in conflict-affected and resource-constrained contexts generate compounded impacts, where weakened infrastructure, limited

state capacity, and fragile social systems interact and intensify both human suffering and economic losses. The region should establish platforms for sharing knowledge, resources, and optimal practices. Regional organizations such as the Arab League and the Gulf Cooperation Council (GCC) can play a key role in fostering cooperation and coordinating joint initiatives under the Arab Strategy for Disaster Risk Reduction 2030.

15. Strengthening Legal and Institutional Frameworks: Effective DRM requires clear legal and institutional frameworks that define roles, responsibilities, and accountability. Countries should enact and enforce disaster management laws that mandate risk reduction, preparedness, and responses, as establishing these dedicated authorities with adequate funding can improve coordination and implementation. In particular, policies and regulatory frameworks should mandate energy and water-efficient standards, require the use of reflective building materials to minimize heat absorption, and promote the integration of green roofs and vertical greening systems in order to support the development of climate-smart buildings.

23.5 Conclusions

The MENA region, with its rich cultural heritages and complex socio-political landscapes, is also one of the most vulnerable areas in the world to a wide array of natural and human-induced hazards. Once known for its hot and dry climate, the MENA region has experienced an increase in heavy rainfall events leading to floods in the recent years. Over the last few years, the region faced disasters including the Dubai floodings in 2024, the Syria–Turkey earthquakes in 2023, the Libyan Dam failure in 2003, and the Beirut explosions in 2020, along with conflicts in Syria, Yemen, and Gaza to name a few. These disasters drew world attention (Francis et al., 2025). This book explored the major disasters that have affected the jurisdiction, providing a historical account of significant events and their impacts. By examining past disasters, we can identify recurring themes and lessons learned that can be used to inform current and future risk management strategies. Secondly, the chapter examines the types of hazards, crises and disasters that are most prevalent in the region. This includes an analysis of both natural hazards, such as earthquakes and droughts, and human-induced crises, such as conflicts and economic instability. Understanding the prevalence and distribution of these hazards is essential for developing targeted mitigation and preparedness measures.

We noted the impacts of these disasters on communities, infrastructure, and the economy. Disasters in the MENA region have far-reaching consequences, often exacerbating existing vulnerabilities and inequalities. For instance, prolonged droughts can lead to food insecurity and displacement, while earthquakes can destroy critical infrastructure and disrupt economic activities. By examining these impacts, we can better understand the cascading effects of disasters and the need for integrated approaches to disaster risk reduction and emergency management.

The MENA region's disaster risk profile is further complicated by its socio-political context. Many countries in the region face challenges related to governance, resource management and political instability, which can hinder effective disaster response and recovery efforts. We noted the absence of a centralized authority, weak inter-agency coordination, and limited public engagement in several countries. Additionally, rapid urbanization and population growth have increased exposure to hazards, particularly in densely populated cities with inadequate infrastructure. Climate change adds another layer of complexity, as rising temperatures, changing precipitation patterns and sea-level rises are expected to exacerbate existing hazards and create new ones. Disaster management in conflict-affected contexts is characterized by a distinct set of challenges, including technical limitations, fragile or ineffective governance and institutions, severe data gaps and chronic shortages of financial and material resources.

Despite these challenges, the MENA region has made significant strides in disaster risk management. Many countries have established national disaster management frameworks, early warning systems, and community-based initiatives to enhance resilience. International cooperation and partnerships have also played a crucial role in supporting disaster preparedness and response efforts. However, there is still much work to be done to address the underlying drivers of risk and build long-term resilience.

The MENA region faces significant challenges in disaster, crisis, and emergency management, but these challenges also present opportunities for innovation and improvement. By implementing the recommendations outlined in this chapter, the region can reduce disaster-related deaths, enhance risk management policies, strengthen community engagement, and leverage technology to build resilience. A proactive, inclusive, and collaborative approach is essential for addressing the complex and evolving risks faced by the MENA region. Ultimately, these efforts will contribute to safer, more resilient communities, and a sustainable future for the region.

References

Alhosani, N., Hussein, K., Al-Areeq, A.M., Al Aghbari, A.A., Elkamali, M., Alsumaiti, T., Almurshidi, A.M.G., Sharif, H.O. and Abdalati, W. (2024). From sandstorms to deluges: Assessing flood risks in Al Ain City, United Arab Emirates. *Geomatics, Natural Hazards and Risk, 15*(1): 2429735.

Elkhalfi, O., Chaabita, R., Ghoujdam, M., Zehraoui, K., El Alaoui, H., & Belhaj, I. (2025). The impact of climate change on food security in the Middle East and North Africa: Challenges and adaptation strategies. *Journal of Agriculture and Food Research, 21*, Article 101963.

Francis, D., Fonseca, R., Nelli, N., Cherif, C., Yarragunta, Y., Zittis, G., & Jan de Vries, A. (2025). From cause to consequence: Examining the historic April 2024 rainstorm in the United Arab Emirates through the lens of climate change. *npj Climate and Atmospheric Science, 8*(1), 183.

Mohammed Qasem, R. H., & Scholz, M. (2025). Climate change impact on resources in the MENA region: A systematic and critical review. *Physics and Chemistry of the Earth, Parts a/b/c, 139*, Article 103936.

The manufacturer's authorised representative in the EU is Springer Nature Customer Service Centre GmbH, Europaplatz 3, 69115 Heidelberg, Germany. If you have any concerns regarding our products, please contact ProductSafety@springernature.com

Printed and bound by CPI Group (UK) Ltd, Croydon, CR0 4YY
17/09/2026
02218437-0015